T0331623

Twelve Landmarks of Twentieth-Century Analysis

The striking theorems showcased in this book are among the most profound results of twentieth-century analysis. The authors' original approach combines rigorous mathematical proofs with commentary on the underlying ideas to provide a rich insight into these landmarks in mathematics. Results ranging from the proof of Littlewood's conjecture to the Banach–Tarski paradox have been selected for their mathematical beauty as well as their educative value and historical role. Placing each theorem in historical perspective, the authors paint a coherent picture of modern analysis and its development, whilst maintaining mathematical rigour with the provision of complete proofs, alternative proofs, worked examples, and more than 150 exercises and solution hints.

This edition extends the original French edition of 2009 with a new chapter on partitions, including the Hardy–Ramanujan theorem, and a significant expansion of the existing chapter on the corona problem.

Twelve Landmarks of Twentieth-Century Analysis

D. CHOIMET
Lycée du Parc, Lyon

H. QUEFFÉLEC
Université de Lille

Illustrated by
MICHAËL MONERAU

Translated from the French by
DANIÈLE GIBBONS and GREG GIBBONS

CAMBRIDGE
UNIVERSITY PRESS

Shaftesbury Road, Cambridge CB2 8EA, United Kingdom

One Liberty Plaza, 20th Floor, New York, NY 10006, USA

477 Williamstown Road, Port Melbourne, VIC 3207, Australia

314–321, 3rd Floor, Plot 3, Splendor Forum, Jasola District Centre, New Delhi – 110025, India

103 Penang Road, #05–06/07, Visioncrest Commercial, Singapore 238467

Cambridge University Press is part of Cambridge University Press & Assessment, a department of the University of Cambridge.

We share the University's mission to contribute to society through the pursuit of education, learning and research at the highest international levels of excellence.

www.cambridge.org
Information on this title: www.cambridge.org/9781107059450

Originally published in French as *Analyse mathématique. Grands théorèmes du vingtième siècle* by Calvage et Mounet, 2009

© Calvage & Mounet, Paris, 2009

First published in English by Cambridge University Press & Assessment 2015
English translation © Cambridge University Press & Assessment 2015

First published 2015

A catalogue record for this publication is available from the British Library

Library of Congress Cataloging-in-Publication data
Choimet, Denis.
[Analyse mathématique. English]
Twelve landmarks of twentieth-century analysis / D. Choimet, Lycée du Parc, Lyon, H. Queffélec, Université de Lille ;
illustrated by Michaël Monerau ;
translated from the French by Danièle Gibbons and Greg Gibbons.
pages cm
Originally published in French as: Analyse mathématique. Grands théorèmes du vingtième siècle (Montrouge (Hauts-de-Seine) : Calvage et Mounet, 2009).
ISBN 978-1-107-05945-0
1. Mathematical analysis. 2. Harmonic analysis. 3. Banach algebras.
I. Queffélec, Hervé. II. Title.
QA300.C45413 2015
515–dc23
2014050264

ISBN   978-1-107-05945-0   Hardback
ISBN   978-1-107-65034-3   Paperback

to our students

Contents

Foreword

Analysis... the word is dangerous. Mention it at a dinner party, and depending on your guests, it will bring to mind lab coats and test tubes, or couches and psychoanalysts, or perhaps again those experts that unveil the subtleties of an economical or political crisis. Clarify that you are referring to mathematical analysis and the image will change: former students will then recall memories of derivatives and integrals, and no doubt remind you that it was much easier to calculate the former than the latter... But perhaps one might ask you: Mathematical analysis, no doubt it's all very nice, but what's its point? In fact, what are you analysing?

The book of Denis Choimet and Hervé Queffélec provides brilliant and profound answers to these questions in a most agreeable manner. We follow the evolution of analysis throughout the twentieth century, from the founding fathers Hardy and Littlewood, to the creators of spaces Wiener and Banach and up through contemporaries such as Lennart Carleson. The historical perspective helps us understand the motivation behind the problems, and the naturalness of their solutions. Moreover, analysis is shown clearly for what it is: a discipline situated in the heart of mathematics, indissolubly linked to arithmetic and number theory, to combinatorics, to probability theory, to logic, to geometry... Its objective is hence to serve mathematics and consequently all of the sciences, and thereby each and every one of us.

I have the pleasure of knowing Denis and Hervé. Hence I can assert that their knowledge of analysis can be qualified as encyclopaedic. However, they were not attempting to write an encyclopaedia, and the roots of their work can be found more in Cambridge than in Paris, Warsaw or Moscow. This wise approach allowed them to explore multiple directions right up to the most recent results, while maintaining the profound unity of a very reasonably formatted book, providing constant encouragement to the reader.

Reaching the end, the reader will lay down the book (close at hand, because there are works that demand to be re-read) with the satisfaction of now having a better understanding of analysis. He will also wish to congratulate Denis Choimet and Hervé Queffélec for their collaboration, which illustrates the connectedness of mathematics and of the community of mathematicians. Whether we study them in "classes préparatoires" or in a university, the mathematics stay equally fascinating. Let us not disfigure them by zebra-striping boundaries.

But time for a break from lyricism, to make way for mathematics. Happy reading to all! You are in for a real treat.

Gilles Godefroy, September 2014.

Preface

This book has a history: it was born after the encounter of two professors from different generations, on the occasion of a series of mathematics seminars organised by the younger of the two at the Lycée Clemenceau in Nantes, in the early part of the years 2000 onwards. The prime objective of these seminars was to allow the professors of this establishment to keep a certain mathematical awareness that the sustained rhythm of preparing students for competitive entrance exams did not always facilitate. The seminars took place roughly once a month, and lasted an hour and a half. Over the years, the professors were joined by an increasing number of students from their classes; a vocation for mathematics was born for many of these, possibly in part due to this initiative. Both authors gave half a dozen talks at these seminars, on themes of their choosing, with a strong emphasis (but not exclusively) on classical analysis.

After the nomination of one of us to Lyon, we thought it would be interesting to assemble and write up these talks in more detail, and to find a connection between them. It seemed to us that a good starting point would be the 1911 paper of Littlewood (Chapter 1), which is at the same time the founding point of what we today call Tauberian theorems, and the beginning of the famous collaboration between Hardy and Littlewood that spanned 35 years, until Hardy's death in 1947. This collaboration produced a large number of remarkable discoveries, not the least of which was that of Ramanujan. The magnificent work of Hardy and Ramanujan on the asymptotic behaviour of the partition function is in fact the subject of an entire chapter (Chapter 8).

Some of these discoveries are explained in detail, in addition to the converse of Abel's radial theorem (Chapter 1) – from the functional equation (approximated or not) of the Jacobi θ_0 function and its applications – via Diophantine approximations and continued fractions, to exponential sums and the close study of the "other" function of Riemann (Chapters 7 and 9), and in passing the asymptotic behaviour of the partition function (Chapter 8). Important extensions of this work include Wiener's Tauberian theorem (Chapter 2), the Tauberian theorems of Ikehara and Newman (Chapter 3), which are precursors of Banach algebras and Gelfand theory (Chapter 11), the latter giving rise to the corona problem, brilliantly solved by Carleson in 1962, shortly after his characterisation of interpolation sequences (Chapter 12). Beurling, the supervisor of Carleson (among others), provided a description of the invariant subspaces under the shift operator, a jewel of functional analysis of the twentieth century: this completes the long Chapter 12. Another extension of the study of sums of squares in Chapter 9 is the Littlewood conjecture about the L^1-norm of exponential sums, which was only resolved in 1981 (Chapter 10). A good half of the book thus pays tribute to the English school of analysis and, in passing, to the Swedish and American schools.

A second main theme starts from the work of the Polish school in the 1930s, in particular that of Stefan Banach. The spaces that today are given his name have been the subject of innumerable studies; one of their specific properties, complementation, is described in Chapter 13. This school highly prized the works of the French, in particular those of Baire and Lebesgue. These in turn are well represented in this book, through the study of the generic properties of derivative functions (Chapter 4), generic properties in the domain of probability theory (Chapter 5, which acknowledges the contributions of the Russian school, with Kolmogorov and Khintchine), and finally the paradoxical properties in measure theory (Chapter 6 on the paradox of Banach–Tarski).

All of these works are profound and difficult, but they deserve to be better known and popularised throughout the mathematical community, both from an historical and a scientific point of view. This has been our ambition.

A few words on the style of this book: we did not seek to write a text for highly skilled specialists, thus we were not ashamed to provide many reminders and lots of details and heuristic explanations, and to provide an historical perspective. Nor did we try to write a book for skilled generalists, thus we were not ashamed to provide complete and rigorous proofs, even if very difficult. Therefore, depending on the themes we study, our book spans multiple levels: some portions are at a graduate level, others are at an advanced undergraduate level, the average being somewhere between the two. We thought it useful to extend each chapter with a dozen or so exercises, as a complement to the main text or as an incentive for the reader to continue his reflections. These exercises do not have detailed solutions, but we hope to have provided sufficient references and hints for a reasonably courageous and interested reader to tackle them.

We hope that this book will serve a large audience, even if only now and then: we are thinking of our colleagues, as well as graduate students or *amateurs* of mathematics and beauty (*amateurs* is to be understood as Jean-Pierre Kahane would say).

Our thanks go to our friend Rached Mneimné, whose enthusiasm, openness and efficiency allowed this atypical book to be published, and to Gilles Godefroy, who was kind enough to write a friendly foreword. We sincerely thank our colleagues and former students who accepted reading in depth certain chapters and providing us with precise and constructive feedback: Walter Appel, Frédéric Bayart, Nicolas Bonnotte, Rémi Catellier, Vincent Clapiès, Jean-François Deldon, Quentin Dufour, Jordane Granier, Jérémy Guéré, Denis Jourdan, Xavier Lamy, Stéphane Malek, Thomas Ortiz, Marc Pauly, Michel Staïner, Carl Tipler. We also address warm thanks to the staff at CUP who trusted us and helped us with great kindness and professionalism during the final steps of this translation: Emma Collison, Clare Dennison, Katherine Law, Roger Astley.

Special thanks must be addressed to Bruno Calado and Michaël Monerau. Bruno proofread (in record time) the whole of the manuscript, flushed out an incalculable number of misprints, and proposed a multitude of interesting improvements: without him, the book would not be as polished as we wanted. Michaël not only read in detail several chapters, but also provided 20 or so magnificent diagrams which greatly help in reading and understanding the underlying text, at times extremely technical. It was truly to our benefit that they put their competence at our service.

Last but not least, we could have contemplated translating this new edition by ourselves, producing no doubt masses of "Frenglish". In any case, we totally underestimated the effort involved in translating 500 pages of serious mathematics! As luck would have it, we were introduced to Danièle and Greg Gibbons, both fluent in French, English and ... mathematics, thus able not only to read our text, but to understand what we were talking about. They did an enormous and excellent job of translating; and for the choice of certain technical terms, it was a true collaboration and a pleasure to discuss with them. Our warmest thanks go to them here.

We welcome with great interest your remarks.

1

The Littlewood Tauberian theorem

1.1 Introduction

In 1897, the Austrian mathematician Alfred Tauber published a short article on the convergence of numerical series [173], which can be summarised as follows.

Let $\sum a_n$ be a convergent series of complex numbers, with $\sum_{n=0}^{\infty} a_n = \ell$. A theorem of Abel [1] states that

$$f(x) = \sum_{n=0}^{\infty} a_n x^n \to \ell \text{ as } x \nearrow 1. \tag{1.1}$$

A theorem of Kronecker [116] states that

$$\frac{1}{n} \sum_{k=1}^{n} k a_k \to 0. \tag{1.2}$$

The converse of these two theorems is false: neither of the two conditions (1) nor (2) is sufficient to imply the convergence of the series $\sum a_n$. However, if **both** conditions are satisfied *simultaneously*, the series $\sum a_n$ converges, giving the following theorem.

1.1.1 Theorem *A necessary and sufficient condition for $\sum a_n$ to converge (with sum ℓ) is that:*

(1) $f(x) = \sum_{n=0}^{\infty} a_n x^n \to \ell$ *as* $x \nearrow 1$,
(2) $\frac{1}{n} \sum_{k=1}^{n} k a_k \to 0$.

The proof of Theorem 1.1.1 follows that of the following special case.

1

1.1.2 Theorem $f(x) \to \ell$ *as* $x \nearrow 1$, *and* $na_n \to 0$ *implies* $\sum_{n=0}^{\infty} a_n = \ell$.

A few remarks on this article: the theorem of Abel cited above gave rise to *Abelian theorems* [113], that is, theorems of the form

If $\sum a_n$ *is a convergent complex series with sum* ℓ, *and* $(b_{n,x})$ *an infinite rectangular matrix indexed by* $\mathbb{N} \times X$, *where* X *is a set with an associated point at infinity satisfying*

$$b_{n,x} \xrightarrow[x \to \infty]{} 1 \ for \ every \ n \in \mathbb{N},$$

then

$$f(x) = \sum_{n=0}^{\infty} a_n b_{n,x}$$

is defined for $x \in X$ *and*

$$f(x) \to \ell \ as \ x \to \infty.$$

Such a theorem generalises the theorem of Abel, where we have $X = [0, 1[$, the point at infinity being 1, and $b_{n,x} = x^n$. It also generalises the case $b_{n,x} = \left(1 - \dfrac{n}{x+1}\right)^+$ and $X = \mathbb{N}$, the point at infinity being ∞, which corresponds to

$$\sum_{n=0}^{\infty} a_n b_{n,x} = \frac{S_0 + \cdots + S_x}{x+1},$$

where S_n is the partial sum of index n of the series $\sum a_n$. The corresponding Abelian theorem is none other than the theorem of Cauchy–Cesàro.

The theorem of Kronecker is now referred to as the *lemma* of Kronecker [153]. For the proof of Theorem 1.1.2, we proceed as follows. With the inequality

$$1 - x^n \leqslant n(1 - x) \text{ if } 0 \leqslant x < 1, \tag{1.3}$$

we show that $S_n - f\left(1 - \dfrac{1}{n}\right) \to 0$, where $S_n = a_0 + \cdots + a_n$.

It is essentially Theorem 1.1.2 which has passed on to posterity. Despite its elegance, it remains relatively superficial, because of the highly restrictive hypothesis ($na_n \to 0$) and because of the limited use of the other hypothesis: we make use of $f\left(1 - \dfrac{1}{n}\right) \to \ell$ while in fact we have $f(x) \to \ell$ as $x \nearrow 1$. Nonetheless, at least two jewels of theorems can be considered as direct descendants of Tauber's theorem: the following results due to L. Fejér and A. Zygmund.

1.1.3 Theorem [Fejér]   *Let D be the unit open disk, J a Jordan curve with interior Ω, and $f : D \to \Omega$ a conformal mapping that can be extended to a homomorphism (also denoted) f from $\overline{D}$ over $\overline{\Omega} = \Omega \cup J$. Then the Taylor series of f converges uniformly on $\overline{D}$.*

The proof begins by showing that if $f(z) = \sum_{n=0}^{\infty} a_n z^n$, then the area of Ω is $\pi \sum_{n=1}^{\infty} n|a_n|^2$. We thus know that $\sum_{n=1}^{\infty} n|a_n|^2 < \infty$, which implies a Tauberian-style condition $\frac{1}{n} \sum_{k=1}^{n} k|a_k| \to 0$. Next, setting $S_n(\theta) = \sum_{j=0}^{n} a_j e^{ij\theta}$ and $r_n = 1 - \frac{1}{n}$, we use Tauber's method to show that $S_n(\theta) - f(r_n e^{i\theta}) \to 0$ uniformly with respect to θ.

1.1.4 Theorem [Zygmund [195]]   *We consider the trigonometric series $\sum_{n=1}^{\infty} (a_n \cos nx + b_n \sin nx)$ satisfying a Tauberian condition*

$$\lim_{N \to \infty} \frac{1}{N} \sum_{n=1}^{N} n\rho_n = 0, \text{ with } \rho_n = |a_n| + |b_n|. \tag{1.4}$$

Then, the primitive series $F(x) = \sum_{n=1}^{\infty} \dfrac{a_n \sin nx - b_n \cos nx}{n}$ converges normally on $\mathbb{R}$. Moreover, setting $N_h = \left[\dfrac{1}{|h|} \right]$, we have

$$\frac{F(x+h) - F(x)}{h} - \sum_{n=1}^{N_h} (a_n \cos nx + b_n \sin nx) \to 0 \text{ when } |h| \searrow 0,$$

uniformly with respect to x. The hypothesis (1.4) is verified for a lacunary series $\sum_{k=1}^{\infty} (\alpha_k \cos n_k x + \beta_k \sin n_k x)$, with

$$\frac{n_{k+1}}{n_k} \geqslant q > 1 \text{ and } |\alpha_k| + |\beta_k| \to 0.$$

In other words, under the hypothesis (1.4), we obtain a *point-by-point* result of differentiation term by term: the derivative series of F converges at a point x_0 with sum ℓ if and only if the function F is differentiable at x_0, and in this case $F'(x_0) = \ell$. For example, the real-valued function $F(x) = \sum_{k=1}^{\infty} \dfrac{2^{-k}}{\sqrt{k}} \sin(2^k x)$ is almost everywhere non-differentiable, because the lacunary series $\sum_{k=1}^{\infty} \dfrac{\cos(2^k x)}{\sqrt{k}}$, where the squares of the coefficients are not summable, is almost everywhere divergent (see [15], Vol. 2, p. 242) and

$\dfrac{1}{\sqrt{k}} \to 0$. However, F is differentiable on a non-countable set of points because it belongs to the *little Zygmund class* (see Exercise 7.4).[1]

Here is the proof: define

$$T_n = \sum_{j=1}^{n} j\rho_j \text{ and } F_n(x) = \frac{a_n \sin nx - b_n \cos nx}{n}.$$

A first Abel transformation gives

$$\sum_{n=1}^{N} \frac{\rho_n}{n} = \sum_{n=1}^{N} \frac{T_n - T_{n-1}}{n^2}$$

$$= \frac{T_N}{N^2} + \sum_{n=1}^{N-1} T_n(n^{-2} - (n+1)^{-2})$$

$$\ll N^{-1} + \sum_{n=1}^{N-1} \frac{n}{n^3} = O(1),$$

where the notation $A \ll B$ means that $A \leqslant \lambda B$, where λ is a positive constant. This proves normal convergence. Moreover, Taylor provides the estimate

$$\frac{F_n(x+h) - F_n(x)}{h} = a_n \cos nx + b_n \sin nx + O(n\rho_n|h|), \qquad (1.5)$$

with O being uniform with respect to all the parameters. From this,

$$\frac{F(x+h) - F(x)}{h} - \sum_{n=1}^{N_h} (a_n \cos nx + b_n \sin nx)$$

$$= \sum_{n=1}^{N_h} \left(\frac{F_n(x+h) - F_n(x)}{h} - F_n'(x) \right) + \sum_{n > N_h} \frac{F_n(x+h) - F_n(x)}{h},$$

---

[1] We even have

$$F(x+h) - F(x) = O\left(h\sqrt{\log \frac{1}{h}}\right) \text{ when } h \searrow 0,$$

uniformly with respect to x, which is better than the general estimate

$$G(x+h) - G(x) = o\left(h \ln \frac{1}{h}\right)$$

for G in the little Zygmund class.

from which, using (1.5) and the fact that $\| F_n \|_\infty \leqslant \frac{\rho_n}{n}$:

$$\left| \frac{F(x+h) - F(x)}{h} - \sum_{n=1}^{N_h} (a_n \cos nx + b_n \sin nx) \right|$$

$$\ll |h| \sum_{n=1}^{N_h} n\rho_n + \frac{1}{|h|} \sum_{n > N_h} \frac{\rho_n}{n}. \quad (1.6)$$

A further Abel transformation gives

$$\sum_{n > N_h} \frac{\rho_n}{n} = \sum_{n > N_h} \frac{n\rho_n}{n^2} = \sum_{n > N_h} \frac{T_n - T_{n-1}}{n^2}$$

$$= \sum_{n > N_h} T_n (n^{-2} - (n+1)^{-2}) - \frac{T_{N_h}}{(N_h + 1)^2},$$

so that[2]

$$\sum_{n > N_h} \frac{\rho_n}{n} = o \left(\sum_{n > N_h} \frac{n}{n^3} \right) + o(N_h^{-1}) = o(N_h^{-1}).$$

Finally, referring back to (1.6):

$$\left| \frac{F(x+h) - F(x)}{h} - \sum_{n=1}^{N_h} (a_n \cos nx + b_n \sin nx) \right| \ll |h| \sum_{n=1}^{N_h} n\rho_n + o \left(\frac{1}{|h| N_h} \right)$$

$$= o(|h| N_h) + o \left(\frac{1}{|h| N_h} \right)$$

$$= o(1),$$

with o being uniform.

The lacunary case corresponds to $a_n = \alpha_k$ when $n = n_k$, $a_n = 0$ otherwise, and similarly for b_n. Let $\varepsilon > 0$ and $N \geqslant 1$ be fixed, then p and k_0 be indices $\geqslant 1$ such that $n_p \leqslant N < n_{p+1}$ and

$$\gamma_k \leqslant \varepsilon \text{ for } k \geqslant k_0, \text{ where } \gamma_k = |\alpha_k| + |\beta_k|.$$

---

[2] This time, we use the full force of the hypothesis $T_n = o(n)$.

Then,[3]

$$\sum_{n=1}^{N} n\rho_n = \sum_{k=1}^{p} n_k \gamma_k$$

$$= \sum_{k=1}^{k_0-1} n_k \gamma_k + \sum_{k=k_0}^{p} n_k \gamma_k$$

$$\leqslant \sum_{k=1}^{k_0-1} n_k \gamma_k + \varepsilon n_p \left(1 + q^{-1} + q^{-2} + \cdots\right)$$

$$\leqslant C_\varepsilon + N\varepsilon \left(1 - q^{-1}\right)^{-1}$$

and hence

$$\varlimsup_{N\to\infty} N^{-1} \sum_{n=1}^{N} n\rho_n \leqslant \varepsilon \left(1 - q^{-1}\right)^{-1} \text{ for every } \varepsilon > 0,$$

so that

$$\frac{1}{N} \sum_{n=1}^{N} n\rho_n \to 0$$

as stated.

The aim of this chapter is to analyse in detail the enormous progress realised by Littlewood in 1911, when, in Tauber's Theorem 1.1.2, he replaced the hypothesis $na_n \to 0$ by na_n bounded, which Hardy had done the year before using the method of Cesàro summation (see Theorem 1.2.6). Littlewood's proof is nonetheless incredibly more elaborate than that of Hardy, and one can wonder why. In fact, Exercise 1.8 provides an indication: supposing that $f(x)$ has a limit when $x \nearrow 1$ is *a priori* a much weaker supposition than that made by Hardy.

Furthermore, when Tauber proved his two theorems, obviously he did not consider them as *conditional converses* of Abelian theorems, that is to say, theorems of the form

If

$$f(x) = \sum_{n=0}^{\infty} a_n b_{n,x} \to \ell \text{ as } x \to \infty$$

---

[3] We use the bound, valid for $1 \leqslant k \leqslant p$:

$$n_k \leqslant \frac{n_{k+1}}{q} \leqslant \frac{n_{k+2}}{q^2} \leqslant \cdots \leqslant \frac{n_p}{q^{p-k}}.$$

and if as well $(a_n)_{n \geqslant 0}$ *verifies certain additional conditions, of smallness or lacunarity for example,*[4] *then* $\sum a_n$ *converges with sum* ℓ.

The study of such converse theorems is precisely what Hardy and Littlewood undertook after 1911. They proposed naming them Tauberian theorems, in honour of Tauber and his seminal result. Tauber's result would today be considered as a *remark*, by the standards of twentieth-century publications, which does not preclude a certain sense of depth and technical difficulty, as evidenced by the papers of Tauber and others.

The subject of power series of a single variable is often considered a little old-fashioned, with the single issue of determining the radius of convergence (for which there are a number of rules, each as boring as the others). In Section 1.4 we see that it is in fact much more rich and complex, as soon as we approach the circle of convergence. Littlewood's theorem already bears witness to this, and the subject poses problems that are open to this day!

We start by examining the state of the art in 1911, before Littlewood's paper, considered as the starting point of his 30-year collaboration with G. H. Hardy.

1.2 State of the art in 1911

In the following, we will consistently use the definitions below:

$$f(x) = \sum_{n=0}^{\infty} a_n x^n = (1-x) \sum_{n=0}^{\infty} S_n x^n = (1-x)^2 \sum_{n=0}^{\infty} \sigma_n x^n,$$

where $S_n = a_0 + \cdots + a_n$ and $\sigma_n = S_0 + \cdots + S_n$.

We suppose (without loss of generality) that our power series have radius of convergence 1 (or in any case $\geqslant 1$). Before Littlewood, the principal results relating the behaviour of S_n, σ_n and $f(x)$ when $x \nearrow 1$ were as follows.

1.2.1 Theorem [Cauchy mean (1821)] *If* $S_n \to \ell$, *then* $\dfrac{\sigma_n}{n} \to \ell$. *The converse is false* ($S_n = u^n$, $|u| = 1$, $u \neq 1$).

1.2.2 Theorem [Abel continuity (1826)] *If* $S_n \to \ell$, *then* $f(x) \to \ell$. *The converse is false* ($a_n = u^n$, $|u| = 1$, $u \neq 1$).

1.2.3 Theorem [Frobenius continuity (1880)] *If* $\dfrac{S_0 + \cdots + S_n}{n} = \dfrac{\sigma_n}{n} \to \ell$, *then* $f(x) \to \ell$. *The converse is false* ($S_n = nu^n$, $|u| = 1$, $u \neq 1$).

---

[4] For Tauber, this corresponds to the condition $\dfrac{1}{n} \sum_{k=0}^{n} k a_k \to 0$.

This result is more general than that of Abel because of the Cauchy mean theorem.

As for Cesàro, he obtained the following results, which extend those of Cauchy (and in the process reprove a theorem of Mertens).

1.2.4 Theorem [Cesàro multiplication (1890)]   (*1*) *If $a_n \to a$ and $b_n \to b$, then $\frac{1}{n}(a_0 b_n + \cdots + a_n b_0) \to ab$.*
(2) *If $\sum a_n$ and $\sum b_n$ converge respectively to A and B, and if $c_n = \sum_{i+j=n} a_i b_j$ and $C_n = c_0 + \cdots + c_n$, then $\dfrac{C_0 + \cdots + C_n}{n} \to AB$. In particular, if $C_n \to C$, we necessarily have $C = AB$ (theorem of Mertens).*

Statement (1) is in fact a simple improvement of Theorem 1.2.1.

As we have already mentioned, Tauber [113, 124] had already established the following conditional converse of Abel's theorem.

1.2.5 Theorem [Tauber (1897)]   *If $f(x) \to \ell$ and $na_n \to 0$ (or if only $\frac{1}{n}\sum_{j=1}^{n} j a_j \to 0$), then $S_n \to \ell$.*

This theorem, the first in a long line, is a gem ("remarkable", according to Littlewood) even if somewhat superficial. It is the ancestor of Fejér's theorem, which can also be shown with the simple identity:

$$\frac{1}{n} \sum_{j=1}^{n} j a_j e^{ij\theta} = S_n(\theta) - \frac{1}{n} \sum_{j=0}^{n-1} S_j(\theta),$$

where $S_n(\theta) = \sum_{k=0}^{n} a_k e^{ik\theta}$. The second term on the righthand side converges uniformly to $f(e^{i\theta})$, after another theorem of Fejér. The first term is bounded in absolute value by $\frac{1}{n}\sum_{j=1}^{n} j|a_j|$, hence it tends uniformly to 0. As a result, $S_n(\theta) \to f(e^{i\theta})$ uniformly, without the necessity of invoking a Tauberian-style argument. However, it is difficult to give a convincing example of an application of Theorem 1.2.5 (possibly the convergence of the series $\sum \dfrac{(-1)^n |\sin n|}{n \ln n}$). In the same vein we have the following two results.

1.2.6 Theorem [Hardy (1910)]   *If $\dfrac{\sigma_n}{n} \to \ell$, and in addition we suppose na_n bounded, then $S_n \to \ell$.*

While it could still be considered "elementary" this theorem is nonetheless more difficult, and possibly more useful, than that of Tauber, as it has applications to Fourier series (functions of bounded variation). Naturally, Hardy posed the question whether this converse remained true for the method of Abel, that is if, in Tauber's theorem, we can replace $na_n \to 0$ by na_n bounded. We shall see Littlewood's response later. Landau, in 1910, gave an improvement to Hardy's theorem: for a_n real, the one-sided condition $na_n \geqslant -C$ is sufficient to improve $\frac{\sigma_n}{n} \to \ell$ to $S_n \to \ell$.

1.2.7 Theorem [Fatou (1905) [113, 157]] *Suppose that f admits an analytic continuation in the neighbourhood of $z = 1$, and that $a_n \to 0$. Then, $S_n \to f(1)$.*

This theorem was completely satisfactory, because it was clearly optimal: if $a_n = u^n$, with $|u| = 1$ and $u \neq 1$, $f(z) = \sum_{n=0}^{\infty}(uz)^n = \dfrac{1}{1 - uz}$ can be analytically continued around 1, however S_n diverges. This was not the case with the theorem of Tauber (or even that of Hardy). Keeping in mind the eternal counter-example seen previously ($a_n = u^n$), one could imagine a proposition of the form

$$\text{if } f(x) \to \ell \text{ and } a_n \to 0, \text{ then } S_n \to \ell.$$

We will see later Littlewood's answer (in the negative) to this question.

Another subject in the air in 1911 was the more general case of the Dirichlet series $\sum a_n e^{-\lambda_n x}$, with $\lambda_n \nearrow +\infty$ (in fact, this is Littlewood's framework) and the case of the power series corresponding to $\lambda_n = n$ via the change of variable $x = e^{-\varepsilon}$: $\sum a_n x^n = \sum a_n e^{-n\varepsilon}$. For these series, Landau proved the following generalisation of Tauber's theorem.

1.2.8 Theorem [Landau (1907)] *Set $\mu_n = \lambda_n - \lambda_{n-1} > 0$. If $f(\varepsilon) = \sum_{n=0}^{\infty} a_n e^{-\lambda_n \varepsilon} \to \ell$ as $\varepsilon \searrow 0$, and in addition $a_n = o\left(\dfrac{\mu_n}{\lambda_n}\right)$, then $S_n = a_0 + \cdots + a_n \to \ell$.*

Up to a change in notation, the proof is the same as for the theorem of Tauber: $S_N - f\left(\dfrac{1}{\lambda_N}\right) \to 0$, using the inequality

$$(\lambda_n - \lambda_{n-1})e^{-\lambda_n \varepsilon} \leqslant \int_{\lambda_{n-1}}^{\lambda_n} e^{-t\varepsilon} dt.$$

We now make a detailed analysis of Littlewood's article and its contributions.

1.3 Analysis of Littlewood's 1911 article

In 1911, John Edensor Littlewood was 26 years old: he was thus a beginner in the world of mathematics, as opposed to Godfrey Harold Hardy, who at 34 was already a confirmed and recognised mathematician. However, one can say that this article marks a triumphant entrance of the author into the big league. It contains (among others) four inspiring sections, that we will subsequently examine one by one.

(1) An analysis of Tauber's proof, and a non-trivial counter-example illustrating to what extent the conditions $na_n \to 0$ and $|na_n| \leqslant C$ are different.

(2) The tour de force of the article: the *affirmative* answer to Hardy's question, in a much more general context than that of power series – Dirichlet series. The proof was much more difficult than anything shown previously (see Exercise 1.8 for a heuristic justification).

(3) A proof of the *optimality* of the result obtained, again in a very general framework, with (prophetic) estimations of independent interest of sums of imaginary exponentials.

(4) A work plan, executed successfully afterwards, that gave birth to the field of Tauberian theorems for which, a century later, J. Korevaar [113] published his monumental *Tauberian Theory, a Century of Developments*.

1.3.1 A non-trivial example

When we analyse the proof of Tauber ($na_n \to 0$), we see that the essential point is as follows: $S_N - f(x) \to 0$ if $N \to +\infty$ and $x \nearrow 1$ in such a manner that $C_1 \leqslant N(1-x) \leqslant C_2$, where C_1 and C_2 are positive constants. It follows that by designating F the cluster set in $\mathbb{C} \cup \{\infty\}$ of $f(x)$, when $x \nearrow 1$, and S the cluster set of S_N, when $N \to +\infty$, we have:

$$\text{if } na_n \to 0, \text{ then } F = S. \tag{1.7}$$

This statement is evidently a generalisation of Tauber's theorem; later (Hadwiger *et al.*) it was shown that (at least in the case where $f(x)$ and S_N are bounded):

$$h(S, F) \leqslant H \varlimsup_{n \to +\infty} |na_n|, \tag{1.8}$$

where H is a numerical constant (the Hadwiger constant) and h is the Hausdorff distance between the compact sets S and F. But this result does not at all imply that $S_N \to \ell$ if $f(x) \to \ell$ and $na_n = O(1)$, and S and F can be very different. Littlewood examined the following case, in which α is a non-zero real number, which we suppose positive to simplify the ideas:

$$f(x) = \sum_{n=1}^{\infty} \frac{x^n}{n^{1+i\alpha}}, \; S_N = \sum_{n=1}^{N} \frac{1}{n^{1+i\alpha}}, \; a_n = \frac{1}{n^{1+i\alpha}}.$$

Littlewood established the following proposition, using the previous definitions of F and S.

1.3.2 Proposition *For the example above, if w is the value of the Riemann zeta function ζ at the point $1 + i\alpha$, we have:*

(1) $F = \left\{ z / |z - w| = |\Gamma(-i\alpha)| = \dfrac{1}{\alpha} \sqrt{\dfrac{\pi\alpha}{\sinh(\pi\alpha)}} \right\},$

(2) $S = \left\{ z / |z - w| = \dfrac{1}{\alpha} \right\}.$

We thus see that $f(x)$ and S_N describe a (circular) orbit around the same sun $w = \zeta(1 + i\alpha)$, but the orbit of S_N is *external* to the orbit of f, and in a certain sense S_N escapes from the influence of f. Moreover, the respective radii R_F and R_S of the orbits satisfy

$$\frac{R_S}{R_F} = \sqrt{\frac{\sinh(\pi\alpha)}{\pi\alpha}} \to +\infty \text{ as } \alpha \to +\infty,$$

even if the difference $R_S - R_F$ stays bounded when α varies within $]0, +\infty[$, in accordance with (1.8), given that $n|a_n| = 1$ for all values of n and α. See Figure 1.1.

Littlewood states that (2) is well known and that (1) (there is in fact an error in sign) can be found in the book *Le calcul des résidus* by Lindelöf (p. 139). Here is a proof, based on a classical formula for the comparison of series/integrals for C^1 complex-valued functions, where $\{t\} = t - [t]$ designates the fractional part of a real number t:

$$\sum_{n=2}^{N} \varphi(n) = \int_{1}^{N} \varphi(t) \, dt + \int_{1}^{N} \{t\} \varphi'(t) \, dt, \tag{1.9}$$

with N an integer $\geqslant 2$ and φ a function C^1 from $[1, N]$ to $\mathbb{C}$ [149, 153]. Setting $x = e^{-\varepsilon}$ for $0 < x < 1$ and applying (1.9) to the function $\varphi(t) = t^{-1-i\alpha} e^{-\varepsilon t}$,

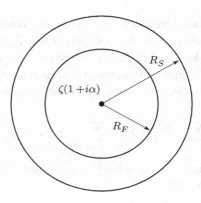

Figure 1.1

we obtain for $f(x) = \sum_{n=1}^{\infty} x^n n^{-1-i\alpha}$ the representation

$$f(x) - x = \sum_{n=2}^{\infty} x^n n^{-1-i\alpha}$$

$$= \int_1^{+\infty} t^{-1-i\alpha} e^{-\varepsilon t} dt - \int_1^{+\infty} \{t\} \left((1 + i\alpha) t^{-2-i\alpha} \right.$$

$$\left. + \varepsilon\, t^{-1-i\alpha} \right) e^{-\varepsilon t} dt.$$

The first integral becomes, after integration by parts:

$$\left[\frac{t^{-i\alpha}}{-i\alpha} e^{-\varepsilon t} \right]_1^{+\infty} - \frac{1}{i\alpha} \int_1^{+\infty} t^{-i\alpha} (\varepsilon e^{-\varepsilon t})\, dt$$

$$= \frac{e^{-\varepsilon}}{i\alpha} - \frac{\varepsilon^{i\alpha}}{i\alpha} \int_\varepsilon^{+\infty} u^{-i\alpha} e^{-u} du$$

$$= \frac{1}{i\alpha} - \frac{\varepsilon^{i\alpha}}{i\alpha} \Gamma(1 - i\alpha) + o(1) = \frac{1}{i\alpha} + \varepsilon^{i\alpha}\, \Gamma(-i\alpha) + o(1)$$

as $\varepsilon \searrow 0$. For the second integral, we observe that

$$\left| \int_1^{+\infty} \{t\}\, \varepsilon t^{-1-i\alpha} e^{-\varepsilon t} dt \right| \leqslant \int_1^{+\infty} \varepsilon e^{-\varepsilon t} \frac{dt}{t} = \varepsilon \int_\varepsilon^{+\infty} e^{-u} \frac{du}{u}$$

$$\leqslant \varepsilon \left(\ln \frac{1}{\varepsilon} + \int_1^{+\infty} e^{-u} du \right)$$

$$\leqslant \varepsilon \ln \frac{1}{\varepsilon} + \varepsilon = o(1),$$

while

$$\int_1^{+\infty} \{t\} t^{-2-i\alpha} e^{-\varepsilon t} dt \to \int_1^{+\infty} \{t\} t^{-2-i\alpha} dt \text{ as } \varepsilon \searrow 0.$$

In summary,

$$f(x) - x = \varepsilon^{i\alpha}\,\Gamma(-i\alpha) + \frac{1}{i\alpha} - (1 + i\alpha)\int_1^{+\infty}\{t\}t^{-2-i\alpha}dt + o(1),$$

and hence

$$f(x) = f(e^{-\varepsilon}) = \varepsilon^{i\alpha}\,\Gamma(-i\alpha) + \zeta(1 + i\alpha) + o(1), \qquad (1.10)$$

because we shall see that $1 + \frac{1}{i\alpha} - (1 + i\alpha)\int_1^{+\infty}\{t\}t^{-2-i\alpha}dt$ is one of the many avatars of $\zeta(1 + i\alpha)$.

An elegant use of the following three functional equations satisfied by the function Γ:

$$\Gamma(1 + u) = u\Gamma(u), \ \Gamma(u)\Gamma(1 - u) = \frac{\pi}{\sin(\pi u)}, \ \Gamma(\overline{u}) = \overline{\Gamma(u)}$$

allows us to compute the absolute value of $\Gamma(-i\alpha)$ and thus prove (1): setting $z = -i\alpha$, we have

$$\frac{\pi}{\sin(\pi z)} = \Gamma(z)\Gamma(1 - z) = -z\Gamma(z)\Gamma(-z) = -z\Gamma(z)\Gamma(\overline{z}) = -z\Gamma(z)\overline{\Gamma(z)},$$

hence $\dfrac{i\pi}{\sinh(\pi\alpha)} = i\alpha|\Gamma(-i\alpha)|^2$, so that

$$|\Gamma(-i\alpha)|^2 = \frac{\pi}{\alpha\sinh(\pi\alpha)} = \frac{1}{\alpha^2}\frac{\pi\alpha}{\sinh(\pi\alpha)}.$$

Similarly, (1.9) gives

$$\sum_{n=2}^{N} n^{-1-i\alpha} = \int_1^N t^{-1-i\alpha}dt - (1 + i\alpha)\int_1^N \{t\}t^{-2-i\alpha}dt$$

and

$$\sum_{n=2}^{N} n^{-s} = \int_1^N t^{-s}dt - s\int_1^N \{t\}t^{-s-1}dt \text{ if Re } s > 1.$$

Letting $N \to +\infty$ in the second equality, we obtain

$$\zeta(s) = 1 + \frac{1}{s - 1} - s\int_1^{+\infty}\{t\}t^{-s-1}dt,$$

the second term giving an analytic continuation of ζ for Re $s > 0$ and $s \neq 1$. The first equality thus gives

$$S_N = \frac{N^{-i\alpha} - 1}{-i\alpha} + 1 - (1 + i\alpha)\int_1^{+\infty}\{t\}t^{-2-i\alpha}dt + o(1),$$

or again

$$S_N = \sum_{n=1}^{N} n^{-1-i\alpha} = \frac{N^{-i\alpha}}{-i\alpha} + \zeta(1+i\alpha) + o(1). \qquad (1.11)$$

This proves statement (2) of Proposition 1.3.2. If we compare (1.10) and (1.11), we easily see that $f(x)$ and S_N both describe circular orbits around $\zeta(1+i\alpha)$, but with radii $|\Gamma(-i\alpha)|$ and $\frac{1}{\alpha}$, which are different.

1.3.3 The tour de force of the article

Today, we have identified the difficulty in Littlewood's Theorem 1.3.4 (known as Theorem B in the article [124], to be described below). We used to work in the rigid context of a single power series $\sum a_n x^n$; Littlewood had the audacity and imagination to plunge the situation into a space E of functions on $[0, 1]$ by considering $\sum a_n x^n \varphi(x^n)$, $\varphi \in E$; he then profited from the flexibility of using these functions in E.

But first, we make the following statement.

1.3.4 Theorem [Littlewood (1911)] *Let $f(x) = \sum_{n=0}^{\infty} a_n x^n$, defined for $|x| < 1$. Suppose that $f(x) \to \ell$ as $x \nearrow 1$ and, in addition, $n|a_n| \leqslant C$. Then, $S_n = a_0 + \cdots + a_n \to \ell$.*

The works of Karamata (1930), Wiener (1932), and Wielandt (1952) have given a more-or-less definitive form to the proof, which is composed of two steps [113].

Step 1: linearisation. This step is an effortless (but useful) *self-reinforcement* of the hypothesis; it is possible here (with the substitution $x \mapsto x^r$) that we use the hypothesis $f(x) \to \ell$ less superficially than Tauber. It consists of showing that, for $\mathcal{P}$ the space of polynomials, we automatically have

$$\sum_{n=0}^{\infty} a_n P(x^n) \to \ell P(1) \text{ as } x \nearrow 1, \text{ for all } P \in \mathcal{P} \text{ such that } P(0) = 0. \quad (1.12)$$

Step 2: approximation. This step only works because of the Tauberian condition $|na_n| \leqslant C$. It consists of observing that $S_N = \sum_{n=0}^{N} a_n$ can be written as

$$S_N = \sum_{n=0}^{\infty} a_n g(x_N^n) = \phi(x_N),$$

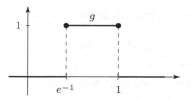

Figure 1.2

where g is the characteristic function of $[e^{-1}, 1]$ (Figure 1.2), $x_N = e^{-1/N}$ and $\phi(x) = \sum_{n=0}^{\infty} a_n g(x^n)$, and then demonstrating, using the approximation theorem of Weierstrass, that we still have

$$\phi(x) = \sum_{n=0}^{\infty} a_n g(x^n) \to \ell \text{ as } x \nearrow 1, \tag{1.13}$$

which completes the proof, as $S_N = \phi(x_N)$.

Let us elaborate on the above. The assertion (1.12) holds for the monomials $x^r, r \in \mathbb{N}^*$, and is equivalent to the statement

$$\sum_{n=0}^{\infty} a_n P(x^n) \to \ell, \text{ if } P(0) = 0 \text{ and } P(1) = 1.$$

We parametrise the polynomials fixing 0 and 1 by setting $P(x) = x + x(1 - x) Q(x)$, where $Q \in \mathcal{P}$ is arbitrary. Accordingly, we set

$$g(x) = x + x(1 - x)h(x),$$

with

$$h(x) = \begin{cases} -\dfrac{1}{1 - x} & \text{if } 0 \leqslant x < e^{-1}, \\[2mm] \dfrac{1}{x} & \text{if } e^{-1} \leqslant x \leqslant 1. \end{cases}$$

The assertions (1.12) and (1.13) become, respectively,

$$\sum_{n=1}^{\infty} a_n x^n (1 - x^n) Q(x^n) \to 0 \text{ for all } Q \in \mathcal{P} \tag{1.14}$$

and

$$\sum_{n=1}^{\infty} a_n x^n (1 - x^n) h(x^n) \to 0. \tag{1.15}$$

To go from (1.14) to (1.15), we must control the "error"

$$E(x) = \sum_{n=1}^{\infty} a_n x^n (1 - x^n) \left(h(x^n) - Q(x^n) \right).$$

However, the hypothesis and the inequality

$$1 - x^n \leqslant n(1 - x) \text{ for } 0 \leqslant x \leqslant 1$$

(already important for Tauber, see (1.3)) give

$$|E(x)| \leqslant \sum_{n=1}^{\infty} \frac{C}{n} x^n n(1 - x) \left| h(x^n) - Q(x^n) \right|,$$

so that

$$\varlimsup_{x \nearrow 1} |E(x)| \leqslant C \varlimsup_{x \nearrow 1} \sum_{n=1}^{\infty} x^n (1 - x) \left| h(x^n) - Q(x^n) \right|$$

$$= C \int_0^1 |h(t) - Q(t)| \, dt,$$

because if $\varphi : [0, 1] \rightarrow \mathbb{C}$ is piecewise continuous, using as points of subdivision the x^n, we have

$$\sum_{n=1}^{\infty} x^n (1 - x) \varphi(x^n) \rightarrow \int_0^1 \varphi(t) \, dt \text{ as } x \nearrow 1.$$

The reader is invited to consult [113] for details. We conclude here, by adjusting the polynomial Q so that the L^1-norm of $h - Q$ becomes small.

And how did Littlewood, the inventor of all this, figure this out 20 or 40 years earlier? Well, much as we have not invented much about transcendence since Hermite, he did much the same. In fact, he considered the more general case of Dirichlet series $\sum a_n e^{-\lambda_n x}$, $x > 0$. In order to lighten the presentation, we transcribe his work in the context of power series and analyse at the same time what was discovered and what was missed. There are three steps along the way.

Step 1: Abel transformation. Write $x = e^{-\varepsilon}$, $\varepsilon > 0$, $f(x) = \sum_{n=0}^{\infty} a_n x^n$. Then, we have

$$f(x) = (1 - x) \sum_{n=0}^{\infty} S_n x^n = \varepsilon \int_0^{+\infty} e^{-\varepsilon t} S(t) \, dt \rightarrow \ell \text{ as } x \nearrow 1, \quad (1.16)$$

with $S(t) = \sum_{n \leqslant t} a_n$. This step is not absolutely necessary, and is in fact somewhat penalising: instead of leading directly to $S_n \rightarrow \ell$, we only obtain $\dfrac{S_0 + \cdots + S_{n-1}}{n} \rightarrow \ell$, but can then conclude using Theorem 1.2.6 of Hardy.

Step 2: linearisation or, more accurately, "monomisation". This step was the great discovery of Littlewood; it was revolutionary and underlies almost all the Tauberian theorems of the twentieth century. It consists of showing that (1.16) implies

$$\varepsilon^{r+1} \int_0^{+\infty} t^r S(t) e^{-\varepsilon t} dt \to \ell\, r!, \text{ for all } r \in \mathbb{N}. \tag{1.17}$$

Littlewood thus introduces by force an extra degree of freedom (the parameter r), which will prove to be decisive. But the introduction of r *is not for free*: to obtain (1.17), we need to use a theorem (Theorem A of Littlewood), which was in fact already known (Hadamard, Kneser, etc.), and is presented below.

1.3.5 Theorem *Let $\phi : \mathbb{R}_+ \to \mathbb{C}$ be C^∞ such that $\phi(x) \to s$ as $x \to +\infty$, and let $\phi', \phi'', \ldots$ be bounded on $\mathbb{R}_+$. Then, $\phi'(x), \phi''(x), \ldots$ tend to 0 as $x \to +\infty$.*

Moreover, today we know that the parameter r is misplaced: it would be better to write

$$\varepsilon \int_0^{+\infty} e^{-\varepsilon r t} S(t)\, dt \to \ell \int_0^{+\infty} e^{-rt} dt \text{ as } \varepsilon \searrow 0, \text{ for all } r > 0,$$

which we obtain effortlessly from (1.16) by changing ε to εr.

Step 3: approximation and large values of r. If Littlewood had thought of linearising (1.17), he would have obtained

$$\varepsilon \int_0^{+\infty} P(\varepsilon t) S(t) e^{-\varepsilon t} dt \to \ell \int_0^{+\infty} P(t) e^{-t} dt \text{ as } \varepsilon \searrow 0, \text{ for all } P \in \mathcal{P}.$$

He could then have used the Weierstrass approximation to finish. However, he missed that linearisation. Nonetheless, he succeeded (with underlying peak functions) using only monomials t^r as follows.[5]

We can suppose that $\ell = 0$. The condition $n a_n = O(1)$ implies that $S(t)$ oscillates slowly.[6] If $S(t)$ does not tend to 0 when $t \to +\infty$, we create special values of ε tending to 0 such that, if r is very large, $\varepsilon^{r+1} \int_0^{+\infty} t^r S(t) e^{-\varepsilon t} dt$ is not close to zero as it should be, after (1.17). The reasoning requires some rather painful estimations of the integrals

---

[5] A bit like trying to prove the theorem of Merten, if $\int_0^1 t^r f(t)\, dt = 0$ for all $r \in \mathbb{N}$, then $f = 0$, without linearising.

[6] This idea of slow oscillation is a second revolutionary idea, but in fact was already present in Theorem 1.2.6 of Hardy in 1910, and abundantly adapted by Karamata, Pitt *et al.*

$$\int_0^r t^r \ln \frac{r}{t} e^{-t} dt \text{ and } \int_r^{+\infty} t^r \ln \frac{t}{r} e^{-t} dt,$$

both $O(r^r e^{-r}) = O(r!)$. But Littlewood succeeds with these estimations and hence concludes the proof of his Theorem B.

1.3.6 Optimality of the Tauberian condition $a_n = O(n^{-1})$

The optimality of the Tauberian condition corresponds to Theorem C of Littlewood's paper (the great predators leave nothing but clean carcasses for passing jackals...), which also shows the optimality of the condition $na_n = O(1)$ in Hardy's theorem of 1910, and is as follows.

1.3.7 Theorem *Let $(\varphi_n)_{n \geqslant 1}$ be a sequence of positive real numbers that tend to $+\infty$. Then there exists a sequence $(a_n)_{n \geqslant 0}$ of complex numbers such that:*

(1) $|a_n| \leqslant \frac{\varphi_n}{n}$ for all $n \geqslant 1$,

(2) $\dfrac{S_0 + \cdots + S_n}{n} \to 0$ and hence (by Frobenius):

$$f(x) = \sum_{n=0}^{\infty} a_n x^n \to 0 \text{ as } x \nearrow 1,$$

(3) S_n does not have a limit when $n \to +\infty$.

In the proof of this theorem, we will modify Littlewood's example slightly in the choice of complex numbers a_n (even if everything can be found in his text). Set

$$\Phi(n) = \sum_{j=1}^{n} \frac{\varphi_j}{j}$$

(for convenience taking $\Phi(0) = 0$) and adjust a_n so that we have

$$S_0 = 0 \text{ and } S_n = e^{i\Phi(n)} \text{ if } n \geqslant 1,$$

in other words

$$a_0 = 0, a_1 = e^{i\Phi(1)}, \ a_n = e^{i\Phi(n)} - e^{i\Phi(n-1)} \text{ if } n \geqslant 2.$$

It is useful to first see (keep in mind the example $\varphi_n = \ln n$) that we can suppose, without loss of generality, that $\varphi_n \leqslant \sqrt{n}$ and $n^{-1}\varphi_n$ decreases to 0, with the aid of the following lemma.

1.3.8 Lemma *There exists a sequence $(\psi_n)_{n \geqslant 1}$ of positive real numbers such that*

(1) $\psi_n \leqslant \min(\varphi_n, \sqrt{n})$,
(2) $\psi_n \to +\infty$,
(3) $n^{-1}\psi_n$ *decreases to* 0.

Proof Set $\varphi_n' = \min(\varphi_n, \sqrt{n})$ and $c = \min\limits_{n \geqslant 1} \varphi_n' > 0$, and consider ψ_n defined by

$$\frac{\psi_n}{n} = \min_{1 \leqslant k \leqslant n} \frac{\varphi_k'}{k}.$$

By construction, $n^{-1}\psi_n$ is decreasing; set $\lambda_n = [\sqrt{n}]$ (the integer part of $\sqrt{n}$), and distinguish two cases for the values of k:

- if $1 \leqslant k \leqslant \lambda_n$,

$$n\frac{\varphi_k'}{k} \geqslant \frac{nc}{\sqrt{n}} = c\sqrt{n};$$

- if $\lambda_n < k \leqslant n$,

$$n\frac{\varphi_k'}{k} \geqslant \varphi_k' \geqslant \inf_{j > \lambda_n} \varphi_j' =: \varphi_n'',$$

which tends to $+\infty$ with n.

Thus,

$$\psi_n \geqslant \min(c\sqrt{n}, \varphi_n'') \text{ and } \psi_n \to +\infty.$$

Finally, $\dfrac{\psi_n}{n} \leqslant \dfrac{\varphi_n'}{n}$, and so we have

$$\psi_n \leqslant \varphi_n' = \min(\varphi_n, \sqrt{n}). \qquad \square$$

Now, if we know how to find a counter-example such that $|a_n| \leqslant n^{-1}\psi_n$, this counter-example will also verify $|a_n| \leqslant n^{-1}\varphi_n$. In what follows, we thus suppose that

$$\varphi_n \leqslant \sqrt{n} \text{ and } n^{-1}\varphi_n \text{ decreases to } 0.$$

We then see that

$$|a_n| \leqslant \Phi(n) - \Phi(n-1) = \frac{\varphi_n}{n} \text{ if } n \geqslant 2,$$

$$\Phi(n) \to +\infty$$

and

$$\Phi(n) - \Phi(n - 1) = \frac{\varphi_n}{n} \text{ decreases to } 0.$$

We know classically [152] that the cluster set of the sequence $(S_n)_{n \geqslant 0}$ is the entire unit circle, and in particular this sequence diverges. The only thing left to show is assertion (2) of Theorem 1.3.7, that is:

$$\sigma_N = \sum_{n=1}^{N} e^{i\Phi(n)} = o(N).$$

For this, Littlewood, anticipating the work of van der Corput (1922–1937), is clearly guided by the analogy with the integral

$$\int_1^N e^{i\Phi(x)} dx = \int_1^N \frac{(e^{i\Phi})'}{i\Phi'} dx = \left[\frac{e^{i\Phi}}{i\Phi'} \right]_1^N - \int_1^N e^{i\Phi} d\left(\frac{1}{i\Phi'} \right),$$

wherein

$$\left| \int_1^N e^{i\Phi(x)} dx \right| \leqslant \frac{1}{\Phi'(N)} + \frac{1}{\Phi'(1)} + \int_1^N \left| d\left(\frac{1}{\Phi'} \right) \right|,$$

which is interesting when Φ' is positive and for example decreasing (so that $\frac{1}{\Phi'}$ is increasing). We obtain in this case:

$$\begin{aligned}
\left| \int_1^N e^{i\Phi(x)} dx \right| &\leqslant \frac{1}{\Phi'(N)} + \frac{1}{\Phi'(1)} + \left| \int_1^N d\left(\frac{1}{\Phi'} \right) \right| \\
&= \frac{1}{\Phi'(N)} + \frac{1}{\Phi'(1)} + \frac{1}{\Phi'(N)} - \frac{1}{\Phi'(1)} \\
&= \frac{2}{\Phi'(N)}.
\end{aligned}$$

By analogy, Littlewood would like to write

$$e^{i\Phi(n)} = \frac{\Delta e^{i\Phi(n)}}{i\Delta\Phi(n)},$$

where for a sequence $(u_n)_{n \geqslant 0}$ the derivative is replaced by the first difference $\Delta u_n = u_n - u_{n-1}$. The catch is that the preceding formula is false, but not by much. If we set $h_n = \Phi(n) - \Phi(n-1) = \frac{\varphi_n}{n}$ (the term h_n decreases to 0 by hypothesis), we have

$$\Delta e^{i\Phi(n)} = e^{i\Phi(n)} - e^{i(\Phi(n)-h_n)} = e^{i\Phi(n)}(1 - e^{-ih_n}),$$

thus

$$e^{i\Phi(n)} = \frac{\Delta e^{i\Phi(n)}}{1 - e^{-ih_n}}. \tag{1.18}$$

Moreover, $h_n \to 0$, so that

$$1 - e^{-ih_n} = ih_n + O(h_n^2) = ih_n \left(1 + O(h_n)\right),$$

whereby

$$e^{i\Phi(n)} = \frac{\Delta e^{i\Phi(n)}}{i\Delta\Phi(n)\left(1 + O(h_n)\right)} = \frac{\Delta e^{i\Phi(n)}}{i\Delta\Phi(n)} + O(\Delta e^{i\Phi(n)}).$$

As

$$\left|\Delta e^{i\Phi(n)}\right| \leqslant \Phi(n) - \Phi(n-1) = h_n,$$

we thus have

$$e^{i\Phi(n)} = \frac{\Delta e^{i\Phi(n)}}{i\Delta\Phi(n)} + O(h_n). \tag{1.19}$$

The estimate (1.19) is sufficient to establish the following proposition, very interesting in itself.

1.3.9 Proposition [Littlewood]   *If h_n decreases to 0, we have the inequality*

$$\left|\sum_{n=1}^{N} e^{i\Phi(n)}\right| = |\sigma_N| \ll \frac{1}{h_N} + \sum_{n=1}^{N} h_n = \frac{1}{h_N} + \Phi(N), \tag{1.20}$$

the notation $A \ll B$ signifying that $A \leqslant \lambda B$, where λ is some positive constant.

To show (1.20) starting from (1.19), it is sufficient to replace the integration by parts with an Abel transformation, where the growth of $\frac{1}{\Phi'}$ is replaced by that of $\frac{1}{h_n}$, and the sum $\sum_{n=1}^{N} h_n$ comes from the error term in the estimate (1.19).

We can see that (1.20) already implies some non-trivial results; for example, if $\Phi(n) = (\ln n)^\alpha$ with $\alpha > 1$, we obtain

$$\left|\sum_{n=1}^{N} e^{i\Phi(n)}\right| \ll \frac{N}{(\ln N)^{\alpha-1}} + O(\ln N)^\alpha \ll \frac{N}{(\ln N)^{\alpha-1}},$$

and then

$$\frac{1}{N}\sum_{n=1}^{N} e^{i\Phi(n)} \to 0.$$

The sequence $(\Phi(n))_{n \geqslant 1}$ is thus equidistributed modulo one, a result usually attributed to Koksma in the 1930s. The result is false for $\alpha = 1$, and more generally if $\Phi(n) = a \ln n$ with $a \in \mathbb{R}^*$, we have

$$N^{-1} \sum_{n=1}^{N} e^{i\Phi(n)} = N^{-1} \sum_{n=1}^{N} n^{ia} = N^{ia} \frac{1}{N} \sum_{n=1}^{N} \left(\frac{n}{N}\right)^{ia}$$

$$\sim N^{ia} \int_{0}^{1} t^{ia} dt,$$

so that

$$\left| N^{-1} \sum_{n=1}^{N} e^{i\Phi(n)} \right| \to \left| \frac{1}{1+ia} \right| = \frac{1}{\sqrt{1+a^2}} \in \,]0, 1[.$$

Nonetheless, the bound (1.20) is somewhat imprecise as the error term $\sum_{n=1}^{N} h_n$ can be too large.[7] If, for example, $\Phi(n) = n^\alpha$ with $0 < \alpha < 1$, the inequality (1.20) gives ($h_n \simeq n^{\alpha-1}$):

$$|\sigma_N| \ll N^{1-\alpha} + \sum_{n=1}^{N} n^{\alpha-1} \ll N^{1-\alpha} + N^\alpha,$$

in other words $|\sigma_N| \ll N^\alpha$ if $\frac{1}{2} < \alpha < 1$, while in fact the correct order of magnitude is $|\sigma_N| \ll N^{1-\alpha}$. However, in the book of N. Bary ([15], Vol. 2, pp. 167–171), there is a lemma now famous among number theorists, obtained by Kuzmin (with geometrical methods) and then by Landau (with analytic methods) in the 1930s, some 20 years after Littlewood. Here it is.

1.3.10 Lemma [Kuzmin–Landau, cf. [15]]   *Suppose that*

$$\pi > h_1 \geqslant h_2 \geqslant \cdots \geqslant h_N > 0,$$

where, as above, $h_n = \Phi(n) - \Phi(n-1)$. Then we have the following inequality:

$$\left| \sum_{n=1}^{N} e^{i\Phi(n)} \right| \leqslant \frac{C}{h_N}, \tag{1.21}$$

where C is a positive absolute constant.

The proof consists of improving the passage from (1.18) to (1.19), by taking more care of the term $(1 - e^{-ih_n})^{-1}$; we have

$$\frac{1}{1 - e^{-ih_n}} = \frac{e^{ih_n/2}}{e^{ih_n/2} - e^{-ih_n/2}} = \frac{\cos(h_n/2) + i\sin(h_n/2)}{2i\sin(h_n/2)}$$

$$= \frac{1}{2i} \cot \frac{h_n}{2} + \frac{1}{2},$$

---

[7] Not for Littlewood, for whom $h_n = n^{-1}\varphi_n$ is hardly greater than n^{-1}, so that $\sum_{n=1}^{N} h_n$ is barely greater than $\ln N$.

so that

$$e^{i\Phi(n)} = \frac{\Delta e^{i\Phi(n)}}{2i} \cot\frac{h_n}{2} + \frac{1}{2}\Delta e^{i\Phi(n)},$$

and an Abel transformation gives exactly (1.21), due to the fortunate property of the cotangent function being decreasing on $]0, \pi[$. The inequality (1.21) gives, for example,

$$\left|\sum_{n=1}^{N} e^{in^\alpha}\right| \ll N^{1-\alpha} \text{ if } 0 < \alpha < 1,$$

and thus can be seen as an improvement over Littlewood's method.

We come back to (1.20) in order to complete the proof of Theorem C. We have, since $h_n = \frac{\varphi_n}{n}$ and $\varphi_n \leqslant \sqrt{n}$:

$$|\sigma_N| = \left|\sum_{n=1}^{N} e^{i\Phi(n)}\right| \ll \frac{N}{\varphi_N} + \sum_{n=1}^{N}\frac{1}{\sqrt{n}} \ll \frac{N}{\varphi_N} + \sqrt{N},$$

so that

$$\left|\frac{\sigma_N}{N}\right| \ll \frac{1}{\varphi_N} + \frac{1}{\sqrt{N}},$$

and $\frac{\sigma_N}{N} \to 0$ since $\varphi_N \to +\infty$.

Before continuing, here is a helpful remark on the more general context of Dirichlet series in which Littlewood worked.

Let $(\lambda_n)_{n \geqslant 0}$ be an increasing sequence of positive real numbers, with limit $+\infty$, and $(a_n)_{n \geqslant 0}$ a sequence of complex numbers. Set

$$S(t) = \sum_{\lambda_n \leqslant t} a_n$$

and

$$f(x) = \sum_{n=0}^{\infty} a_n e^{-\lambda_n x} = x\int_0^{+\infty} e^{-xt} S(t)\, dt.$$

Also

$$\mu_n = \lambda_n - \lambda_{n-1} \text{ and } \nu_n = \frac{\mu_n}{\lambda_n} = 1 - \frac{\lambda_{n-1}}{\lambda_n}.$$

Then, from Littlewood, we obtain the following results.

1.3.11 Theorem *Suppose that*

$$f(x) = \sum_{n=0}^{\infty} a_n e^{-\lambda_n x}$$

exists for all x > 0, and that:

(1) $f(x) \to \ell$ *as* $x \searrow 0$;

(2) $|a_n| \leqslant C \dfrac{\mu_n}{\lambda_n}$, *where C is a positive constant.*

Then, if $\dfrac{\lambda_{n+1}}{\lambda_n} \to 1$, *we have*

$$S_n = a_0 + a_1 + \cdots + a_n \to \ell.$$

1.3.12 Theorem   *The Tauberian condition* $a_n = O(\lambda_n^{-1} \mu_n)$ *of the preceding theorem is optimal in the following sense (supposing again that* $v_n \to 0$*). If* $\varphi_n \to +\infty$, *there exists* $(a_n)_{n \geqslant 1}$ *such that*

(1) $a_n = O\left(\varphi_n \dfrac{\mu_n}{\lambda_n} \right) = O(\varphi_n v_n)$;

(2) $\dfrac{\mu_1 S_1 + \cdots + \mu_n S_n}{\mu_1 + \cdots + \mu_n} \to 0$ *where* $S_n = a_0 + \cdots + a_n$, *and hence*[8]

$$f(x) = \sum_{n=0}^{\infty} a_n e^{-\lambda_n x} \to 0 \text{ as } x \searrow 0;$$

(3) S_n *does not have a limit when* $n \to +\infty$.

The proof is almost identical to the above, with the difference that we are manipulating *weighted* exponential sums: we do not repeat the argument here. It suffices to perform an Abel transformation and to know that if $a_n > 0$ and

$$S_n = a_0 + \cdots + a_n \to +\infty,$$

then the series $\sum \dfrac{a_n}{S_n}$ is divergent.

1.3.13 Collaboration and follow-on work

The work of Littlewood was not notable for its elegance (does mathematics always have to be a beauty contest?), but his talent and creativity shine through, as well as a certain style: that of an ardent fan of "hard" analysis, under whose feet conjectures don't grow. Unsurprisingly, this paper caught Hardy's

---

[8]   According to a result of Hardy (see [73] or [77], Vol. VI, p. 311) whereby if, for all $x > 0$, $S_n e^{-\lambda_n x} \to 0$ as $n \to +\infty$, and

$$\frac{\mu_1 S_1 + \cdots + \mu_n S_n}{\mu_1 + \cdots + \mu_n} \to \ell,$$

then $f(x) = \sum_{n=0}^{\infty} a_n e^{-\lambda_n x}$ exists for all $x > 0$ and $f(x) \to \ell$ as $x \searrow 0$.

attention, and he sought to collaborate with this brilliant young mathematician who had solved his problem for him.

In fact, this collaboration had already started in 1911. When one digs a bit deeper, it is apparent that the equality $F = S$ of (1.7) is due to Hardy, as well as the non-trivial example $\sum_{n=1}^{\infty} \dfrac{x^n}{n^{1+i\alpha}}$. And there are similar examples in the paper, even though the principal Theorems B and C truly belong to Littlewood, who can be considered (even more than Tauber) to be the father of Tauberian theorems. Theorem B has passed to posterity under the name of the Hardy–Littlewood theorem, actually under a slightly less restrictive hypothesis when the a_n are real numbers: the bilateral estimate $n|a_n| \leqslant C$ can be replaced by a unilateral estimate $na_n \geqslant -C$, for example. But this refinement (sometimes useful, see Exercise 1.12) is infinitely less difficult than the passage from $na_n \to 0$ to $n|a_n| \leqslant C$.

As for follow-on work, it was obvious that Littlewood's results raised a host of questions; here are three among hundreds.

Question 1. Theorems 1.3.11 and 1.3.12 are established under the hypothesis

$$\frac{\lambda_{n+1}}{\lambda_n} \to 1.$$

What happens if we omit this condition? Littlewood claimed to be "practically certain" that if $\dfrac{\lambda_{n+1}}{\lambda_n} \geqslant q > 1$, then no Tauberian condition is necessary for the implication

$$\sum_{n=0}^{\infty} a_n e^{-\lambda_n x} \xrightarrow[x \searrow 0]{} \ell \Rightarrow \sum_{n=0}^{N} a_n \to \ell.$$

His intuition was correct: this is the theorem of high indices that he proved in 1926 with Hardy, variants of which reappear in a very recent work [72]. The theorem of high indices was completed in 1966 by W. Rudin [162]: if $\underline{\lim}_{n \to +\infty} \dfrac{\lambda_{n+1}}{\lambda_n} = 1$, we *must have* a Tauberian condition, because an example exists where $\sum_{n=0}^{\infty} a_n e^{-\lambda_n x} \to \ell$ and $\sum a_n$ diverges. As for Theorem 1.3.11, Ananda-Rau showed in 1928 [113] that it remains true when we remove the hypothesis $\dfrac{\lambda_{n+1}}{\lambda_n} \to 1$, but keep the *bilateral* Tauberian condition $|a_n| \leqslant C \mu_n \lambda_n^{-1}$.

Question 2. In the general form given by Littlewood:

$$x \int_0^{+\infty} S(t) e^{-xt} dt \xrightarrow[x \searrow 0]{} \ell \text{ and } a_n = O(\star) \Rightarrow S(t) \xrightarrow[t \to +\infty]{} \ell,$$

can we replace the exponential kernel by another? That is to say, can we have

$$x \int_0^{+\infty} N(xt)S(t)\,dt \xrightarrow[x \searrow 0]{} \ell \text{ and } a_n = O(\star) \text{ (or } S(t) = \ldots) \Rightarrow S(t) \xrightarrow[t \to +\infty]{} \ell,$$
(1.22)

or at least

$$t^{-1} \int_0^t S(u)\,du \to \ell \text{ as } t \to +\infty,$$

where the "kernel" N is a positive function in $L^1(\mathbb{R}_+)$ such that[9]

$$\int_0^{+\infty} N(t)\,dt = 1?$$

And would a good choice of N lead to a proof of the prime number theorem? The response (*affirmative for the prime number theorem*) was given by N. Wiener [186] (see also Chapter 2 of the present book) in his monumental 100-page article: the implication (1.22) holds if and only if $\int_0^{+\infty} N(t)t^{i\xi}\,dt \neq 0$ for all real numbers ξ. This is true for $N(t) = e^{-t}$, as

$$\int_0^{+\infty} e^{-t}t^{i\xi}\,dt = \Gamma(1 + i\xi) \neq 0!.$$

It is also true for the kernel $N(t) = -\dfrac{d}{dt}\left(\dfrac{t}{e^t - 1}\right)$ associated with the Lambert summation method $\sum a_n \dfrac{nx}{e^{nx} - 1}$, because (see [113] or Exercise 2.8):

$$\int_0^{+\infty} N(t)t^{i\xi}\,dt = i\xi\,\Gamma(1 + i\xi)\,\zeta(1 + i\xi).$$

The Tauberian theorem associated with the Lambert summation method, that is,

$$\sum_{n=1}^{\infty} a_n \frac{nx}{e^{nx} - 1} \xrightarrow[x \searrow 0]{} \ell \text{ and } na_n \geqslant -C \Rightarrow \sum_{n=1}^{N} a_n \to \ell,$$

implies the convergence of the series $\sum n^{-1}(\Lambda(n) - 1)$, where Λ is the von Mangoldt function.

Using Kronecker's lemma, this convergence implies

$$\sum_{n \leqslant x} \Lambda(n) \sim x \text{ as } x \to +\infty,$$

---

[9] This is so that the integral in (1.22) appears as a barycentre of the values S.

a statement which is notoriously and easily equivalent to the prime number theorem (see Exercise 1.12). Note that the initial proof by Hardy–Littlewood used even more than the prime number theorem!

Question 3. (L. Carleson). The hypothesis $f(x) \underset{x \nearrow 1}{\longrightarrow} \ell$, where $f(x) = \sum_{n=0}^{\infty} a_n x^n$, implies that $f(x)$ does not oscillate very much when $x \nearrow 1$. Take the even stronger hypothesis that f is of bounded variation on $[0, 1[$: can we conclude that $S_n = a_0 + \cdots + a_n \to \ell$ under a less restrictive hypothesis than $a_n = O(n^{-1})$? The response (negative) was given by H. Shapiro in 1965, and even better by P. B. Kennedy and P. Szüsz [106]: if $\varphi_n \to +\infty$, there exists a power series $f(x) = \sum_{n=0}^{\infty} a_n x^n$ such that

  (i) $|a_n| \leqslant \dfrac{\varphi_n}{n}$ if $n \geqslant 1$,
  (ii) f is increasing and bounded on $[0, 1[$,
  (iii) $\sum a_n$ diverges.

This constitutes a reinforcement of Theorem C of Littlewood (see Exercise 1.11). There are many more examples, but the impact made by the famous 1911 paper of Littlewood on his collaboration with Hardy and subsequent analytical research was enormous. It is shown on the one hand by the complete works of these two great mathematicians, and on the other hand by the title and contents of the work cited previously [113]: *Tauberian Theory, a Century of Developments*.

1.4 Appendix: Power series

The theory of power series is often considered somewhat brutal. To the series $\sum a_n z^n$, we associate a radius of convergence R that is a rough function of the coefficients a_n:

$$\frac{1}{R} = \varlimsup_{n \to +\infty} |a_n|^{1/n}.$$

It is almost as if we were raising $|a_n|$ to the power zero – but a number raised to the power zero is in fact 1! This explains why we often find $R = 1$, for example for the totally different series $a_n = e^{\sqrt{n}}$ or $a_n = e^{-\sqrt{n}}$. We often include the diagram of Figure 1.3 in the search for R, the question marks indicating where we cannot draw any conclusion.

Can *anything* really happen on the circle of uncertainty $|z| = R$? If so, then we can select an arbitrary set E of this circle, and then find a power series $\sum a_n z^n$ with radius of convergence R, that converges on E but diverges on the

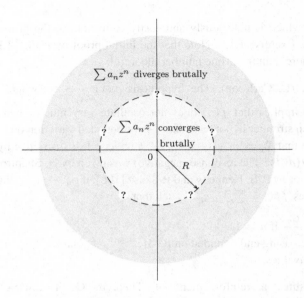

$|z| < R \Rightarrow \sum a_n z^n$ converges brutally
$|z| > R \Rightarrow \sum a_n z^n$ diverges brutally
$|z| = R \Rightarrow$ anything can happen

Figure 1.3

complement of E in the circle; for example, we could take $R = 1$, $E = \{1\}$. The series exists, but good luck in finding it (see Exercise 1.5).

A moment of reflection shows that E cannot be totally arbitrary: it must be a Borel subset of the circle, and more precisely an $F_{\sigma\delta}$ (a countable intersection of countable unions of closed sets).

We talk more about E in the following theorem (where $R = 1$ and $\mathbb{T}$ designates the unit circle $|z| = 1$).

1.4.1 Theorem [See [114], in particular pp. 90–91] *Let $E \subset \mathbb{T}$ be the set of points of convergence on the boundary of a power series with radius of convergence* 1. *Then:*

(1) E is necessarily an $F_{\sigma\delta}$,
(2) E can be any arbitrary F_σ,
(3) E cannot be any arbitrary G_δ [Lukasenko].

Point (1) is easy: the set E of points in a topological space X of simple convergence of a sequence $(f_n)_{n\geqslant 0}$ of continuous functions from X to $\mathbb{C}$ (here, $f_n(t) = \sum_{j=0}^n a_j e^{ijt}$) is always an $F_{\sigma\delta}$, because

$$E = \bigcap_{p \in \mathbb{N}^*} \bigcup_{N \in \mathbb{N}} \bigcap_{n \geqslant N} \left\{ t \in X / |f_n(t) - f_N(t)| \leqslant \frac{1}{p} \right\} =: \bigcap_{p \in \mathbb{N}^*} \bigcup_{N \in \mathbb{N}} F_{N,p}.$$

Point (2) is difficult, even if the case $E = \{a\}$ is covered by Theorem 1.4.2 below (see Exercise 1.5). Finally, for point (3), the example is simple ($E = \mathbb{T} \setminus \bigcup_{n \geqslant 1} F_n$, F_n closed but with empty interior, $m(^cE) < 1$, $m(I \cap {}^cE) > 0$ for all infinite arcs I), but the proof is difficult and uses the famous theorem of Carleson from 1966, solution of an old conjecture of Lusin, which states that the Fourier series of a square-summable function converges almost everywhere. We also have the following result.

1.4.2 Theorem   *(1) If $\sum a_n z^n$ converges uniformly for $|z| \leqslant 1$, then $\sum_{n=0}^{\infty} |a_n|^2 < \infty$.*
(2) Conversely, if $\sum_{n=0}^{\infty} |a_n|^2 < \infty$, the series $\sum a_n e^{int}$ converges almost everywhere [Carleson, 1966].
(3) If $\sum_{n=0}^{\infty} |a_n|^2 = \infty$ and if $|a_n|$ is decreasing, we can find signs $\varepsilon_n = \pm 1$ such that $\sum a_n \varepsilon_n e^{int}$ diverges everywhere [Dvoretzky–Erdős, 1956].

Result (1) is an easy consequence of Parseval's identity [153]; result (2) is extremely difficult [7]; finally, result (3) is inventive, but elementary [55]: we can suppose that $|a_1| \leqslant 1$ and $a_n \to 0$. Let $p_1 = 1$, and by induction define integers $p_1 < \cdots < p_k < p_{k+1} < \cdots$ as follows: p_{k+1} is the smallest integer strictly superior to p_k such that

$$\sum_{n \in B_k} |a_n| > 1,$$

where B_k is the block of integers $]\!] p_k, p_{k+1}]\!]$. We then have

$$1 < \sum_{n \in B_k} |a_n| \leqslant 2 \text{ and } \sum_{k=1}^{\infty} \frac{1}{p_{k+1} - p_k} = \infty. \tag{1.23}$$

In fact, $\sum_{p_k < n < p_{k+1}} |a_n| \leqslant 1$ and $|a_{p_{k+1}}| \leqslant |a_1| \leqslant 1$. Moreover,

$$\sum_{n \in B_k} |a_n|^2 \leqslant |a_{p_k}| \sum_{n \in B_k} |a_n| \leqslant 2|a_{p_k}|,$$

hence $\sum_{k=1}^{\infty} |a_{p_k}| = \infty$. But

$$2 \geqslant \sum_{n \in B_k} |a_n| \geqslant (p_{k+1} - p_k)|a_{p_{k+1}}|,$$

so that $\dfrac{1}{p_{k+1} - p_k} \geqslant \dfrac{|a_{p_{k+1}}|}{2}$, which proves (1.23). Set

$$\ell_k = \frac{1}{2(p_{k+1} - p_k)} \text{ for } k \geqslant 1.$$

As $\sum_{k=1}^{\infty} \ell_k = \infty$, it is possible, by placing them end to end, to find closed arcs I_k, of length ℓ_k, such that each point of $\mathbb{T}$ belongs to an infinite number of I_k. We adjust the signs ε_n (by block) so that $\left| \sum_{n \in B_k} \varepsilon_n a_n e^{int} \right|$ is large at $t = t_k$, the centre of I_k. For example, let us take

$$\left| \sum_{n \in B_k} \varepsilon_n a_n e^{int} \right| \geqslant \frac{1}{2} \sum_{n \in B_k} |a_n|,$$

which is possible because (see Exercise 1.1)

$$\sup_{\varepsilon_n = \pm 1} \left| \sum_{n=1}^{N} \varepsilon_n z_n \right| \geqslant \frac{1}{2} \sum_{n=1}^{N} |z_n|.$$

Next, the absolute value above cannot vary much when t remains in the tiny interval I_k centred on t_k:

$$\left| \sum_{n \in B_k} \varepsilon_n a_n e^{int} \right| \geqslant \frac{1}{4} \sum_{n \in B_k} |a_n| > \frac{1}{4} \text{ for all } t \in I_k. \tag{1.24}$$

Indeed, if we set $P(t) = \sum_{n \in B_k} \varepsilon_n a_n e^{int}$ and $Q(t) = e^{-ip_k t} P(t)$, we have

$$\left| |P(t)| - |P(t_k)| \right| = \left| |Q(t)| - |Q(t_k)| \right| \leqslant |Q(t) - Q(t_k)|$$

$$= \left| \sum_{n \in B_k} \varepsilon_n a_n \left(e^{i(n - p_k)t} - e^{i(n - p_k)t_k} \right) \right|$$

$$\leqslant \sum_{n \in B_k} |a_n| (n - p_k) |t - t_k|$$

$$\leqslant \sum_{n \in B_k} |a_n| (p_{k+1} - p_k) \frac{1}{4(p_{k+1} - p_k)}$$

$$\leqslant \frac{1}{4} \sum_{n \in B_k} |a_n|,$$

so that, if $t \in I_k$:

$$\left| \sum_{n \in B_k} \varepsilon_n a_n e^{int} \right| = |P(t)|$$

$$\geqslant |P(t_k)| - \frac{1}{4} \sum_{n \in B_k} |a_n|$$

$$\geqslant \left(\frac{1}{2} - \frac{1}{4} \right) \sum_{n \in B_k} |a_n|$$

$$\geqslant \frac{1}{4} \sum_{n \in B_k} |a_n| > \frac{1}{4},$$

by (1.23), which gives (1.24).

We can now conclude. As each real number t belongs to an infinity of I_k, we observe by invoking (1.24) that the series $\sum \varepsilon_n a_n e^{int}$ diverges, as required, for all real numbers t.

If we now consider the case of uniform convergence within the entire disk of convergence, we have the following result [108, 151].

1.4.3 Theorem [Kisliakov, 1981] *Let $(a_n)_{n \geqslant 0}$ be a square-summable complex sequence $\sum_{n=0}^{\infty} |a_n|^2 < \infty$. Then there exists a complex sequence $(b_n)_{n \geqslant 0}$ such that $|b_n| \geqslant |a_n|$ for all $n \geqslant 0$ and $\sum b_n e^{int}$ converges uniformly on $\mathbb{R}$.*

The proof uses a reformulation, due to Vinogradov, of the theorem of Carleson.

This appendix and the presentation of Littlewood's theorem have hopefully shown two points. First of all, the subject is not at all old-fashioned (a complete description of the sets E of "convergence at the boundary" described in Theorem 1.4.1 is still not known in 2014). Moreover, the behaviour of a power series in a neighbourhood of its circle of convergence can be simultaneously complicated and rich in information about the sequence $(a_n)_{n \geqslant 0}$ itself (*transition convergence–divergence*).

Exercises

1.1. Let $n \geqslant 2$ be a fixed integer, $z_1, \ldots, z_n$ complex numbers, and

$$M = \sup \left| \sum_{k=1}^{n} \varepsilon_k z_k \right|,$$

where the least upper bound is taken over all n-tuples $(\varepsilon_1, \ldots, \varepsilon_n)$ belonging to $\{-1, 1\}^n$.

(a) Show that

$$\sum_{k=1}^{n} |\operatorname{Re}(z_k)| \leqslant M.$$

(b) Show that

$$M \geqslant \frac{1}{2} \sum_{k=1}^{n} |z_k|.$$

1.2. We keep here the notation of Section 1.3.1.

(a) Show the result of Hardy: $F = S$ if $na_n \to 0$.

(b) Consider the statement: "there exists a numerical constant $\lambda > 0$ such that, for all power series $\sum a_n x^n$ verifying $\sup_{n \geqslant 0} |na_n| \leqslant 1$, we have $S \subset \lambda \, \overline{\text{co}}(F \cup \{0\})$", where $\overline{\text{co}}$ designates the closed convex hull. Is this statement true or false?

1.3. Let $\sum a_n z^n$ be a power series satisfying

$$a_0 + \cdots + a_n = O(1) \text{ and } a_{n+1} - a_n = O(n^{-2}).$$

We set

$$S_n(\theta) = \sum_{k=0}^{n} a_k e^{ik\theta} \text{ for } 0 < |\theta| \leqslant \pi.$$

The letter C denotes a numerical constant, which can be different from one question to the next.

(a) With the aid of an Abel transformation, show that

$$\left| S_p(\theta) \right| \leqslant C \left(1 + p|\theta| \right).$$

(b) Using an *additional* Abel transformation, show that if $1 \leqslant p < q$,

$$\left| S_p(\theta) - S_q(\theta) \right| \leqslant \frac{C}{p|\theta|}.$$

(c) Show that the partial sums of our power series are uniformly bounded in the unit disk.

1.4. Properties of the Hardy series. Let w be a complex number with positive real part, and

$$f(z) = \sum_{n=1}^{\infty} \frac{z^n}{n^w},$$

where $z \in D$, the open unit disk.

(a) Show that

$$\frac{1}{n^w} = \frac{1}{\Gamma(w)} \int_0^{+\infty} t^{w-1} e^{-nt} dt \text{ for all } n \geqslant 1.$$

(b) Show that for $z \in D$,

$$f(z) = \frac{z}{\Gamma(w)} \int_0^{+\infty} \frac{t^{w-1}}{e^t - z} dt.$$

What are the singular points of the power series $\sum n^{-w} z^n$ on its circle of convergence $|z| = 1$?

(c) Let $u \in \mathbb{C}$, u of modulus 1, $u \neq 1$. Show (in several different ways, if possible) that the series $\sum u^n n^{-w}$ converges. What can be said when $u = 1$?

(d) In what follows, we suppose (Hardy series) that $w = 1 + i\alpha$, $\alpha \in \mathbb{R}^*$. We propose to establish a few remarkable properties of the corresponding power series.

  (i) Show that $\sum u^n n^{-w}$ converges for $|u| = 1$, $u \neq 1$ and diverges *with bounded partial sums* (which hence excludes $\sum u^n n^{-1}$) for $u = 1$. This provides a simple example, compared with the more complicated one of Körner [114].

  (ii) Show that (see Exercise 1.3) the partial sums of the series are uniformly bounded on the unit disk.

  (iii) Let $(\varepsilon_n)_{n \geqslant 1}$ be a sequence decreasing to 0, such that

$$\sum_{n=1}^{\infty} n^{-1} \varepsilon_n = +\infty.$$

Show that the power series $\sum \varepsilon_n n^{-w} z^n$ converges uniformly, but not normally, in the unit disk.

  (iv) Show that

$$\lim_{\varepsilon \searrow 0} \sum_{n=1}^{\infty} \frac{1}{n^{w+\varepsilon}}$$

exists, and identify this limit.

1.5. The aim of this exercise is to show that every finite subset of the unit circle $\mathbb{T}$ is a set of convergence at the boundary of a power series with radius of convergence 1.

(a) Show that there exists a power series $\sum a_n z^n$, with radius of convergence 1, with $a_n \to 0$, that diverges everywhere on $\mathbb{T}$.

(b) Select distinct $u_1, \ldots, u_p \in \mathbb{T}$. We define the sequence $(b_n)_{n \geqslant 0}$ as follows:

$$\sum_{n=0}^{\infty} b_n z^n = (z - u_1) \times \cdots \times (z - u_p) \sum_{n=0}^{\infty} a_n z^n.$$

Show that, for this new power series with radius of convergence 1, the set of convergence at the boundary is $\{u_1, \ldots, u_p\}$.

1.6. Variant of the theorem of Littlewood. Suppose that $f(x) = \sum_{n=1}^{\infty} a_n x^n$ exists for $|x| < 1$ and remains bounded when $x \nearrow 1$, $(a_n)_{n \geqslant 1}$ satisfying the Tauberian condition $|n^{1-\alpha} a_n| \leqslant C$, with α and C positive constants. We set

$$g(x) = \frac{f(x)}{x} = \sum_{n=1}^{\infty} a_n x^{n-1} \text{ for } |x| < 1.$$

(a) Show that if $0 < x < 1$, then

$$\sum_{n=1}^{\infty} a_n n^{-\alpha} x^{n-1} = \frac{1}{\Gamma(\alpha)} \int_0^{+\infty} t^{\alpha-1} e^{-t} g(xe^{-t}) \, dt.$$

(b) Show that if $|g(x)| \leqslant M$ for $0 < x < 1$, then

$$\left| \sum_{n=1}^{\infty} a_n n^{-\alpha} x^{n-1} \right| \leqslant M \text{ for } 0 < x < 1.$$

(c) Show that the series $\sum n^{-\alpha} a_n$ converges.

1.7. Let $\varphi : \mathbb{R} \to \mathbb{C}$ be 1-periodic, satisfying a Hölder condition of order $\alpha > \frac{1}{2}$, meaning

$$|\varphi(u) - \varphi(v)| \leqslant M|u - v|^{\alpha} \text{ for } u, v \in \mathbb{R},$$

with mean zero ($\int_0^1 \varphi(t) \, dt = 0$) and Fourier series[10] $\varphi(t) = \sum_{k \neq 0} c_k e^{2i\pi kt}$. Let a be a "badly approximated" irrational real number,[11] that is to say, there exists a positive constant C and $N \geqslant 1$ such that

$$\left| a - \frac{p}{q} \right| \geqslant \frac{C}{q^N} \text{ for } p \in \mathbb{Z} \text{ and } q \in \mathbb{N}^*.$$

Furthermore, let q and r be two integers such that $0 \leqslant r < q$.
(a) For $0 \leqslant x < 1$, set

$$f(x) = \sum_{n=1}^{\infty} \frac{\varphi((nq+r)a)}{n} x^n.$$

Show that

$$f(x) = -\sum_{k \neq 0} c_k e^{2i\pi kra} \log(1 - xe^{2i\pi kqa}).$$

---

[10] According to a theorem of Bernstein, the Fourier series of φ is absolutely convergent. See Exercise 10.1.

[11] For example, all algebraic irrationals are badly approximated, as is π (see [60]).

(b) Show that there exists a positive constant C' such that

$$\left|\log(1 - xe^{2i\pi kqa})\right| \leqslant C' \ln(1 + |k|) \text{ for all } k \in \mathbb{Z}^*,$$

and then that $f(x)$ has a limit ℓ when $x \nearrow 1$.

(c) Show that the series $\sum \dfrac{\varphi((nq+r)a)}{n}$ converges. Show in particular that the series $\sum(-1)^n \dfrac{|\sin n|}{n}$ converges. What can be said about the series $\sum u^n \dfrac{|\sin n|}{n}$ for $|u| = 1$ and $u \neq 1$?

1.8. Series that are Abel-summable but never (C, k)-summable, after Hardy [76]

(a) Define the sequence $(a_n)_{n \geqslant 0}$ by

$$\exp\left(\frac{1}{1+z}\right) = \sum_{n=0}^{\infty} a_n z^n \text{ for } |z| < 1.$$

Show that $\sum_{n=0}^{\infty} a_n x^n \to \ell$ as $x \nearrow 1$, but none of the iterated Cesàro means of $S_n = \sum_{j=0}^{n} a_j$ converge.

(b) Set

$$f(x) = \sum_{n=1}^{\infty} (-1)^n e^{c\sqrt{n}} x^n \text{ for } |x| < 1$$

(where c is a fixed positive constant). It is clear that none of the iterated Cesàro means of $S_n = \sum_{j=1}^{n} a_j$, where $a_n = (-1)^n e^{c\sqrt{n}}$, converge; but we propose to show that $f(x) \to \ell$ as $x \nearrow 1$. For $b \geqslant 0$ and $0 \leqslant x < 1$, set

$$f_b(x) = \sum_{n=1}^{\infty} (-1)^n n^b x^n.$$

(i) Show that if $t > 0$,

$$\sum_{n=1}^{\infty} (-1)^n e^{-nt} = -\frac{1}{e^t + 1} = \phi(t),$$

where ϕ can be extended to an analytic function on $|\operatorname{Im} t| < \pi$, and bounded on $|\operatorname{Im} t| \leqslant \dfrac{\pi}{2}$.

(ii) Let $k \in \mathbb{N}$. Show that $|f_k(x)| \leqslant A k!$ if $0 \leqslant x < 1$ (with A a numerical constant) and that $f_k(x) \to \ell(k)$ as $x \nearrow 1$.

(iii) Show that

$$|f_b(x)| \leqslant B(1+b)^{1+b} \text{ if } b \geqslant 0 \text{ and } 0 \leqslant x < 1$$

(where B is a numerical constant) and that

$$f_b(x) \to \ell(b) \text{ as } x \nearrow 1.$$

(iv) Show that

$$f(x) = \sum_{k=0}^{\infty} \frac{c^k}{k!} f_{k/2}(x)$$

and conclude.

1.9. The notations are those of Section 1.2. Prove *Hardy's following generalisation* of the theorem of Frobenius: if $\frac{\sigma_n}{n} \to \ell$ and if k is a fixed non-zero positive integer, then $\sum_{n=0}^{\infty} a_n x^{n^k} \to \ell$ as $x \nearrow 1$.

1.10. Let $(\lambda_n)_{n \geqslant 1}$ be a sequence of positive real numbers, increasing to $+\infty$, with

$$\frac{\mu_n}{\lambda_n} \to 0 \text{ and } \varphi_n := n \frac{\mu_n}{\lambda_n} \to +\infty,$$

where $\mu_n = \lambda_n - \lambda_{n-1}$ (for example, $\lambda_n = e^{n^\alpha}$, $0 < \alpha < 1$). Show that there exists a Dirichlet series $\sum a_n e^{-\lambda_n x}$ such that $\frac{\sigma_n}{n} \to \ell$, but such that $\sum_{n=1}^{\infty} a_n e^{-\lambda_n x}$ does not have a limit when $x \searrow 0$. Thus, the theorem of Hardy from Exercise 1.9 for the exponents $\lambda_n = n^k$ cannot be extended to all exponents.

1.11. An example of Kennedy and Szüsz. Let $(\varphi(n))_{n \geqslant 1}$ be a sequence of positive real numbers increasing to $+\infty$, and

$$(c_k)_{k \geqslant 1}, \ (d_k)_{k \geqslant 1}, \ (n_k)_{k \geqslant 1}$$

three increasing sequences of non-zero positive integers, to be constructed.
(a) Consider the intervals of integers $I_k = [\![d_k, d_k + 2n_k[\![= I_k^+ \cup I_k^-$, with $I_k^+ = [\![d_k, d_k + n_k[\![, \ I_k^- = [\![d_k + n_k, d_k + 2n_k[\![$, as well as the polynomials

$$f_k(x) = c_k \left(\sum_{n \in I_k^+} \frac{x^n}{n} - \sum_{n \in I_k^-} \frac{x^n}{n} \right) = c_k \sum_{n \in I_k^+} \left(\frac{x^n}{n} - \frac{x^{n+n_k}}{n+n_k} \right).$$

Show that the f_k are increasing on $[0, 1]$, and that

$$\sum_{n \in I_k^+} \frac{c_k}{n} \geqslant \frac{c_k n_k}{d_k + n_k} \geqslant \frac{1}{2}$$

if, for example, $d_k = c_k n_k$ (a choice that we will adopt in what follows). Show that then $f_k(1) \leqslant \frac{1}{c_k}$.

(b) If the c_k are given, show that we can select non-zero positive integers n_k ($k \geqslant 1$) in such a way that

$$\varphi(n_k) \geqslant c_k \text{ et } n_{k+1}c_{k+1} \geqslant n_k(c_k + 2) \text{ for all } k \geqslant 1.$$

Thus the intervals I_k above are non-overlapping.

(c) Next, we define, for $0 \leqslant x < 1$, $f(x) = \sum_{n=0}^{\infty} a_n x^n$ by $f(x) = \sum_{k=1}^{\infty} f_k(x)$, which defines the a_n, with $|a_n| \leqslant 1$. Show that, for an appropriate choice of c_k, the power series $\sum_{n=0}^{\infty} a_n x^n$ thus defined has the following three properties:

   (i) $|a_n| \leqslant \dfrac{\varphi(n)}{n}$ if $n \geqslant 1$;

   (ii) f is increasing, bounded on $[0, 1[$, and $f(x) \to \ell$ as $x \nearrow 1$;

   (iii) $\sum a_n$ diverges, as $\displaystyle\sum_{n \in I_k^+} a_n \geqslant \dfrac{1}{2}$.

1.12. Let Λ be the von Mangoldt function, $d(n)$ the number of divisors of the integer $n \geqslant 1$, and set $u_n = \dfrac{\Lambda(n) - 1}{n}$. The sequence $(u_n)_{n \geqslant 1}$ verifies the unilateral estimate $nu_n \geqslant -1$. We also set

$$f(x) = (1 - x) \sum_{n=1}^{\infty} u_n \frac{nx^n}{1 - x^n} \text{ for } |x| < 1.$$

(a) Show that

$$f(x) = (1 - x) \sum_{n=1}^{\infty} (\ln n - d(n))x^n \text{ for } |x| < 1.$$

(b) Using the estimates (see [174])

$$\sum_{n=1}^{N} d(n) = N \ln N + (2\gamma - 1)N + O(\sqrt{N})$$

and

$$\sum_{n=1}^{N} \ln n = N \ln N - N + O(\ln N),$$

show that $f(x) \to -2\gamma$ as $x \nearrow 1$, where γ is the Euler constant.

(c) Show that $\sum u_n$ converges to -2γ. Deduce that

$$\sum_{n \leqslant x} \Lambda(n) \sim x \text{ as } x \to +\infty,$$

which is equivalent to the prime number theorem.

1.13. Optimality of the Abel non-tangential theorem. We denote by c_0 the Banach space of complex sequences $a = (a_n)_{n \geqslant 0}$ with limit zero, equipped with its usual norm $\|a\| = \sup\limits_{n \geqslant 0} |a_n|$. Let

$$D = \{z \in \mathbb{C} / |z| < 1\}$$

and $E \subset D$ "tangent at 1 to the unit circle", that is to say, such that

$$\sup_{z \in E} \frac{|1 - z|}{1 - |z|} = +\infty.$$

(a) Fix $z \in D$. Let $L_z : c_0 \to \mathbb{C}$ be the linear form defined by

$$L_z(a) = (1 - z) \sum_{n=0}^{\infty} a_n z^n.$$

Show that

$$\|L_z\| = \frac{|1 - z|}{1 - |z|}.$$

(b) Show that there exists $a \in c_0$ such that $\sup\limits_{z \in E} |L_z(a)| = +\infty$.

(c) Let a be as in the preceding question. Define $(b_n)_{n \geqslant 0}$ by

$$\sum_{n=0}^{\infty} b_n z^n = (1 - z) \sum_{n=0}^{\infty} a_n z^n \text{ for } |z| < 1.$$

Show that the power series $\sum b_n z^n$ converges for $z = 1$, but does not converge uniformly on E.

1.14. Generalisation of the Littlewood theorem. Let a be a positive real number, and $f(x) = \sum_{n=0}^{\infty} a_n x^{n^a}$, which we suppose to exist for $0 \leqslant x < 1$. Suppose that $f(x) \to \ell$ as $x \nearrow 1$, and that $a_n = O(n^{-1})$ when $n \to \infty$. Show that the series with general term a_n converges, and that its sum is ℓ.

1.15. Around the theorem of Zygmund. Let $(\varepsilon_k)_{k \geqslant 1}$ be a sequence of positive real numbers with limit zero, and

$$f(x) = \sum_{k=1}^{\infty} \varepsilon_k 2^{-k} \sin\left(2^k x\right).$$

Show that f is almost everywhere differentiable if and only if $\sum_{k=1}^{\infty} \varepsilon_k^2 < +\infty$, and it is everywhere differentiable if and only if $\sum_{k=1}^{\infty} \varepsilon_k < +\infty$.

2

The Wiener Tauberian theorem

2.1 Introduction

In 1932, Norbert Wiener published a monumental 100-page article entitled "Tauberian theorems" [186] in the prestigious journal *Annals of Mathematics*. The 38-year-old Wiener was already a world-renowned mathematician: he provided a model for Brownian motion between 1920 and 1924, and made fundamental contributions in such domains as harmonic analysis, potential theory, etc.

The same year, he was invited by Besicovitch and Hardy to give a series of lectures in Cambridge; the subject he chose was Fourier transforms and their applications. Rather than an exhaustive overview, Wiener's presentation was more a sample of his own contributions to the subject. He concentrated on three themes: Plancherel's L^2 theory, his own unified Tauberian theory based on his article [186], and harmonic analysis. The whole was published in 1933 as a book [187]. In this chapter, we concentrate on the second theme.

We start by introducing a few definitions and notations. As usual, we denote by $L^1(\mathbb{R})$ the space of Lebesgue-measurable functions $f : \mathbb{R} \to \mathbb{C}$, such that

$$\|f\|_1 := \int_{\mathbb{R}} |f(t)|dt < \infty.$$

If $f \in L^1(\mathbb{R})$ and $a \in \mathbb{R}$, the translate f_a of f is defined by

$$f_a(t) = f(t - a) \text{ for } t \in \mathbb{R}.$$

A subspace H of $L^1(\mathbb{R})$ is said to be *translation-invariant* if $f_a \in H$ for all $f \in H$ and $a \in \mathbb{R}$.

Wiener's contribution to Tauberian theory consists essentially of the following two theorems, in which, as in all that follows, the integral symbol without bounds represents $\int_{\mathbb{R}}$.

2.1.1 Theorem [approximation theorem]   *Let $f \in L^1(\mathbb{R})$ and V be the subspace of $L^1(\mathbb{R})$ generated by the translates f_a of f ($a \in \mathbb{R}$). Then V is dense in $L^1(\mathbb{R})$ if and only if the Fourier transform of f is zero-free on $\mathbb{R}$.*

2.1.2 Theorem [general Tauberian theorem]   *Let $K_1 \in L^1(\mathbb{R})$.*

(1) If $\widehat{K_1}(x) \neq 0$ for all $x \in \mathbb{R}$, and if $g \in L^\infty(\mathbb{R})$ and $\ell \in \mathbb{C}$ satisfy

$$\int K_1(x - t)g(t)\, dt \to \ell \int K_1(t)\, dt \text{ as } x \to +\infty, \qquad (2.1)$$

then, for, all $K_2 \in L^1(\mathbb{R})$,

$$\int K_2(x - t)g(t)\, dt \to \ell \int K_2(t)\, dt \text{ as } x \to +\infty. \qquad (2.2)$$

(2) If there exists an $x_0 \in \mathbb{R}$ such that $\widehat{K_1}(x_0) = 0$, then there exist $g \in L^\infty(\mathbb{R})$, $\ell \in \mathbb{C}$ and $K_2 \in L^1(\mathbb{R})$ that satisfy (2.1) but not (2.2).

We first take a brief look at these two theorems. It is easy to see that the first implies point (1) of the second: let $K_1 \in L^1(\mathbb{R})$ be such that $\widehat{K_1}$ is zero-free, g and ℓ satisfying (2.1). The set W of $K_2 \in L^1(\mathbb{R})$ verifying (2.2) is a subspace of $L^1(\mathbb{R})$, translation-invariant, containing K_1 and closed in $L^1(\mathbb{R})$.

To justify this, let $(F_p)_{p \geqslant 0}$ be a sequence of functions in W converging to $F \in L^1(\mathbb{R})$. Then,

$$\left| \int F(x - t)g(t)\, dt - \ell \int F(t)\, dt \right|$$

$$\leqslant \left| \int \big(F(x-t) - F_p(x-t)\big) g(t)\, dt \right| + \left| \int F_p(x-t)g(t)\, dt - \ell \int F_p(t)\, dt \right|$$

$$+ \left| \ell \int \big(F_p(t) - F(t)\big)\, dt \right|$$

$$\leqslant (\|g\|_\infty + |\ell|)\, \|F - F_p\|_1 + \left| \int F_p(x - t)g(t)\, dt - \ell \int F_p(t)\, dt \right|.$$

Hence

$$\varlimsup_{x \to +\infty} \left| \int F(x - t)g(t)\, dt - \ell \int F(t)\, dt \right| \leqslant (\|g\|_\infty + |\ell|)\, \|F - F_p\|_1$$

for all p, so that $F \in W$ by letting p tend to $+\infty$. In conclusion, W is at the same time dense (Theorem 2.1.1) and closed in $L^1(\mathbb{R})$, so that $W = L^1(\mathbb{R})$.

Suppose now that there exists an $x_0 \in \mathbb{R}$ such that $\widehat{K_1}(x_0) = 0$, and select $g(t) = e^{ix_0 t}$. We remark that

$$\int K_1(x - t)e^{ix_0 t}\, dt = e^{ix_0 x}\, \widehat{K_1}(x_0) = 0,$$

so that (2.1) holds (with $\ell = 0$). If we then select $K_2 \in L^1(\mathbb{R})$ such that $\widehat{K_2}(x_0) \neq 0$, such as for example a Gaussian kernel $K_2(t) = e^{-t^2}$, then

$$\int K_2(x - t)e^{ix_0t}\,dt = e^{ix_0x}\,\widehat{K_2}(x_0)$$

does not tend to 0 as $x \to +\infty$.

In contrast with the Tauberian theorems of Tauber and Littlewood, the Tauberian nature of Wiener's theorem is not completely evident. To understand it, consider $g \in L^\infty(\mathbb{R})$ and $K_1 \in L^1(\mathbb{R})$ such that $\int K_1(t)\,dt = 1$. For all x, $\int K_1(x - t)g(t)\,dt$ appears as a mean of the values of g, where $g(t)$ is weighted by $K_1(x - t)$. Under the *Tauberian* hypothesis of a non-vanishing Fourier transform of K_1, the ultimate would be for (2.1) to imply $g(t) \to \ell$ as $t \to +\infty$. This, however, is **not** the conclusion of Wiener's theorem, which allows us only to replace the kernel K_1 with an arbitrary kernel K_2 in $L^1(\mathbb{R})$. In general, we need to "pay a bit more" to come to a satisfying conclusion: first of all, make a judicious choice of K_2, and then have some extra information on the behaviour of g towards $+\infty$ (for example, relatively slow oscillations). We return to this in Section 2.4.

In this chapter, after a brief overview of Fourier transforms, we analyse the original proof of Wiener's approximation Theorem 2.1.1, and then the application of the general Tauberian Theorem 2.1.2 to Littlewood's theorem on power series. We then present the elementary proof given by Newman (1975) to the lemma of Wiener concerning absolutely convergent Fourier series, which is one of the crucial ingredients in Wiener's proof. Finally, we see how Gelfand theory provides an essentially algebraic proof of the approximation theorem.

2.2 A brief overview of Fourier transforms

In this section, we summarise without proof the main results of the theory of Fourier transforms in $L^1(\mathbb{R})$ (for details, see [39] or [117]).

We start with a result from the theory of Lebesgue integrals that will be of constant use, the *continuity of translation in* $L^1(\mathbb{R})$: for each $f \in L^1(\mathbb{R})$, the mapping

$$\tau_f : \mathbb{R} \to L^1(\mathbb{R}), \ a \mapsto f_a$$

is continuous. If f is continuous and compactly supported, the result is due to the uniform continuity of f. The general case follows because the set of compactly supported continuous functions is dense in $L^1(\mathbb{R})$.

The Fourier transform of $f \in L^1(\mathbb{R})$ is the function

$$\widehat{f} : \mathbb{R} \to \mathbb{C}, \ x \mapsto \int f(t) e^{-ixt} dt.$$

The function $\widehat{f}$ is continuous, and tends to 0 at $\pm\infty$ (Riemann–Lebesgue lemma). In the case where $\widehat{f} \in L^1(\mathbb{R})$, we have, almost everywhere (and everywhere if f is continuous):

$$f(t) = \frac{1}{2\pi} \int \widehat{f}(x) e^{itx} dx$$

(inversion theorem). In particular:

(1) the Fourier transformation $f \mapsto \widehat{f}$ is injective;
(2) if f and $\widehat{f}$ are in $L^1(\mathbb{R})$ and if f is even, then $\widehat{f}$ is even and $\widehat{\widehat{f}} = 2\pi f$ almost everywhere.

From an operational point of view, the Fourier transformation has two essential properties.

(1) Translations are transformed into multiplication by a character[1] and conversely: if $a \in \mathbb{R}$,

$$\widehat{f_a}(x) = e^{-iax} \widehat{f}(x) \ \text{ and } \ \widehat{e^{iat} f(t)}(x) = \left(\widehat{f}\right)_a(x).$$

(2) Convolution products are transformed into ordinary products: if f and g are in $L^1(\mathbb{R})$, then

$$\widehat{f * g} = \widehat{f} \cdot \widehat{g},$$

where

$$f * g(x) = \int f(x - t) g(t) \, dt$$

exists for almost all x and defines an element of $L^1(\mathbb{R})$.

We also have the equality

$$\int \widehat{f}(t) g(t) \, dt = \int f(t) \widehat{g}(t) \, dt,$$

valid for all $f, g \in L^1(\mathbb{R})$, which follows immediately from Fubini's theorem.

---

[1] A character is a continuous group homomorphism from $\mathbb{R}$ to $\mathbb{T} = \{z \in \mathbb{C} / |z| = 1\}$; in other words, a function of the type $x \mapsto e^{iax}, a \in \mathbb{R}$.

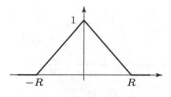

Figure 2.1

Here is an example of the computation of a Fourier transform that will be useful in what follows. Let R be a positive real number, and k_R the triangle function (Figure 2.1) defined by

$$k_R(x) = \max\left(0, 1 - \left|\frac{x}{R}\right|\right).$$

Its Fourier transform can easily be calculated:

$$\widehat{k_R}(x) = Rh(Rx), \text{ with } h(x) = \left(\frac{\sin(x/2)}{x/2}\right)^2.$$

As $\widehat{k_R} \in L^1(\mathbb{R})$, by the inversion theorem, $k_R = \widehat{K_R}$, where

$$K_R(x) = \frac{1}{2\pi}\widehat{k_R}(x) = \frac{R}{2\pi}h(Rx).$$

The family of functions $(K_R)_{R>0}$ is known as the *Fejér kernel*. An essential point is that the Fejér kernel is an *approximate identity*, that is to say, it verifies the following conditions:

$$\begin{cases} \bullet \ K_R \geqslant 0, \\[2mm] \bullet \ \displaystyle\int K_R(x)\,dx = \widehat{K_R}(0) = k_R(0) = 1, \\[2mm] \bullet \ \text{for all } \delta > 0, \ \lim_{R \to +\infty}\int_{|x|\geqslant\delta} K_R(x)\,dx = 0. \end{cases}$$

This last condition means that when R increases, the graph of K_R concentrates most of its mass in a neighbourhood of the origin. This has an important consequence: if $f \in L^1(\mathbb{R})$, then

$$f * K_R \to f \text{ in } L^1(\mathbb{R}), \text{ as } R \to +\infty.$$

Another set of functions will also be very useful: the trapezoid functions. If $a < b < c < d$ are real numbers, we define $\varphi_{a,b,c,d}$ (or simply φ if there is no risk of confusion) as the function with values 0 outside $[a, d]$, 1 on $[b, c]$, and linear on each of the two segments $[a, b]$ and $[c, d]$. See Figure 2.2.

In the special case where $(a, b, c, d) = (-e - \eta, -e, e, e + \eta)$, with $e, \eta > 0$, the Fourier transform of φ is easy to calculate. One can proceed with

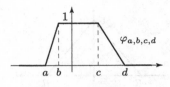

Figure 2.2

a brute-force computation, or more cleverly note that φ is the difference of two triangle functions:

$$\varphi(x) = \left(1 + \frac{e}{\eta}\right)k_{e+\eta} - \frac{e}{\eta}k_e.$$

From this,

$$\widehat{\varphi}(x) = \left(1 + \frac{e}{\eta}\right)(e+\eta)h\left((e+\eta)x\right) - \frac{e}{\eta}eh(ex)$$

$$= \frac{4}{\eta x^2}\left(\sin^2\frac{(e+\eta)x}{2} - \sin^2\frac{ex}{2}\right),$$

so that finally

$$\widehat{\varphi}(x) = \frac{2}{\eta x^2}\left(\cos(ex) - \cos\left((e+\eta)x\right)\right). \tag{2.3}$$

The preceding results can easily be extended to $L^1(\mathbb{R}^d)$ functions. In this case, the Fourier transform is defined by the formula

$$\widehat{f}(x) = \int f(t)e^{-i\langle x,t\rangle}dt,$$

where $\langle\,\cdot\,,\cdot\,\rangle$ denotes the usual inner product of $\mathbb{R}^d$. The inversion theorem then becomes

$$f(t) = \frac{1}{(2\pi)^d}\int \widehat{f}(x)e^{i\langle t,x\rangle}dx.$$

2.3 Wiener's original proof

In this section, we present Wiener's original proof of Theorem 2.1.1. We have tried to highlight the many ingenious ideas contained therein. However, for reasons of readability, we have not hesitated to use a more modern algebraic terminology, that of linear algebra and of ideals. The algebraic concepts themselves were clearly present in Wiener's work: in addition to studying analysis under Hardy at Cambridge, Wiener had learned group theory from Landau in Göttingen.

2.3.1 The ideas of the proof of the approximation theorem

We continue with the notations of Theorem 2.1.1: that is, f is a fixed element of $L^1(\mathbb{R})$ and V denotes the subspace of $L^1(\mathbb{R})$ generated by the translates of f.

It is easy to see that a necessary condition for the subspace V to be dense in $L^1(\mathbb{R})$ is that $\widehat{f}$ is zero-free. Indeed, if the Fourier transform of f is zero at some x, we have successively:

- $\widehat{f_a}(x) = e^{-ixa}\widehat{f}(x) = 0$ for all $a \in \mathbb{R}$;
- $\widehat{v}(x) = 0$ for all $v \in V$ (by linearity);
- $\widehat{w}(x) = 0$ for all w belonging to the closure of V in $L^1(\mathbb{R})$, as the linear mapping $w \mapsto \widehat{w}(x)$ is continuous on $L^1(\mathbb{R})$.

But as a result, V cannot be dense in $L^1(\mathbb{R})$.

The converse is much more delicate. Suppose that $\widehat{f}$ is zero-free. We must show that we can approximate an arbitrary element of $L^1(\mathbb{R})$ by a linear combination of a finite number of translates of f. However, among the functions that are susceptible to being thus approximated, we are sure to find elements of the *principal ideal* of $L^1(\mathbb{R})$ generated by f, that is, functions of the form $f * u$, with $u \in L^1(\mathbb{R})$. Indeed, at least in the case where f and u are continuous and u is compactly supported, we have, for $\varepsilon > 0$:

$$\varepsilon \sum_{n\in\mathbb{Z}} u(n\varepsilon) f_{n\varepsilon}(x) = \varepsilon \sum_{n\in\mathbb{Z}} f(x-n\varepsilon) u(n\varepsilon) \xrightarrow[\varepsilon\to 0]{} \int f(x-t)u(t)\,dt = f * u(x).$$

In fact, the following theorem shows that the closure of V has a more robust algebraic structure than V.

2.3.2 Theorem *Let W be a subspace of $L^1(\mathbb{R})$ that is closed and translation-invariant. Then W is an ideal of $L^1(\mathbb{R})$, in the sense that $w * g \in W$ for $w \in W$ and $g \in L^1(\mathbb{R})$.*

Proof Let $w \in W$, $g \in L^1(\mathbb{R})$ and $\varepsilon > 0$. There exists a function $h : \mathbb{R} \to \mathbb{C}$ continuous and compactly supported such that $\|g - h\|_1 \leqslant \varepsilon$. Then,

$$\|w * g - w * h\|_1 \leqslant \|w\|_1 \|g - h\|_1 \leqslant \varepsilon \|w\|_1.$$

Thus it suffices to prove the theorem for the case of g a continuous compactly supported function. Let $[a, b]$ be an interval that contains the support of g, and $x_0 < x_1 < \cdots < x_n$ a subdivision of $[a, b]$.

Finally, set

$$k(x) = \sum_{i=0}^{n-1} w(x - x_i) \int_{x_i}^{x_{i+1}} g(t)\, dt.$$

It is clear that $k \in W$. Moreover,

$$\|w * g - k\|_1 = \int \left| \sum_{i=0}^{n-1} \int_{x_i}^{x_{i+1}} (w(x - t) - w(x - x_i))\, g(t)\, dt \right| dx$$

$$\leqslant \sum_{i=0}^{n-1} \int \int_{x_i}^{x_{i+1}} |w(x - t) - w(x - x_i)|\, |g(t)|\, dt\, dx$$

$$\leqslant \sum_{i=0}^{n-1} \int_{x_i}^{x_{i+1}} \|w_t - w_{x_i}\|_1 |g(t)|\, dt$$

$$= \sum_{i=0}^{n-1} \int_{x_i}^{x_{i+1}} \|w - w_{x_i - t}\|_1 |g(t)|\, dt.$$

Finally, by the continuity of translation in $L^1(\mathbb{R})$, we can choose the subdivision $(x_i)_{0 \leqslant i \leqslant n}$ in such a way that $\|w - w_{x_i - t}\|_1 \leqslant \varepsilon$ for all $i \in [\![0, n - 1]\!]$ and all $t \in [x_i, x_{i+1}]$. Hence $\|w * g - k\|_1 \leqslant \varepsilon \|g\|_1$, which completes the proof. $\qquad \square$

2.3.3 Remark Conversely, any closed ideal I of $L^1(\mathbb{R})$ is translation-invariant. In fact, if $f \in I$ and if $(g_n)_{n \geqslant 0}$ is an approximate identity, we have, for $a \in \mathbb{R}$,

$$(f * g_n)_a = f * (g_n)_a \in I.$$

On the contrary,

$$(f * g_n)_a = f_a * g_n \to f_a \text{ in } L^1(\mathbb{R}),$$

hence $f_a \in I$.

In particular, the closure of V in $L^1(\mathbb{R})$ contains the ideal generated by f. If we were optimists, we could try to show that the principal ideal of $L^1(\mathbb{R})$ generated by f is *equal* to $L^1(\mathbb{R})$ itself, which would imply that for all $g \in L^1(\mathbb{R})$, there exists an $h \in L^1(\mathbb{R})$ such that

$$g = f * h. \tag{2.4}$$

By the injectivity of the Fourier transformation, this last condition is equivalent to saying that

$$\frac{\widehat{g}}{\widehat{f}} \text{ is a Fourier transform.}$$

Now, if ever

$$\frac{1}{\widehat{f}} \text{ was a Fourier transform,} \tag{2.5}$$

say $\dfrac{1}{\widehat{f}} = \widehat{h}$, $h \in L^1(\mathbb{R})$, we would have

$$\frac{\widehat{g}}{\widehat{f}} = \widehat{g * h},$$

and we would be done! Unfortunately, (2.5) is not possible, because of the Riemann–Lebesgue lemma.

Even though Wiener could not get the desired result on the real line, he could succeed on the unit circle $\mathbb{T} = \{z \in \mathbb{C}/|z| = 1\}$, where the obstruction of Riemann–Lebesgue disappears because of the compactness of $\mathbb{T}$. Rather than inverting Fourier transforms, he managed to invert absolutely convergent Fourier series: this is Wiener's lemma (see Theorem 2.3.8).

The second part of his work[2] consists of showing that a Fourier transform with a sufficiently concentrated support is the sum of an absolutely convergent Fourier series. This second ingredient via Wiener's lemma allows us to obtain (2.4) for $\widehat{g}$ compactly supported. To conclude, we only need to show that the set of functions in $L^1(\mathbb{R})$ whose Fourier transform is compactly supported is dense in $L^1(\mathbb{R})$.

2.3.4 Further properties of the Fourier transform

In this section, we prove two useful results.

- The first relates the absolute convergence of the Fourier series to the integrability of the Fourier transform, for functions with sufficiently concentrated support.
- The second shows that the set of functions whose Fourier transform is compactly supported is dense in $L^1(\mathbb{R})$.

---

[2] That we present first in order to be coherent!

2.3.5 Proposition   *Let* $\varepsilon \in \ ]0, \pi[$ *and* f *be a continuous function whose support is contained in* $[-\pi + \varepsilon, \pi - \varepsilon]$. *Then,*

$$\sum_{n \in \mathbb{Z}} |\widehat{f}(n)| < \infty \ \Leftrightarrow \ \int |\widehat{f}(t)| dt < \infty.$$

More precisely, there exist positive constants $\alpha(\varepsilon)$ *and* $\beta(\varepsilon)$ independent *of* f *such that*

$$\alpha(\varepsilon) \int |\widehat{f}(t)| dt \leqslant \sum_{n \in \mathbb{Z}} |\widehat{f}(n)| \leqslant \beta(\varepsilon) \int |\widehat{f}(t)| dt. \tag{2.6}$$

Proof   The idea of the proof is to use a function φ such that $f = f\varphi$, the function φ being smooth enough so that its Fourier transform tends to 0 sufficiently rapidly towards $\pm\infty$.

Here, Wiener introduces the function

$$\varphi = \varphi_{-\pi, -\pi+\varepsilon, \pi-\varepsilon, \pi},$$

which is zero outside $[-\pi, \pi]$, equals 1 on $[-\pi + \varepsilon, \pi - \varepsilon]$, and is linear on each of the two intervals $[-\pi, -\pi + \varepsilon]$ and $[\pi - \varepsilon, \pi]$. See Figure 2.3.

The Fourier transform of the function φ is easily calculated using the formula (2.3):

$$\widehat{\varphi}(x) = \frac{2}{x^2 \varepsilon} \left(\cos\left((\pi - \varepsilon)x\right) - \cos(\pi x) \right).$$

We note that $\widehat{\varphi} \in L^1(\mathbb{R})$. According to the inversion theorem, $\varphi = \widehat{\psi}$, where

$$\psi(x) = \frac{1}{2\pi} \widehat{\varphi}(x) = \frac{1}{\pi x^2 \varepsilon} \left(\cos\left((\pi - \varepsilon)x\right) - \cos(\pi x) \right).$$

Thus there exists a constant[3] $M > 0$ such that

$$(1 + x^2)|\psi(x)| \leqslant M \text{ for all } x \in \mathbb{R}.$$

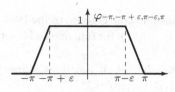

Figure 2.3

---

[3] Depending of course on ε.

Letting $\psi_n(t) = \psi(t - n)$ for $n \in \mathbb{Z}$, we obtain

$$\widehat{f}(n) = \int f(t)\varphi(t)e^{-int}dt = \int f(t)\widehat{\psi}_n(t)\,dt = \int \widehat{f}(t)\psi_n(t)\,dt$$

and hence, using the Beppo Levi theorem,

$$\sum_{n \in \mathbb{Z}} |\widehat{f}(n)| \leqslant \sum_{n \in \mathbb{Z}} \int |\widehat{f}(t)| \frac{M}{1 + (t - n)^2} dt = \int |\widehat{f}(t)| \sum_{n \in \mathbb{Z}} \frac{M}{1 + (t - n)^2} dt.$$

As the function $t \mapsto \sum_{n \in \mathbb{Z}} \dfrac{M}{1 + (t - n)^2}$ is continuous and 1-periodic, it is bounded, which establishes the second bound in (2.6).

To prove the first bound, it will be convenient to identify f with the 2π-periodic function equal to it on $[-\pi, \pi]$. The Fourier coefficients of f are thus

$$c_n(f) = \frac{1}{2\pi} \int_{-\pi}^{\pi} f(t)e^{-int}dt = \frac{1}{2\pi} \widehat{f}(n),$$

with $\widehat{f}$ still denoting the Fourier transform of f, considered now as a function with support in $[-\pi, \pi]$.

Parseval's identity then gives

$$\begin{aligned}
\widehat{f}(t) &= \int f(x)e^{-ixt}dx = \int_{-\pi}^{\pi} f(x)\overline{\varphi(x)e^{ixt}}dx \\
&= 2\pi \sum_{n \in \mathbb{Z}} c_n(f)\overline{c_n(\chi)}, \text{ where } \chi(x) = \varphi(x)e^{ixt} \\
&= 2\pi \sum_{n \in \mathbb{Z}} \frac{1}{2\pi} \widehat{f}(n)\overline{\frac{1}{2\pi} \widehat{\varphi}(n - t)} \\
&= \sum_{n \in \mathbb{Z}} \widehat{f}(n)\overline{\psi(n - t)}, \text{ as } \widehat{\varphi} = 2\pi\psi.
\end{aligned}$$

Using again the Beppo Levi theorem, we have

$$\int |\widehat{f}(t)|dt \leqslant \sum_{n \in \mathbb{Z}} |\widehat{f}(n)| \int |\psi(n - t)|dt = \left(\int |\psi(t)|dt \right) \left(\sum_{n \in \mathbb{Z}} |\widehat{f}(n)| \right),$$

which proves the first inequality of (2.6). $\qquad\square$

2.3.6 Proposition *The set of functions in $L^1(\mathbb{R})$ whose Fourier transform is compactly supported is dense in $L^1(\mathbb{R})$.*

Proof The inspiration comes from the case of 2π-periodic functions. Recall that if $f : \mathbb{R} \to \mathbb{C}$ is locally integrable and 2π-periodic, we define its Fourier coefficients by the formula

$$\widehat{f}(n) = \frac{1}{2\pi} \int_0^{2\pi} f(t)e^{-int}\,dt \text{ for } n \in \mathbb{Z}.$$

The Fourier transform of f is then the function $\widehat{f} : \mathbb{Z} \to \mathbb{C}$. The partial sums of the Fourier series of f are defined as

$$S_n(f)(x) = \sum_{k=-n}^{n} \widehat{f}(k)e^{ikx}.$$

Each $S_n(f)$ is thus a trigonometric polynomial, in other words a function whose Fourier transform is of finite support (which, for $\mathbb{Z}$ with its discrete topology, is equivalent to being compact). In general, the sequence $(S_n(f))_{n\in\mathbb{Z}}$ does not converge to f in $L^1([0, 2\pi])$. Nonetheless, things improve if we replace the $S_n(f)$ by their Cesàro means

$$\sigma_n(f) = \frac{1}{n+1} \sum_{k=0}^{n} S_k(f).$$

In this case, $\sigma_n(f) \to f$ in $L^1([0, 2\pi])$.

Finally, note the following important formula:

$$\sigma_n(f)(x) = f * K_n(x) = \frac{1}{2\pi} \int_0^{2\pi} f(x-t)K_n(t)\,dt,$$

where $(K_n)_{n \geqslant 0}$ is the Fejér kernel, defined by

$$K_n(x) = \sum_{k=-n}^{n} \left(1 - \frac{|k|}{n+1}\right) e^{ikx}.$$

The Fourier transform of K_n can be represented (with k the abscissa, $\widehat{K_n}(k)$ the ordinate) by a triangle of height 1, more and more spread out as n grows (Figure 2.4).

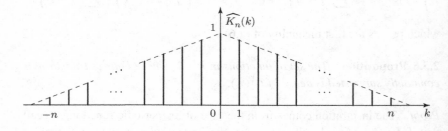

Figure 2.4

We will adapt this idea to the case of functions defined on $\mathbb{R}$, by using[4] the continuous Fejér kernel $(K_R)_{R>0}$. This terminology is justified by the fact that the Fourier transform of K_R, this time on the real line, is indeed a triangle function.

Let $g \in L^1(\mathbb{R})$. For each $R > 0$, set $g_R = g * K_R$. We then have that $\widehat{g_R} = \widehat{g} \cdot \widehat{K_R} = \widehat{g} \cdot k_R$, and therefore the Fourier transform of $\widehat{g_R}$ is compactly supported. But we already know that $g_R \to g$ in L^1 as $R \to +\infty$, which completes the proof. $\qquad\square$

2.3.7 Absolutely convergent Fourier series

The heart of Wiener's proof is based on a study of what is today known as the Wiener algebra, to be examined thoroughly in Chapter 11. It consists of the set W of functions

$$f : \mathbb{R} \to \mathbb{C}, x \mapsto \sum_{n \in \mathbb{Z}} c_n e^{inx}$$

where the c_n are complex numbers such that $\sum_{n \in \mathbb{Z}} |c_n| < \infty$. The c_n are unique; they are in fact the Fourier coefficients of f:

$$c_n = \frac{1}{2\pi} \int_0^{2\pi} f(t) e^{-int} dt.$$

The set W is an algebra of complex-valued functions. This is clear for addition and scalar multiplication, but requires some explanation for the product: if f and g are two elements of W such that $f(x) = \sum_{n \in \mathbb{Z}} c_n e^{inx}$ and $g(x) = \sum_{n \in \mathbb{Z}} d_n e^{inx}$, then

$$(fg)(x) = \sum_{n \in \mathbb{Z}} a_n e^{inx} \text{ with } a_n = \sum_{k \in \mathbb{Z}} c_k d_{n-k}.$$

The formula

$$\|f\|_W = \sum_{n \in \mathbb{Z}} |c_n|$$

defines a norm on W. What Wiener does not say (but uses implicitly) is that $(W, \|\cdot\|_W)$ is a Banach space, isometric to the space $\ell^1(\mathbb{Z})$ of summable sequences of complex numbers indexed by $\mathbb{Z}$. Moreover, as we have

$$\|fg\|_W \leqslant \|f\|_W \|g\|_W,$$

we are dealing with a *Banach algebra*.

Wiener's objective is to prove the following statement.

---

[4] See Section 2.2.

2.3.8 Theorem [Wiener's lemma]   *Let $f \in W$ such that $f(x) \neq 0$ for all x. Then $1/f \in W$.*

In order to establish his lemma, Wiener tackled the problem in the following manner.

- First, he showed the result under an additional highly restrictive condition on $\|f\|_W$ and $|c_0|$. Here, we see the first inklings of the theory of Banach algebras, with the use of the Neumann series[5] $\sum a^n$.
- To go beyond this condition, the idea is to next localise the function f in a neighbourhood of each point $x \in [-\pi, \pi]$, multiplying it by a function in W with a small support centred on x, and constructing a *local inverse* of f at x. A crucial point of the proof is to show that by localising an element g of W in a neighbourhood of one of its zeros, we can force the norm $\|g\|_W$ to be arbitrarily small.
- Finally, we "glue together" all these localised inverses using an argument based on a *partition of unity*.

2.3.9 Proposition   *Let $f \in W$, $f(x) = \sum_{n \in \mathbb{Z}} c_n e^{inx}$ such that $\|f\|_W < 2|c_0|$. Then $1/f \in W$.*

Proof   Note that the condition

$$\sum_{n \neq 0} |c_n| = \|f\|_W - |c_0| < |c_0|$$

implies that f can never be zero. If we write $f = c_0(1 - g)$, then $g \in W$ satisfies

$$\|g\|_W = \frac{\|f\|_W}{|c_0|} - 1 < 1,$$

so that $1 - g$ is invertible in the Banach algebra W. It is thus the same for f. $\square$

Before continuing, we introduce a convenient convention: if $f : \mathbb{R} \to \mathbb{C}$ is a function whose support is contained in an interval $[\alpha, \beta]$ such that $\beta - \alpha < 2\pi$, then f will often be identified in what follows with the 2π-periodic function g that coincides with f on $[\alpha, \alpha + 2\pi[$. See Figure 2.5.

Whenever the function g is an element of W, we write in a somewhat abusive manner $\|f\|_W$ for $\|g\|_W$.

Using the trapezoid functions $\varphi_{a,b,c,d}$ (defined in Section 2.2), we prove a fundamental property of W, namely its *local* character, first brought to light by Wiener.

---

[5] See Chapter 11, p. 331.

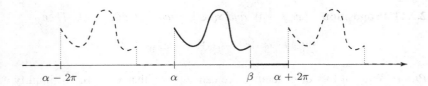

Figure 2.5

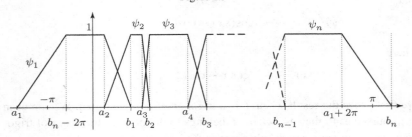

Figure 2.6

2.3.10 Proposition *Let* $f : \mathbb{R} \to \mathbb{C}$ *be a* 2π*-periodic function such that, for each* $x \in \mathbb{R}$, *there exists a* $g_x \in W$ *such that* f *and* g_x *coincide in a neighbourhood of* x. *Then* $f \in W$.

Proof Using the compactness of $[-\pi, \pi]$, we can cover this interval with a finite number of intervals $]a_1, b_1[, \ldots,]a_n, b_n[$, of length $< 2\pi$, such that f coincides on each $[a_k, b_k]$ with an element f_k of W. In addition, we can suppose that

$$a_1 < -\pi < b_n - 2\pi < a_2 < b_1 < a_3 < b_2 < \ldots$$
$$\ldots < a_{n-1} < b_{n-2} < a_n < b_{n-1} < a_1 + 2\pi < \pi < b_n.$$

In this case, set $b_0 = b_n - 2\pi$, $a_{n+1} = a_1 + 2\pi$ and

$$\psi_k = \varphi_{a_k, b_{k-1}, a_{k+1}, b_k} \text{ for } 1 \leqslant k \leqslant n,$$

ψ_k considered as a 2π-periodic function that is an element of W because ψ_k is continuous and piecewise C^1. See Figure 2.6.

The function $\psi = \sum_{k=1}^{n} \psi_k$ is everywhere[6] equal to 1, so that

$$f = f\psi = \sum_{k=1}^{n} f\psi_k = \sum_{k=1}^{n} f_k \psi_k \in W,$$

because W is an algebra. $\qquad\square$

---

[6] If you prefer, the ψ_k form a partition of unity, relative to the open covering $(]a_k, b_k[)_{1 \leqslant k \leqslant n}$ of $[-\pi, \pi]$.

2.3.11 Proposition   *Let $f \in W$ and $x_0 \in \mathbb{R}$ such that $f(x_0) = 0$. Then*

$$\lim_{\varepsilon \to 0} \left\| f \varphi_{x_0 - 2\varepsilon, x_0 - \varepsilon, x_0 + \varepsilon, x_0 + 2\varepsilon} \right\|_W = 0.$$

Proof   Without loss of generality, we can suppose that $x_0 = 0$. To simplify, we denote by φ_ε the function $\varphi_{-2\varepsilon, -\varepsilon, \varepsilon, 2\varepsilon}$. We can easily calculate, using (2.3), that

$$\widehat{\varphi_\varepsilon}(x) = \frac{2}{x^2 \varepsilon} (\cos(\varepsilon x) - \cos(2\varepsilon x)) = \varepsilon r(\varepsilon x),$$

where

$$r(x) = \frac{2}{x^2} (\cos x - \cos(2x)) \in L^1(\mathbb{R}).$$

We first prove the result when f is a trigonometric polynomial. We can even limit ourselves to the case where f belongs to the basis of the space of trigonometric polynomials with value zero at 0 formed by functions $t \mapsto e_m(t) - 1$ where $e_m(t) = e^{imt}$, m a non-zero integer.

Let $\varepsilon \in \,]0, 1]$. The support of $f\varphi_\varepsilon$ is thus contained in

$$[-2, 2] = [-\pi + \varepsilon_0, \pi - \varepsilon_0], \text{ where } \varepsilon_0 = \pi - 2.$$

In addition,

$$\widehat{f \varphi_\varepsilon}(x) = \widehat{\varphi_\varepsilon}(x - m) - \widehat{\varphi_\varepsilon}(x),$$

so that

$$\int |\widehat{f\varphi_\varepsilon}(x)| dx = \varepsilon \int |r(\varepsilon(x - m)) - r(\varepsilon x)| \, dx = \int |r(u - m\varepsilon) - r(u)| du \to 0$$

as $\varepsilon \searrow 0$, by continuity of translation in L^1.

Moreover, by Proposition 2.3.5, we have

$$\|f\varphi_\varepsilon\|_W \leqslant \beta(\varepsilon_0) \int |\widehat{f\varphi_\varepsilon}(x)| dx. \tag{2.7}$$

The result follows immediately.

From this, and the density of trigonometric polynomials in W, we now prove the general case. More precisely, if $f \in W$ and $\eta > 0$, there exists a trigonometric polynomial p (a partial sum of f) such that $\|f - p\|_W \leqslant \eta$. In particular,

$$|p(0)| = |f(0) - p(0)| \leqslant \|f - p\|_\infty \leqslant \|f - p\|_W \leqslant \eta.$$

Thus, setting $q = p - p(0)$, we obtain

$$\|f\varphi_\varepsilon\|_W = \|(f - p)\varphi_\varepsilon + p(0)\varphi_\varepsilon + q\varphi_\varepsilon\|_W \leqslant 2\eta\|\varphi_\varepsilon\|_W + \|q\varphi_\varepsilon\|_W.$$

Moreover, if again we suppose that $\varepsilon \in]0, 1]$, by Proposition 2.3.5 we have

$$\|\varphi_\varepsilon\|_W \leqslant \beta(\varepsilon_0) \int |\widehat{\varphi_\varepsilon}(x)| dx = \beta(\varepsilon_0) \|r\|_1,$$

so that

$$\varlimsup_{\varepsilon \to 0} \|f\varphi_\varepsilon\|_W \leqslant 2\eta\beta(\varepsilon_0)\|r\|_1.$$

As η is arbitrary, the proof is complete. $\qquad\qquad\square$

2.3.12 Remark We can make the inequality (2.7) a bit more precise by noting that the function r is C^∞ on $\mathbb{R}$ and that $r' \in L^1(\mathbb{R})$. Therefore, by choosing $m = 1$, so that $f(t) = e^{it} - 1$, we see that

$$\int |\widehat{f\varphi_\varepsilon}(x)| dx = \varepsilon \int \left| \int_0^1 r'(u - t\varepsilon) \, dt \right| du \leqslant \varepsilon \int_0^1 \int |r'(u - t\varepsilon)| \, du \, dt = \varepsilon \|r'\|_1.$$

We then deduce from Proposition 2.3.5 the existence of a positive constant $C_0 = \beta(\varepsilon_0)\|r'\|_1$ such that, for all $\varepsilon \in]0, 1]$, we can write

$$e^{it} - 1 = \sum_{n \in \mathbb{Z}} c_n(\varepsilon) e^{int}$$

for $|t| \leqslant \varepsilon$, with

$$\sum_{n \in \mathbb{Z}} |c_n(\varepsilon)| \leqslant C_0 \varepsilon.$$

In reality, the hypothesis $\varepsilon \leqslant 1$ is naive and the result holds for all $\varepsilon > 0$, if we replace C_0 by $\max(C_0, 2)$. In fact, if $\varepsilon > 1$, we write

$$e^{it} - 1 = a_0 + a_1 e^{it}, \text{ with } a_0 = -a_1 = -1,$$

and we have $|a_0| + |a_1| = 2 \leqslant 2\varepsilon$.

For an application of this result, see Exercise 2.11.

2.3.13 Proposition *Let $f \in W$ and $x_0 \in \mathbb{R}$ such that $f(x_0) \neq 0$. There exists a neighbourhood V of x_0 and $g \in W$ such that $f(x)g(x) = 1$ for all $x \in V$.*

Proof We can suppose that $x_0 = 0$. For each $\varepsilon > 0$, set

$$f_\varepsilon = f(0) + (f - f(0))\varphi_{-2\varepsilon, -\varepsilon, \varepsilon, 2\varepsilon} =: f(0) + g_\varepsilon.$$

On the one hand, we have

$$\|f_\varepsilon\|_W \leqslant |f(0)| + \|g_\varepsilon\|_W,$$

and on the other hand,

$$\left| \frac{1}{2\pi} \int_{-\pi}^{\pi} f_\varepsilon(t)\, dt \right| \geqslant |f(0)| - \left| \frac{1}{2\pi} \int_{-\pi}^{\pi} g_\varepsilon(t)\, dt \right| \geqslant |f(0)| - \|g_\varepsilon\|_W.$$

If ε is chosen such that

$$|f(0)| + \|g_\varepsilon\|_W < 2\left(|f(0)| - \|g_\varepsilon\|_W\right),$$

then f_ε will be invertible in W because of Proposition 2.3.9. But this is definitely possible because of Proposition 2.3.11, by which

$$\|g_\varepsilon\|_W \to 0 \text{ as } \varepsilon \searrow 0.$$

As f and f_ε coincide on a neighbourhood of 0, the proposition is shown. □

Wiener's lemma, stated in Theorem 2.3.8, follows immediately from Propositions 2.3.10 and 2.3.13.

2.3.14 Conclusion of the proof of the approximation theorem

Let $f \in L^1(\mathbb{R})$ such that $\widehat{f}$ is zero-free, and $g \in L^1(\mathbb{R})$ such that $\widehat{g}$ is compactly supported: $\operatorname{supp}\widehat{g} \subset [-a, a]$ $(a > 0)$. We show that g belongs to the ideal of $L^1(\mathbb{R})$ generated by f. We thus deduce that the closure of the subspace of $L^1(\mathbb{R})$ generated by the translates of f contains all the functions whose Fourier transform is compactly supported (Theorem 2.3.2), and hence is equal to $L^1(\mathbb{R})$ (Proposition 2.3.6). The proof is then complete.

To start, we suppose that $a < \pi$, fix a real number R in $]a, \pi[$, and set $f_R = f * K_R$, where $(K_R)_{R>0}$ is the Fejér kernel. Then, $\widehat{f_R} = \widehat{f} \cdot \widehat{K_R}$ has support $[-R, R] \subset \,]-\pi, \pi[$, and is non-zero on $]-R, R[$. Consider the function

$$\psi : \,]-\pi, \pi[\to \mathbb{C}, \quad x \mapsto \begin{cases} \dfrac{\widehat{g}(x)}{\widehat{f_R}(x)} & \text{if } x \in \,]-R, R[, \\ 0 & \text{otherwise,} \end{cases}$$

where ψ could also be taken as a 2π-periodic function defined on $\mathbb{R}$.

Since

$$\int |\widehat{\widehat{g}}| = 2\pi \int |g| < \infty,$$

we know that $\widehat{g} \in W$ because of Proposition 2.3.5. Similarly, $\widehat{f_R} \in W$. But then, by Proposition 2.3.13, ψ coincides in a neighbourhood of each point of $]-R, R[$ with an element of W. This is obviously the case in the neighbourhood of any point in $[-\pi, -R] \cup [R, \pi]$: ψ is zero in a neighbourhood of such a point. Thus, using Proposition 2.3.10, $\psi \in W$. Applying once again Proposition 2.3.5, we conclude that

$$\int |\widehat{\psi}| < \infty.$$

According to the inversion theorem, there thus exists an $h \in L^1(\mathbb{R})$ such that $\psi = \widehat{h}$. But then, $\widehat{g}(x) = \widehat{f_R}(x)\widehat{h}(x)$ for all $x \in \mathbb{R}$: this is clear on $]-R, R[$, and also elsewhere! By the injectivity of the Fourier transform,

$$g = f_R * h = f * (K_R * h),$$

which completes the proof.

In the general case (as previously, we suppose $\widehat{g}$ is compactly supported), set $g_R(x) = Rg(Rx)$ for all $R > 0$. Then, $\widehat{g_R}(x) = \widehat{g}\left(\dfrac{x}{R}\right)$, and by choosing R sufficiently small, the support of $\widehat{g_R}$ is contained in $]-\pi, \pi[$. By applying the previous result to $t \mapsto Rf(Rt)$ instead of f, we obtain the existence of $h \in L^1(\mathbb{R})$ such that

$$Rg(Rx) = \int Rf(Rt)\,h(x-t)\,dt = \int f(u)\,h\left(x - \frac{u}{R}\right) du,$$

so that $g = f * v$, where $v(t) = \dfrac{1}{R}h\left(\dfrac{t}{R}\right)$.

2.4 Application to Littlewood's theorem

In his book [187], Wiener gives a first application of his "general Tauberian theorem" regarding the Littlewood theorem on power series that we studied at length in Chapter 1. Let us see how this method, appropriately applied to the context of summability methods for divergent series, allows a short and elegant proof of Littlewood's difficult result.

2.4.1 The Tauberian theorem of Pitt

A function $f : \mathbb{R} \to \mathbb{C}$ is said to be *slowly oscillating at* $+\infty$ if, for all $\varepsilon > 0$, there exist $A, \delta > 0$ such that if $y, x \geqslant A$ and $|y - x| \leqslant \delta$, then $|f(y) - f(x)| \leqslant \varepsilon$.

For example, all uniformly continuous functions are slowly oscillating at $+\infty$. We find a less evident example in the proof of Littlewood's theorem.

Pitt's theorem, stated below, is a more "Tauberian-like" variant of Wiener's Tauberian theorem.

2.4.2 Theorem   *Let $g \in L^\infty(\mathbb{R})$ be slowly oscillating at $+\infty$ and $K \in L^1(\mathbb{R})$ whose Fourier transform is zero-free; suppose there is a constant $\ell \in \mathbb{C}$ such that*

$$\int K(x - t)g(t)\,dt \to \ell \int K(t)\,dt \text{ as } x \to +\infty.$$

Then $g(t) \to \ell$ as $t \to +\infty$.

Proof   We can suppose that g is real and ℓ is zero. If ever g does not tend to 0 at $+\infty$, there exist $\eta > 0$ and a sequence $(u_n)_{n \geqslant 0}$ going to $+\infty$ such that, say, $g(u_n) \geqslant 2\eta$. As g is slowly oscillating, there also exists a $\delta > 0$ such that, for n sufficiently large and $|x - u_n| \leqslant \delta$, we have $g(x) \geqslant \eta$.

Wiener's Tauberian theorem allows us to replace the function K in the hypothesis by any other kernel in $L^1(\mathbb{R})$: for example, the Fejér kernel $(K_R)_{R>0}$, which has the good taste to be non-negative, and an approximate identity (see the proof of Proposition 2.3.6). Thus, for n large enough and $R > 0$, we have

$$\int K_R(u_n - t)g(t)\,dt = \int_{u_n - \delta}^{u_n + \delta} K_R(u_n - t)g(t)\,dt + \int_{|t - u_n| \geqslant \delta} K_R(u_n - t)g(t)\,dt$$

$$\geqslant \eta \int_{-\delta}^{\delta} K_R(u)\,du - \|g\|_\infty \int_{|u| \geqslant \delta} K_R(u)\,du.$$

As this last lower bound goes to η as $R \to +\infty$, we can choose $R > 0$ such that this bound is $\geqslant \dfrac{\eta}{2}$. For this choice of R,

$$\varliminf_{n \to +\infty} \int K_R(u_n - t)g(t)\,dt \geqslant \frac{\eta}{2},$$

which contradicts the hypothesis

$$\int K_R(x - t)g(t)\,dt \to 0 \text{ as } x \to +\infty. \qquad \square$$

2.4.3 Summability methods

We now examine how Pitt's theorem can be used to prove Tauberian theorems related to summability methods. These methods allow us to assign a "sum" to series that are divergent in the usual sense. Take a C^1 function $f : \mathbb{R}_+ \to \mathbb{C}$, such that $f(0) = 1$, and let $(a_n)_{n \geqslant 0}$ be a complex sequence. Suppose that, for each $t > 0$, the series $\sum a_n f(tn)$ is convergent. We say that the series $\sum a_n$ is *f-convergent* if there exists an $\ell \in \mathbb{C}$ such that

$$\sum_{n=0}^{\infty} a_n f(tn) \to \ell \text{ as } t \searrow 0.$$

In this case, it is reasonable to call ℓ the f-*sum* of the series $\sum a_n$, whether or not it converges in the usual sense.

Here are a few examples.

(1) If $f(t) = e^{-t}$, the f-convergence is called Abel convergence. Under the change of variable $x = e^{-t}$, this is exactly the same as studying

$$\lim_{x \nearrow 1} \sum_{n=0}^{\infty} a_n x^n.$$

(2) If $f(t) = \dfrac{te^{-t}}{1 - e^{-t}}$ for $t > 0$ and $f(0) = 1$, the f-convergence is called Lambert convergence (see Exercise 2.8).

(3) By extending our definition to series of the form $\sum_{n=0}^{\infty} a_n f(t\lambda_n)$, where $(\lambda_n)_{n \geqslant 0}$ is an increasing sequence tending to $+\infty$, we encompass the case of Dirichlet series $\sum a_n e^{-\lambda_n s}$.

Of course, we hope that this method will allow us to sum more series than the usual method. In other words:

ordinary convergence $\Rightarrow$ f-convergence (with the same sum).

The converse provides an inexhaustible source of Tauberian theorems, for which the theorems of Wiener and Pitt provide invaluable assistance, for reasons that we now explain.

Set $s(t) = \sum_{k \leqslant t} a_k$ and $s_n = \sum_{k=0}^{n} a_k$. For $t > 0$, we have (assuming all desired convergences take place):

$$\sum_{n=0}^{\infty} a_n f(tn) = \sum_{n=0}^{\infty} s_n \left(f(tn) - f\left(t(n+1)\right) \right) \text{ (Abel transformation)}$$

$$= -t \sum_{n=0}^{\infty} s_n \int_{n}^{n+1} f'(tx)\, dx$$

$$= -t \sum_{n=0}^{\infty} \int_{n}^{n+1} s(x) f'(tx)\, dx$$

$$= -t \int_{0}^{+\infty} s(x) f'(tx)\, dx.$$

We would like to transform this "multiplicative convolution" into an additive one, which we obtain through a double change of variables. If we write $t = e^{-y}$ and $x = e^u$, we obtain

$$\sum_{n=0}^{\infty} a_n f(tn) = -\int e^{u-y} s(e^u) f'(e^{u-y}) \, du = \int K(y-u) g(u) \, du,$$

where

$$g(u) = s(e^u) \quad \text{and} \quad K(u) = -e^{-u} f'(e^{-u}).$$

If we manage to show that the function g is bounded and slowly oscillating at $+\infty$, that $K \in L^1(\mathbb{R})$ and that

$$\widehat{K}(x) = -\int e^{-u} f'(e^{-u}) e^{-ixu} \, du = -\int_0^{+\infty} f'(v) v^{ix} \, dv \neq 0$$

for all x, we can conclude, using Pitt's theorem, that if

$$\sum_{n=0}^{\infty} a_n f(tn) \to \ell \quad \text{as } t \searrow 0,$$

then

$$\sum_{n=0}^{\infty} a_n = \ell.$$

We now apply these ideas to the theorem of Littlewood.

2.4.4 Theorem *Let $\sum a_n x^n$ be a power series with radius of convergence $\geqslant 1$ such that*

$$h(x) = \sum_{n=0}^{\infty} a_n x^n \to \ell \quad \text{as } x \nearrow 1.$$

Suppose in addition that $n|a_n| \leqslant C$ for all $n \geqslant 0$. Then the series $\sum a_n$ converges and its sum is ℓ.

Proof We keep the preceding notations. Here, $f(t) = e^{-t}$, so that

$$K(t) = e^{-t-e^{-t}} \in L^1(\mathbb{R}).$$

First we verify that the sequence $(s_n)_{n \geqslant 0}$, and hence also the function g, is bounded.[7] For $N \geqslant 1$ and $x \in [0, 1[$, we have

$$|s_N - h(x)| \leqslant \sum_{n=0}^{N} |a_n|(1 - x^n) + \sum_{n=N+1}^{\infty} |a_n| x^n$$

$$\leqslant (1 - x) \sum_{n=0}^{N} n|a_n| + \frac{C}{N} \sum_{n=N+1}^{\infty} x^n$$

$$\leqslant CN(1 - x) + \frac{C}{N(1 - x)}.$$

---

[7] This argument is already present in [124, 173].

Choosing $x = 1 - N^{-1}$, this gives the result because the function h is bounded on $[0, 1[$.

Next, let us check that the Fourier transform of K is zero-free:

$$\widehat{K}(x) = \int_0^{+\infty} v^{ix} e^{-v} dv = \Gamma(1 + ix) \neq 0 \text{ (see Exercise 2.3).}$$

Finally, for $v > u > 0$, we have[8]

$$|g(v) - g(u)| = \left| \sum_{e^u < k \leqslant e^v} a_k \right|$$

$$\leqslant C \sum_{e^u < k \leqslant e^v} \frac{1}{k}$$

$$\leqslant C \frac{e^v - e^u + 1}{e^u} = C(e^{v-u} + e^{-u} - 1),$$

which shows that the function g is slowly oscillating at $+\infty$. This proves Littlewood's theorem.   □

2.5 Newman's proof of the Wiener lemma

Before continuing, let us note[9] that the Gelfand theory of Banach algebras provides an additional proof of Wiener's lemma (Theorem 2.3.8). This was certainly the first historical success of this theory. In 1975, D. J. Newman [133] gave an alternative proof, also disarming in its simplicity, but much more elementary because it only uses the Neumann series $\sum a^n$ (with a an element of a Banach algebra). It is this proof that we explain below.

Let $f \in W$, $f(x) = \sum_{n \in \mathbb{Z}} c_n e^{inx}$. Then,

$$\|f\|_\infty \leqslant \sum_{n \in \mathbb{Z}} |c_n| = \|f\|_W.$$

In the other direction, we have the following inequality:

2.5.1 Lemma   *If $f \in W$ is C^1, then $\|f\|_W \leqslant \|f\|_\infty + 2\|f'\|_\infty$.*

Proof   First of all, $|c_0| = |\widehat{f}(0)| \leqslant \|f\|_\infty$. Moreover, from the Cauchy–Schwarz inequality and Parseval's identity, we have

---

[8] Because the interval $]e^u, e^v]$ contains at most $e^v - e^u + 1$ integers.
[9] The reader is referred to Chapter 11 of this book.

$$\sum_{n \neq 0} |c_n| = \sum_{n \neq 0} |nc_n| \frac{1}{|n|}$$

$$\leqslant \left(2 \sum_{n=1}^{\infty} \frac{1}{n^2} \right)^{1/2} \left(\sum_{n \neq 0} |nc_n|^2 \right)^{1/2}$$

$$= \frac{\pi}{\sqrt{3}} \|f'\|_2 \leqslant 2 \|f'\|_\infty. \qquad \square$$

We proceed with the proof of Wiener's lemma. Let $f \in W$, $f(x) = \sum_{n \in \mathbb{Z}} c_n e^{inx}$, such that $f(x) \neq 0$ for all x. We must show that $\frac{1}{f} \in W$. Let us fix p a partial sum of f, to be specified later. We would really like to write

$$\frac{1}{f} = \frac{1}{p \left(1 - \dfrac{p-f}{p} \right)} = \sum_{n=1}^{\infty} \frac{(p-f)^{n-1}}{p^n}. \qquad (2.8)$$

We arrange things so that the series $\sum \dfrac{(p-f)^{n-1}}{p^n}$ is well-defined and converges, say to an element g, in the Banach algebra W. We thus obtain $fg = 1$ through the formal calculation of (2.8), henceforth properly justified.

We can of course suppose that $|f(x)| \geqslant 1$ for all x, and choose p such that $\|f - p\|_W \leqslant \frac{1}{3}$. Then, for all x,

$$|p(x)| \geqslant |f(x)| - |p(x) - f(x)| \geqslant 1 - \|p - f\|_\infty \geqslant 1 - \|p - f\|_W \geqslant \frac{2}{3},$$
$$(2.9)$$

which ensures that p is zero-free. Moreover,

$$\|(p-f)^{n-1}\|_W \leqslant \|p - f\|_W^{n-1} \leqslant \frac{1}{3^{n-1}}.$$

Finally, the function $\frac{1}{p}$ is 2π-periodic and C^1, and thus belongs to W.

We now bound $\left\| \dfrac{1}{p^n} \right\|_W$. First of all,

$$\left\| \frac{1}{p^n} \right\|_\infty \leqslant \left(\frac{3}{2} \right)^n,$$

from (2.9). Also,

$$\left\| \left(\frac{1}{p^n} \right)' \right\|_\infty = \left\| \frac{np'}{p^{n+1}} \right\|_\infty \leqslant n \|p'\|_\infty \left(\frac{3}{2} \right)^{n+1}.$$

Using Lemma 2.5.1,

$$\left\| \frac{(p-f)^{n-1}}{p^n} \right\|_W \leqslant \|(p-f)^{n-1}\|_W \left\| \frac{1}{p^n} \right\|_W$$

$$\leqslant \|(p-f)^{n-1}\|_W \left(\left\| \frac{1}{p^n} \right\|_\infty + 2 \left\| \left(\frac{1}{p^n} \right)' \right\|_\infty \right)$$

$$\leqslant \frac{1}{3^{n-1}} \left(\left(\frac{3}{2} \right)^n + 2n \|p'\|_\infty \left(\frac{3}{2} \right)^{n+1} \right)$$

$$= O\left(\frac{n}{2^n} \right),$$

which completes the proof.

2.6 Proof of Wiener's theorem using Gelfand theory

In this section, we assume a familiarity with the general properties of Banach algebras, as presented in Chapter 11.

2.6.1 Addition of a unit to $L^1(\mathbb{R})$

The Banach algebra $L^1(\mathbb{R})$ has a serious drawback: it does not have a unit element. We remedy this with a formal method, in order to be able to use the full artillery of Gelfand theory. We annex a supplementary element e to $L^1(\mathbb{R})$, and set

$$\mathcal{L} = \{\lambda e + f, \ \lambda \in \mathbb{C} \text{ and } f \in L^1(\mathbb{R})\}.$$

With operations defined by

$$\begin{cases} (\lambda e + f) + (\mu e + g) &= (\lambda + \mu)e + (f + g), \\ \alpha(\lambda e + f) &= \alpha\lambda e + \alpha f, \\ (\lambda e + f)(\mu e + g) &= \lambda\mu e + (\lambda g + \mu f + f * g) \end{cases}$$

and with a norm defined by

$$\|\lambda e + f\| = |\lambda| + \|f\|_1,$$

our $\mathcal{L}$ becomes a Banach algebra, with unit e. As the previously defined internal multiplication extends the convolution product in $L^1(\mathbb{R})$, we continue to write it as $*$.

2.6.2 Remark If we want to effectively construct $\mathcal{L}$, we could set $\mathcal{L} = \mathbb{C} \times L^1$, and write e for $(1, 0)$, λ for $(\lambda, 0)$ and f for $(0, f)$. An alternative solution is to embed L^1 in the Banach algebra of complex measures on $\mathbb{R}$ and for e take the Dirac measure at 0.

In order to take advantage of Gelfand theory, we need to know the spectrum of $\mathcal{L}$. An element φ of the spectrum satisfies in particular $\varphi(e) = 1$. It is thus completely determined by its restriction to L^1, which is a multiplicative functional. Conversely, if ψ is a multiplicative functional on L^1, the formula

$$\phi(\lambda e + f) = \lambda + \psi(f)$$

defines an element of the spectrum of $\mathcal{L}$. The spectrum of $\mathcal{L}$ is thus fully elucidated by the following theorem.

2.6.3 Theorem   *The non-zero multiplicative functionals on L^1 are exactly the maps of the form*

$$\varphi_\alpha : L^1(\mathbb{R}) \to \mathbb{C}, \ f \mapsto \widehat{f}(\alpha),$$

with $\alpha \in \mathbb{R}$.

Proof   Let φ be a non-zero multiplicative functional on L^1. In particular, φ is continuous on L^1 (see Proposition 11.2.4 of Chapter 11). Thus, there exists a $u \in L^\infty(\mathbb{R})$ such that

$$\varphi(f) = \int f(t)u(t)\, dt \text{ for all } f \in L^1.$$

The equality $\varphi(f * g) = \varphi(f)\varphi(g)$ can therefore be written

$$\int u(t) \int f(t - x)g(x)\, dx\, dt = \varphi(f)\varphi(g).$$

As $|u(t)f(t - x)g(x)| \leqslant \|u\|_\infty |f(t-x)g(x)| \in L^1(\mathbb{R}^2)$, by Fubini's theorem we have

$$\int g(x) \int u(t)f(t - x)\, dt\, dx = \varphi(f)\varphi(g),$$

which can be rewritten as

$$\int g(x)\varphi(f_x)\, dx = \varphi(f)\varphi(g) = \int g(x)u(x)\varphi(f)\, dx.$$

As this is true for all $g \in L^1(\mathbb{R})$, we have $\varphi(f_x) = \varphi(f)u(x)$ almost everywhere.

Now select f so that $\varphi(f) \neq 0$. The continuity of the function $x \mapsto f_x$, as well as that of φ, proves that, up to a modification of u on a negligible set, we can suppose that u is continuous, and the above equality holds everywhere.

We can then write, for all $(x, y) \in \mathbb{R}^2$,

$$\varphi(f)u(x + y) = \varphi(f_{x+y}) = \varphi(f_x)u(y) = \varphi(f)u(x)u(y).$$

So the non-zero continuous function u satisfies

$$u(x + y) = u(x)u(y) \text{ for all } x, y \in \mathbb{R}.$$

Thus there exists a $z \in \mathbb{C}$ such that $u(x) = e^{zx}$, for all $x \in \mathbb{R}$. As u is bounded, z is a pure imaginary: $z = -i\alpha$ with $\alpha \in \mathbb{R}$, which gives

$$\varphi(f) = \int f(x)e^{-i\alpha x}dx = \widehat{f}(\alpha).$$

The converse is evident. $\qquad\qquad\qquad\qquad\qquad\qquad\qquad\qquad\qquad\square$

We have thus shown that the spectrum of $\mathcal{L}$ consists of the maps

$$\psi_\alpha : \mathcal{L} \to \mathbb{C}, \ \lambda e + f \mapsto \lambda + \widehat{f}(\alpha),$$

with α in $\mathbb{R}$, and also of the projection

$$\pi : \mathcal{L} \to \mathbb{C}, \ \lambda e + f \mapsto \lambda.$$

An element $\lambda e + f$ of $\mathcal{L}$ is thus invertible if and only if $\lambda \neq 0$ and $\widehat{f}$ is never equal to $-\lambda$.

2.6.4 An algebraic proof of Wiener's theorem

Let $f \in L^1(\mathbb{R})$ such that $\widehat{f}$ is zero-free. To prove Wiener's approximation theorem, we have already seen that it is sufficient to show that each function $g \in L^1(\mathbb{R})$ whose Fourier transform is compactly supported is in the ideal generated by f. In other words, for such a function g, we want to find $h \in L^1(\mathbb{R})$ such that $g = f * h$. Essentially, we want to find the inverse of f, but the problem is precisely that f is not invertible in $\mathcal{L}$!

We will see how the well-established Gelfand theory shatters the difficulty in just a few lines. Let $k \in L^1(\mathbb{R})$, to be chosen later. For $x \in \mathbb{R}$, let $\widetilde{f}(x) = \overline{f(-x)}$. Then $\widehat{f * \widetilde{f}} = |\widehat{f}|^2 > 0$. According to the description of the spectrum of $\mathcal{L}$, the element $e + f * \widetilde{f} - k$ is invertible in $\mathcal{L}$ if and only if $-|\widehat{f}|^2 + \widehat{k}$ is never equal to 1; this will certainly be the case if

$$\widehat{k} \leqslant 1. \qquad\qquad\qquad\qquad\qquad\qquad\qquad\qquad (2.10)$$

If condition (2.10) is satisfied, there will be an $i \in \mathcal{L}$ such that

$$(e + f * \widetilde{f} - k) * i = e,$$

so that $g = (e + f * \tilde{f} - k) * i * g$. If we can choose k so that in addition

$$(e - k) * g = 0, \qquad (2.11)$$

or equivalently $g = k*g$, we will have $g = f*\tilde{f}*i*g$, and setting $h = \tilde{f}*i*g$, which as required is in $L^1(\mathbb{R})$ (given that $L^1(\mathbb{R})$ is an ideal of $\mathcal{L}$), we will have found our baby! By the injectivity of the Fourier transform, condition (2.11) is equivalent to

$$\hat{g} = \hat{k}\,\hat{g}. \qquad (2.12)$$

Ultimately, to ensure conditions (2.10) and (2.11), it suffices to choose for k a function whose Fourier transform is real-valued, with values in $[0, 1]$ and equal to 1 on the support of $\hat{g}$, which is definitely possible because this support is compact. Indeed, each function in the Schwartz space of rapidly decaying C^∞ functions is a Fourier transform [39]. The algebraic proof of Wiener's approximation theorem is thus complete.

Exercises

2.1. Let $f : \mathbb{R} \to \mathbb{C}$ be C^1 and compactly supported. Using Cauchy–Schwarz inequality and Plancherel's theorem, show that $\hat{f} \in L^1(\mathbb{R})$. Note that we can thus, without computation, establish the existence of functions in $L^1(\mathbb{R})$ whose Fourier transform is compactly supported.

2.2. Let $f : \mathbb{R}^d \to \mathbb{C}$ be a compactly supported C^k function, with $k > \dfrac{d}{2}$.

(a) Let $P \in \mathbb{C}[X_1, \ldots, X_d]$ be of degree $\leqslant k$, $P = \sum_{\alpha \in \mathbb{N}^d} c_\alpha X^\alpha$, with the standard notation $X^\alpha = X_1^{\alpha_1} \times \cdots \times X_d^{\alpha_d}$ if α is the multi-index $(\alpha_1, \ldots, \alpha_d)$. We define the differential operator

$$P(\partial_1, \ldots, \partial_d) = \sum_\alpha c_\alpha \partial_1^{\alpha_1} \cdots \partial_d^{\alpha_d}.$$

Show that, for $x \in \mathbb{R}^n$,

$$\widehat{P(\partial_1, \ldots, \partial_d)(f)}(x) = P(ix)\hat{f}(x).$$

(b) Show that $\hat{f} \in L^1(\mathbb{R}^d)$, and then that f is a Fourier transform of a function in $L^1(\mathbb{R}^d)$.

(c) Extend Wiener's theorem to $\mathbb{R}^d$.

2.3. Recall that

$$\Gamma(s) = \int_0^{+\infty} t^{s-1}e^{-t}dt, \text{ for } \mathrm{Re}\, s > 0.$$

Suppose that there exists an $s \in \mathbb{C}$ such that $\operatorname{Re} s > 0$ and $\Gamma(s) = 0$.

(a) Show that for all polynomials P with complex coefficients,

$$\int_0^1 (-\ln v)^{s-1} P(v) \, dv = 0.$$

(b) Show that this leads to a contradiction.

Thus, the function Γ is zero-free in the right half-plane.

2.4. Let $f : \mathbb{R}_+ \to \mathbb{C}$ be continuous. Show that f is slowly oscillating at $+\infty$ if and only if f is uniformly continuous.

2.5. This exercise illustrates an additional proof of Wiener's approximation theorem using Gelfand theory (see Chapter 11 if necessary).

Let $A(\mathbb{R})$ be the set of Fourier transforms of functions in $L^1(\mathbb{R})$. For $f = \widehat{g} \in A(\mathbb{R})$, define $\|f\| = \|g\|_1 = \int |g(t)| dt$. Equipped with this norm, $A(\mathbb{R})$ is a Banach algebra, isometric to $L^1(\mathbb{R})$.

Let E be a compact subset of $\mathbb{R}$, and $A(E)$ the algebra of restrictions to E of elements of $A(\mathbb{R})$. If $I(E)$ denotes the ideal of functions of $A(\mathbb{R})$ that are zero on E, the algebra $A(E)$ is isomorphic to the quotient $A(\mathbb{R})/I(E)$, and we can equip $A(E)$ with the quotient norm defined by $\|f_{|E}\|_E = \inf_{h \in I(E)} \|f + h\|$. Then $A(E)$ is a unitary Banach algebra.[10]

(a) Show that the spectrum of $A(E)$ can be identified with E, in the sense that the characters of $A(E)$ are the evaluations at the points of E.

(b) Let $f \in A(\mathbb{R})$ be zero-free on E. Show that there exists a $g \in A(\mathbb{R})$ such that $fg = 1$ on E.

(c) Let $F \in L^1(\mathbb{R})$ such that $f = \widehat{F}$ is zero-free on $\mathbb{R}$. Show that the ideal of $L^1(\mathbb{R})$ generated by F contains all the functions of $L^1(\mathbb{R})$ whose Fourier transform is compactly supported. This again proves Wiener's approximation theorem.

2.6. Recall that the Fourier transform on $L^1(\mathbb{R}) \cap L^2(\mathbb{R})$ can be extended to an isometry[11] from $L^2(\mathbb{R})$ *onto* $L^2(\mathbb{R})$ [160]. If $f \in L^2(\mathbb{R})$, $\widehat{f}$ is thus, in particular, defined almost everywhere. For f again in $L^2(\mathbb{R})$, let us denote by V the subspace of $L^2(\mathbb{R})$ generated by the translates of f.

(a) Show that $g \in L^2(\mathbb{R})$ is in $V^\perp$ if and only if $\widehat{g} \cdot \overline{\widehat{f}} = 0$.

(b) Show that V is dense in $L^2(\mathbb{R})$ if and only if the set of zeros of $\widehat{f}$ has measure zero.

---

[10] Unlike $A(\mathbb{R})$.

[11] In reality, up to a multiplicative constant.

2.7. Denote by W^+ the Hardy–Wiener algebra of functions from $\mathbb{T}$ to $\mathbb{C}$ of the form

$$z \mapsto \sum_{n=0}^{\infty} c_n z^n, \text{ with } \sum |c_n| < \infty,$$

and H^2 the Hardy space of functions in $L^2(\mathbb{T})$ such that

$$\widehat{f}(n) = \frac{1}{2\pi} \int_0^{2\pi} f(e^{it}) e^{-int} dt = 0 \text{ for } n < 0.$$

Let E be a compact subset of the circle. The compact set E is said to be W^+-*determinant* if any element f of W^+ such that $f_{|E} = 0$ is zero; it is a *generator* for H^2 if for all functions $g \in H^2$ such that $\widehat{g}(n) \neq 0$ for all $n \geqslant 0$, the family $(g_a)_{a \in E}$ is complete in H^2.

(a) Show that if E is W^+-determinant, then it is a generator.

(b) Show the converse.

(c) Carleson has shown that there exist W^+-determinant sets of measure zero. What can you conclude?

2.8. The Hardy and Littlewood Tauberian theorem relative to Lambert summability. Let $(a_n)_{n \geqslant 1}$ be a complex sequence such that

$$\sum_{n=1}^{\infty} a_n \frac{nte^{-nt}}{1 - e^{-nt}} \to \ell \in \mathbb{C} \text{ as } t \searrow 0.$$

Suppose in addition that $n|a_n| \leqslant C$ for all $n \geqslant 1$, and we will prove that $\sum_{n=1}^{\infty} a_n = \ell$.

(a) For $v > 0$, let $f(v) = \dfrac{ve^{-v}}{1 - e^{-v}}$. Show that for $t > 0$ and $x \in \mathbb{R}$,

$$-\int_0^{+\infty} f'(v) v^{t+ix} dv = (t + ix)\zeta(t + 1 + ix)\Gamma(t + 1 + ix),$$

with $\zeta(s) = \sum_{n=1}^{\infty} \dfrac{1}{n^s}$ for $\operatorname{Re} s > 1$ and $\Gamma(s) = \int_0^{+\infty} t^{s-1} e^{-t} dt$ for $\operatorname{Re} s > 0$.

(b) Show that there is a constant $K > 0$ such that

$$\left| \sum_{n=1}^{N} a_n \right| \leqslant K \text{ for } N \geqslant 1.$$

(c) Conclude.

2.9. The Liouville function. Let $\lambda : \mathbb{N}^* \to \mathbb{R}$ be the Liouville function, that is, the completely multiplicative function such that $\lambda(1) = 1$ and $\lambda(p) = -1$ for all primes p.

(a) Show that $\sum_{d|n} \lambda(d) = 1$ if $n \geqslant 1$ is a perfect square, and 0 otherwise.

(b) Let $u_n = \dfrac{\lambda(n)}{n}$. Show that if $|x| < 1$, then

$$\sum_{n=1}^{\infty} n u_n \frac{x^n}{1 - x^n} = \sum_{n=1}^{\infty} x^{n^2} = L(x).$$

(c) Show that $(1 - x)L(x) \to 0$ as $x \to 1$. Using the Hardy–Littlewood theorem relative to Lambert summability (Exercise 2.8), show that the series $\sum u_n$ converges, and that its sum is zero.

(d) Let $m \in \mathbb{N}^*$ be fixed. Show that the series $\sum u_{mn}$ converges and that its sum is zero. However, the u_n are not all zero.

(e) Let $\sum v_n$ be an *absolutely* convergent series such that $\sum_{n=1}^{\infty} v_{mn} = 0$ for all integers $m \geqslant 1$. Show that all of the v_n are zero.

2.10. The Erdős–Feller–Pollard renewal theorem [41]. Let $(X_n)_{n \geqslant 1}$ be a sequence of random variables with values in $\mathbb{Z}$, independent, identically distributed and not almost surely constant. For $n \geqslant 1$, let $S_n = X_1 + \cdots + X_n$; the sequence $(S_n)_{n \geqslant 1}$ is the random walk associated with the common law of the X_n. We also define

$$p_0 = 1 \quad \text{and} \quad p_n = \mathbf{P}(S_n = 0) \text{ for } n \geqslant 1,$$

$$T = \inf\{n \geqslant 1 / S_n = 0\}$$

and

$$q_0 = 0 \quad \text{and} \quad q_n = \mathbf{P}(T = n) \text{ for } n \geqslant 1.$$

We say that the random walk is
- *aperiodic* if the GCD of the integers $n \geqslant 1$ such that $p_n > 0$ is 1,
- *recurrent* if

$$\mathbf{P}(T < \infty) = \sum_{n=1}^{\infty} q_n = 1,$$

- *positive recurrent* if it is recurrent and in addition verifies

$$\mathbf{E}(T) = \sum_{n=1}^{\infty} n q_n =: m < \infty.$$

(a) Establish the *renewal equation*

$$p_n = \sum_{k=1}^{n} q_k p_{n-k} \text{ for } n \geqslant 1$$

and

$$P = 1 + PQ,$$

where

$$P(z) = \sum_{n=0}^{\infty} p_n z^n \text{ and } Q(z) = \sum_{n=1}^{\infty} q_n z^n \text{ for } |z| < 1.$$

(b) Define φ, the characteristic function of X_1, by

$$\varphi(t) = \mathbf{E}(e^{itX_1}) = \sum_{k \in \mathbb{Z}} P(X_1 = k)e^{ikt}.$$

Show that

$$p_n = \frac{1}{2\pi} \int_{-\pi}^{\pi} \varphi(t)^n dt.$$

Deduce that $p_n \to 0$ when $n \to +\infty$.

Show then that the random walk associated with X_1 such that

$$\mathbf{P}(X_1 = -1) = \mathbf{P}(X_1 = 1) = \frac{1}{2}\mathbf{P}(X_1 = 0) = \frac{1}{4}$$

is at the same time aperiodic and recurrent.

In what follows, we consider *more generally* two sequences $(p_n)_{n \geqslant 0}$ and $(q_n)_{n \geqslant 0}$ of non-negative real numbers satisfying the renewal equation, with $(p_n)_{n \geqslant 0}$ aperiodic, such that

$$p_0 = 1, \ q_0 = 0, \ p_n \leqslant 1, \ \sum_{n=1}^{\infty} q_n = 1 \text{ and } m := \sum_{n=1}^{\infty} nq_n < \infty.$$

(c) Denote by D (resp. $\overline{D}$) the open (resp. closed) unit disk of $\mathbb{C}$ and W^+ the algebra of functions $f : \overline{D} \to \mathbb{C}$ such that there exists a complex sequence $(c_n)_{n \geqslant 0}$ satisfying

$$\|f\|_{W^+} := \sum_{n=0}^{\infty} |c_n| < \infty \text{ and } f(z) = \sum_{n=0}^{\infty} c_n z^n \text{ for } |z| \leqslant 1.$$

Also, set

$$r_n = \sum_{k > n} q_k \text{ for } n \geqslant 0.$$

Show that we define an element R of W^+ by setting

$$R(z) = \sum_{n=0}^{\infty} r_n z^n \text{ for } |z| \leqslant 1.$$

Show also that

$$(1 - z)R(z) = 1 - Q(z) \text{ for } |z| \leqslant 1, \text{ and } R(1) = m.$$

(d) Show that the GCD of the integers $n \geqslant 1$ such that $q_n > 0$ is equal to 1.
(e) Show that the function R is zero-free on $\overline{D}$.
(f) Show (or assume) that, equipped with the norm $\| \cdot \|_{W^+}$, the algebra W^+ is a Banach algebra whose spectrum coincides with $\overline{D}$ (see Chapter 11 if necessary). By noting that

$$(1 - z)P(z)R(z) = 1 \text{ for } |z| < 1,$$

show that the function

$$D \to \mathbb{C}, \ z \mapsto (1 - z)P(z)$$

has an extension to $\overline{D}$ which is an element of W^+. Deduce that $(p_n)_{n \geqslant 0}$ converges to $\dfrac{1}{m}$.
(g) Can there exist, on the additive group $\mathbb{Z}$, random walks that are at the same time aperiodic and *positive* recurrent? (Aperiodic and recurrent is possible, as we have seen in part (b).)

2.11. Let A be a finite subset of $\mathbb{Z}$. For $(s, t) \in \mathbb{T}^2$, define

$$d(s, t) = \sup_{k \in A} |s^k - t^k| \text{ and } \overline{d}(s, t) = \sup_{f \in \mathcal{P}_A, \|f\|_\infty \leqslant 1} |f(s) - f(t)|,$$

where $\mathcal{P}_A$ is the set of trigonometric polynomials with spectrum in A, that is to say, the functions of the form

$$f : \mathbb{T} \to \mathbb{C}, \ t \mapsto \sum_{k \in A} c_k t^k.$$

Then d and $\overline{d}$ are semi-metrics on $\mathbb{T}$. Independently, Bourgain and Rodriguez-Piazza showed the "equivalence" of d and $\overline{d}$: there exist positive constants C and C' such that

$$Cd \leqslant \overline{d} \leqslant C'd.$$

This equivalence has applications in harmonic analysis (see [120], Chapter 13). This exercise outlines Bourgain's proof of this result. Recall (Remark 2.3.12) that there exists a positive constant C_0 such that, for all $\varepsilon > 0$, we have, for $|\alpha| \leqslant \varepsilon$:

$$e^{i\alpha} - 1 = \sum_{n \in \mathbb{Z}} a_n(\varepsilon)e^{in\alpha}, \text{ with } \sum_{n \in \mathbb{Z}} |a_n(\varepsilon)| \leqslant C_0\varepsilon. \qquad (\star)$$

(a) Verify that $d \leqslant \overline{d}$.

(b) In what follows, we intend to show the existence of $C' > 0$ such that $\overline{d} \leqslant C'd$. For this, fix s and t in $\mathbb{T}$. Show that without loss of generality, we can assume that $s = 1$.

(c) Set $\delta = d(1, t)$ and write $t^k = e^{i\alpha_k}$ for $k \in A$, with $|\alpha_k| \leqslant \pi$. Show that $|\alpha_k| \leqslant \dfrac{\pi}{2}\delta =: \varepsilon$.

(d) Let $f \in \mathcal{P}_A$ such that $\|f\|_\infty \leqslant 1$. Using $(\star)$, show that

$$f(t) - f(1) = \sum_{n \in \mathbb{Z}} a_n(\varepsilon)f(t^n).$$

(e) Conclude.

2.12. We want to show that for any finite family $(f_1, \ldots, f_p)$ of functions in $L^1(\mathbb{R})$, the ideal I of $L^1(\mathbb{R})$ generated by the f_j is distinct from $L^1(\mathbb{R})$.[12]

(a) Let $f \in L^1(\mathbb{R})$. Show that there exists a function $r : \mathbb{R} \to \mathbb{R}$, continuous and even, tending to 0 at $\pm\infty$, dominating $\widehat{f}$, non-increasing and convex on $\mathbb{R}_+$.

(b) Show that there exists a $g \in L^1(\mathbb{R})$ such that $\widehat{g} = r$.

(c) Let $u \in L^1(\mathbb{R})$ be defined by $u(x) = e^{-|x|}$ for $x \in \mathbb{R}$. By part (b), there exist $g_1, \ldots, g_p$ in $L^1(\mathbb{R})$ so that $|\widehat{f_j}| \leqslant \widehat{g_j}$ for $1 \leqslant j \leqslant p$. Set $g = g_1 + \cdots + g_p + u \in L^1(\mathbb{R})$. Show that $g \notin I$.

---

[12] As we will see, a less algebraic variant of this result is the following. Let h be a positive, continuous function that tends to zero as $|x| \to +\infty$. Then there exists a $g \in L^1(\mathbb{R})$ such that $h \leqslant \widehat{g}$. This means that, in the Riemann–Lebesgue lemma, the convergence to zero can be arbitrarily slow.

3

The Newman Tauberian theorem

3.1 Introduction

The famous prime number theorem (PNT for short), proved independently (in the same year 1896) by J. Hadamard and Ch. de la Vallée-Poussin [174], states that

$$\pi(x) \underset{x \to +\infty}{\sim} \frac{x}{\ln x},$$

where $\pi(x)$ denotes the number of primes equal to or less than x. The initial proof of this theorem required a deep understanding of the properties of the Riemann ζ function, in particular the crucial fact that

$$\zeta(1 + it) \neq 0 \text{ for all } t \in \mathbb{R}^*, \tag{3.1}$$

among other things. In contrast, the 1931 Tauberian theorem of Wiener–Ikehara [174] that we will state and prove in this chapter (Theorem 3.5.2) required only minimum knowledge of the function ζ (knowing (3.1) was enough), but in counterpart required a deep familiarity with the Fourier transform. A major strength of this theorem is the weakness of its hypotheses. Still today, this approach to the PNT (and its generalisation to prime numbers in an arithmetic progression, following Dirichlet) is one of the most direct.[1]

However, in the 1980s, that is to say quite recently in the scale of the "popularization" of mathematics, D. J. Newman gave a new proof of this theorem, under more restrictive hypotheses, naturally verified in the application to the PNT. Here, Fourier analysis was replaced by elementary complex analysis. This proof eventually became quite well known; nonetheless, we thought it useful to include it here, insisting on the exact contribution of Newman, often poorly explained in works on these topics.[2] We also give some applications of this result in addition to the PNT.

---

[1] See also the proof of H. Daboussi, given in [175].

[2] With the exception of [89] – see also [192].

In the following proof of Newman's lemma, we will see the triumph of the old-fashioned point of view of improper Riemann integrals, like that of $\frac{\sin t}{t}$ over $]0, \infty[$. The symbol $\int_0^{+\infty} f(t)\,dt$ represents the limit of $\int_0^T f(t)\,dt$ when $T \to +\infty$. The value $I(T) = \int_0^T f(t)\,dt$ may have a limit as $T \to +\infty$ even if f does not have an absolutely convergent Lebesgue integral (this is often the case). This can be true if $I(T)$ does not *oscillate* too much, in contrast (for example) with $J(T) = \int_0^T \sin t\,dt$, for which

$$\varliminf_{T \to +\infty} J(T) = 0 \text{ and } \varlimsup_{T \to +\infty} J(T) = 2.$$

It is this absence of oscillation ($\varliminf = \varlimsup$) that is the basis of Tauberian theorems, including the PNT. In fact, we know from the time of Tchebycheff [174] that if $Q(x) = \frac{\pi(x)}{\frac{x}{\ln x}}$, we have

$$\varliminf_{x \to +\infty} Q(x) > 0 \text{ and } \varlimsup_{x \to +\infty} Q(x) < \infty.$$

The PNT, which will be proved later in this chapter, states more precisely that

$$\varliminf_{x \to +\infty} Q(x) = \varlimsup_{x \to +\infty} Q(x) = 1.$$

By convention, we say that a complex function defined on a *closed subset* E of the complex plane is holomorphic on E if it can be extended to a holomorphic function (in the usual sense) over an open subset of $\mathbb{C}$ containing E.

3.2 Newman's lemma

Newman's main contribution is to be found in this lemma[3] and especially in its proof [134, 135].

3.2.1 Lemma [Newman] *Let $f : \mathbb{R}^+ \to \mathbb{C}$ be a measurable and bounded function, and let*

$$F(z) = \int_0^{+\infty} f(t)e^{-zt}\,dt$$

be its Laplace transform, defined on the open half-plane

$$\{z \in \mathbb{C} / \operatorname{Re} z > 0\} =: \Omega.$$

Suppose that F is holomorphic on $\overline{\Omega}$, and continue to use F to name its analytic extension to an open set $V \supset \overline{\Omega}$. Then the integral $\int_0^{+\infty} f(t)\,dt$ converges, and has value $F(0)$.

---

[3] Also attributed to Ingham.

Proof Without loss of generality, we can suppose that $\|f\|_\infty = 1$ and $F(0) = 0$. Indeed, the first point is trivial; for the second point, consider the function

$$h(t) = f(t) - F(0)e^{-t},$$

whose Laplace transform is, for $z \in \Omega$,

$$H(z) = F(z) - \frac{F(0)}{1+z}.$$

The function H is holomorphic on the open set $W := V \setminus \{-1\}$, h is bounded on $\mathbb{R}_+$, $H(0) = F(0) - F(0) = 0$. If we know how to deal with this case, we can deduce that

$$\int_0^T h(t)\,dt \to 0 \text{ as } T \to +\infty,$$

or again

$$\int_0^T f(t)\,dt - F(0)\int_0^T e^{-t}dt \to 0.$$

But $\int_0^T e^{-t}dt = 1 - e^{-T} \to 1$, so that $\int_0^T f(t)\,dt \to F(0)$.

We thus suppose, in what follows, that $F(0) = 0$. The advantage is that the function $\dfrac{F(z)}{z}$ is holomorphic on V.

Now define the truncated Laplace transform

$$F_T(z) = \int_0^T f(t)e^{-zt}dt,$$

which is clearly an entire function. We want to show that

$$F_T(0) - F(0) = F_T(0) \to 0 \text{ as } T \to +\infty.$$

In this context, the use of an integral representation of $F_T(0) - F(0)$ comes to mind.

For this, fix $R > 0$ and let Γ be the circle with centre 0 and radius R, traced out once in the counter-clockwise direction (Figure 3.1). Let Γ^+ (resp. Γ^-) be the portion of this path contained in the open right (resp. left) half-plane of the complex plane. Let S be the vertical segment with origin iR and finishing at $-iR$.

Cauchy's integral formula then gives

$$F_T(0) = \frac{1}{2i\pi}\int_{\Gamma^+}\frac{F_T(z)}{z}dz + \frac{1}{2i\pi}\int_{\Gamma^-}\frac{F_T(z)}{z}dz. \tag{3.2}$$

We cannot write a similar formula for F, which is holomorphic on V, but not necessarily in a neighbourhood of Γ. Nonetheless, as the function $\dfrac{F(z)}{z}$ is

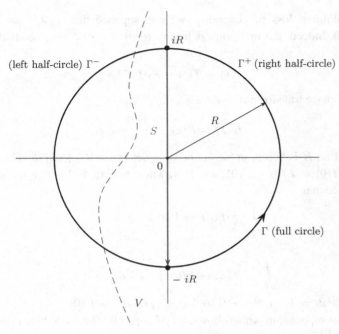

Figure 3.1

holomorphic on V and as $\Gamma^+ \cup S$ as well as its interior (namely the set of points having non-zero winding number with respect to this closed curve) are contained in V, Cauchy's theorem gives

$$0 = \frac{1}{2i\pi} \int_{\Gamma^+} \frac{F(z)}{z}\,dz + \frac{1}{2i\pi} \int_S \frac{F(z)}{z}\,dz. \qquad (3.3)$$

By subtracting (3.2) and (3.3) term by term, we obtain the desired integral representation:

$$F_T(0) = \frac{1}{2i\pi} \int_{\Gamma^+} \frac{F_T(z) - F(z)}{z}\,dz + \frac{1}{2i\pi} \int_{\Gamma^-} \frac{F_T(z)}{z}\,dz$$
$$- \frac{1}{2i\pi} \int_S \frac{F(z)}{z}\,dz.$$

Recall now the simple but fundamental estimation

$$\left| \int_\gamma \varphi(z)\,dz \right| \leqslant L(\gamma) \sup_{z \in \gamma} |\varphi(z)|, \qquad (3.4)$$

where $L(\gamma)$ is the length of the curve γ.

In order to use this estimation, observe that, since $|f(t)| \leqslant 1$, we have

$$|F_T(z) - F(z)| \leqslant \int_T^{+\infty} |e^{-tz}| dt = \frac{e^{-T \operatorname{Re} z}}{\operatorname{Re} z} \text{ for } z \in \Gamma^+, \qquad (3.5)$$

so that

$$\left| \frac{F_T(z) - F(z)}{z} \right| \leqslant \frac{|z|}{\operatorname{Re} z} \frac{e^{-T \operatorname{Re} z}}{R^2} \text{ for these same } z. \qquad (3.6)$$

In this inequality, the enemy is the term $\dfrac{|z|}{\operatorname{Re} z}$. If it was not there, we could easily bound $I^+(T)$, even without the term $e^{-T \operatorname{Re} z}$.

For $z \in \Gamma^-$, we have

$$|F_T(z)| \leqslant \int_0^T e^{-T \operatorname{Re} z} dt = \frac{e^{-T \operatorname{Re} z} - 1}{-\operatorname{Re} z} = \frac{e^{T | \operatorname{Re} z|} - 1}{|\operatorname{Re} z|} \leqslant \frac{e^{T | \operatorname{Re} z|}}{|\operatorname{Re} z|},$$

so that

$$\left| \frac{F_T(z)}{z} \right| \leqslant \frac{|z|}{|\operatorname{Re} z|} \frac{e^{T | \operatorname{Re} z|}}{R^2} \text{ for } z \in \Gamma^-.$$

The factor $\dfrac{|z|}{\operatorname{Re} z}$ still torments us, in the Racinian[4] sense of the word:

> *Britannicus le gêne, Albine, et chaque jour*
> *Je sens que je deviens importune à mon tour.*

The factor $e^{T | \operatorname{Re} z|}$ is also troublesome, even though it provided us with a margin of security in the inequality (3.6). Thus, a first idea (classical and not at all due to Newman) is to replace $F(z)$ and $F_T(z)$, respectively, by $F(z)e^{Tz}$ and $F_T(z)e^{Tz}$, which have the same values at 0 as the initial functions. Then we obtain

$$F_T(0) = \frac{1}{2i\pi} \int_{\Gamma^+} \frac{F_T(z) - F(z)}{z} e^{Tz} dz + \frac{1}{2i\pi} \int_{\Gamma^-} \frac{F_T(z)}{z} e^{Tz} dz$$
$$- \frac{1}{2i\pi} \int_S \frac{F(z)}{z} e^{Tz} dz,$$

with the estimations

$$\left| \frac{F_T(z) - F(z)}{z} e^{Tz} \right| \leqslant \frac{|z|}{\operatorname{Re} z} \frac{1}{R^2} \text{ for } z \in \Gamma^+$$

and

$$\left| \frac{F_T(z)}{z} e^{Tz} \right| \leqslant \frac{|z|}{|\operatorname{Re} z|} \frac{1}{R^2} \text{ for } z \in \Gamma^-.$$

---

[4] See [154], *Britannicus*, Act 1, Scene 1.

The factor $\dfrac{|z|}{\operatorname{Re} z}$ still causes trouble. And here, to get rid of this factor and particularly its denominator, Newman introduced a *simple but decisive idea*: replace

$$F(z)e^{Tz} \text{ and } F_T(z)e^{Tz},$$

respectively, by

$$F(z)e^{Tz}\Big(1 + \frac{z^2}{R^2}\Big) \text{ and } F_T(z)e^{Tz}\Big(1 + \frac{z^2}{R^2}\Big),$$

which have the same value at 0 as the initial functions. By setting

$$G_T(z) = \frac{F_T(z) - F(z)}{z} e^{Tz}\Big(1 + \frac{z^2}{R^2}\Big)$$

and

$$H_T(z) = \frac{F_T(z)}{z} e^{Tz}\Big(1 + \frac{z^2}{R^2}\Big),$$

we thus obtain

$$F_T(0) = \underbrace{\frac{1}{2i\pi} \int_{\Gamma^+} G_T(z)\,dz}_{:=I^+(T)} + \underbrace{\frac{1}{2i\pi} \int_{\Gamma^-} H_T(z)\,dz}_{:=I^-(T)}$$

$$- \underbrace{\frac{1}{2i\pi} \int_S \frac{F(z)}{z} e^{Tz}\Big(1 + \frac{z^2}{R^2}\Big)\,dz}_{:=I_S(T)}.$$

Now, since

$$\frac{1}{z} = \frac{\bar{z}}{R^2} \text{ for } z \in \Gamma,$$

we have, for $z \in \Gamma^+$:

$$|G_T(z)| = |F(z) - F_T(z)|e^{T\operatorname{Re} z}\left| \frac{\bar{z}}{R^2} + \frac{z}{R^2} \right|$$

so that, remembering (3.5):

$$|G_T(z)| \leqslant \frac{e^{-T\operatorname{Re} z}}{\operatorname{Re} z} e^{T\operatorname{Re} z} \frac{2\operatorname{Re} z}{R^2} = \frac{2}{R^2} \text{ for } z \in \Gamma^+, \qquad (3.7)$$

and similarly

$$|H_T(z)| \leqslant \frac{2}{R^2} \text{ for } z \in \Gamma^-. \qquad (3.8)$$

The rest of the proof can then be split into three simple steps.

Step 1. $|I^+(T)| \leqslant \dfrac{1}{R}$. In fact, (3.4) and (3.7) give

$$|I^+(T)| \leqslant \frac{1}{2\pi} \pi R \frac{2}{R^2} = \frac{1}{R}.$$

Step 2. $|I^-(T)| \leqslant \dfrac{1}{R}$. This time we apply (3.4) and (3.8).

Step 3. $I_S(T) \to 0$ as $T \to +\infty$. Indeed,

$$I_S(T) = -\frac{1}{2i\pi} \int_{-R}^{R} \frac{F(iy)}{y} \left(1 - \frac{y^2}{R^2} \right) e^{iTy} dy,$$

and this quantity tends to 0 for R fixed, when $T \to +\infty$, according to the Riemann–Lebesgue lemma.

As $|F_T(0)| \leqslant |I^+(T)| + |I^-(T)| + |I_S(T)|$, we can conclude from the three preceding steps that

$$\varlimsup_{T \to +\infty} |F_T(0)| \leqslant \frac{2}{R}.$$

As R is arbitrarily large, this shows that

$$\lim_{T \to +\infty} F_T(0) = 0,$$

which concludes the proof of Newman's lemma. □

3.3 The Newman Tauberian theorem

We present two versions of this theorem, one for integrals and the other for series. Even if these statements hold the upper hand, keep in mind that most of the work was accomplished in the proof of Newman's lemma [134, 135].

3.3.1 Theorem [Newman's theorem for integrals] *Let $f : \mathbb{R}^+ \to \mathbb{R}^+$ be a non-decreasing function, and a, c positive real numbers. Suppose that the Laplace transform of f,*

$$F(z) = \int_0^{+\infty} f(t) e^{-zt} dt,$$

is defined on the open half-plane $\operatorname{Re} z > a$ *and that, more precisely:*

(i) *$f(t)e^{-at}$ is bounded on $\mathbb{R}^+$;*

(ii) *$F(z) - \dfrac{c}{z - a}$ has a holomorphic extension G on the closed half-plane $\operatorname{Re} z \geqslant a$.*

Then, we have

$$f(t) \sim ce^{at} \text{ as } t \to +\infty. \tag{3.9}$$

We now present the version of the theorem for Dirichlet series. A good reference for these series is the text of Hardy and Riesz [86].

3.3.2 Theorem [Newman's theorem for series]   *Let $(a_n)_{n \geqslant 1}$ be a sequence of non-negative real numbers, and a, c be positive real numbers. Suppose that the Dirichlet series*

$$\Phi(s) = \sum_{n=1}^{\infty} a_n n^{-s}$$

is defined on the open half-plane $\operatorname{Re} s > a$ *and that, more precisely, with*

$$A(x) = \sum_{n \leqslant x} a_n \text{ for } x \geqslant 0,$$

the following properties are verified:

(*i*) *$A(x)x^{-a}$ is bounded on $\mathbb{R}^+$;*
(*ii*) *$\Phi(s) - \dfrac{c}{s-a}$ has a holomorphic extension G on the closed half-plane $\operatorname{Re} s \geqslant a$.*

Then we have

$$A(x) \sim \frac{c}{a} x^a \text{ as } x \to +\infty.$$

3.3.3 Remarks (1) Ikehara's[5] Theorems 3.5.1 and 3.5.2, stated further on, have respectively the same conclusions as Theorems 3.3.1 and 3.3.2, but under the following weaker hypotheses:

f is non-decreasing, $F(z) - \dfrac{c}{z-a}$ has a continuous extension on $\operatorname{Re} z \geqslant a$

and

$a_n \geqslant 0$, $\Phi(s) - \dfrac{c}{s-a}$ has a continuous extension on $\operatorname{Re} s \geqslant a$.

A brief glance at these different statements shows us that Ikehara's theorems are "better" than those of Newman. Indeed, the weaker hypotheses – both on the function f and on its Laplace transform F (or on the sequence (a_n) and its generating series Φ) – lead to the same conclusions.

What then is the interest of Theorems 3.3.1 and 3.3.2? The answer is as follows: their proof is (a bit) more simple, and at least one of the additional hypotheses will naturally be verified in the applications to number theory that we have in mind. For the comparison of the proofs, we quote Newman promoting his theorem: "The proof of Ikehara's theorem requires

---

[5] Ikehara was a student of N. Wiener.

analysis à la Fourier". One could slyly object that the proof of his theorem requires "analysis à la Cauchy", which amounts to almost the same thing. To say "à la Fourier" means

$$\frac{1}{2\pi} \int_0^{2\pi} f(e^{it})\, dt\,;$$

whereas to say "à la Cauchy" means

$$\frac{1}{2i\pi} \int_{\mathbb{T}} f(z)\frac{dz}{z}\,.$$

But to criticise is easy, to create is more difficult. . .

(2) The series version of Newman's theorem is an immediate consequence of the integral version. To see this, set

$$A(x) = \sum_{n \leqslant x} a_n \text{ and } f(t) = A(e^t),$$

and note that

$$\Phi(s) = \sum_{n=1}^{\infty} a_n n^{-s} = s \int_1^{+\infty} A(x) x^{-s-1} dx = s \int_0^{+\infty} f(t)e^{-st} dt.$$

$$(3.10)$$

Indeed, the first integral in (3.10) gives, if $\operatorname{Re} s > a$,

$$s \sum_{n=1}^{\infty} \int_n^{n+1} (a_1 + \cdots + a_n) x^{-s-1} dx = \sum_{n=1}^{\infty} (a_1 + \cdots + a_n)(n^{-s} - (n+1)^{-s}),$$

or again

$$\sum_{n=1}^{\infty} (a_1 + \cdots + a_n) n^{-s} - \sum_{n=2}^{\infty} (a_1 + \cdots + a_{n-1}) n^{-s} = \sum_{n=1}^{\infty} a_n n^{-s}.$$

The change of variable $x = e^t$ then gives the second integral.

Next, set

$$F(s) = \int_0^{+\infty} f(t)e^{-st} dt = \frac{\Phi(s)}{s} \text{ for } \operatorname{Re} s > a.$$

If $\Phi(s) = \dfrac{c}{s-a} + \psi(s)$ with ψ holomorphic on $\operatorname{Re} s \geqslant a$, we see that

$$F(s) = \frac{c}{s(s-a)} + \frac{\psi(s)}{s}$$

$$= \frac{c}{a}\Big(\frac{1}{s-a} - \frac{1}{s}\Big) + \frac{\psi(s)}{s}$$

$$= \frac{c}{a}\frac{1}{s-a} + \chi(s),$$

with χ holomorphic on $\text{Re}\, s \geqslant a$. As the a_n are non-negative, f is non-decreasing, and since $\sum_{n \leqslant x} a_n \underset{x \to +\infty}{=} O(x^a)$ by hypothesis, $f(t) \underset{t \to +\infty}{=} O(e^{at})$. The integral version of Newman's Theorem 3.3.1 then gives $f(t) \underset{t \to +\infty}{\sim} \frac{c}{a} e^{at}$, that is to say, $A(x) \underset{x \to +\infty}{\sim} \frac{c}{a} x^a$. The same evidently holds true for Ikehara's theorem.

(3) The Tauberian condition here is the fact that f is non-decreasing, and $f(t)e^{-at}$ is bounded.

(4) In his article [134] about Theorem 3.3.1, Newman wrote: "Actually, this theorem dates back to Ingham…"; however, a study of the article by Ingham [94] convinced us that both the statement and the methods used were more "à la Ikehara" and "à la Fourier", so it seems fair to attribute Theorem 3.3.1 to Newman alone.

Let us now show how Newman's Lemma 3.2.1 implies Theorem 3.3.1 in a standard Tauberian manner. Under the hypotheses of this theorem, we have, for $\text{Re}\, z > a$:

$$\int_0^{+\infty} \left(f(t)e^{-at} - c \right) e^{-(z-a)t} dt = F(z) - \frac{c}{z-a},$$

with a holomorphic extension to the closed set $\text{Re}\, z \geqslant a$, or equivalently, by setting $z - a = w$,

$$\int_0^{+\infty} \left(f(t)e^{-at} - c \right) e^{-wt} dt = F(w+a) - cw^{-1} \text{ for } \text{Re}\, w > 0,$$

with a holomorphic extension to the closed set $\text{Re}\, w \geqslant 0$. By Newman's lemma:[6]

$$\int_0^{+\infty} \left(f(t)e^{-at} - c \right) dt \text{ converges.} \tag{3.11}$$

To obtain (3.9) from (3.11), we "make right" the error that says, since the integral converges in (3.11), the general term must tend to 0, that is to say, $f(t)e^{-at} - c \to 0$ as $t \to +\infty$ (thus (3.9)).

For this, set

$$I(T) = \int_0^T \left(f(t)e^{-at} - c \right) dt \text{ for } T > 0,$$

and let $\varepsilon > 0$. As f is non-decreasing, we have

$$I(T + \varepsilon) - I(T) = \int_T^{T+\varepsilon} \left(f(t)e^{-at} - c \right) dt$$
$$\geqslant \int_T^{T+\varepsilon} \left(f(T)e^{-at} - c \right) dt$$

---

[6] Note that $f(t)e^{-at} - c$ is bounded on $\mathbb{R}_+$.

$$= f(T) \int_T^{T+\varepsilon} e^{-at} dt - c\varepsilon$$

$$= f(T)e^{-aT} \frac{1 - e^{-a\varepsilon}}{a} - c\varepsilon,$$

so that

$$e^{-aT} f(T) \leqslant \frac{ca\varepsilon}{1 - e^{-a\varepsilon}} + \frac{a}{1 - e^{-a\varepsilon}} \big(I(T + \varepsilon) - I(T)\big).$$

As $I(T+\varepsilon) - I(T) \underset{T \to +\infty}{\to} 0$ by (3.11), taking the upper limit when $T \to +\infty$ in this inequality, with ε fixed, gives

$$\varlimsup_{T \to +\infty} e^{-aT} f(T) \leqslant \frac{ca\varepsilon}{1 - e^{-a\varepsilon}}.$$

Letting ε tend to 0, we obtain

$$\varlimsup_{T \to +\infty} e^{-aT} f(T) \leqslant c.$$

Similarly, by writing

$$I(T) - I(T - \varepsilon) \leqslant \int_{T-\varepsilon}^T \big(f(T)e^{-at} - c\big) \, dt,$$

we obtain

$$f(T)e^{-aT} \geqslant \frac{ca\varepsilon}{e^{a\varepsilon} - 1} + \frac{a}{e^{a\varepsilon} - 1} \big(I(T) - I(T - \varepsilon)\big),$$

and hence

$$\varliminf_{T \to +\infty} e^{-aT} f(T) \geqslant \frac{ca\varepsilon}{e^{a\varepsilon} - 1}$$

and

$$\varliminf_{T \to +\infty} e^{-aT} f(T) \geqslant c,$$

so that, finally,

$$e^{-aT} f(T) \to c \text{ as } t \to +\infty,$$

which justifies our initial statement, and proves Theorem 3.3.1.

3.4 Applications

We present two applications of Newman's theorems. The first, due to P. Erdős and his collaborators [58], is about the study of a curious recursive sequence. The second is none other than the PNT.

3.4.1 Theorem [Erdős *et al.*] *Let* q *be an integer* $\geqslant 2$, *and* $r_1, \ldots, r_q$ *positive real numbers with sum strictly greater than* 1. *Next, select integers* $m_1, \ldots, m_q \geqslant 2$, *such that* $\dfrac{\ln m_j}{\ln m_k}$ *is irrational for at least one pair* (j, k), *and consider the recursive sequence* $(a_n)_{n \geqslant 0}$ *defined by* $a_0 = 1$ *and*

$$a_n = \sum_{j=1}^{q} r_j a_{[n/m_j]} \, for \, n \geqslant 1,$$

where $[\cdot]$ *is the integer part. Then:*

(1) the equation $\sum_{j=1}^{q} r_j m_j^{-a} = 1$ *has a single solution* $a > 0$;

(2) there exists a constant $c_0 > 0$ *such that* $a_n \sim c_0 n^a$, *when* $n \to +\infty$;

(3) in particular, if $a_n = a_{[n/2]} + a_{[n/3]} + a_{[n/6]}$, *we have*

$$a_n \sim \frac{12}{\ln 432} n.$$

Proof (1) The function $\varphi(x) = \sum_{j=1}^{q} r_j m_j^{-x}$ is continuous, decreasing on $[0, +\infty[$, and $\varphi(0) = \sum_{j=1}^{q} r_j > 1$ while $\varphi(+\infty) = 0$, which shows the existence and uniqueness of a. Note that $(a_n)_{n \geqslant 0}$ is *non-decreasing* (easy induction), and that

$$a_n \leqslant C n^a \text{ for } n \geqslant 1, \tag{3.12}$$

for some positive constant C. Indeed, select C such that (3.12) holds when $n \leqslant \max(m_1, \ldots, m_q)$. If $n > \max(m_1, \ldots, m_q)$ and if the bound (3.12) holds for all integers $< n$, we see that

$$a_n = \sum_{j=1}^{q} r_j a_{[n/m_j]} \leqslant \sum_{j=1}^{q} r_j C \left(\frac{n}{m_j} \right)^a = C n^a \sum_{j=1}^{q} r_j m_j^{-a} = C n^a.$$

Next (but we will not use this result) an easy proof by induction shows that

$$a_n \geqslant (n + 1)^a \text{ for } n \geqslant 0.$$

Thus the sequence $(n^{-a} a_n)_{n \geqslant 1}$ has an upper bound, and 1 as a lower bound; it remains to see if the sequence actually has a limit, which is incomparably more difficult.

Figure 3.2 represents $n^{-1} a_n$ as a function of n for $1 \leqslant n \leqslant 10\,000$; the sequence $(n^{-1} a_n)_{n \geqslant 1}$ gives the impression that it oscillates indefinitely... Beware of drawing conclusions from numerical evidence!

Moreover, the article of Erdős [58] is called "A very slowly convergent sequence" and theory shows that the rate of convergence is $\dfrac{1}{\ln n}$ (see also [59]).

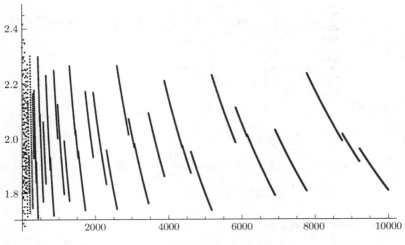

Figure 3.2

(2) To prove the second item of Theorem 3.4.1, we proceed as follows: set $a(x) = a_{[x]}$ for $x \geqslant 0$. The *recursive relation* on the a_n can be transformed into the following *functional equation*:

$$a(x) = \begin{cases} 1 & \text{if } 0 \leqslant x < 1, \\ \displaystyle\sum_{j=1}^{q} r_j\, a\Big(\frac{x}{m_j}\Big) & \text{if } x \geqslant 1, \end{cases}$$

essentially because, if $x \geqslant 1$ verifies $[x] = n$, then $\left[\dfrac{x}{m_j}\right] = \left[\dfrac{n}{m_j}\right]$. We then set

$$\Phi(s) = s \int_1^{+\infty} a(x) x^{-s-1} dx \text{ for } \mathrm{Re}\, s > a.$$

This definition is correct according to (3.12), since

$$|a(x) x^{-s-1}| \leqslant C x^{a - \mathrm{Re}\, s - 1} \text{ for } x \geqslant 1.$$

The functional equation satisfied by $a(x)$ thus gives an explicit expression for Φ. In fact:

$$\Phi(s) = \int_1^{+\infty} \sum_{j=1}^{q} r_j\, a\Big(\frac{x}{m_j}\Big) s x^{-s-1} dx$$

$$= \sum_{j=1}^{q} r_j \int_{1/m_j}^{+\infty} a(y) s (m_j y)^{-s-1} m_j\, dy$$

$$= \sum_{j=1}^{q} r_j m_j^{-s} \int_{1/m_j}^{1} s y^{-s-1} dy + \sum_{j=1}^{q} r_j m_j^{-s} \int_{1}^{+\infty} a(y) s y^{-s-1} dy$$

$$= \sum_{j=1}^{q} r_j m_j^{-s} (m_j^s - 1) + \sum_{j=1}^{q} r_j m_j^{-s} \Phi(s),$$

so that

$$\Phi(s) = \frac{\sum_{j=1}^{q} r_j (1 - m_j^{-s})}{1 - \sum_{j=1}^{q} r_j m_j^{-s}} =: \frac{P(s)}{Q(s)} \text{ for Re } s > a, \qquad (3.13)$$

which is correct because

$$\left| \sum_{j=1}^{q} r_j m_j^{-s} \right| \leqslant \sum_{j=1}^{q} r_j m_j^{-\operatorname{Re} s} < \sum_{j=1}^{q} r_j m_j^{-a} = 1,$$

so that

$$\sum_{j=1}^{q} r_j m_j^{-s} \neq 1 \text{ if Re } s > a.$$

However, the arithmetical hypothesis on the m_j allows us to show even more.[7] In fact,

$$\sum_{j=1}^{q} r_j m_j^{-s} \neq 1 \text{ for Re } s \geqslant a \text{ and } s \neq a. \qquad (3.14)$$

Indeed, if $\sum_{j=1}^{q} r_j m_j^{-s} = 1$ with $s = a + iy$, $y \in \mathbb{R}^*$, by taking the real parts, we find that

$$\sum_{j=1}^{q} r_j m_j^{-a} \cos(y \ln m_j) = 1,$$

so that by point (1) of Theorem 3.4.1, $\cos(y \ln m_j) = 1$ for all j, and hence

$$y \ln m_j = 2k_j \pi, \text{ with } k_j \in \mathbb{Z}.$$

As $y \neq 0$, we must have $\dfrac{\ln m_j}{\ln m_k} \in \mathbb{Q}$ for all j and k, which contradicts the hypothesis.

---

[7] It is a variant of Lemma 3.4.8 about the non-vanishing of the ζ function on the line Re $s = 1$, due to Hadamard and de la Vallée-Poussin.

Now, (3.13) shows that $\frac{P}{Q}$ has a simple pole at a, with residue

$$c = \frac{P(a)}{Q'(a)} = \frac{-1 + \sum_{j=1}^{q} r_j}{\sum_{j=1}^{q} r_j m_j^{-a} \ln m_j}. \tag{3.15}$$

This shows that $\Phi(s) - \frac{c}{s-a}$ has a holomorphic extension to the closed half-plane $\operatorname{Re} s \geqslant a$. Moreover, from the second equality of (3.10), if we set

$$b(x) = \sum_{n \leqslant x} (a_n - a_{n-1}),$$

we thus obtain

$$\sum_{n=1}^{\infty} \frac{a_n - a_{n-1}}{n^s} = s \int_{1}^{+\infty} b(x) x^{-s-1} dx \text{ for } \operatorname{Re} s > a.$$

But $b(x) = a_{[x]} - a_0 = a(x) - 1$ for $x \geqslant 1$, so that

$$\sum_{n=1}^{\infty} \frac{a_n - a_{n-1}}{n^s} = s \int_{1}^{+\infty} a(x) x^{-s-1} dx - 1 = \Phi(s) - 1 \text{ for } \operatorname{Re} s > a.$$

As the sequence $(a_n)_{n \geqslant 0}$ is non-decreasing, the sequence $(a_n - a_{n-1})_{n \geqslant 1}$ is non-negative, and Newman's Theorem 3.3.2 for series can be applied, after we note that

$$b(x) = a(x) - 1 = O(x^a) \text{ as } x \to +\infty,$$

according to (3.12). We obtain, with c given by (3.15):

$$b(x) \sim \frac{c}{a} x^a,$$

which implies

$$a_n \sim \frac{c}{a} n^a.$$

This proves point (2) of Theorem 3.4.1, with $c_0 = \frac{c}{a}$.

(3) As for the third point, it is a simple numerical application corresponding to

$$r_1 = r_2 = r_3 = 1, \ a = 1, \ m_1 = 2, \ m_2 = 3, \ m_3 = 6.$$

The expression for c from (3.15) thus gives

$$c_0 = c = \frac{3 - 1}{\frac{1}{2} \ln 2 + \frac{1}{3} \ln 3 + \frac{1}{6} \ln 6} = \frac{12}{3 \ln 2 + 2 \ln 3 + \ln 6} = \frac{12}{\ln 432}. \qquad \square$$

3.4.2 Remark The proof of Erdős [59] used Wiener's Tauberian theorem and the renewal equation ([161], p. 229). The arithmetical hypothesis on the m_j corresponds to the fact that they are not all powers of some integer $d \geqslant 2$. If we have $m_j = d^{u_j}$ with u_j integer, the sequence $(n^{-a}a_n)_{n \geqslant 1}$ does not converge, but its cluster set is an interval; for example, if $a_0 = 1$ and $a_n = 2a_{[n/3]} + 3a_{[n/9]}$ for $n \geqslant 1$, the cluster set of the sequence $(n^{-1}a_n)_{n \geqslant 1}$ is the segment $\left[\dfrac{3}{2}, \dfrac{9}{2} \right]$ (see Exercise 3.3).

We now come to the fundamental application of Newman's theorem to the PNT. As before, $[x]$ denotes the integer part of a real number x, and (m, n) the greatest common divisor (GCD) of the positive integers m, n. Finally, p will always be a prime number.

Recall now the statement of the famous prime number theorem.

3.4.3 Theorem [prime number theorem (PNT)] *We have*

$$\pi(x) \sim \frac{x}{\ln x} \ as \ x \to +\infty,$$

where $\pi(x)$ is the number of primes less than or equal to x. Equivalently, if p_n is the nth prime number, we have

$$p_n \sim n \ln n \ as \ n \to +\infty.$$

Before showing how Newman's Tauberian theorem can be used to prove the PNT, it is useful to restate some equivalent forms of the PNT, which are due to the famous identity discovered by Euler around 1737:

$$\sum_{n=1}^{\infty} n^{-s} = \prod_p (1 - p^{-s})^{-1} \text{ for all real } s > 1, \tag{3.16}$$

where p runs over the set of prime numbers. This identity has a useful generalisation [110, 135, 174, 177] as follows: we say that a function $f : \mathbb{N}^* \to \mathbb{C}$ is *multiplicative* if

$$f(1) = 1 \text{ and } \big((m, n) = 1 \Rightarrow f(mn) = f(m)f(n)\big).$$

The function f is said to be *completely (or totally) multiplicative* if

$$f(1) = 1 \text{ and } f(mn) = f(m)f(n) \text{ for all } m, n \geqslant 1.$$

Also, f is said to be *summable* if $\sum_{n=1}^{\infty} |f(n)| < \infty$. For example, the Euler totient function φ is multiplicative, and the function $n \mapsto n^{-s}$ is completely multiplicative for all values $s \in \mathbb{C}$. With these definitions, we can state the following result.

3.4.4 Lemma   (*1*) *If f is a multiplicative and summable function, then we have*

$$\sum_{n=1}^{\infty} f(n) = \prod_p \left(1 + \sum_{k=1}^{\infty} f(p^k)\right).$$

(*2*) *If f is a completely multiplicative and summable function, then we have*

$$\sum_{n=1}^{\infty} f(n) = \prod_p \frac{1}{1 - f(p)}.$$

(See Exercises 3.4 and 3.5 for some applications.)

To quote Hlawka *et al.* [89] on the identity (3.16):[8] "The door to the Prime Number Theorem was opened not by Euclid's, but by Euler's proof of the infinity of the set of primes". Recall that indeed the identity (3.16) and the divergence of the harmonic series imply the divergence of the series $\sum \frac{1}{p}$, and in particular the infinitude of the sequence of prime numbers. Following Euler, Riemann introduced his famous function ζ *in the complex field*:

$$\zeta(s) = \sum_{n=1}^{\infty} n^{-s} \text{ where } s \in \mathbb{C}, \text{ Re } s > 1,$$

for which we also have

$$\sum_{n=1}^{\infty} n^{-s} = \prod_{j=1}^{\infty} (1 - p_j^{-s})^{-1} \text{ for Re } s > 1$$

(here we number each prime: $p_1 = 2$, $p_2 = 3$, $p_3 = 5$, etc.). We conclude from this equality that the function ζ is never zero in the half-plane $\text{Re } s > 1$. Moreover, we have (formally)

$$\frac{1}{\zeta(s)} = \prod_{j=1}^{\infty} (1 - p_j^{-s}) = 1 - \sum_{j_1} p_{j_1}^{-s} + \sum_{j_1 < j_2} (p_{j_1} p_{j_2})^{-s} + \ldots,$$

or indeed

$$\frac{1}{\zeta(s)} = \sum_{n=1}^{\infty} \frac{\mu(n)}{n^s},$$

where μ is the *Möbius function* defined by

$$\mu(n) = \begin{cases} 1 & \text{if } n = 1, \\ (-1)^k & \text{if } n \text{ is the product of } k \text{ distinct prime numbers,} \\ 0 & \text{otherwise.} \end{cases}$$

---

[8] According to certain number theorists, we have (if lucky!) about one good idea per century in this domain.

We set

$$M(x) = \sum_{n \leqslant x} \mu(n) \text{ for } x \geqslant 1.$$

We also have, formally,

$$\ln \zeta(s) = -\sum_{j=1}^{\infty} \ln(1 - p_j^{-s}),$$

so that, by taking derivatives and changing the sign,

$$-\frac{\zeta'(s)}{\zeta(s)} = \sum_{n=1}^{\infty} \frac{\Lambda(n)}{n^s},$$

where Λ is the *von Mangoldt function* defined by

$$\Lambda(n) = \begin{cases} \ln p & \text{if } n = p^k, k \geqslant 1, \\ 0 & \text{otherwise.} \end{cases}$$

Finally, we set

$$\psi(x) = \sum_{n \leqslant x} \Lambda(n) \text{ for } x \geqslant 1.$$

We can easily provide rigorous arguments for the preceding heuristics,[9] and show that we have effectively, for $\operatorname{Re} s > 1$, the formal equalities used to introduce μ and Λ, as well as their corresponding "summatory functions", M and ψ

These functions allow us to give two equivalent formulations (in the sense of *elementary* equivalent) of the PNT, as follows.

3.4.5 Proposition *The following are equivalent:*

(i) $\pi(x) \sim \dfrac{x}{\ln x}$,

(ii) $\psi(x) \sim x$.

3.4.6 Proposition *The following are equivalent:*

(i) $\pi(x) \sim \dfrac{x}{\ln x}$,

(ii) $M(x) = o(x)$.

---

[9] Without which there would be no reason to introduce the famous functions of Möbius and von Mangoldt!

Let us accept these propositions for now (see Exercise 3.6) and see how *each of them* leads to a proof of the PNT, once we know the following properties of the function ζ [135, 174, 177].

3.4.7 Lemma *The function ζ has a meromorphic extension to the half-plane* $\operatorname{Re} s > 0$, *with a unique simple pole of residue* 1 *at the point* $s = 1$.

3.4.8 Lemma [Hadamard–de la Vallée-Poussin] *The function ζ thus extended is never zero on the line* $\operatorname{Re} s = 1$:

$$\zeta(1 + iy) \neq 0 \text{ for all } y \in \mathbb{R}^*.$$

3.4.9 First proof of the prime number theorem

The proof consists of showing that $M(x) = o(x)$. Set

$$a_n = 1 + \mu(n) \text{ for } n \geqslant 1,$$

where μ is the Möbius function. Then the two preceding lemmas imply that, for $\operatorname{Re} s > 1$,

$$\sum_{n=1}^{\infty} a_n n^{-s} = \zeta(s) + \frac{1}{\zeta(s)} = \frac{1}{s-1} + g(s),$$

where g is holomorphic on the closed set $\operatorname{Re} s \geqslant 1$. Note that the a_n are non-negative and that

$$\sum_{n \leqslant x} a_n \leqslant 2x \text{ for } x \geqslant 1.$$

We can thus apply Newman's Tauberian theorem to obtain

$$\sum_{n \leqslant x} \left(1 + \mu(n)\right) = [x] + M(x) = x + O(1) + M(x) \sim x,$$

so that

$$M(x) + x + O(1) = x + o(x),$$

hence $M(x) = o(x)$.

3.4.10 Second proof of the prime number theorem

This time, we show that $\psi(x) \sim x$. Set $a_n = \Lambda(n)$, for $n \geqslant 1$. It is clear that $\Lambda(n) \leqslant \ln n$, so $\sum_{n=1}^{\infty} \Lambda(n) n^{-s}$ is defined for $\operatorname{Re} s > 1$, and as previously seen, is equal to

$$-\frac{\zeta'(s)}{\zeta(s)} = \frac{1}{s-1} + h(s),$$

where h is holomorphic on the closed half-plane $\operatorname{Re} s \geqslant 1$, according to Lemmas 3.4.7 and 3.4.8.

If we knew that $\psi(x) = O(x)$, Newman's Tauberian theorem could be applied, and we could conclude that $\psi(x) \sim x$, which implies the PNT using Proposition 3.4.5.

It is here that we uncover the *main weakness* of this theorem: before we can show an equivalence of the type $f(t) \sim ce^{at}$ when we have information about the Laplace transform F of f, we must already know that $f(t) = O(e^{at})$, which is not always "for free"! The Tauberian Theorem 3.5.2 of Ikehara is *exempt from this weakness*, and in a way provides the shortest proof of the PNT.

Of course, an expert in analytic number theory would claim that the estimation $\psi(x) = O(x)$ comes almost for free, as it was obtained (in a non-trivial manner) by Tchebycheff [174] around 1850. We can thus say that Newman's theorem leads automatically from Tchebycheff to the PNT.

3.4.11 Remarks   (1) The first proof may seem simpler, because the estimate

$$M(x) = o(x)$$

is a direct consequence of the theorem of Newman, as the inequality $|M(x)| \leqslant x$ is evident. But what we gain on the one hand, we lose on the other! Indeed, once we know that $\psi(x) = O(x)$, from Newman we deduce $\psi(x) \sim x$, and hence the PNT using fairly simple arguments, detailed in Exercise 3.6. It is, however, much less immediate (except for experts) to use the information $M(x) = o(x)$ to derive the PNT. For details, see [174], pp. 38–44 (see also [110, 177]).

(2) Define the *logarithmic integral* function as

$$\operatorname{li}(x) = \int_2^x \frac{dt}{\ln t}.$$

Successive integration by parts shows that for each integer $k \geqslant 1$, we have

$$\operatorname{li}(x) \underset{x \to +\infty}{=} \frac{x}{\ln x} + \frac{x}{\ln^2 x} + \frac{2x}{\ln^3 x} + \cdots + \frac{(k-1)! \, x}{\ln^k x} + o\left(\frac{x}{\ln^k x}\right). \tag{3.17}$$

However, it turns out that $\operatorname{li}(x)$ is a much better approximation to $\pi(x)$ than the function $\dfrac{x}{\ln x}$. Indeed, even if the proofs of Hadamard and de la Vallée-Poussin (which used the function ζ and Lemma 3.4.8 intensively) were much more complicated than the proof shown here, they gave a deeper result, namely the existence of a positive constant c such that

$$\pi(x) = \text{li}(x) + O\left(xe^{-c\sqrt{\ln x}}\right) \text{ as } x \to +\infty. \qquad (3.18)$$

Now, if we put together (3.17) and (3.18), we obtain a magnificent asymptotic expansion at infinity of the function π:

$$\pi(x) = \frac{x}{\ln x} + \sum_{j=2}^{k} \frac{(j-1)!\,x}{\ln^j x} + o\left(\frac{x}{\ln^k x}\right),$$

for each integer $k \geq 1$.

(3) Hardy showed in 1914 that the ζ function has an infinite number of zeros on the "critical" line $\text{Re}\,s = \frac{1}{2}$. Subsequently, progress in understanding the zeros of ζ and in a better error term for the asymptotic expansion (3.17) has been infinitesimal; see, for example, the 1958 result of Vinogradov [174]:

$$\pi(x) = \text{li}(x) + O\left(x \exp\left(-c\frac{(\ln x)^{3/5}}{(\ln\ln x)^{1/5}}\right)\right).$$

(4) If the famous Riemann hypothesis was true, we would have [174]

$$\pi(x) = \text{li}(x) + O(x^{1/2}\ln x) \text{ and } \psi(x) = x + O(x^{1/2}\ln^2 x).$$

(5) "Elementary" proofs of the PNT exist, that is to say, proofs that do not rely on holomorphic functions and the ζ function. One such approach was developed by Selberg (and independently by Erdős) around 1947–1948, for which Selberg was awarded the Fields Medal in 1950. However, to quote Hlawka *et al.* ([89], p. 114) again: "As the price of limiting itself to elementary analysis, the proof of Selberg and Erdős demands manipulative skill of high order, rich in tricks, together with a far from transparent thought process".

3.5 The theorems of Ikehara and Delange

We start by giving the statement of Ikehara's theorem, which we have already mentioned, as well as its generalisation by Delange in the case of a multiple pole or a "conical singularity".[10]

If $(a_n)_{n\geq 0}$ is a complex sequence, we use again its *summatory function*, defined by

$$A(x) = \sum_{n \leq x} a_n.$$

---

[10] The definition is contained in the statement of Theorem 3.5.3.

3.5.1 Theorem [Ikehara's theorem for integrals]   *Let* $f : \mathbb{R}^+ \to \mathbb{R}^+$ *be a non-decreasing function, and* a, c *be positive real numbers. We suppose that the Laplace transform of* f,

$$F(z) = \int_0^{+\infty} f(t)e^{-zt}dt,$$

is defined on the open half-plane $\mathrm{Re}\, z > a$ *and that* $F(z) - \dfrac{c}{z-a}$ *has a continuous extension* G *on the closed half-plane* $\mathrm{Re}\, z \geqslant a$. *Then,*

$$f(t) \sim ce^{at} \text{ as } t \to +\infty.$$

3.5.2 Theorem [Ikehara's theorem for series]   *Let* $\Phi(s) = \sum_{n=1}^{\infty} a_n n^{-s}$ *be a Dirichlet series with non-negative coefficients, convergent if* $\mathrm{Re}\, s > a > 0$. *Suppose that there exists a constant* $c > 0$ *such that* $\Phi(s) - \dfrac{c}{s-a}$ *has a continuous extension* G *to the closed half-plane* $\mathrm{Re}\, s \geqslant a$. *Then,*

$$A(x) \sim \frac{c}{a}x^a \text{ as } x \to +\infty.$$

3.5.3 Theorem [Delange's theorem for series]   *Let* $\Phi(s) = \sum_{n=1}^{\infty} a_n n^{-s}$ *be a Dirichlet series with non-negative coefficients, convergent if* $\mathrm{Re}\, s > a > 0$. *Suppose that:*

(1) Φ *has a continuous extension to the closed half-plane* $\mathrm{Re}\, s \geqslant a$ *minus the point* a;

(2) in a right-side neighbourhood[11] of a, *we have*

$$\Phi(s) = (s - a)^{-\omega}g(s) + h(s),$$

where ω *is not a negative integer,* $(s - a)^{-\omega}$ *is taken to be a holomorphic determination of this function,* g *and* h *are holomorphic at* a, *and* $g(a) \neq 0$.

Then,

$$A(x) \sim \frac{g(a)}{a\Gamma(\omega)}x^a(\ln x)^{\omega-1} \text{ as } x \to +\infty.$$

We present below the proof of Ikehara's theorem, but we refer the reader to [47] for Delange's theorem, which will be useful for Exercise 3.5. As in the case of the Newman–Ikehara theorems, we could state a similar theorem of Delange for integrals, but we leave this as an exercise for the reader.

---

[11] By a *right-side neighbourhood* of a we mean the intersection of a neighbourhood of a and the half-plane $\mathrm{Re}\, s > a$.

It is in fact the integral formulation of Ikehara's Theorem 3.5.1 that we now prove.

Proof We begin with a few remarks on the proof of the Fourier inversion formula. Recall[12] that the Fourier transform $\widehat{u}$ of a function $u \in L^1(\mathbb{R})$ is defined by

$$\widehat{u}(t) = \int e^{-itx} u(x)\, dx \text{ for } x \in \mathbb{R}.$$

In this equality, as always in this book, the integral without limits is extended over $\mathbb{R}$. The Fourier inversion formula states that, if $\widehat{u}$ is also in $L^1(\mathbb{R})$, we have, almost everywhere (everywhere if u is continuous):

$$u(x) = \frac{1}{2\pi} \int e^{ixt} \widehat{u}(t)\, dt. \tag{3.19}$$

The proof of (3.19) can be broken down into two steps.

Step 1 establishes a fundamental identity (a simple consequence of Fubini's theorem), in which g is an arbitrary even element of $L^1(\mathbb{R})$ and $h = \frac{1}{2\pi}\widehat{g}$:

$$\frac{1}{2\pi} \int e^{ixt} \widehat{u}(t) g(t)\, dt = u * h(x) = \int u(x - y) h(y)\, dy \text{ for } x \in \mathbb{R}. \tag{3.20}$$

Step 2 makes use of step 1, by applying (3.20) to a sequence of functions $(g_n)_{n \geqslant 1}$ uniformly bounded, simply convergent to 1 and such that, if $h_n = \frac{1}{2\pi}\widehat{g}_n$, the sequence $(h_n)_{n \geqslant 1}$ tends weakly to the Dirac measure δ_0 at the origin, in the following sense:[13] $h_n \in L^1(\mathbb{R})$ and, for all functions $v : \mathbb{R} \to \mathbb{C}$ continuous and with limit zero at $\pm\infty$,

$$\int h_n(y) v(y)\, dy \to v(0).$$

An adequate passage to the limit in the equality

$$\frac{1}{2\pi} \int e^{ixt} \widehat{u}(t) g_n(t)\, dt = \int u(x - y) h_n(y)\, dy$$

would then give (3.19), at least if u is continuous and tends to 0 at infinity. A possible choice for $(g_n)_{n \geqslant 1}$ is the following: consider the "triangle" function

$$g : \mathbb{R} \to \mathbb{R}, \ x \mapsto \max(1 - |x|, 0).$$

---

[12] See Chapter 2.
[13] We consider here the complex measures $d\mu_n(y) = h_n(y)\, dy$ as elements of the dual of the space $C_0(\mathbb{R})$ of continuous functions from $\mathbb{R}$ to $\mathbb{C}$ with limit zero at $\pm\infty$.

We thus have $g(0) = 1$ and

$$\widehat{g}(t) = \frac{2(1 - \cos t)}{t^2},$$

so, if we set $h(t) := \frac{1}{2\pi}\widehat{g}(t)$, we obtain

$$\int h(t)\,dt = 1.$$

Finally, we set $g_n(x) = g\left(\frac{x}{n}\right)$, so that $h_n(t) := \frac{1}{2\pi}\widehat{g_n}(t) = nh(nt)$.
The functions g_n in this example have interesting properties:

- g_n has compact support $[-n, n]$;
- h_n is positive and has integral 1;
- for each fixed $\delta > 0$, we have

$$R_n(\delta) := \int_{|t|>\delta} h_n(t)\,dt = \int_{|u|>n\delta} h(u)\,du \to 0 \text{ as } n \to +\infty.$$

We thus identify $(h_n)_{n\geqslant 1}$ as an approximation of the identity (Fejér kernel), in the sense of Chapter 2.

The proof of Ikehara's theorem is modelled on the proof of the Fourier inversion theorem. There again, as a first step, we establish an identity similar to (3.20); as a second step, we use this identity with the sequence $(g_n)_{n\geqslant 1}$ defined above.

Consider now a non-decreasing function $f : \mathbb{R}^+ \to \mathbb{R}^+$ whose Laplace transform F verifies the hypothesis of Theorem 3.5.1. The function G will be that mentioned in the statement of the theorem.

Step 1. We begin by extending f as 0 for negative real numbers. We then select $\varepsilon > 0$, and set $\varphi(x) = f(x)e^{-ax}$ and

$$\psi_\varepsilon(x) = \varphi(x)e^{-\varepsilon x} - ce^{-\varepsilon x}\mathbf{1}_{\mathbb{R}_+}(x) \text{ for } x \in \mathbb{R}.$$

Applying our hypothesis to $s = a + \varepsilon + it$, we obtain, for $t \in \mathbb{R}$:

$$G(a + \varepsilon + it) = F(a + \varepsilon + it) - \frac{c}{\varepsilon + it}$$

$$= \int_0^{+\infty} (\varphi(x) - c)\,e^{-(\varepsilon+it)x}dx$$

$$= \int \psi_\varepsilon(x)e^{-itx}dx = \widehat{\psi_\varepsilon}(t).$$

Now the identity (3.20) applied to $\psi_\varepsilon \in L^1(\mathbb{R})$ gives us, for $x \in \mathbb{R}$ and $n \geqslant 1$, and $h_n = \dfrac{1}{2\pi} \widehat{g}_n$ as before:

$$
\begin{aligned}
\frac{1}{2\pi} \int e^{ixt} G(a + \varepsilon + it) g_n(t)\, dt &= \frac{1}{2\pi} \int e^{ixt} \widehat{\psi_\varepsilon}(t) g_n(t)\, dt \\
&= \int \psi_\varepsilon(y) h_n(x - y)\, dy \\
&= \int \varphi(y) e^{-\varepsilon y} h_n(x - y)\, dy \\
&\quad - c \int_0^{+\infty} e^{-\varepsilon y} h_n(x - y)\, dy.
\end{aligned}
$$

First, let ε tend to 0 in this equality; by uniform convergence of $G(a + \varepsilon + it)$ to $G(a + it)$ on the (compact) support of g_n when $\varepsilon \searrow 0$, the left-hand term tends to

$$
\frac{1}{2\pi} \int e^{ixt} G(a + it) g_n(t)\, dt.
$$

On the contrary, by monotone convergence (φ and h_n are positive), we have

$$
\int \varphi(y) e^{-\varepsilon y} h_n(x - y)\, dy \underset{\varepsilon \searrow 0}{\longrightarrow} \int \varphi(y) h_n(x - y)\, dy
$$

and

$$
\int_0^{+\infty} e^{-\varepsilon y} h_n(x - y)\, dy \underset{\varepsilon \searrow 0}{\longrightarrow} \int_0^{+\infty} h_n(x - y)\, dy = \int_{-\infty}^x h_n(y)\, dy.
$$

We thus obtain the finiteness of the integral $\int \varphi(y) h_n(x - y)\, dy$, as well as the equation

$$
\frac{1}{2\pi} \int e^{ixt} G(a + it) g_n(t)\, dt = \int \varphi(y) h_n(x - y)\, dy - c \int_{-\infty}^x h_n(y)\, dy.
$$

Now let x tend to $+\infty$. The left-hand term tends to zero by the Riemann–Lebesgue lemma, whereas

$$
\int_{-\infty}^x h_n(y)\, dy \to \int h_n(y)\, dy = 1.
$$

We have thus shown the following relation:

$$
\int \varphi(y) h_n(x - y)\, dy \to c \text{ as } x \to +\infty, \tag{3.21}
$$

which is in fact the essential information that can be derived from the hypothesis.

Step 2. We now let n tend to $+\infty$ in (3.21). If we could permute the limits in n and x, this would give formally:

$$
\begin{aligned}
c &= \lim_{n \to +\infty} \lim_{x \to +\infty} \int \varphi(y) h_n(x - y) \, dy \\
&= \lim_{x \to +\infty} \lim_{n \to +\infty} \int \varphi(y) h_n(x - y) \, dy \\
&= \lim_{x \to +\infty} \lim_{n \to +\infty} \int \varphi(x - y) h_n(y) \, dy \\
&= \lim_{x \to +\infty} \varphi(x),
\end{aligned}
$$

which is in fact the desired result. We justify this formal permutation with a standard Tauberian argument based on the non-decreasing nature of f, but which nonetheless requires a minimum of care. Fix $\delta > 0$, and set

$$
F_n(x) = \int \varphi(x - y) h_n(y) \, dy,
$$

as well as

$$
L = \varlimsup_{x \to +\infty} \varphi(x) \quad \text{and} \quad \ell = \varliminf_{x \to +\infty} \varphi(x).
$$

As is usual in this kind of reasoning, we show that $L \leqslant c$, and then that $\ell \geqslant c$, which will complete the proof. We start by showing the following *a priori* estimation:

$$
\varphi(x) \leqslant e^{2a\delta} \frac{F_n(x + \delta)}{I_n(\delta)} \quad \text{for } x \geqslant 0, \tag{3.22}
$$

where we have set

$$
I_n(\delta) = \int_{-\delta}^{\delta} h_n(y) \, dy = 1 - R_n(\delta) > 0.
$$

Indeed, as f is non-decreasing and f and h_n are non-negative, we have

$$
\begin{aligned}
F_n(x + \delta) &= \int f(x + \delta - y) e^{-a(x+\delta-y)} h_n(y) \, dy \\
&\geqslant \int_{-\delta}^{\delta} f(x + \delta - y) e^{-a(x+\delta-y)} h_n(y) \, dy \\
&\geqslant \int_{-\delta}^{\delta} f(x) e^{-a(x+2\delta)} h_n(y) \, dy \\
&= \varphi(x) e^{-2a\delta} I_n(\delta),
\end{aligned}
$$

which proves (3.22). Passing to the $\varlimsup$ as $x \to +\infty$ in this inequality, we obtain, using (3.21):

$$L \leqslant \frac{ce^{2a\delta}}{I_n(\delta)}.$$

We now successively let n tend to $+\infty$, and then δ tend to 0. We obtain $L \leqslant ce^{2a\delta}$, and then $L \leqslant c$. In particular, we find that *the hypothesis of Newman's theorem is satisfied*: φ is bounded by a constant M on $\mathbb{R}^+$ (hence on $\mathbb{R}$).

We can now show the *a priori* lower bound

$$\varphi(x) \geqslant \frac{\big(F_n(x-\delta) - M\big(1 - I_n(\delta)\big)\big)e^{-2a\delta}}{I_n(\delta)}. \tag{3.23}$$

Indeed,

$$F_n(x-\delta) = \int \varphi(x - \delta - y)h_n(y)\,dy$$

$$\leqslant \int_{-\delta}^{\delta} \varphi(x - \delta - y)h_n(y)\,dy + M\int_{|y|>\delta} h_n(y)\,dy$$

$$= \int_{-\delta}^{\delta} f(x - \delta - y)e^{-a(x-\delta-y)}h_n(y)\,dy + M\big(1 - I_n(\delta)\big)$$

$$\leqslant f(x)e^{-a(x-2\delta)}I_n(\delta) + M\big(1 - I_n(\delta)\big)$$

$$= \varphi(x)e^{2a\delta}I_n(\delta) + M\big(1 - I_n(\delta)\big),$$

which proves (3.23). Passing to the $\varliminf$ as $x \to +\infty$, we obtain this time

$$\ell \geqslant \frac{\big(c - M\big(1 - I_n(\delta)\big)\big)e^{-2a\delta}}{I_n(\delta)}.$$

By successively letting n tend to $+\infty$ and then δ to 0, we get $\ell \geqslant ce^{-2a\delta}$ and then $\ell \geqslant c$, which completes the proof of Ikehara's Theorem 3.5.1. $\quad\square$

Exercises

3.1. For each integer $n \geqslant 1$, let $P(n)$ be the least common multiple (LCM) of $1, 2, \ldots, n$.
(a) Show that $\ln P(n) = \psi(n)$, where ψ is the von Mangoldt function.
(b) Show that the radius of convergence of the power series $\sum P(n)z^n$ is $1/e$.

3.2. For each integer $n \geqslant 1$, let $d(n)$ be the number of divisors $\geqslant 1$ of n. We recall [174] that

$$\varlimsup_{n\to+\infty} \frac{\ln d(n)\ln\ln n}{\ln n} = \ln 2,$$

and in particular $d(n) \leqslant C_\varepsilon n^\varepsilon$ for all $\varepsilon > 0$. Let $(\lambda_n)_{n \geqslant 1}$ be an increasing sequence of positive integers. For $n \geqslant 1$, let P_n be the LCM of $\lambda_1, \ldots, \lambda_n$.

(a) Show that $d(P_n) \geqslant n$.

(b) Give the best possible lower bound for P_n, and in particular show that $\sum_{n=1}^\infty P_n^{-\delta} < \infty$ for all $\delta > 0$.

3.3. In the recursive sequence of Erdős (Section 3.4), we suppose that the quotient $\dfrac{\ln m_j}{\ln m_k}$ is always rational.

(a) Show that there exists an integer d such that $m_j = d^{u_j}$ for all j, where the u_j are integers.

(b) What is the recurrence relation verified by $b_p = a_{d^p}$?

(c) Show that if $d^p \leqslant n < d^{p+1}$ ($p \in \mathbb{N}$), then $a_n = b_p$.

(d) Does the sequence $\left(\dfrac{a_n}{n^a} \right)_{n \geqslant 1}$ converge? What can be said about its cluster set? The example

$$a_n = 2a_{\lfloor n/3 \rfloor} + 3a_{\lfloor n/9 \rfloor}$$

can be considered (among others).

3.4. Let φ be the Euler totient function and, for $n \in \mathbb{N}$, $r(n)$ the number of solutions of the equation $\varphi(m) = n$.

(a) Show that $\lim_{m \to +\infty} \varphi(m) = +\infty$.

(b) Show that $r(n) < \infty$ for all $n \in \mathbb{N}$.

(c) Let $s \in \mathbb{C}$ such that $\operatorname{Re} s > 1$. Show that the series $\sum r(n) n^{-s}$ converges, and that

$$\sum_{n=1}^\infty \frac{r(n)}{n^s} = \sum_{m=1}^\infty \varphi(m)^{-s}.$$

(d) Under the same hypotheses, prove the following equality, in which the product is over the prime numbers:

$$f(s) := \sum_{n=1}^\infty \frac{r(n)}{n^s} = \prod_p \left(1 + \frac{1}{(p^s - 1)(1 - p^{-1})^s} \right).$$

(e) Show that f is meromorphic on $\operatorname{Re} s > 0$, with a unique simple pole at 1, with residue

$$\rho = \prod_p \left(1 + \frac{1}{p(p-1)} \right) = \frac{\zeta(2)\zeta(3)}{\zeta(6)} = \frac{315}{2\pi^4} \zeta(3).$$

(f) Find an equivalent to $\sum_{n \leqslant x} r(n)$ when $x \to +\infty$.

3.5. A result of Ramanujan.[14] Let q be a positive real number. For $x \in [0, 1[$, we set

$$g_q(x) = \sum_{k=0}^{\infty} (k+1)^q x^k$$

and consider, for $\mathrm{Re}\, s > 1$, the Dirichlet series

$$f(s) = \sum_{n=1}^{\infty} \frac{d(n)^q}{n^s}.$$

(a) Show that $f(s) = \prod_p g_q(p^{-s})$.

(b) Show that $f(s) = \zeta(s)^{2^q} \Phi(s)$, where Φ is an absolutely convergent Dirichlet series for $\mathrm{Re}\, s > \frac{1}{2}$, and $\Phi(1) \neq 0$.

(c) Conclude that

$$\sum_{n \leqslant x} d(n)^q \sim \frac{\Phi(1)}{\Gamma(2^q)} x (\ln x)^{2^q - 1} \quad \text{as } x \to +\infty.$$

3.6. Let ψ be the von Mangoldt function.

(a) Show (see Exercise 3.1) that if $x \geqslant 1$,

$$\psi(x) = \sum_{p^k \leqslant x} \ln p = \sum_{p \leqslant x} \left[\frac{\ln x}{\ln p} \right] \ln p \leqslant \pi(x) \ln x.$$

(b) Show that if $1 \leqslant y \leqslant x$,

$$\pi(x) \leqslant y + \sum_{y < p \leqslant x} \frac{\ln p}{\ln y} \left[\frac{\ln x}{\ln p} \right] \leqslant y + \frac{1}{\ln y} \psi(x).$$

(c) Prove Proposition 3.4.5 by judicious choice of y as a function of x.

3.7. An integer $n \geqslant 1$ is said to be *squarefull* if $n = 1$ or if, with p being a prime number,

$$p \mid n \Rightarrow p^2 \mid n.$$

Let E be the set of squarefull integers and a_n the indicator function of E, that is to say, the sequence which is equal to 1 on E and 0 otherwise. For $\mathrm{Re}\, s > 1$, we set

$$\Phi(s) = \sum_{n=1}^{\infty} \frac{a_n}{n^s}.$$

---

[14] The reader can learn much more about Srinivasa Ramanujan by consulting the extraordinary book [79].

(a) Show that, p being a prime number (as always), we have

$$\Phi(s) = \prod_p \frac{1 - p^{-s} + p^{-2s}}{1 - p^{-s}} = \frac{\zeta(2s)\zeta(3s)}{\zeta(6s)},$$

and that the result remains true for $\mathrm{Re}\, s > \frac{1}{2}$.

(b) Using Ikehara's theorem, show that

$$\sum_{n \leqslant x} a_n \sim \frac{\zeta\left(\frac{3}{2}\right)}{\zeta(3)} \sqrt{x} \text{ as } x \to +\infty.$$

4

Generic properties of derivative functions

The purpose of this chapter is to bring out appropriate mathematical concepts to express whether a subset of $\mathbb{R}$ (to begin with) is *small* or *large*. The notion of smallness that we would like to define is subject to three conditions.

(1) Heredity: any subset of a small set must also be small.
(2) Stability under countable union (any countable union of small sets is also small).
(3) No interval $[a, b]$ (with $a < b$) is small.

A subset of $\mathbb{R}$ will be *large* if its complement is *small*. If $P(x)$ is an assertion depending on a real number x, we say that P is *generic* (or *typical*) if $P(x)$ is true for x belonging to a large subset of $\mathbb{R}$.

Here, among others, are three possible points of view.

- Cardinality: the small sets are those that are finite or countable.
- Measure: the small sets are those that are negligible in the sense of Lebesgue.
- Category: the small sets are those that are of first category in the sense of Baire.

In what follows, we will leave the first point of view aside, in order to compare the other two notions in a specific situation: the study of the points of continuity of derivative functions. For all this chapter, a good reference is [32].

4.1 Measure and category

4.1.1 Measure

A subset A of $\mathbb{R}$ is said to be *negligible in the sense of Lebesgue* if for all $\varepsilon > 0$, there exists a sequence $(I_n)_{n \geqslant 0}$ of open intervals such that

$$A \subset \bigcup_{n \geqslant 0} I_n \text{ and } \sum_{n=0}^{\infty} \lambda(I_n) \leqslant \varepsilon,$$

where $\lambda(I_n)$ denotes the length of I_n.

We can verify that this definition satisfies our specifications. This is easy for conditions (1) and (2), but a bit more delicate for (3).[1] Any countable subset of $\mathbb{R}$ and the Cantor middle third set (which is uncountable) are very simple examples of negligible subsets of $\mathbb{R}$.

4.1.2 Category

Our second definition makes sense in any topological space X. The prototype for a small subset of X in the sense of category is a closed set with an empty interior. As this definition is too unstable, we are led to modify it as follows: a subset of X is said to be *of first category*, or *meagre*, if it is contained in a countable union of closed subsets of X whose interiors are empty. A subset of X is said to be *residual* if its complement is of first category. In the sense of category, the small sets are those of first category, and the large sets the residual ones.

Of course, the smallness thus defined is stable by taking subsets or countable unions. What meaning do we give to condition (3) in this general context? A reasonable solution would be to require that no open ball can be of first category. In general, this last condition does not hold; for example, every open ball of $\mathbb{Q}$ is a countable union of non-isolated points, hence of first category. But as soon as we restrict our study to complete metric spaces, the situation becomes much more manageable.

4.1.3 Theorem [Baire] *If $(F_n)_{n \geqslant 0}$ is a sequence of closed subsets of a complete metric space, all with empty interiors, then $\bigcup_{n \geqslant 0} F_n$ has an empty interior.*

Proof We refer to [152]. □

We can paraphrase Baire's theorem by saying that a set of first category in a complete metric space has an empty interior, or alternatively that a residual set is dense. Any countable subset of $\mathbb{R}$, as well as the Cantor middle third set (as an example of a closed set with empty interior), are of first category. It is more difficult to give examples of subsets of $\mathbb{R}$ that are negligible in the sense of

---

[1] *Hint:* using the compactness of $[a, b]$, reduce this to the case of finite coverings.

Lebesgue, but are nonetheless residual. This is for example the case for the set of Liouville numbers (see Exercise 4.1).

4.2 Functions of Baire class one

In a topological space, we call an F_σ-set (resp. a G_δ-set) any countable union of closed sets (resp. any countable intersection of open sets). Of course, the complement of an F_σ-set is a G_δ-set, and vice versa. In addition, we denote by $\delta(A)$ the diameter of a non-empty subset A of $\mathbb{R}$:

$$\delta(A) = \sup_{(x,y) \in A \times A} |x - y| \in \mathbb{R}_+ \cup \{+\infty\}.$$

4.2.1 Points of discontinuity of a function

Consider a function $f : [0, 1] \to \mathbb{R}$. Given $x_0 \in \mathbb{R}$, we call the *oscillation of* f *at* x_0 the non-negative extended real number defined by

$$\omega(x_0) = \lim_{\eta \to 0^+} \delta\big(f([x_0 - \eta, x_0 + \eta] \cap [0, 1])\big)$$

(where the limit is monotone). By definition, $\omega(x_0) = 0$ if and only if f is continuous at x_0. The following proposition, simple to verify, shows that the oscillation is an upper semi-continuous function.

4.2.2 Proposition *For all $\varepsilon > 0$, the set $\{x \in [0, 1] / \omega(x) < \varepsilon\}$ is an open subset of the segment $[0, 1]$.*

4.2.3 Corollary *The set of points of continuity of f is a G_δ-set of the segment* $[0, 1]$.

Proof This set is no other than

$$\bigcap_{n \geqslant 1} \left\{ x \in [0, 1] / \omega(x) < \frac{1}{n} \right\}. \qquad \square$$

4.2.4 Case of functions of Baire class one

A function $f : [0, 1] \to \mathbb{R}$ is said to be *of Baire class one* if it is the pointwise limit of a sequence of continuous functions on $[0, 1]$. We know that continuity is preserved by uniform convergence. The following theorem (also due to Baire) shows that we still have something to say even if the convergence is only pointwise.

4.2.5 Proposition   *If* $f : [0, 1] \to \mathbb{R}$ *is Baire class one, then the set of its points of discontinuity is of first category. In particular, the set of points of continuity of* f *is dense in* $[0, 1]$.

Proof   Let $f : [0, 1] \to \mathbb{R}$ be Baire class one, and $(f_n)_{n \geqslant 1}$ a sequence of continuous functions from $[0, 1]$ to $\mathbb{R}$ that converges pointwise to f. We already know that the set of points of discontinuity of f is an F_σ of $[0, 1]$. It remains to show that this F_σ has an empty interior. From what precedes and by Baire's theorem, it is sufficient to show that, for all $\varepsilon > 0$, the closed set

$$D := \{x \in [0, 1] \,/\, \omega(x) \geqslant 7\varepsilon\}$$

has an empty interior. Suppose that this is not the case, and set, for all $n \geqslant 0$,

$$E_n = \left\{x \in [0, 1] \,/\, |f_p(x) - f_q(x)| \leqslant \varepsilon \text{ for } p, q \geqslant n\right\}.$$

Clearly,

$$D = \bigcup_{n \geqslant 1} \underbrace{D \cap E_n}_{\text{closed subset of } \mathbb{R}},$$

as, for each $x \in [0, 1]$, $f_k(x) \to f(x)$ as $k \to +\infty$. By Baire's theorem, there exists an $n \geqslant 1$ such that $D \cap E_n$ contains an open non-empty interval I. Set $x_0 \in I$. For $x \in I$, we have, by passing to the limit,

$$|f_n(x) - f(x)| \leqslant \varepsilon.$$

Now, the function f_n is continuous at x_0: there exists an $\eta > 0$ such that, for all $x \in I$,

$$|x - x_0| \leqslant \eta \Rightarrow |f_n(x) - f_n(x_0)| \leqslant \varepsilon.$$

Thus, for all $x \in I$ such that $|x - x_0| \leqslant \eta$, we have

$$|f(x) - f(x_0)| \leqslant |f(x) - f_n(x)| + |f_n(x) - f_n(x_0)| + |f_n(x_0) - f(x_0)| \leqslant 3\varepsilon,$$

which contradicts the fact that $\omega(x_0) \geqslant 7\varepsilon$.   □

4.2.6 Corollary   *Let* $f : [0, 1] \to \mathbb{R}$ *be differentiable on* $\mathbb{R}$. *Then,* f' *is Baire class one.*

Proof   The function f' is the pointwise limit of the sequence $(g_n)_{n \geqslant 1}$ of continuous functions on $[0, 1]$, defined by

$$g_n(x) = \begin{cases} n\left(f(x + \frac{1}{n}) - f(x)\right) & \text{if } 0 \leqslant x \leqslant 1 - \frac{1}{n}, \\ n\left(f(1) - f(1 - \frac{1}{n})\right) & \text{if } 1 - \frac{1}{n} < x \leqslant 1. \end{cases}$$

□

4.3 The set of points of discontinuity of derivative functions

The purpose of this section is to evaluate the set of points of discontinuity of a derivative function, first in the sense of category, and then in the sense of measure. As we will see, the two points of view can be quite opposite: the set of points of discontinuity of a derivative is always small from the point of view of category, but, in a generic sense, is large from the point of view of measure.

4.3.1 Characterisation of the set of points of discontinuity of a derivative

Given a function $f : [0, 1] \to \mathbb{R}$, let C_f (resp. D_f) be the set of its points of continuity (resp. discontinuity).

4.3.2 Theorem   *Let F be a subset of $[0, 1]$. The following propositions are equivalent:*

(i) there exists an $f : [0, 1] \to \mathbb{R}$, differentiable such that $F = D_{f'}$;
(ii) F is an F_σ-set of first category.

4.3.3 Remark   We shall see in the course of the proof that in assertion (i), the function f can be chosen with a *bounded derivative*.

Proof   The implication (i) $\Rightarrow$ (ii) has already been shown. For the converse, first suppose that F is a closed subset of $[0, 1]$, with empty interior and containing the points 0 and 1. The set $\Omega = [0, 1] \setminus F$ is thus an open subset of $\mathbb{R}$. Let $\omega =]a, b[$ be a fixed connected component of Ω, and consider the function

$$\varphi : \omega \to \mathbb{R}, \ x \mapsto (x - a)^2 \sin \frac{1}{x - a}.$$

The function φ is differentiable on ω, and for $x \in \omega$:

$$\varphi'(x) = 2(x - a) \sin \frac{1}{x - a} - \cos \frac{1}{x - a}.$$

The derivative function φ' is thus bounded on ω (with $\|\varphi'\|_\infty \leqslant 3$), and φ' takes the values -1, 0 and 1 in any right neighbourhood of a. We can thus select a real c_ω such that $0 < c_\omega < \dfrac{b - a}{2}$ and $\varphi'(a + c_\omega) = 0$. Define then the function

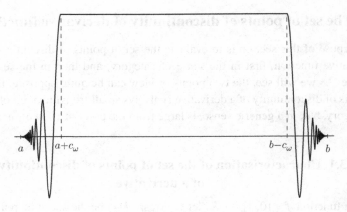

Figure 4.1

$$f_\omega : \omega \to \mathbb{R}, x \mapsto \begin{cases} (x-a)^2 \sin \dfrac{1}{x-a} & \text{if } a < x < a + c_\omega, \\[2mm] c_\omega^2 \sin \dfrac{1}{c_\omega} & \text{if } a + c_\omega \leqslant x \leqslant b - c_\omega, \\[2mm] (b-x)^2 \sin \dfrac{1}{b-x} & \text{if } b - c_\omega < x < b \end{cases}$$

and, finally,

$$f : [0, 1] \to \mathbb{R}, x \mapsto \begin{cases} 0 & \text{if } x \in F, \\ f_\omega(x) & \text{if } x \text{ is in the component } \omega \text{ of } \Omega. \end{cases}$$

See Figure 4.1. We will show that f is differentiable on $[0, 1]$, and that $D_{f'} = F$.

It is clear that f is C^1 on the open set Ω. Let $x \in F$, $x < 1$; let us show that f is right-differentiable at x, with $f_d'(x) = 0$.

Let $t \in \,]x, 1]$. To begin with, suppose that t belongs to Ω, and let $\omega = \,]a, b[$ be the connected component containing it. There are three possible cases:

- if $a < t < a + c_\omega$, then

$$|f(t) - f(x)| = (t-a)^2 \left| \sin \frac{1}{t-a} \right| \leqslant (t-a)^2 \leqslant (t-x)^2 ;$$

- if $a + c_\omega \leqslant t \leqslant b - c_\omega$, then

$$|f(t) - f(x)| = c_\omega^2 \left| \sin \frac{1}{c_\omega} \right| \leqslant c_\omega^2 \leqslant (t-x)^2 ;$$

- finally, if $b - c_\omega < t < b$, then

$$|f(t) - f(x)| = (b-t)^2 \left| \sin \frac{1}{b-t} \right| \leqslant (b-t)^2 \leqslant c_\omega^2 \leqslant (t-x)^2.$$

The same bound evidently holds true if $t \in F$. We thus have

$$|f(t) - f(x)| \leqslant (t - x)^2 \text{ for all } t \in \,]x, 1],$$

which proves the result. Similarly, one can show that f is left-differentiable for each $x \in F$ such that $x > 0$, with $f'_g(x) = 0$, which shows that f is differentiable for each point x of F, with $f'(x) = 0$.

Time to conclude: f is thus differentiable on $[0, 1]$, and its derivative is bounded, with the more precise estimate

$$\|f'\|_\infty \leqslant 3. \tag{4.1}$$

Moreover, if $x \in F$, as Ω is dense in $[0, 1]$, each neighbourhood of x contains an endpoint of a connected component of Ω, and hence some points where the derivative f' takes the values -1 and 1; the function f' is thus discontinuous at x. This proves that $D_{f'} = F$.

4.3.4 Remark In the case where $0 \notin F$ (for example), it is sufficient to modify the definition of f on the connected component containing 0 to ensure that f' is continuous at 0.

Now we consider the general case: let $F = \bigcup_{n \geqslant 0} F_n$ be an F_σ-set of $[0, 1]$ with empty interior. We can obviously suppose that the sequence $(F_n)_{n \geqslant 0}$ is non-decreasing. By Theorem 4.3.2 (and in particular the estimate (4.1)), we can find, for each $n \geqslant 0$, a function $f_n : [0, 1] \to \mathbb{R}$, differentiable on $[0, 1]$ and satisfying

$$D_{f'_n} = F_n, \; \|f_n\|_\infty \leqslant 1 \text{ and } \|f'_n\|_\infty \leqslant 3.$$

Then we set

$$f = \sum_{n=0}^{\infty} \frac{f_n}{5^n}.$$

The function f is differentiable on $[0, 1]$, and

$$f' = \sum_{n=0}^{\infty} \frac{f'_n}{5^n}.$$

By uniform convergence, f' is continuous on $\Omega := [0, 1] \setminus F$. Select now $x \in F$, and $N \geqslant 0$ minimal such that $x \in F_N$. We have

$$\left\| \sum_{n > N} \frac{f'_n}{5^n} \right\|_\infty \leqslant 3 \sum_{n > N} \frac{1}{5^n} = \frac{3}{4.5^N},$$

so that $\sum_{n>N} \frac{f_n'}{5^n}$ oscillates at x by at most $\frac{3}{2.5^N}$. On the contrary, the func-

tion $\sum_{n<N} \frac{f_n'}{5^n}$ is continuous at x, and the oscillation of $\frac{f_N'}{5^N}$ at x is greater

than $\frac{2}{5^N}$ by construction of f_N, as f_N' takes on the values -1 and 1 in every

neighbourhood of x. The function f' is thus certainly discontinuous at x. In
conclusion, we have $D_{f'} = F$.   □

4.3.5 Discontinuity almost everywhere of the generic bounded derivative

In this section, we call $B\Delta$ the set of bounded derivatives on $[0, 1]$. A function
$f : [0, 1] \to \mathbb{R}$ is thus an element of $B\Delta$ if f is bounded and if there exists
a differentiable function $F : [0, 1] \to \mathbb{R}$ such that $F' = f$. We are going to
study the Lebesgue measure (denoted by λ) of the set of points of continuity
of a generic element of $B\Delta$. To achieve this, we again make intensive (and
legitimate) use of Baire's theorem.

4.3.6 Proposition $(B\Delta, \|\cdot\|_\infty)$ *is a Banach space.*

Proof   It is clear that $B\Delta$ is a linear subspace of the space $\mathcal{B}$ of bounded functions from $[0, 1]$ to $\mathbb{R}$. Let us show that $B\Delta$ is closed in the Banach space
$(\mathcal{B}, \|\cdot\|_\infty)$. Let $(f_n)_{n\geqslant 1}$ be a sequence of points of $B\Delta$ that converges uniformly to $f \in \mathcal{B}$. Let F_n be the primitive of f_n that is null at 0. Then, the
sequence of real numbers $\big(F_n(0)\big)_{n\geqslant 1}$ is convergent, and the sequence $(F_n')_{n\geqslant 1}$
converges uniformly on $[0, 1]$. According to a standard theorem on the dif-
ferentiation of uniform limits, $(F_n)_{n\geqslant 1}$ converges uniformly on $[0, 1]$ to a
primitive of f. Thus, $f \in B\Delta$.   □

We can now state the main result.

4.3.7 Theorem [Bruckner–Petruska [33]]   *The generic bounded derivative is
discontinuous almost everywhere. In other words: the set of $f \in B\Delta$ such that
$\lambda(C_f) = 0$ is residual.*

Proof   The proof proceeds in several steps.

(1) *Approximation by functions that are discontinuous on a dense subset.* Let
   I be a non-empty open interval contained in $[0, 1]$. Denote by C_I the set of
   elements of $B\Delta$ that are continuous on I. The set C_I is a closed subset of
   $(B\Delta, \|\cdot\|_\infty)$; moreover, given $f \in C_I$ and $c \in I$, by Theorem 4.3.2 we can

find an element g of $B\Delta$, with arbitrarily small norm, and discontinuous exactly at c. The function $f + g$ is no longer an element of C_I, which shows that C_I has an empty interior. By Baire's theorem, the union of the C_I, for I taken over the (countable) set of open intervals with rational endpoints contained in $[0, 1]$, is of first category in $B\Delta$. Its complement, which is none other than the set of elements of $B\Delta$ discontinuous on a dense subset of $[0, 1]$, is thus residual (and, in particular, dense in $B\Delta$).

(2) Given $\delta > 0$, call A_δ the set of $f \in B\Delta$ such that $\lambda(C_f) \geqslant \delta$. *Let us show that A_δ is closed in* $(B\Delta, \|\cdot\|_\infty)$. For this, let $(f_n)_{n\geqslant 0}$ be a sequence of elements of A_δ converging to $g \in B\Delta$, and set $B_n = \bigcup_{p\geqslant n} C_{f_p}$ for $n \geqslant 0$. The sequence $(B_n)_{n\geqslant 0}$ is thus non-increasing, and each of its elements has measure greater than δ. As a result,

$$\lambda\left(\bigcap_{n=0}^{\infty} B_n\right) \geqslant \delta$$

(this works because $[0, 1]$ has finite measure). Now, if $x \in \bigcap_{n=0}^{\infty} B_n$, x belongs to an infinite number of C_{f_p}. Thanks to the uniform convergence of the sequence $(f_p)_{p\geqslant 0}$ to g, we can conclude that g is continuous at x. Therefore, $\bigcap_{n=0}^{\infty} B_n \subset C_g$, so that $\lambda(C_g) \geqslant \delta$; in other words, $g \in A_\delta$.

(3) *Let us show finally that A_δ has empty interior in $B\Delta$.* For this, choose $f \in A_\delta$. According to the first step, we can find an element g of $B\Delta$, arbitrarily close to f, and such that $E := D_g$ is dense in $[0, 1]$. As the Lebesgue measure is regular, there exists G, a G_δ-set of $[0, 1]$ containing E, such that $\lambda(G \setminus E) = 0$. But then, $F := [0, 1] \setminus G$ is an F_σ-set of $[0, 1]$ with empty interior. By Theorem 4.3.2, there exists a function $h \in B\Delta$, with norm arbitrarily small, such that $D_h = F$. Let us show that

$$C_{g+h} = G \setminus E.$$

- First, if $x \in G \setminus E$, then g and h are continuous at x, hence so is $g + h$.
- Next, if $x \in E$, then g is discontinuous at x and h is continuous at x, so that $g + h$ is discontinuous at x.
- Finally, if $x \in F$, then g is continuous at x and h is discontinuous at x, so that $g + h$ is discontinuous at x.

Now, as $G \setminus E$ has measure zero, we have $g + h \notin A_\delta$. As $g + h$ is arbitrarily close to f, this completes the proof of the third point.

By applying Baire's theorem one more time, $\bigcup_{n\geqslant 1} A_{1/n}$ is of first category in $B\Delta$, so its complement is residual. It suffices then to remark that this last set is none other than the set of elements of $B\Delta$ that are discontinuous almost everywhere. $\qquad\square$

4.4 Differentiable functions that are nowhere monotonic

In this last section, we will use Baire theory to show the existence of functions that are differentiable on [0, 1] but are not monotonic on any interval with non-empty interior contained in [0, 1]. Yet again, Baire's theorem will overcome the difficulties only thanks to an additional ingredient, a construction due to the Romanian mathematician Dimitrie Pompeiu [148].

As a derivative, continuous or not, satisfies the intermediate value property (the image of an interval is an interval), such a function is certainly to be found within the set of functions that are differentiable, non-constant, and whose derivatives are zero on a dense subset of [0, 1]. However, it is not at all evident that this set is non-empty!

Once we have overcome this difficulty, the idea [184] is then to use a method that *a priori* is most surprising, but has proven useful in the past: to show that a set non-empty, it suffices to prove that it is large.[2] We will thus show that, generically, a bounded derivative that is zero on a dense subset of [0, 1] does not have a constant sign on any interval contained in [0, 1]. This is in contrast to common practice, where functions are always piecewise monotonic.

In what follows, we will consider, without loss of generality, real-valued functions defined on the segment [0, 1].

4.4.1 The functions of Pompeiu type

We will say that a function $f : [0, 1] \to \mathbb{R}$ is *of Pompeiu type* if it is differentiable, *non-constant*, and if f' is zero on a dense subset of [0, 1].

4.4.2 Theorem   *There exist functions of Pompeiu type.*

Proof Let $(a_n)_{n \geqslant 1}$ be a sequence of positive real numbers such that $\sum_{n=1}^{\infty} a_n < \infty$ and let $(r_n)_{n \geqslant 1}$ be an enumeration of the rational numbers of [0, 1]. For all $x \in [0, 1]$, we set

---

[2] One of the first successes of this idea was the non-constructive proof by Cantor of the existence of transcendental numbers: $\mathbb{R}$ is not countable, whereas the set of algebraic numbers is!

$$f(x) = \sum_{n=1}^{\infty} a_n (x - r_n)^{1/3}.$$

It is clear that the function f is increasing on $[0, 1]$. To study its differentiability, we adopt the following notations: for $n \geqslant 1, x \in [0, 1]$ and $h \in \mathbb{R}^*$, set

$$u_n(x, h) = \frac{a_n}{h} \left((x + h - r_n)^{1/3} - (x - r_n)^{1/3} \right),$$

$$v_n(x) = \frac{a_n}{3(x - r_n)^{2/3}}$$

(with the convention $0^{-1} = +\infty$) and

$$V(x) = \sum_{n=1}^{\infty} v_n(x).$$

We also set

$$E = \{r_n, n \geqslant 1\} \text{ and } F = \{x \in [0, 1] / V(x) = +\infty\}.$$

4.4.3 Lemma $0 \leqslant u_n(x, h) \leqslant 4v_n(x)$.

Proof   We have $u_n(x, h) \geqslant 0$ because the map $x \mapsto (x - r_n)^{1/3}$ is nondecreasing. The other inequality is evident if $x = r_n$. Otherwise,

$$u_n(x, h) = \frac{a_n}{h} \left((x + h - r_n)^{1/3} - (x - r_n)^{1/3} \right)$$

$$= \frac{a_n}{(x + h - r_n)^{2/3} + (x + h - r_n)^{1/3}(x - r_n)^{1/3} + (x - r_n)^{2/3}}$$

$$\leqslant \frac{4a_n}{3(x - r_n)^{2/3}} = 4v_n(x),$$

since $\alpha^2 + \alpha\beta + \beta^2 \geqslant \frac{3}{4}\beta^2$ for $\alpha, \beta \in \mathbb{R}$. $\square$

Next, it is clear that $E \subset F$. In particular, F is dense in $[0, 1]$. Now select $x \in [0, 1]$, and consider two cases.

- If $x \notin F$, then according to the preceding lemma and an argument of uniform (or dominated) convergence, f is differentiable at x and

$$f'(x) = \sum_{n=1}^{\infty} v_n(x) = V(x).$$

- If $x \in F$, then we have, for $h \neq 0$ and $N \geqslant 1$,

$$\frac{f(x + h) - f(x)}{h} \geqslant \sum_{n=1}^{N} u_n(x, h) \text{ since } u_n(x, h) \geqslant 0,$$

so that[3]

$$\lim_{h \to 0} \frac{f(x+h) - f(x)}{h} \geqslant \sum_{n=1}^{N} v_n(x)$$

and hence, by letting $N \to +\infty$:

$$\lim_{h \to 0} \frac{f(x+h) - f(x)}{h} \geqslant \sum_{n=1}^{\infty} v_n(x) = +\infty.$$

Thus, $f'(x) = +\infty$.

Ultimately, f induces an increasing homeomorphism – that we still call f – from $[0, 1]$ onto a segment $[a, b]$ satisfying

$$f'(x) = \begin{cases} V(x) > 0 & \text{if } x \notin F, \\ +\infty & \text{if } x \in F. \end{cases}$$

The inverse bijection $g = f^{-1}$ is hence differentiable on $[a, b]$, its derivative is zero exactly on $f(F)$, which is a dense subset of $[a, b]$ since F is dense in $[0, 1]$. To complete the proof, we only need to compose g with an appropriate diffeomorphism. $\qquad\square$

4.4.4 Remarks   (1) We have

$$\int_0^1 V(x)\, dx = \sum_{n=1}^{\infty} a_n \big((1 - r_n)^{1/3} - (0 - r_n)^{1/3}\big) < \infty.$$

Therefore, F is negligible in the sense of Lebesgue. Nonetheless, F is not countable: in fact, g' is certainly zero at each of its points of continuity. Thus, the set $C_{g'}$ of points of continuity of g' is contained in $f(F)$. At the same time, because g', as a derivative, is of Baire class one, $C_{g'}$ is a residual subset of $[a, b]$ and hence not countable (see Exercise 4.5).
(2) The function g' is bounded. To see this, let $x \in [a, b]$. If $x \in f(F)$, then $g'(x) = 0$. Otherwise, x can be written as $f(a)$, with $a \in [0, 1] \setminus F$. We thus have $g'(x) = \dfrac{1}{f'(a)}$. But

$$f'(a) = \sum_{n=1}^{\infty} \frac{a_n}{3(a - r_n)^{2/3}} \geqslant \frac{a_1}{3(a - r_1)^{2/3}} \geqslant \frac{a_1}{3}.$$

As a result, $0 \leqslant g'(x) \leqslant \dfrac{3}{a_1}$.

---

[3] Whether x is in E or in $F \setminus E$.

(3) Starting with a function of Pompeiu type on $[0, 1]$, we easily obtain one on $]0, 1[$ (by restriction), and then on $\mathbb{R}$ (by composition with a diffeomorphism). Hence, with $x_0 \in [0, 1]$ fixed, we can construct by translation and then restriction a function of Pompeiu type on $[0, 1]$ whose derivative at x_0 is not zero. This will be useful in what follows.

4.4.5 Differentiable functions that are nowhere monotonic

In what follows, as was the case in Section 4.3.5, we call $B\Delta$ the Banach space of bounded derivatives on $[0, 1]$, and $B\Delta_0$ the set of elements of $B\Delta$ that are zero on a dense subset of $[0, 1]$. We start by establishing some additional results about $B\Delta$ and $B\Delta_0$.

4.4.6 Proposition *The set of zeros of an element f of $B\Delta$ is a G_δ-set of $[0, 1]$.*

Proof This is more generally true for any function $f : [0, 1] \to \mathbb{R}$ of Baire class one. Indeed, we can select a sequence $(f_n)_{n \geqslant 1}$ of continuous functions that converges pointwise to f on $[0, 1]$. In this case, given $x \in [0, 1]$, x is a zero of f if and only if there is a subsequence[4] of the sequence $\big(f_n(x)\big)_{n \geqslant 1}$ that converges to 0. Thus, the set of zeros of f is none other than

$$\bigcap_{p=1}^{\infty} \bigcap_{q=1}^{\infty} \bigcup_{n=q}^{\infty} |f_n|^{-1} \left(\left] -\frac{1}{p}, \frac{1}{p} \right[\right),$$

which is indeed a G_δ-set of $[0, 1]$. $\qquad \square$

4.4.7 Proposition $(B\Delta_0, \| \cdot \|_\infty)$ *is a Banach space.*

Proof We must show that $B\Delta_0$ is closed in $(B\Delta, \| \cdot \|_\infty)$ and that it is a linear space. We start with the first point. Let $(f_n)_{n \geqslant 1}$ be a sequence of elements of $B\Delta_0$ that converges uniformly to $f \in B\Delta$. For each $n \geqslant 1$, let Z_n be the set of zeros of f_n in $[0, 1]$. As we saw above, this is a G_δ-set that is dense in $[0, 1]$. According to Baire's theorem, this is also the case for $Z_\infty := \bigcap_{n=1}^{\infty} Z_n$, which is contained in the set of zeros of f. This last set is thus dense in $[0, 1]$. By reusing the same argument, we can show that $B\Delta_0$ is in reality *a linear subspace of $B\Delta$*, which is not at all evident! $\qquad \square$

---

[4] We take a subsequence in order to make a G_δ-set appear.

The existence of functions of Pompeiu type also leads to this:

$B \Delta_0$ *contains functions that are not identically zero.*

Now let $(I_n)_{n \geqslant 1}$ be an enumeration of the segments contained in $[0, 1]$ with rational endpoints, and not reduced to a point. Let $B \Delta_0'$ be the set of elements of $B \Delta_0$ that have constant sign on at least one interval with non-empty interior contained in $[0, 1]$. We have of course

$$B \Delta_0' = \bigcup_{n=1}^{\infty} (E_n \cup F_n),$$

where E_n (resp. F_n) is the set of elements of $B \Delta_0$ which are $\geqslant 0$ (resp. $\leqslant 0$) on I_n.

4.4.8 Proposition   *The sets E_n and F_n are closed subsets of $B \Delta_0$ with empty interior.*

Proof   It is clear that E_n is closed in $B \Delta_0$. Let $f \in E_n$ and $\varepsilon > 0$. As $f \in B \Delta_0$, there exists an $x_0 \in I_n$ such that $f(x_0) = 0$. Moreover, we can construct "à la Pompeiu" an element g of $B \Delta_0$ satisfying $g(x_0) < 0$ and $\|g\|_\infty \leqslant \varepsilon$. Then set $h = f + g$. As the sum of two elements of $B \Delta_0$, the function $h \in B \Delta_0$. Moreover, $\|f - h\|_\infty \leqslant \varepsilon$. But $h \notin E_n$, since $h(x_0) < 0$. $\square$

According to Baire's theorem, $B \Delta_0'$ is of first category, and hence with empty interior in $B \Delta_0$. As a result, $B \Delta_0 \setminus B \Delta_0'$ is non-empty. We can thus state the following result.

4.4.9 Theorem   *There exists a function f, differentiable on $[0, 1]$, whose derivative is not of constant sign on any interval with non-empty interior contained in $[0, 1]$.*

4.4.10 Remark   The first example of a differentiable function that is nowhere monotonic was given by Köpcke (1889). Later, many simpler constructions of such functions were given (see, for example, [105]).

Exercises

4.1. An irrational number x is said to be a *Liouville number* if for all $n \geqslant 1$, there exists a rational $\dfrac{p}{q}$ ($p \in \mathbb{Z}, q \geqslant 2$) such that

$$\left| x - \frac{p}{q} \right| < \frac{1}{q^n}. \tag{4.2}$$

This means that x is well approximated by the rationals.

(a) Show that if x is a Liouville number and $n \geqslant 1$, there exist in reality infinitely many pairs (p, q) satisfying (4.2).

(b) Using the mean value theorem, show that if x is an algebraic real number of degree $d \geqslant 2$, there exists a numerical constant $M > 0$ such that $\left| x - \dfrac{p}{q} \right| \geqslant \dfrac{M}{q^d}$ for every rational $\dfrac{p}{q}$. Conclude that every Liouville number is transcendental.

(c) Show that the set $\mathcal{L}$ of Liouville numbers is a residual subset of $\mathbb{R}$.

(d) Show that $\mathcal{L}$ is negligible in the sense of Lebesgue. *Hint:* bound the measure of $\mathcal{L} \cap [-N, N]$ for $N \geqslant 1$.

4.2. Every real number is the sum of two Liouville numbers

(a) Show that if X and Y are two residual subsets of $\mathbb{R}$, then $X \cap Y \neq \varnothing$.

(b) Deduce that $\mathbb{R} = \mathcal{L} + \mathcal{L}$.

4.3. Select a fixed real number $\alpha \in \,]0, 1[$.

(a) Construct a closed subset of $[0, 1]$ with empty interior and with measure α.

(b) Construct a function $f : [0, 1] \to \mathbb{R}$, differentiable on $[0, 1]$, with bounded derivative, such that f' is discontinuous almost everywhere.

4.4. Show that there exists no function $f : [0, 1] \to \mathbb{R}$ continuous exactly on $[0, 1] \cap \mathbb{Q}$.

4.5. Let (X, d) be a complete metric space without any isolated point. Show that every residual subset of X is uncountable.

4.6. Consider the function

$$f : [0, 1] \to \mathbb{R}, t \mapsto \begin{cases} \sin \dfrac{1}{t} & \text{if } t \neq 0, \\ 0 & \text{if } t = 0. \end{cases}$$

Show that $f \in B\Delta$, but that $f^2 \notin B\Delta$. Thus, $B\Delta$ is not an algebra.

4.7. Show that a subset D of $\mathbb{R}$ is the set of points of discontinuity of a function $f : \mathbb{R} \to \mathbb{R}$ if and only if D is an F_σ-set.

4.8. This exercise aims to present an unexpected application of Baire's theorem.[5]

(a) Let (E, d) and (F, δ) be two metric spaces, the second being complete. Let $X \subset E$, and $f : X \to F$ be a continuous function. For each $x \in \overline{X}$, we set

---

[5] This exercise was inspired by Gilles Godefroy.

$$\omega(x) = \lim_{r \searrow 0} \delta\big(f(B(x,r) \cap X)\big) \in \overline{\mathbb{R}},$$

where the symbol $\delta(A)$ denotes, as on p. 105, the diameter of the set A. Finally, let Y be the set of $x \in \overline{X}$ such that $\omega(x) = 0$.

(i) Show that if $\varepsilon > 0$, the set $\{x \in \overline{X}/\omega(x) < \varepsilon\}$ is an open subset of $\overline{X}$.

(ii) Show that Y is a G_δ-set of E.

(iii) Show that the function f has a continuous extension to Y.

(b) Suppose now that f is a homeomorphism. Show that $Y = X$.

(c) Show that if two *normed* spaces are homeomorphic, then they are simultaneously complete.[6]

4.9. Our purpose is to show that there exist normed Baire spaces which are not complete.

(a) Let E be a normed space, and F a non-meagre subspace of E. Show that F is dense, and a Baire space.

(b) Show that if E is an infinite-dimensional normed space, E can be written as a countable union of an increasing sequence of proper subspaces (*hint:* use an algebraic basis of E), and conclude.

4.10. Nowhere analytic functions. Given a C^∞ function $f : \mathbb{R} \to \mathbb{C}$ and a fixed real number x, denote by $R(f, x)$ the radius of convergence of the Taylor series of f at the point x.

(a) First, we intend to construct an explicit example of a function f such that $R(f, x) = 0$ for all $x \in \mathbb{R}$, following a method due to Bernal-Gonzàlez [17].

(i) Let $(c_n)_{n \geqslant 1}$ be a sequence of non-negative real numbers. Define the sequence $(b_n)_{n \geqslant 1}$ by

$$b_1 = 2 + c_1 \text{ and } b_k = 2 + c_k + \sum_{j=1}^{k-1} b_j^{k+1-j} \text{ for } k \geqslant 2,$$

and the function $f : \mathbb{R} \to \mathbb{C}$ by

$$f(x) = \sum_{k=1}^{\infty} b_k^{1-k} e^{ib_k x}.$$

Show that f is C^∞ on $\mathbb{R}$.

(ii) Show that $|f'(x)| \geqslant c_1$ for $x \in \mathbb{R}$.

(iii) More generally, show that if $n \geqslant 2$ and $x \in \mathbb{R}$, $|f^{(n)}(x)| \geqslant c_n$.

(iv) Conclude.

---

[6] Which is notoriously false in the context of metric spaces.

(b) We now consider the space E of C^∞ functions from $\mathbb{R}$ to $\mathbb{C}$, equipped with the family of semi-norms $(p_k)_{k \geqslant 0}$ defined by

$$p_k(f) = \sup_{0 \leqslant i \leqslant k, |x| \leqslant k} |f^{(i)}(x)|.$$

The semi-norms p_k define on E the topology of a Fréchet space.[7] A basis of neighbourhoods of 0 in E is formed by the $V_{n,\varepsilon}$ ($n \in \mathbb{N}$, $\varepsilon > 0$), where

$$V_{n,\varepsilon} = \{f \in E / p_n(f) \leqslant \varepsilon\}.$$

The topology of E is derived from the complete metric defined by

$$d(f, g) = \sum_{k=0}^{\infty} 2^{-k} \frac{p_k(f - g)}{1 + p_k(f - g)}.$$

Finally, if $(f_n)_{n \geqslant 0}$ is a sequence of elements of E, saying $(f_n)_{n \geqslant 0}$ converges to $f \in E$ means that for each $k \geqslant 0$, the sequence $(f_n^{(k)})_{n \geqslant 0}$ converges uniformly to $f^{(k)}$ on every compact subset of $\mathbb{R}$ (for details, see [153]).

(i) Given integers $b, p \geqslant 1$, we set

$$F(b, p) = \left\{ f \in E \ \middle| \ \begin{array}{l} \text{there exists } x \in [-p, p] \text{ such that} \\ |f^{(n)}(x)| \leqslant b^n n! \text{ for all } n \geqslant 0 \end{array} \right\}.$$

Show that $F(b, p)$ is a closed subset of E with empty interior.

(ii) Show that the set A of the functions $f \in E$ such that $R(f, x) = 0$ for all $x \in \mathbb{R}$ is a residual subset of E, and that each function of E is the sum of two functions of A. We thus obtain an enhancement of a theorem of Morgenstern (1950), which can be found, for example, in the excellent book [51].

---

[7] A Fréchet space is a topological vector space whose topology can be defined by a distance d such that the metric space (E, d) is complete.

5

Probability theory and existence theorems

5.1 Introduction

The "dummies" in probability will tell you: "Probability theory, it's just a minor sideline of measure theory, where the measure is positive and of total mass one". *Nothing could be more wrong*. It is true that Kolmogorov's probability theory is a magnificent application of Lebesgue's theory and of convergence theorems of integrals.[1] However, this domain of mathematics has its own vocabulary and way of thinking, and a specific set of tools: independence, conditioning, filtration, stochastic integrals, etc. Moreover, while probability theory can in many ways be considered as applied mathematics, it is also extremely useful in a number of domains of pure mathematics: functional analysis and number theory, among others. It also gives rise to *existence proofs* that are very difficult to obtain by other means (for example, Dvoretzky's theorem on the geometry of Banach spaces, or results on ultraflat polynomials in Fourier analysis). We will attempt to illustrate this in the present chapter, with three fairly elementary examples.

(1) Khintchine's inequalities and three of their applications.
(2) Gaussian random variables and an application to *Dvoretzky's theorem* in a very particular case.
(3) The method of Bourgain selectors and its application to the *combinatorial dimension* of subsets of a lattice.

Our probability spaces will always be denoted by $(\Omega, \mathcal{A}, \mathbf{P})$. We assume a familiarity with the notions of random variables, expectation (denoted by $\mathbf{E}$), variance (denoted by V), characteristic functions and independence, as well as with the inequalities of Markov and Tchebycheff. Let us recall finally that:

---

[1] Which indeed were invented for other reasons than to study the limit of a sequence of the type $\int_0^n \left(1 + \frac{x}{n}\right)^n e^{-2x} dx$!.

- $|F|$ is the cardinality of a finite set F;
- a *Rademacher* sequence (finite or infinite) is a sequence (ε_n) of independent and identically distributed random variables (abbreviated i.i.d.), with

$$\mathbf{P}(\varepsilon_n = 1) = \mathbf{P}(\varepsilon_n = -1) = \frac{1}{2} \; ;$$

- a *Gaussian standard sequence* (finite or infinite) is a sequence (g_n) of random variables i.i.d. with common density

$$\frac{1}{\sqrt{2\pi}} e^{-x^2/2},$$

or equivalently with common characteristic function

$$\mathbf{E}(e^{itg_n}) = e^{-\frac{t^2}{2}} \, .$$

A fundamental property of these standard Gaussian sequences is as follows.

5.1.1 Theorem [Hilbertian stability theorem] *Let $(g_1, \ldots, g_n)$ be a Gaussian standard sequence, and $a = (a_1, \ldots, a_n) \in \mathbb{R}^n$. Then the random variables*

$$X := \sum_{j=1}^{n} a_j g_j \quad and \quad Y := \|a\|_2 g_1$$

have the same distribution, where $\|a\|_2$ is the standard Euclidean norm of a:

$$\|a\|_2 = \left(\sum_{j=1}^{n} a_j^2 \right)^{1/2}.$$

Proof Indeed, X and Y have the same characteristic function: for all $t \in \mathbb{R}$,

$$\mathbf{E}(e^{itX}) = \prod_{j=1}^{n} \mathbf{E}(e^{ita_j g_j}) = \prod_{j=1}^{n} e^{-\frac{t^2}{2} a_j^2} = e^{-\frac{t^2}{2} \|a\|_2^2} = \mathbf{E}(e^{itY}). \qquad \square$$

5.2 Khintchine's inequalities and applications

Let $(\varepsilon_n)_{n \geqslant 1}$ be a Rademacher sequence, defined on a probability space $(\Omega, \mathcal{A}, \mathbf{P})$, which will play only a passive role. These random variables are orthogonal in the Hilbert space $L^2(\Omega, \mathcal{A}, \mathbf{P})$, since if $i \neq j$, we have

$$\mathbf{E}(\varepsilon_i \varepsilon_j) = \mathbf{E}(\varepsilon_i)\mathbf{E}(\varepsilon_j) = 0.$$

However, their independence implies that they are much more than orthogonal, in opposition – for example – to the case of the imaginary exponentials e^{int} of harmonic analysis. This *super-orthogonality* leads to the following inequalities [120], in which the L^p-norms refer to $(\Omega, \mathcal{A}, \mathbf{P})$, that is to say

$$\|X\|_p = \left(\mathbf{E}(|X|^p)\right)^{1/p},$$

meaning that these norms are all equivalent on the *infinite*-dimensional space generated by the ε_j.

5.2.1 Theorem [Khintchine's inequalities]   *Let $a_1, \ldots, a_N \in \mathbb{C}$ and*

$$X = \sum_{j=1}^{N} a_j \varepsilon_j.$$

Then we have

$$\|X\|_p \leqslant \sqrt{p}\,\|X\|_2 \text{ for } 2 \leqslant p < \infty \tag{5.1}$$

and

$$\|X\|_1 \geqslant \frac{1}{\sqrt{2}}\,\|X\|_2. \tag{5.2}$$

A detailed proof is given, for example in [120], but we will not fully reproduce it here. We limit ourselves to indicating that (5.1) can be shown first when the a_j are real numbers and $p = 2q$ is an even integer. Thereafter, a brute force computation succeeds: starting with the equality $|X|^p = X^p$ and using the multinomial identity, we expand, we integrate, and the independence generates enough null terms so that, up to a factor q^q, we recover the multinomial expansion of $(a_1^2 + \cdots + a_N^2)^q = \|X\|_2^{2q}$.

More precisely:

$$|X|^p = X^p = \sum \frac{p!}{\alpha_1! \times \cdots \times \alpha_N!} a_1^{\alpha_1} \times \cdots \times a_N^{\alpha_N}\, \varepsilon_1^{\alpha_1} \times \cdots \times \varepsilon_N^{\alpha_N},$$

where the sum is taken over the N-tuples of non-negative integers with sum p. Taking the expectation of both sides, we have

$$\mathbf{E}(|X|^p) = \sum \frac{p!}{\alpha_1! \times \cdots \times \alpha_N!} a_1^{\alpha_1} \times \cdots \times a_N^{\alpha_N}\, \mathbf{E}(\varepsilon_1^{\alpha_1}) \times \cdots \times \mathbf{E}(\varepsilon_N^{\alpha_N}),$$

but only the terms for which $\alpha_j = 2\beta_j$ is even for $1 \leqslant j \leqslant N$ provide a non-zero contribution, so that

$$\mathbf{E}(|X|^p) = \sum \frac{(2q)!}{(2\beta_1)! \times \cdots \times (2\beta_N)!} a_1^{2\beta_1} \times \cdots \times a_N^{2\beta_N},$$

where this time $\beta_1 + \cdots + \beta_N = q$. However, $(2\beta)! \geqslant 2^\beta \beta!$ for all non-negative integers β, hence

$$\mathbf{E}(|X|^p) \leqslant \sum \frac{(2q)!}{2^{\beta_1 + \cdots + \beta_N} \beta_1! \times \cdots \times \beta_N!} a_1^{2\beta_1} \times \cdots \times a_N^{2\beta_N}$$

$$= \frac{(2q)!}{2^q q!} \sum \frac{q!}{\beta_1! \times \cdots \times \beta_N!} (a_1^2)^{\beta_1} \times \cdots \times (a_N^2)^{\beta_N}$$

$$= \frac{(2q)!}{2^q q!} (a_1^2 + \cdots + a_N^2)^q$$

$$\leqslant q^q \|X\|_2^{2q},$$

because

$$\frac{(2q)!}{q!} = (q+1) \times (q+2) \times \cdots \times (2q) \leqslant (2q) \times (2q) \times \cdots \times (2q) = 2^q q^q,$$

so that

$$\|X\|_p \leqslant \sqrt{q} \, \|X\|_2 = \sqrt{\frac{p}{2}} \, \|X\|_2$$

in this case. For arbitrary p, we interpolate via Hölder's inequalities [120].

As for (5.2), it is difficult if we are aiming to have the best constant $\dfrac{1}{\sqrt{2}}$ (see [120] again), but if we are content with any numerical constant, (5.2) is an easy consequence of Hölder's inequalities and the case $p = 4$ in (5.1) (see Exercise 10.8 in Chapter 10).

5.2.2 Application

We present here a first application of Khintchine's inequalities to the *problem of phases* of Salem [15]: let $N \geqslant 2$ be given, $\rho_1, \ldots, \rho_N \in \mathbb{R}_+$ also fixed (the "amplitudes") and $\varphi_1, \ldots, \varphi_N$ real numbers that we will vary (the "phases"). Let f be the sum of the elementary de-phased signals $\rho_j \cos(jt + \varphi_j)$, that is to say:

$$f(t) = \sum_{j=1}^{N} \rho_j \cos(jt + \varphi_j).$$

The function f is of course continuous and 2π-periodic; we would like to minimise its L^∞-norm: $\|f\|_\infty = \sup_{t \in \mathbb{R}} |f(t)|$. We begin with two easy observations.

- First,

$$\|f\|_\infty \leqslant \sum_{j=1}^{N} \rho_j,$$

with equality ($t = 0$) if all the signals are in phase, in other words $\varphi_1 = \cdots = \varphi_N = 0$.

- Second,

$$\|f\|_\infty \geqslant \|f\|_2 = \frac{1}{\sqrt{2}} \Big(\sum_{j=1}^{N} \rho_j^2 \Big)^{1/2},$$

where $\| \cdot \|_2$ refers to the normalised Lebesgue measure on the unit circle $\mathbb{T}$:

$$dm(t) = \frac{dt}{2\pi} \quad \text{on } [0, 2\pi].$$

Let

$$R = \Big(\sum_{j=1}^{N} \rho_j^2 \Big)^{1/2}$$

be the standard Euclidean norm of the vector $(\rho_1, \ldots, \rho_N)$ of amplitudes. The following existence theorem [15] tells us that for certain phases, the inequality $\|f\|_\infty \geqslant \dfrac{R}{\sqrt{2}}$ is (up to a factor $\sqrt{\ln}$) almost the only obstruction to the smallness of $\|f\|_\infty$.

5.2.3 Theorem [Salem's theorem of phases]   (*1*) *There exist phases* $\varphi_1, \ldots,$ *$\varphi_N \in \mathbb{R}$ such that*

$$\|f\|_\infty \leqslant C R \sqrt{\ln N}, \tag{5.3}$$

where C is an absolute constant.
(2) *The result is in general optimal, in the sense that the factor* $\sqrt{\ln N}$ *cannot be replaced by a slower growth factor.*

Theorem 5.2.3 is an existence theorem, which can be obtained by bounding either a minimum (compactness and functions of several variables) or an average value (probability theory). We will thus present two proofs of this theorem.

5.2.4 First proof by compactness

Let p be an even integer $\geqslant 2$, that we will subsequently adjust. For $x = (x_1, \ldots, x_N) \in \mathbb{R}^N$, set

$$f(x, t) = f_x(t) = \sum_{j=1}^{N} \rho_j \cos(jt + x_j) \text{ and } \Phi(x) = \int_{\mathbb{T}} f_x(t)^p \, dm(t).$$

The function Φ is continuous, and 2π-periodic with respect to each variable x_j. Thus, by compactness, it attains its minimum at a point $a = (\varphi_1, \ldots, \varphi_N)$ of $\mathbb{R}^N$, for which we have[2]

$$\Delta\Phi(a) = \sum_{j=1}^{N} \frac{\partial^2 \Phi}{\partial x_j^2}(a) \geqslant 0.$$

Moreover, by differentiating under the integral, we have

$$\frac{\partial \Phi}{\partial x_j}(x) = \int_{\mathbb{T}} pf_x(t)^{p-1}\big(-\rho_j \sin(jt + x_j)\big) \, dm(t),$$

and then

$$\frac{\partial^2 \Phi}{\partial x_j^2}(a) = \int_{\mathbb{T}} p(p-1)f_a(t)^{p-2}\rho_j^2 \sin^2(jt + \varphi_j) \, dm(t)$$

$$- \int_{\mathbb{T}} pf_a(t)^{p-1}\rho_j \cos(jt + \varphi_j) \, dm(t),$$

so that, taking the sum over j and bounding the $\sin^2$ by 1:

$$0 \leqslant \frac{\Delta\Phi(a)}{p} \leqslant \int_{\mathbb{T}} (p-1)f_a(t)^{p-2}R^2 \, dm(t) - \int_{\mathbb{T}} f_a(t)^p \, dm(t).$$

However, according to Hölder's inequality, $\|f_a\|_p = \big(\int_{\mathbb{T}} |f_a|^p\big)^{1/p}$ increases when $p > 0$ increases.[3] Hence, bounding $p - 1$ by p, we obtain[4]

$$\int_{\mathbb{T}} f_a(t)^p \, dm(t) \leqslant pR^2 \int_{\mathbb{T}} f_a(t)^{p-2} \, dm(t) = pR^2 \|f_a\|_{p-2}^{p-2}$$

$$\leqslant pR^2 \|f_a\|_p^{p-2} = pR^2 \Big(\int_{\mathbb{T}} f_a(t)^p \, dm(t)\Big)^{1-2/p}.$$

Finally, after simplifying and taking square roots:

$$\|f_a\|_p \leqslant R\sqrt{p}. \tag{5.4}$$

Note that, even if the proof was called "by compactness", (5.4) has a random character: there exists at least one a such that ... To exploit this *randomness*

---

[2] In reality, each term of the sum is non-negative.

[3] Indeed, if $0 < p < q$, we have, setting $r = \frac{q}{p} > 1$ and $\frac{1}{r} + \frac{1}{r'} = 1$:

$$\int_{\mathbb{T}} |f_a|^p \leqslant \Big(\int_{\mathbb{T}} \big(|f_a|^p\big)^r\Big)^{1/r} \Big(\int_{\mathbb{T}} 1^{r'}\Big)^{1/r'} = \Big(\int_{\mathbb{T}} |f_a|^q\Big)^{p/q}.$$

[4] Including when $p = 2$!

we use Bernstein's inequality [120], a result that can truly be qualified as deterministic. It states that

$$\|f_a'\|_\infty \leqslant N\|f_a\|_\infty. \tag{5.5}$$

Set $M = \|f_a\|_\infty$, and fix a real number t_0 such that $M = |f_a(t_0)|$. Inequality (5.5) then leads to

$$|f_a(t) - f_a(t_0)| \leqslant |t - t_0|\|f_a'\|_\infty \leqslant \frac{1}{2N}NM = \frac{M}{2} \text{ if } |t - t_0| \leqslant \frac{1}{2N},$$

so that, for these same t,

$$|f_a(t)| \geqslant \frac{M}{2}.$$

Hence,

$$\|f_a\|_p^p = \int_{t_0-\pi}^{t_0+\pi} |f_a(t)|^p \frac{dt}{2\pi} \geqslant \int_{t_0-\frac{1}{2N}}^{t_0+\frac{1}{2N}} \left(\frac{M}{2}\right)^p \frac{dt}{2\pi} = \frac{1}{2\pi N}\left(\frac{M}{2}\right)^p,$$

and thus

$$M \leqslant 2(2\pi N)^{1/p}\|f_a\|_p.$$

Then, we select for p the smallest even integer $\geqslant 2$ such that $p \geqslant \ln(2\pi N)$. We obtain the existence of a numerical constant C and an even integer p, such that

$$\|f_a\|_\infty \leqslant 2e\|f_a\|_p \text{ and } p \leqslant C\ln N. \tag{5.6}$$

This is a way to quantify the following classical result: $\|f\|_p \to \|f\|_\infty$ when $p \to \infty$. It is clear that (5.4) and (5.6) imply Salem's result.

5.2.5 Second proof by probability theory

Let $(\varepsilon_1, \ldots, \varepsilon_N)$ be a Rademacher sequence on $(\Omega, \mathcal{A}, \mathbf{P})$. For $\omega \in \Omega$, set

$$f_\omega(t) = \sum_{j=1}^{N} \varepsilon_j(\omega)\rho_j \cos(jt) = \sum_{j=1}^{N} \rho_j \cos\left(jt + \varphi_j(\omega)\right),$$

with

$$\varphi_j(\omega) = \begin{cases} 0 & \text{if } \varepsilon_j(\omega) = 1, \\ \pi & \text{if } \varepsilon_j(\omega) = -1. \end{cases}$$

Fix again an even positive integer $p \geqslant 2$. By Fubini's theorem and Khintchine's inequalities, we have[5]

$$\mathbf{E}\Big(\|f_\omega\|_p^p\Big) = \mathbf{E}\Big(\int_{\mathbb{T}} \Big| \sum_{j=1}^{N} \varepsilon_j(\omega)\rho_j \cos(jt) \Big|^p dm(t)\Big)$$

$$= \int_{\mathbb{T}} \mathbf{E}\Big(\Big| \sum_{j=1}^{N} \varepsilon_j(\omega)\rho_j \cos(jt) \Big|^p\Big) dm(t)$$

$$\leqslant \int_{\mathbb{T}} p^{p/2} \Big(\sum_{j=1}^{N} \rho_j^2 \cos^2(jt) \Big)^{p/2} dm(t)$$

$$\leqslant p^{p/2} R^p, \quad \text{by bounding the } \cos^2 \text{ by } 1.$$

But then, as the average value of the $\|f_\omega\|_p^p$ is bounded by $p^{p/2} R^p$, there exists at least one $\omega_0 \in \Omega$ such that

$$\|f_{\omega_0}\|_p^p \leqslant p^{p/2} R^p,$$

so that

$$\|f_{\omega_0}\|_p \leqslant R\sqrt{p}.$$

The sequence of $(\varphi_j)_{1 \leqslant j \leqslant N} = \big(\varphi_j(\omega_0)\big)_{1 \leqslant j \leqslant N}$ satisfies our requirement, and we find (5.4). We conclude, as before, by using (5.6).

5.2.6 Proof of point (2) of Theorem 5.2.3

Consider the polynomial

$$f(t) = \sum_{k=1}^{n} \cos(2^k t + \psi_k),$$

corresponding to the choice

$$N = 2^n, \quad \rho_j = \begin{cases} 1 & \text{if } j = 2^k \\ 0 & \text{otherwise} \end{cases} \quad \text{and} \quad \psi_k = \varphi_{2^k}.$$

---

[5] By setting $X(\omega) = \|f_\omega\|_p^p = \int_{\mathbb{T}} |f_\omega(t)|^p dm(t)$, we define a random variable which takes on the 2^N values (distinct or not) $\int_{\mathbb{T}} \Big| \sum_{j=1}^{N} \delta_j \rho_j \cos(jt) \Big|^p dm(t)$ with probability 2^{-N}, as $(\delta_1, \ldots, \delta_N)$ runs over $\{-1, 1\}^N$.

The set E of powers of 2 is a set of lacunary integers in the sense of Hadamard, and we thus know [120] that $E \cup -E$ is a Sidon set, that is to say, there exists a numerical constant $c > 0$ such that

$$\left\| \sum_{k=1}^{n} \rho_k \cos(2^k t + \psi_k) \right\|_\infty \geqslant c \sum_{k=1}^{n} \rho_k,$$

for all choices of ρ_k and ψ_k. In this case, we thus have $\|f\|_\infty \geqslant cn$, while $R\sqrt{\ln N} = \sqrt{n}\sqrt{n \ln 2}$. The factor $\sqrt{\ln N}$ is thus unavoidable in the estimate (5.3).

5.2.7 Second application

A second application of Khintchine's inequalities concerns the Banach space $X = L^1(\mu)$, where μ is a positive measure, finite or not, defined on a measurable space T. In what follows, $(\varepsilon_i)_{1 \leqslant i \leqslant N}$ will again be a Rademacher sequence on $(\Omega, \mathcal{A}, \mathbf{P})$.

It is well known that a Hilbert space H verifies the generalised parallelogram identity, which can be expressed as follows:

$$\mathbf{E}\left(\left\| \sum_{i=1}^{N} \varepsilon_i f_i \right\|^2 \right) = \sum_{i=1}^{N} \|f_i\|^2 \tag{5.7}$$

for all $f_1, \ldots, f_N \in H$. Indeed, the left-hand side of (5.7) equals

$$\sum_{1 \leqslant i,j \leqslant N} \langle f_i, f_j \rangle \mathbf{E}(\varepsilon_i \varepsilon_j) = \sum_{i=1}^{N} \langle f_i, f_i \rangle = \sum_{i=1}^{N} \|f_i\|^2.$$

The case $N = 2$ is the familiar parallelogram identity.

If a Banach space Y is isomorphic to the Hilbert space H, in the sense that there exists a continuous linear isomorphism[6] from Y to H, then, for a constant $C \geqslant 1$ and arbitrary $f_1, \ldots, f_N \in Y$, we have

$$C^{-1} \sum_{i=1}^{N} \left\| f_i \right\|^2 \leqslant \mathbf{E}\left(\left\| \sum_{i=1}^{N} \varepsilon_i f_i \right\|^2 \right) \leqslant C \sum_{i=1}^{N} \|f_i\|^2. \tag{5.8}$$

Indeed, if $u : H \to Y$ is an isomorphism and if $f_i = u(g_i)$, where $g_i \in H$, we have

---

[6] And for its inverse, as a result of the Banach open mapping theorem.

$$\mathbf{E}\Big(\Big\|\sum_{i=1}^{N}\varepsilon_{i}f_{i}\Big\|^{2}\Big)=\mathbf{E}\Big(\Big\|u\Big(\sum_{i=1}^{N}\varepsilon_{i}g_{i}\Big)\Big\|^{2}\Big)$$

$$\leqslant\|u\|^{2}\mathbf{E}\Big(\Big\|\sum_{i=1}^{N}\varepsilon_{i}g_{i}\Big\|^{2}\Big)$$

$$=\|u\|^{2}\sum_{i=1}^{N}\|g_{i}\|^{2}$$

$$\leqslant C\sum_{i=1}^{N}\|f_{i}\|^{2},$$

with $C=\|u\|^{2}\|u^{-1}\|^{2}$, and we can similarly prove the leftmost inequality of (5.8).

It is also well known that a space X of type $L^{1}(\mu)$ is very far from being isomorphic to a Hilbert space when dim $X=\infty$. For example (see the exercises of Chapter 13), the space ℓ_{1} does not contain any subspace isomorphic to ℓ_{2}, and conversely. Or again (see [120]), the unit sphere of $L^{1}([0,1])$ does not contain any extremal point, unlike that of a Hilbert space (where the set of extremal points is the unit sphere). Or yet again, ℓ_{1} is not reflexive, contrary to ℓ_{2}. Nonetheless, Khintchine's inequalities will lead to a surprising property of this space: it is "half of a Hilbert space", in the sense that it satisfies the left-hand inequality of (5.8) in the form given below.

5.2.8 Theorem *There exists a numerical constant $C\geqslant 1$ such that, for any space $X=L^{1}(\mu)$, we have*

$$C^{-1}\sum_{i=1}^{N}\|f_{i}\|_{1}^{2}\leqslant\mathbf{E}\Big(\Big\|\sum_{i=1}^{N}\varepsilon_{i}f_{i}\Big\|_{1}^{2}\Big)\tag{5.9}$$

for all $f_{1},\dots,f_{N}\in X$.

5.2.9 Remark In the inequality (5.9), $\|\cdot\|_{1}$ denotes the usual norm on $L^{1}(\mu)$ defined by

$$\|f\|_{1}=\int_{T}|f(t)|d\mu(t).$$

The second portion of this inequality is thus

$$\int_{\Omega}\Big(\int_{T}\Big|\sum_{i=1}^{N}\varepsilon_{i}(\omega)f_{i}(t)\Big|d\mu(t)\Big)^{2}d\mathbf{P}(\omega).$$

The norm $\|\cdot\|_{1}$ is hence relative to the measure μ and to the variable t, whereas the expectation is relative to the probability $\mathbf{P}$ and the variable ω.

Proof   By setting

$$S = \Big\| \sum_{i=1}^{N} \varepsilon_i f_i \Big\|_1,$$

we define a random variable on Ω, satisfying

$$S(\omega) = \int_T \Big| \sum_{i=1}^{N} \varepsilon_i(\omega) f_i(t) \Big| d\mu(t) \ \ \text{for } \omega \in \Omega.$$

The Fubini–Tonelli theorem[7] and Khintchine's inequalities give

$$\mathbf{E}(S) = \int_\Omega S(\omega) d\mathbf{P}(\omega)$$

$$= \int_\Omega \int_T \Big| \sum_{i=1}^{N} \varepsilon_i(\omega) f_i(t) \Big| d\mu(t) d\mathbf{P}(\omega)$$

$$= \int_T \int_\Omega \Big| \sum_{i=1}^{N} \varepsilon_i(\omega) f_i(t) \Big| d\mathbf{P}(\omega) d\mu(t)$$

$$\geqslant \frac{1}{\sqrt{2}} \int_T \Big(\int_\Omega \Big| \sum_{i=1}^{N} \varepsilon_i(\omega) f_i(t) \Big|^2 d\mathbf{P}(\omega) \Big)^{1/2} d\mu(t)$$

$$= \frac{1}{\sqrt{2}} \int_T \Big(\sum_{i=1}^{N} |f_i(t)|^2 \Big)^{1/2} d\mu(t), \ \ \text{by (5.7) applied to } H = \mathbb{C}.$$

At this stage, it is useful to have the following inequality.

5.2.10 Lemma   *We have the inequality*

$$\int_T \Big(\sum_{i=1}^{N} |f_i(t)|^2 \Big)^{1/2} d\mu(t) \geqslant \Big(\sum_{i=1}^{N} \|f_i\|_1^2 \Big)^{1/2}. \tag{5.10}$$

Indeed, the right-hand side of (5.10) can be written

$$\sum_{i=1}^{N} \lambda_i \|f_i\|_1$$

for $\lambda_i \geqslant 0$ such that

$$\sum_{i=1}^{N} \lambda_i^2 = 1,$$

---

[7] Applicable here even if μ is not σ-finite, because the expectation with respect to the ε_i is in fact a finite sum.

namely

$$\lambda_i = \frac{\|f_i\|_1}{\left(\sum_{i=1}^{N} \|f_i\|_1^2\right)^{1/2}}.$$

But then, by Cauchy–Schwarz:

$$\left(\sum_{i=1}^{N} \|f_i\|_1^2\right)^{1/2} = \sum_{i=1}^{N} \lambda_i \int_T |f_i(t)| d\mu(t)$$

$$= \int_T \left(\sum_{i=1}^{N} \lambda_i |f_i(t)|\right) d\mu(t)$$

$$\leqslant \int_T \left(\sum_{i=1}^{N} \lambda_i^2\right)^{1/2} \left(\sum_{i=1}^{N} |f_i(t)|^2\right)^{1/2} d\mu(t)$$

$$= \int_T \left(\sum_{i=1}^{N} |f_i(t)|^2\right)^{1/2} d\mu(t).$$

We thus obtain

$$\mathbf{E}(S) \geqslant \frac{1}{\sqrt{2}} \int_T \left(\sum_{i=1}^{N} |f_i(t)|^2\right)^{1/2} d\mu(t) \geqslant \frac{1}{\sqrt{2}} \left(\sum_{i=1}^{N} \|f_i\|_1^2\right)^{1/2}.$$

The inequality (5.9) follows with $C = 2$, since $\mathbf{E}(S^2) \geqslant \mathbf{E}(S)^2$. $\quad\square$

5.2.11 Remark Here is another way of expressing the inequality (5.9): the space $L^1(\mu)$ is *of cotype 2*. See [120] for more details.

5.2.12 Third application

Here is an application of this result, thus a third application of Khintchine, in which $C_0(\mathbb{R}^d)$ is the Banach space of continuous functions on $\mathbb{R}^d$ vanishing at infinity, equipped with the norm $\|\cdot\|_\infty$.

5.2.13 Corollary *Let d be an integer $\geqslant 1$. Then the Banach spaces $X = L^1(\mathbb{R}^d)$ and $Y = C_0(\mathbb{R}^d)$ are not isomorphic. In particular, the Fourier transform $\Phi : X \to Y$, defined by*

$$\Phi f(x) = \int_{\mathbb{R}^d} e^{-i<x,y>} f(y) \, dy,$$

is not surjective.

Proof   By Theorem 5.2.8, it is sufficient to show that Y is not of cotype 2. For this, consider N functions $f_1, \ldots, f_N \in Y$, with compact and pairwise disjoint supports, such that $\| f_i \|_\infty = 1$ for $1 \leqslant i \leqslant N$. We thus have, for all $\omega \in \Omega$:

$$\left\| \sum_{i=1}^{N} \varepsilon_i(\omega) f_i \right\|_\infty = 1,$$

hence

$$\mathbf{E}\left(\left\| \sum_{i=1}^{N} \varepsilon_i(\omega) f_i \right\|_\infty^2 \right) = 1.$$

On the contrary,

$$\sum_{i=1}^{N} \| f_i \|_\infty^2 = N.$$

Thus, there cannot exist a positive constant C such that we have, *for all $N \geqslant 1$* and all $(f_1, \ldots, f_N) \in Y^N$:

$$C^{-1} \sum_{i=1}^{N} \| f_i \|_\infty^2 \leqslant \mathbf{E}\left(\left\| \sum_{i=1}^{N} \varepsilon_i(\omega) f_i \right\|_\infty^2 \right).$$

This proves the first part of the corollary.[8] Finally, if Φ were surjective, as it is of course injective, it would be an isomorphism of X to Y according to the open mapping theorem, and we have just seen that such an isomorphism does not exist.   □

5.3 Hilbertian subspaces of $L^1([0,1])$

In this section, to simplify, we limit ourselves to the case of *real* normed spaces. We have already mentioned Dvoretzky's theorem, and will deem satisfactory the following somewhat vague statement.

5.3.1 Theorem [Dvoretzky]   *If X is a real normed space of dimension n, X contains a* large *subspace Y, with dimension at least of order $\ln n$, and almost isometric to a Hilbert space.*

---

[8]   If we were very knowledgeable, we could say [120] that X is "weakly sequentially complete" while Y is not. However, this argument turns out to be more complicated than the proof given above using probabilities.

This theorem is a superb illustration of the methods used in this chapter, but it is technically too difficult to be detailed here (see [120]). If dim $X = \infty$, it is false to say that X contains a subspace isomorphic to a Hilbert space (see Exercise 5.7). But it is true, and in a non-trivial manner, for certain spaces *a priori* very far from being Hilbertian.

5.3.2 Theorem   *The space $X = L^1([0, 1])$ contains an infinite-dimensional closed subspace Y isometric to a Hilbert space.*

Proof   Let $(g_n)_{n \geqslant 1}$ be a standard Gaussian sequence defined on the probability space $([0, 1], \mathcal{B}, m)$, where $\mathcal{B}$ is the σ-algebra of Borel sets and m the Lebesgue measure.[9] Note that the g_n are indeed elements of X, because

$$\int_0^1 |g_n(t)| dt = \frac{1}{\sqrt{2\pi}} \int_{\mathbb{R}} |x| e^{-x^2/2} dx = \sqrt{\frac{2}{\pi}}.$$

Furthermore, let H be the real Hilbert space $\ell_2(\mathbb{N}^*)$ of functions

$$f : \mathbb{N}^* \to \mathbb{R} \text{ such that } \|f\|_2^2 := \sum_{n=1}^{\infty} f(n)^2 < \infty,$$

and define the operator $T : H \to X$ as follows:

$$T(a) = \sum_{n=1}^{\infty} a_n g_n, \text{ for all } a = (a_n)_{n \geqslant 1} \in H.$$

To show that T is well-defined, use for example Theorem 5.1.1 on the stability of Gaussian random variables. In fact, if we set

$$S_n = \sum_{j=1}^{n} a_j g_j,$$

we have, for $q > p \geqslant 1$:

$$\|S_q - S_p\|_1 = \left\| \sum_{j=p+1}^{q} a_j g_j \right\|_1 = \left\| \left(\sum_{j=p+1}^{q} a_j^2 \right)^{1/2} g_1 \right\|_1 = \left(\sum_{j=p+1}^{q} a_j^2 \right)^{1/2} \|g_1\|_1,$$

a quantity which tends to 0 when $p, q \to \infty$, as $a \in H$. The sequence $(S_n)_{n \geqslant 1}$ is a Cauchy sequence in X; thus it converges in X to a limit $T(a)$, naturally denoted $\sum_{j=1}^{\infty} a_j g_j$. Moreover, S_n has the same distribution as $(\sum_{j=1}^{n} a_j^2)^{1/2} g_1$, so that by setting

[9] Such a sequence exists, because the preceding probability space is *universal* (see [21], Theorem 20.4, p. 265)

$$\gamma = \|g_1\|_1 = \sqrt{\frac{2}{\pi}},$$

we have

$$\|S_n\|_1 = \gamma \Big(\sum_{j=1}^n a_j^2 \Big)^{1/2}.$$

Passing to the limit in this equality gives

$$\|T(a)\|_1 = \gamma \|a\|_2.$$

The map $S = \frac{1}{\gamma} T$ is thus an isometry from H to the closed subspace Y of X defined by $Y = \operatorname{Im} T = \operatorname{Im} S$. $\qquad\square$

5.3.3 Remarks (1) Gaussian random variables (and Gaussian vectors) will continue to play an essential role in the proof of the general theorem of Dvoretzky [120].

(2) Another application of Gaussian random variables is given in Exercise 5.6.

(3) By using p-stable variables ($1 < p \leqslant 2$) instead of Gaussian variables [120], we can similarly show that $L^1([0, 1])$ contains a closed subspace isometric to the space

$$\ell_p = \Big\{ f : \mathbb{N}^* \to \mathbb{R} / \sum_{n=1}^\infty |f(n)|^p < \infty \Big\}.$$

The result remains true for $p = 1$, but is false for $p > 2$ (see Exercise 5.11).

(4) Even though the space ℓ_1 is of cotype 2, just like $L^1([0, 1])$, it does not contain any closed subspace isomorphic to the Hilbert space ℓ_2 (see Exercise 5.7).

(5) It can be shown [120] that $L^1([0, 1])$ does not contain any infinite-dimensional Hilbert subspace that is *complemented* (see Chapter 13).

5.4 Concentration of binomial distributions and applications

In this section, we fix a real number p in $[0, 1]$, and $n \in \mathbb{N}^*$. Recall that a random variable S_n follows a binomial distribution $\mathcal{B}(n, p)$ if, setting $q = 1 - p$, we have

$$\mathbf{P}(S_n = k) = \binom{n}{k} p^k q^{n-k} \text{ for } 0 \leqslant k \leqslant n.$$

A first spectacular application of these variables to analysis is the construction of the Bernstein polynomials [21] associated with a function f continuous on $[0, 1]$:

$$B_n(p) = \mathbf{E}\left(f\left(\frac{S_n}{n}\right)\right) = \sum_{k=0}^{n} \binom{n}{k} f\left(\frac{k}{n}\right) p^k q^{n-k}.$$

One can show, using only the Tchebycheff inequality,[10] that

$$B_n(p) \to f(p) \text{ as } n \to +\infty, \text{ uniformly on } [0, 1].$$

This is in fact how S. Bernstein discovered these polynomials.

We are now going to study a second type of application [23]. The context is as follows: we have a discrete set[11] D, in which we would like to produce subsets F satisfying a certain property (P), for example, of a functional or combinatorial nature. The bizarre nature of the property (P) makes it difficult to explicitly construct such an F. In certain cases, we can then resort to a random technique called the *Bourgain method of selectors* [120]. It consists of the following: let $(X_i)_{i \in D}$ be a family of independent Bernoulli random variables, indexed by D and defined on a certain space Ω:

$$\mathbf{P}(X_i = 1) = p_i, \ \mathbf{P}(X_i = 0) = 1 - p_i.$$

Having fixed ω in Ω, we then select a random set $F = F_\omega \subset D$ by keeping only those points of D for which $X_i(\omega) = 1$ and rejecting all the others:

$$F_\omega = \{i \in D / X_i(\omega) = 1\}.$$

Hopefully, for a judicious choice of p_i, adapted to the property (P) under consideration, the set F_ω will have this property for a set of values of ω with positive probability, and in particular for at least one ω! The following self-evident truth is thus of crucial importance to the method of selectors:

$$\text{if } A \subset \Omega \text{ and } \mathbf{P}(A) > 0, \text{ then } A \neq \varnothing.$$

Observe that

$$|F_\omega| = \sum_{i \in D} X_i(\omega),$$

thus if $|D| = n$ and if the parameter $p_i = p$ is independent of i, the cardinality of F_ω follows a binomial distribution $\mathcal{B}(n, p)$, which explains the importance of these distributions in this context. Moreover, if S_n is a binomial variable

---

[10] There is an underlying *uniform* weak law of large numbers.
[11] That is to say, without a topology, and in many cases countable.

$\mathcal{B}(n, p)$, this variable has expectation np and has a tendency to be *concentrated* around this expectation, as expressed for example by Tchebycheff's inequality

$$\mathbf{P}(|S_n - np| \geqslant \varepsilon) \leqslant \frac{np(1 - p)}{\varepsilon^2}.$$

This simple fact will be sufficient for us here, even if more precise inequalities of concentration exist, such as those of Bernstein [120].

We now give a definition: fix once and for all an *infinite and countable* set A, and $D = A^2$ the Cartesian product of A by itself. We would like to define [22] the *combinatorial dimension* of an infinite subset F of D, which will be a real number α between 1 and 2. For example, if $\Delta = \{(x, x), x \in A\}$ is the diagonal of D, we would like to say that Δ is a copy of A, so that it is of dimension 1, whereas $F = A \times A$ is of dimension 2. In the general case, we proceed as follows: if s is a positive integer, we call an *s-square* of D any subset of D of the form $Q = A_1 \times A_2$ with $A_j \subset A$ and $|A_j| = s$. We also set

$$\psi_F(s) = \sup_{Q \; s\text{-square}} |F \cap Q|.$$

The combinatorial dimension of F, $\dim F$, is defined to be the best α such that

$$\psi_F(s) \leqslant C s^\alpha \text{ for all } s \geqslant 1.$$

More precisely, we set

$$\overline{\dim} \, F = \varlimsup_{s \to +\infty} \frac{\ln \psi_F(s)}{\ln s} \text{ and } \underline{\dim} \, F = \varliminf_{s \to +\infty} \frac{\ln \psi_F(s)}{\ln s}.$$

The two quantities above are respectively called the upper and lower combinatorial dimensions of F. If they are equal, we denote by $\dim F$ their common value.

For $\alpha = \frac{3}{2}$, for example, we could give [22] explicit examples of F with upper dimension α, but these examples are already quite elaborate. For an arbitrary $\alpha \in \,]1, 2[$, the possibility of resorting to the method of selectors thrills us: we can then obtain the following theorem, whose proof is based on a key lemma in which h denotes a *determinant* function, that is, a function $h : \mathbb{R}^+ \to \mathbb{R}^+$ verifying

- h is increasing and convex,
- $h(0) = 0$,
- $\dfrac{h(s)}{s} \to +\infty$ as $s \to +\infty$,
- there exists an $s_1 \geqslant 0$ such that

$$h(s) \leqslant s^\alpha \text{ for } s \geqslant s_1.$$

We will then apply this lemma to the function $h(s) = s^\alpha$, but the result with an arbitrary h is useful in harmonic analysis [158].

5.4.1 Theorem [Körner–Blei]   *Let h be a determinant function. Then there exists an $F \subset D$ such that*

$$0 < \varlimsup_{s \to \infty} \frac{\psi_F(s)}{h(s)} < \infty.$$

In particular, for all $\alpha \in\ ]1, 2[$, there exists an $F \subset D$ such that $\overline{\dim} F = \alpha$.

The proof is based on the following key lemma.

5.4.2 Lemma   *Let M be a positive integer and A_0 a finite subset of A. Then there exists an integer $p \geqslant M$, a p-square $Q_p = U_p \times U_p \subset D$ with $U_p \cap A_0 = \varnothing$, and a subset F of Q_p such that*

(1) $|F| = \psi_F(p) \geqslant \dfrac{1}{4} h(p)$,
(2) $\psi_F(s) \leqslant Ch(s)$ for $s \geqslant 1$, where $C \geqslant 1$ depends only on h.

Proof   We proceed in two steps.

Step 1 (random). Using selectors we construct an integer $n \geqslant M$, an n-square Q_n with sides disjoint from A_0, and a subset G of Q_n such that

$$\psi_G(n) > \frac{1}{2} h(n) \text{ and } \psi_G(s) \leqslant Ch(s) \text{ for } 1 \leqslant s \leqslant M. \tag{5.11}$$

For this, fix $s_0 \geqslant 1$ such that

$$8s \leqslant h(s) \leqslant s^\alpha \leqslant \frac{s^2}{4} \text{ and } 2s - (2 - \alpha)h(s) \leqslant -1 \text{ if } s \geqslant s_0. \tag{5.12}$$

Even if it means increasing M, we can suppose that $M \geqslant s_0$, which we do in what follows. Then, let n be an integer such that

$$n \geqslant 4M2^{M^2}$$

(this choice will become clearer as we proceed). Fix U_n, a subset of A with cardinality n and disjoint from A_0, and set $Q_n = U_n \times U_n$. Consider selectors $(X_i)_{i \in Q_n}$ defined on Ω, with expectation $\dfrac{h(n)}{n^2}$ and thus with variance

$$V(X_i) = \mathbf{E}(X_i^2) - \mathbf{E}(X_i)^2 \leqslant \mathbf{E}(X_i^2) = \mathbf{E}(X_i) = \frac{h(n)}{n^2}.$$

For $\omega \in \Omega$, let G_ω be the set of $i \in Q_n$ such that $X_i(\omega) = 1$, and let

$$S(\omega) = \sum_{i \in Q_n} X_i(\omega)$$

be the cardinality of G_ω. We have

$$E(S) = \sum_{i \in Q_n} E(X_i) = n^2 \frac{h(n)}{n^2} = h(n),$$

and the Tchebycheff concentration inequality implies

$$\mathbf{P}(|S - h(n)| \geqslant n) =: \mathbf{P}(E_n) \leqslant \frac{V(S)}{n^2} = \frac{1}{n^2} \sum_{i \in Q_n} V(X_i) \leqslant \frac{h(n)}{n^2}.$$

Hence, using (5.12) and the fact that $n \geqslant s_0$:

$$\mathbf{P}(E_n) \leqslant \frac{1}{4}. \tag{5.13}$$

Now fix an s-square K_s contained in Q_n, with $s_0 \leqslant s \leqslant M$. The random variable

$$|G \cap K_s| : \omega \mapsto |G_\omega \cap K_s| = \sum_{i \in K_s} X_i(\omega)$$

follows a binomial distribution $\mathcal{B}\left(s^2, \dfrac{h(n)}{n^2}\right)$, so that

$$\mathbf{P}\left(|G \cap K_s| \geqslant h(s)\right) = \sum_{m=h(s)}^{s^2} \binom{s^2}{m} \left(\frac{h(n)}{n^2}\right)^m \left(1 - \frac{h(n)}{n^2}\right)^{s^2-m}$$

$$\leqslant \left(\frac{h(n)}{n^2}\right)^{h(s)} \sum_{m=h(s)}^{s^2} \binom{s^2}{m}$$

$$\leqslant \left(\frac{h(n)}{n^2}\right)^{h(s)} \sum_{m=0}^{s^2} \binom{s^2}{m}$$

$$= 2^{s^2} \left(\frac{h(n)}{n^2}\right)^{h(s)}.$$

For s fixed, the number of such s-squares contained in Q_n is at most

$$\binom{n}{s}^2 = \left(\frac{n(n-1)\times\cdots\times(n-s+1)}{s!}\right)^2 \leqslant \left(\frac{n^s}{s!}\right)^2 \leqslant n^{2s}.$$

As the probability of a union is less than or equal to the sum of the probabilities, we have

$$\mathbf{P}\left(\sup_{K_s \subset Q_n} \frac{|G \cap K_s|}{h(s)} \geqslant 1\right) \leqslant n^{2s} 2^{s^2} \left(\frac{h(n)}{n^2}\right)^{h(s)} \leqslant \frac{2^{M^2}}{n}.$$

Indeed, since $s_0 \leqslant s \leqslant M$, we have

$$n^{2s}2^{s^2}\left(\frac{h(n)}{n^2}\right)^{h(s)} \leqslant n^{2s}2^{M^2}n^{(\alpha-2)h(s)} = 2^{M^2}n^{2s-(2-\alpha)h(s)} \leqslant \frac{2^{M^2}}{n},$$

according to (5.12).

Note that if K_s is an arbitrary s-square and $\omega \in \Omega$, we have

$$G_\omega \cap K_s = (G_\omega \cap Q_n) \cap K_s = G_\omega \cap (K_s \cap Q_n) \subset G_\omega \cap K'_t,$$

where $t \leqslant s$ and where K'_t is a t-square contained in Q_n. It follows that

$$\mathbf{P}\left(\frac{\psi_G(s)}{h(s)} \geqslant 1\right) = \mathbf{P}\left(\sup_{K_s\ s\text{-square}} \frac{|G \cap K_s|}{h(s)} \geqslant 1\right) \leqslant \frac{2^{M^2}}{n}.$$

Finally, by letting s vary between s_0 and M, and because of the choice of n, we obtain

$$\mathbf{P}\left(\sup_{s_0 \leqslant s \leqslant M} \frac{\psi_G(s)}{h(s)} \geqslant 1\right) =: \mathbf{P}(E'_n) \leqslant \frac{M2^{M^2}}{n} \leqslant \frac{1}{4}. \tag{5.14}$$

If C_n denotes the complement of the event $E_n \cup E'_n$, then the inequalities (5.13) and (5.14) show that

$$\mathbf{P}(C_n) = 1 - \mathbf{P}(E_n \cup E'_n) \geqslant 1 - \mathbf{P}(E_n) - \mathbf{P}(E'_n) \geqslant \frac{1}{2}.$$

In particular, $C_n \neq \varnothing$! This allows us to fix $\omega_0 \in C_n$ and to consider the set G_{ω_0} defined above. We then see that

$$\psi_{G_{\omega_0}}(n) \geqslant |G_{\omega_0} \cap Q_n| = |G_{\omega_0}| = S(\omega_0) > h(n) - n \geqslant \frac{1}{2}h(n),$$

thanks to (5.12), since $n \geqslant s_0$. On the contrary, if $s_0 \leqslant s \leqslant M$, we have

$$\psi_{G_{\omega_0}}(s) \leqslant h(s)$$

thanks to (5.14), since $\omega_0 \notin E'_n$. Finally, if $1 \leqslant s < s_0$ and if K_s is an s-square, we have

$$|G_{\omega_0} \cap K_s| \leqslant |K_s| \leqslant s^2 \leqslant C_0 h(s), \text{ with } C_0 = \sup_{1 \leqslant s \leqslant s_0} \frac{s^2}{h(s)} < \infty.$$

The conditions (5.11) are thus satisfied with $G = G_{\omega_0}$ and $C = \max(1, C_0)$, which completes the proof of step 1.

Observe that, in this step, we made extremely sloppy estimations of the binomial coefficients, and one can reasonably question the efficiency of probabilities here. The response is as follows: we took a huge number n^2 of selectors X_i, independent by definition. This *independence*, along with the largeness of n, is indeed what allows us, in the estimations, to factor out a *large power* of a *small quantity*, that is $\left(\dfrac{h(n)}{n^2}\right)^{h(s)}$, and the presence of this term plays a decisive role. In other words, everything lies in the independence. . .

Step 2 (combinatorial). We correct the set $G = G_{\omega_0}$ obtained in step 1 to obtain a set $F \subset G$, such that

$$|F| = \psi_F(p) \geqslant \frac{1}{4}h(p) \text{ for at least one integer } p \geqslant M$$

and

$$\psi_F(s) \leqslant Ch(s) \text{ for all } s \geqslant 1.$$

For the convenience of the reader, we first rewrite the conditions (5.11) satisfied by G:

$$\psi_G(n) > \frac{1}{2}h(n) \text{ and } \psi_G(s) \leqslant Ch(s) \text{ for } 1 \leqslant s \leqslant M.$$

Denote by p the smallest integer $j \geqslant M$ so that $\psi_G(j) > \frac{1}{2}h(j)$; such an integer exists and is $\leqslant n$, by (5.11) and the choice of n. By definition of $\psi_G(p)$, there exists a p-square $K_p = A_p \times B_p$ such that $|G \cap K_p| > \frac{1}{2}h(p)$. Fix $(a, b) \in G \cap K_p$, and set

$$A'_p = A_p \setminus \{a\}, \ B'_p = B_p \setminus \{b\} \text{ and } F = G \cap (A'_p \times B'_p).$$

In other words, F is derived from G by removing from $G \cap K_p$ both the vertical fibre and the horizontal fibre that intersect at (a, b), thus removing at most $2p$ points from G. It is clear that $\psi_F(p) \leqslant |F|$. Furthermore,

$$\psi_F(p) \geqslant |F \cap K_p| = |F| \geqslant |G \cap K_p| - 2p > \frac{1}{2}h(p) - 2p \geqslant \frac{1}{4}h(p),$$

because $h(p) \geqslant 8p$ after (5.12), since $p \geqslant s_0$. In particular, $\psi_F(p) = |F|$. It remains to estimate $\psi_F(s)$ *for all* $s \geqslant 1$, which will be done in three steps:

- if $1 \leqslant s \leqslant M$, we have $\psi_F(s) \leqslant \psi_G(s) \leqslant Ch(s)$, after (5.11);
- if $M < s \leqslant p - 1$, we have $\psi_F(s) \leqslant \psi_G(s) \leqslant \frac{1}{2}h(s) \leqslant Ch(s)$, by definition of p;
- finally, if $s \geqslant p$ and if K_s is an s-square, we have

$$F \cap K_s \subset F = G \cap K_{p-1},$$

where K_{p-1} is the $(p-1)$-square $A'_p \times B'_p$. Then,

$$|F \cap K_s| \leqslant |G \cap K_{p-1}| \leqslant \frac{1}{2}h(p-1),$$

by definition of p, and

$$\psi_F(s) \leqslant \frac{1}{2}h(p-1) \leqslant \frac{1}{2}h(s).$$

The set F thus satisfies the conditions. This type of reasoning, where combinatorics and determinism intervene and lend a hand to randomness, is frequently encountered when using the method of selectors. $\qquad\square$

We now need to extend this *finite* result to an *infinite* result, which is done using a "gluing lemma". We first need the following definition: let π_1 and π_2 be the projections of D onto the coordinate axes, defined by $\pi_1(x, y) = x$ and $\pi_2(x, y) = y$. We say that the subsets F_j ($j \in \mathbb{N}^*$) of $D = A^2$ are *doubly disjoint* if we have the following property:

$$\pi_1(F_i) \cap \pi_1(F_j) = \varnothing \text{ and } \pi_2(F_i) \cap \pi_2(F_j) = \varnothing, \text{ if } i \neq j.$$

To go from Lemma 5.4.2 to the Körner–Blei Theorem 5.4.1, we use the following simple lemma.

5.4.3 Lemma *Let* $(F_j)_{j \geqslant 1}$ *be a sequence of doubly disjoint subsets of D, such that*

$$\psi_{F_j}(s) \leqslant Ch(s) \text{ for } s, j \geqslant 1,$$

and let $F = \bigcup_{j=1}^{\infty} F_j$. *Then,*

$$\psi_F(s) \leqslant 2Ch(s) \text{ for } s \geqslant 1.$$

Proof Let $Q = A_1 \times A_2$ be an s-square. We set

$$t_{1j} = |\pi_1(F_j) \cap A_1|, \quad t_{2j} = |\pi_2(F_j) \cap A_2| \text{ and } t_j = \max(t_{1j}, t_{2j}).$$

As the $\pi_1(F_j) \cap A_1$ are two-by-two disjoint and contained in A_1, we have

$$\sum_{j=1}^{\infty} t_{1j} \leqslant |A_1| = s,$$

and similarly $\sum_{j=1}^{\infty} t_{2j} \leqslant s$. Moreover, $F_j \cap Q$ is contained in a t_j-square, thus

$$|F_j \cap Q| \leqslant \psi_{F_j}(t_j) \leqslant Ch(t_j).$$

As $F \cap Q = \bigcup_{j \in \mathbb{N}^*}(F_j \cap Q)$, we have

$$|F \cap Q| \leqslant \sum_{j=1}^{\infty} |F_j \cap Q| \leqslant C \sum_{j=1}^{\infty} h(t_j)$$

$$\leqslant C \Big(\sum_{j=1}^{\infty} h(t_{1j}) + \sum_{j=1}^{\infty} h(t_{2j}) \Big)$$

$$\leqslant C \Big(h \Big(\sum_{j=1}^{\infty} t_{1j} \Big) + h \Big(\sum_{j=1}^{\infty} t_{2j} \Big) \Big) \leqslant 2Ch(s).$$

Indeed, h is super-additive, as it is convex and has value zero at zero.[12] As Q is arbitrary, we can conclude that $\psi_F(s) \leqslant 2Ch(s)$.   □

We can now give the proof of the Körner–Blei theorem.

Proof Lemma 5.4.2 allows us to recursively construct an increasing sequence $(p_j)_{j \geqslant 1}$ of positive integers and a sequence $(F_j)_{j \geqslant 1}$ of doubly disjoint sets such that

$$\psi_{F_j}(p_j) \geqslant \frac{1}{4} h(p_j) \text{ and } \psi_{F_j}(s) \leqslant Ch(s) \text{ for } s \geqslant 1.$$

In fact, having constructed $p_1, \ldots, p_j$ and $F_1, \ldots, F_j$, we use Lemma 5.4.2 with

$$M = 1 + p_j \text{ and } A_0 = \bigcup_{i=1}^{j} \left(\pi_1(F_i) \cup \pi_2(F_i) \right)$$

to produce an integer $p_{j+1} \geqslant M$ and a finite set $F_{j+1} \subset D$, doubly disjoint from $F_1, \ldots, F_j$, such that

$$\psi_{F_{j+1}}(p_{j+1}) \geqslant \frac{1}{4} h(p_{j+1}) \text{ and } \psi_{F_{j+1}}(s) \leqslant Ch(s) \text{ for } s \geqslant 1.$$

Then set $F = \bigcup_{j=1}^{\infty} F_j$. Lemma 5.4.3 shows that

$$\overline{\lim_{s \to \infty}} \frac{\psi_F(s)}{h(s)} \leqslant 2C,$$

whereas

$$\psi_F(p_j) \geqslant \psi_{F_j}(p_j) \geqslant \frac{1}{4} h(p_j),$$

hence

$$\overline{\lim_{s \to \infty}} \frac{\psi_F(s)}{h(s)} \geqslant \frac{1}{4}.$$

F is the set we were looking for, and it has upper combinatorial dimension α (that is to say, $\overline{\dim} F = \alpha$) if we take as determinant function the function $h(s) = s^\alpha$.   □

5.4.4 Remark The Körner–Blei theorem has specialised, nonetheless very interesting, applications to the theory of Fourier series, and notably to the p-Sidon sets. We refer the reader to the text [22], which treats many other aspects, in particular the fact that if $1 \leqslant \beta \leqslant \alpha < 2$, we can find F such that

---

[12] To see this, it suffices to combine, for $x, y > 0$, the inequalities $\dfrac{h(x)}{x} \leqslant \dfrac{h(x+y)}{x+y}$ and $\dfrac{h(y)}{y} \leqslant \dfrac{h(x+y)}{x+y}$.

$\underline{\dim} F = \beta$ and $\overline{\dim} F = \alpha$. More recently, Rodriguez-Piazza [158] gave an application to the theory of Sidon sets that more or less closed the study of the "grid" condition for these sets.

Exercises

5.1. Let p be a fixed real number such that $1 \leqslant p < 2$. Let U be the space of power series $f(z) = \sum_{n=0}^{\infty} a_n z^n$, uniformly convergent on the unit disk D, equipped with the norm

$$r(f) = \sup_{N \in \mathbb{N}, z \in D} \left| \sum_{n=0}^{N} a_n z^n \right|.$$

We want to show, by contradiction, that there exists an $f \in U$ such that

$$\sum_{n=0}^{\infty} |a_n|^p = +\infty.$$

We suppose the contrary.

(a) Show that (U, r) is a Banach space.

(b) Show that the map $T : U \to \ell_p$, defined by

$$T\left(\sum_{n=0}^{\infty} a_n z^n \right) = (a_n)_{n \geqslant 0},$$

has a closed graph, and is continuous. Thus there exists a constant $C > 0$ such that

$$\|T(f)\|_p \leqslant Cr(f) \text{ for all } f \in U.$$

(c) Show that this leads to a contradiction by using the Salem theorem of phases. Study the case $p = 2$.

5.2. Uchiyama's theorem [181]. Let N be an integer $\geqslant 1$. For $n \in \mathbb{Z}$, define $e_n \in L^1(\mathbb{T})$ by $e_n(t) = e^{int}$ for $t \in \mathbb{R}$.

(a) Show that there exist signs $\alpha_n = \pm 1$ such that

$$\left\| \sum_{n=1}^{N} \alpha_n e_n \right\|_1 \geqslant \frac{1}{\sqrt{2}} \sqrt{N}.$$

(b) Prove Uchiyama's theorem: there exists a subset A of $\{1, 2, \ldots, N\}$ such that $|A| \geqslant cN$ and

$$\left\| \sum_{n \in A} e_n \right\|_1 \geqslant c\sqrt{N},$$

where c is a numerical constant.

5.3. Multipliers of C in ℓ_1. Let C be the Banach space of continuous 2π-periodic functions from $\mathbb{R}$ to $\mathbb{C}$, equipped with the norm $\| \cdot \|_\infty$. We say that a complex sequence $m = (c_n)_{n \in \mathbb{Z}}$ is a *multiplier* of C if

$$\sum_{n \in \mathbb{Z}} |c_n \widehat{f}(n)| < \infty \text{ for all } f \in C.$$

(a) Show that any element of ℓ_2 is a multiplier of C.
(b) Conversely, let m be a multiplier of C.
   (i) By applying the closed graph theorem to $T : C \to \ell_1$ defined by

$$T(f) = (c_n \widehat{f}(n))_{n \in \mathbb{Z}},$$

show that there exists a constant $M > 0$ such that

$$\sum_{n \in \mathbb{Z}} |c_n \widehat{f}(n)| \leqslant M \|f\|_\infty \text{ for all } f \in C.$$

   (ii) By considering the adjoint operator[13] T^* of T, show (with the notations of Exercise 5.2), that

$$\left\| \sum_{|n| \leqslant N} c_n x_n e_n \right\|_1 \leqslant M \sup_{|n| \leqslant N} |x_n| \text{ for all } (x_n)_{|n| \leqslant N} \in \mathbb{C}^{2N+1},$$

where $M = \|T\| = \|T^*\|$.
   (iii) Conclude that $m \in \ell_2$.

5.4. Give an explicit example of a set $F \subset (\mathbb{N}^*)^2$, with combinatorial dimension $\frac{3}{2}$. We refer to [22]!

5.5. Let d be an integer $\geqslant 2$, and $1 < \alpha < d$. How can we modify the selectors of Section 5.4 to obtain a set $F \subset (\mathbb{N}^*)^d$, of combinatorial dimension α?

5.6. Gaussian model for the measure on the sphere. Let $S = S^{n-1}$ be the unit sphere of the Euclidean space $\mathbb{R}^n$, and μ its normalised area measure, that

---

[13] The operator T^* is the linear operator $(\ell_1)^* = \ell_\infty \to C^*$ defined by

$$(T^*x)(f) \overset{\text{denoted}}{=} \langle T^*x, f \rangle \overset{\text{defined}}{=} x(Tf) \overset{\text{denoted}}{=} \langle x, Tf \rangle = \sum_{n \in \mathbb{Z}} x_n c_n \widehat{f}(n)$$

for $f \in C$ and $x \in (\ell_1)^*$. One can show [118] that T^* is continuous and that $\|T^*\| = \|T\|$.

is to say, the unique Borel probability on S invariant under the action of the orthogonal group $O(n)$. We intend to estimate the integral

$$I_n = \int_S \Big(\sup_{1 \leqslant j \leqslant n} |x_j| \Big) d\mu(x), \text{ where } x = (x_1, \ldots, x_n) \in S.$$

(a) Let $G = (g_1, \ldots, g_n)$ be a standard Gaussian sequence. We set

$$\|G\| = \Big(\sum_{j=1}^{n} g_j^2 \Big)^{1/2}.$$

Show that

$$\int_S f \, d\mu = \mathbf{E}\Big(f\Big(\frac{G}{\|G\|}\Big) \Big) \text{ for all } f \in C(S).$$

(b) By using well-known estimates for Gaussian random variables (see [120], Chapter 8), show that

$$a\sqrt{\frac{\ln n}{n}} \leqslant I_n \leqslant b\sqrt{\frac{\ln n}{n}},$$

where $n \geqslant 2$ and a, b are positive constants.

5.7. We say that a Banach space X, with dual X^*, satisfies Schur's property if each *sequence* $(x_n)_{n \geqslant 0}$ of X weakly convergent to 0 (in the sense that $x^*(x_n) \to 0$ for all $x^* \in X^*$) converges in norm to 0: $\|x_n\| \to 0$. For example, ℓ_1 has this property [120].

(a) Show that Schur's property is passed on to subspaces and to isomorphic spaces.

(b) Show that ℓ_2 does not satisfy Schur's property.

(c) Show that ℓ_1 does not contain any infinite-dimensional subspace isomorphic to a Hilbert space.[14]

5.8. The Kottman constant. The *Kottman constant* of a real infinite-dimensional Banach space X, denoted by $K(X)$, is defined as the least upper bound of the set of real numbers $\alpha > 0$ for which there exists a sequence $(x_n)_{n \geqslant 1}$ of the unit sphere of X, with (x_n) α-distant, in the following sense:

$$\|x_i - x_j\| \geqslant \alpha \text{ if } i \neq j.$$

By Riesz's theorem, we always have $K(X) \geqslant 1$, and even [120] $K(X) > 1$.

(a) Show that $K(\ell_2) = \sqrt{2}$. *Hint:* prove and then use the inequality

$$\sum_{1 \leqslant i,j \leqslant N} \|x_i - x_j\|_2^2 \leqslant \sum_{1 \leqslant i,j \leqslant N} (\|x_i\|_2^2 + \|x_j\|_2^2) \text{ for } x_1, \ldots, x_N \in \ell_2.$$

---

[14] This result can also be proved [120] without using Schur's property for ℓ_1.

(b) Show that $K(\ell_1) = 2$. It can be shown [185] that $K(\ell_p) = 2^{1/p}$ if $1 \leqslant p < \infty$.

(c) Let $p > 2$. By using a Rademacher sequence on $[0, 1]$, show that $K(L^p([0, 1])) \geqslant 2^{1-1/p}$ (we have in fact equality, see [185]).

5.9. A theorem of Erdős by selectors [57]. A subset F of an Abelian additive group is said to be *sum-free* or *free* if, for all $(x, y, z) \in F^3$, $x + y \neq z$. Let

$$A = \{n_1 < \cdots < n_N\} \subset \mathbb{N}^* \subset \mathbb{Z}.$$

We intend to prove (a theorem of Erdős) that A contains a free subset F of cardinality greater than $\dfrac{N}{3}$.

(a) Show that there exists a prime number $p > n_N$, of the form $p = 3k + 2$.

(b) Let $S = \{k + 1, \ldots, 2k + 1\}$. Show that S is free in the additive group $\mathbb{F}_p = \mathbb{Z}/p\mathbb{Z}$, and that $|S| > \dfrac{p - 1}{3}$.

(c) Let t be a random variable uniformly distributed on $\mathbb{F}_p^*$, that is to say:

$$\mathbf{P}(t = j) = \frac{1}{p - 1} \text{ for all } j \in \mathbb{F}_p^*.$$

Set

$$X = X(t) = \sum_{j \in S} \mathbf{1}_{\{t^{-1}j \in A\}}.$$

Show that

$$\mathbf{E}(X) = \frac{N|S|}{p - 1} > \frac{N}{3}.$$

(d) Show that there exists a $t_0 \in \mathbb{F}_p^*$ such that $X(t_0) > \dfrac{N}{3}$, and that $F = A \cap (t_0^{-1} S)$ meets our requirements.

5.10. Let α be a real number such that $\dfrac{1}{2} < \alpha < 1$. We intend to construct a function $f : \mathbb{R} \to \mathbb{C}$, which is 2π-periodic and α-Hölderian,[15] such that

$$\sum_{n=0}^{\infty} n^{\alpha - \frac{1}{2}} |c_n| = +\infty,$$

where (c_n) denotes the sequence of Fourier coefficients of f.

---

[15] In the sense that there exists a $C > 0$ such that $|f(x) - f(y)| \leqslant C|x - y|^\alpha$ for $x, y \in \mathbb{R}$.

(a) Let k be a non-negative integer. Using Salem's theorem of phases, or the results of Chapter 9, construct a trigonometric polynomial f_k such that

$$f_k(x) = \sum_{2^k < n \leqslant 2^{k+1}} c_n e^{inx}, \quad |c_n| = 1, \quad \|f_k\|_\infty \leqslant C 2^{k/2} \sqrt{k+1},$$

where C is a numerical constant.

(b) Show that the function

$$f = \sum_{k=0}^{\infty} \frac{2^{-k(\alpha+1/2)}}{\sqrt{k+1}} f_k$$

answers the question (see Exercise 10.1 in Chapter 10).

5.11. We intend to prove a complement to Theorem 5.3.2.

(a) Let $(I_n)_{n \geqslant 1}$ be a sequence of non-trivial intervals of $[0, 1]$, pairwise disjoint. Denote by f_n the indicator function of I_n, and by Y the closed subspace of $L^1([0, 1])$ generated by the f_n, in other words, the closure of $\mathrm{Vect}(f_n)_{n \geqslant 1}$. Show that Y is isometric to ℓ_1.

(b) Let p be a real number > 2, and $(e_n)_{n \geqslant 1}$ the canonical basis of the Banach space ℓ_p ($e_n(k) = \delta_{n,k}$ for $k \geqslant 1$). By calculating

$$\mathbf{E}\left(\left\| \sum_{n=1}^{N} \varepsilon_n e_n \right\|_p^2 \right),$$

show that ℓ_p is not of cotype 2.

(c) Show that, if $p > 2$, the space $L^1([0, 1])$ does not contain any subspace isomorphic to ℓ_p.

6

The Hausdorff–Banach–Tarski paradoxes

6.1 Introduction

In April 1901, Henri Lebesgue published a note [119] in the *Comptes Rendus de l'Académie des Sciences* in which he defined the measure that now bears his name. This measure is a map λ, defined on the class $\mathcal{L}$ of subsets of $\mathbb{R}^d$ known as measurable,[1] with values in $[0, +\infty]$, and satisfying:

(1) $\lambda\left(\bigcup_{n=0}^{\infty} A_n\right) = \sum_{n=0}^{\infty} \lambda(A_n)$ for all *sequences* $(A_n)_{n \geqslant 0}$ of elements of $\mathcal{L}$ pairwise disjoint (σ-additivity);

(2) $\lambda([0, 1]^d) = 1$ (normalisation);

(3) $\lambda\big(g(A)\big) = \lambda(A)$ if $A \in \mathcal{L}$ and g is an isometry of $\mathbb{R}^d$ (invariance of the measure under the action of the group of isometries).

Naturally, if $d = 1$ (resp. 2, 3), λ is a formalisation of the notion of length (resp. surface, volume). But in the study that follows, it is interesting to place ourselves in a somewhat more general context: that of a group G operating on a non-empty set E. We denote by $\mathcal{P}(E)$ the set of all subsets of E. A positive measure μ on E, defined on a σ-algebra $\mathcal{T} \subset \mathcal{P}(E)$, is called G-*invariant* if

$$gA \in \mathcal{T} \text{ and } \mu(gA) = \mu(A) \text{ for } A \in \mathcal{T} \text{ and } g \in G,$$

when setting $gA = \{ga, a \in A\}$.

Paradoxically, Lebesgue's research, which required a very long time to catch on in France [10], was to immediately inspire the young Polish school, whose figurehead was Stefan Banach. Banach notably started to investigate the existence of a positive measure satisfying the three conditions above, while defined on the whole of $\mathcal{P}(\mathbb{R}^d)$ [12]. This led him, in collaboration with the logician

---

[1] For the definition of the Lebesgue σ-algebra, see [169].

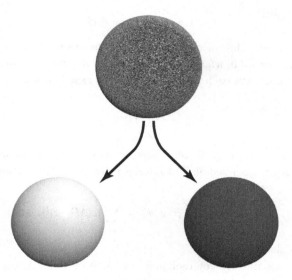

Figure 6.1

A. Tarski, to the discovery in 1924 of the curious phenomena that we will study in this chapter. Without spoiling the charm of the subject, we can say that these paradoxical phenomena only appear, in their most spectacular form, for dimension 3 and over (we will explain why). In dimension 3, they allow us, for example, to chop a closed ball into a finite number of pieces, and then, by reorganising the pieces of the puzzle, to obtain *two* copies of the initial ball – and then as many copies as we want (Figure 6.1). We shall even see that we can do much better.

In reality, the question of the existence of an exhaustive measure was solved in the negative as early as 1905 by Vitali: by doing this, he ruined any hope of a theory of measure ridden of its bristling statements about algebras, σ-algebras and monotone classes.

6.1.1 Proposition   (*1*) *There does not exist a probability measure defined on* $\mathcal{P}(\mathbb{T})$ (*where* $\mathbb{T}$ *is the unit circle of* $\mathbb{C}$) *that is invariant under rotations.*
(*2*) *There does not exist a positive measure* v *defined on* $\mathcal{P}(\mathbb{R})$, *invariant under translations and satisfying* $v([0, 1]) = 1$.

Proof   (1) Vitali's idea was to partition $\mathbb{T}$ into a countable number of sub-sets derived from one another by rotations. Let us call a *rational point* of the circle $\mathbb{T}$ any point of the form $e^{2i\pi x}$, with $x \in \mathbb{Q}$. The set of such points is a subgroup of $\mathbb{T}$, that we will call G. With G, we can associate an equivalence relation on $\mathbb{T}$ defined by

$$x \sim y \Leftrightarrow x^{-1}y \in G.$$

The equivalence class of $x \in \mathbb{T}$ is then xG. The axiom of choice, *whose use will prove crucial in what follows*, guarantees the existence of a subset M of $\mathbb{R}$ that intersects each equivalence class in exactly one point. It follows that

$$\mathbb{T} = \bigcup_{g \in G} (gM),$$

the gM being pairwise disjoint. Now, suppose that there exists a probability measure μ defined on $\mathcal{P}(\mathbb{T})$ that is invariant under rotations. As G is countable, we have

$$1 = \mu(\mathbb{T}) = \sum_{g \in G} \mu(gM) = \sum_{g \in G} \mu(M) \in \{0, +\infty\},$$

which is absurd.

(2) Again, reasoning by contradiction, we suppose that such a measure ν exists. We have the natural bijection at hand:

$$u : [0, 1[\to \mathbb{T}, \ x \mapsto e^{2i\pi x}.$$

For each subset A of $\mathbb{T}$, set $\mu(A) = \nu\big(u^{-1}(A)\big)$. It is clear that μ is a measure defined on $\mathcal{P}(\mathbb{T})$. Moreover, using the invariance under translation, we have, for all $n \geqslant 1$,

$$\nu\left(\left\{ \frac{1}{n} \right\}\right) = \nu(\{1\}),$$

and as $\nu([0, 1])$ is finite, this implies $\nu(\{1\}) = 0$ and thus

$$\mu(\mathbb{T}) = \nu([0, 1[) = \nu([0, 1]) = 1.$$

Finally, we verify that μ is invariant under rotations, which will provide the contradiction because of statement (1). If $\alpha \in [0, 1[$ and $A \subset \mathbb{T}$, by setting $B = u^{-1}(A)$ we have

$$
\begin{aligned}
\mu(e^{2i\pi\alpha} A) &= \mu(\{e^{2i\pi(\alpha+x)}, x \in B\}) \\
&= \mu(\{e^{2i\pi(\alpha+x)}, x \in B \cap [0, 1-\alpha[\}) \\
&\quad + \mu(\{e^{2i\pi(\alpha+x)}, x \in B \cap [1-\alpha, 1[\}) \\
&= \mu(\{e^{2i\pi x}, x \in (B+\alpha) \cap [\alpha, 1[\}) \\
&\quad + \mu(\{e^{2i\pi x}, x \in (B+\alpha-1) \cap [0, \alpha[\}) \\
&= \nu((B+\alpha) \cap [\alpha, 1[) + \nu((B+\alpha-1) \cap [0, \alpha[) \\
&= \nu(B \cap [0, 1-\alpha[) + \nu(B \cap [1-\alpha, 1[) \\
&= \nu(B) = \mu(A). \qquad \square
\end{aligned}
$$

6.1.2 Remarks (1) Thus there exist subsets of $\mathbb{R}$ that are not measurable in the sense of Lebesgue. More precisely, one can show that any Lebesgue-measurable subset of $\mathbb{R}$ with positive measure contains a non-measurable subset (see Exercises 6.1 and 6.3).

(2) Using a very interesting technique known as *Ulam matrices*, Ulam showed that there does not exist a non-zero measure defined on all subsets of $\aleph_1$ (smallest uncountable cardinal) without mass on points: we say that $\aleph_1$ is not a *measurable cardinal*. In the case of the continuum c (that is to say, the cardinality of $\mathbb{R}$), the situation is more complicated: whether the continuum is a measurable cardinal is independent of the axioms of Zermelo–Fraenkel augmented with the axiom of choice (ZFC). In other words, we can add to ZFC the axiom "c is a measurable cardinal", or its negation, and neither lead to a contradiction. On the contrary, as we have seen, it is a theorem within ZFC that there does not exist an extension of the Lebesgue measure to a measure defined on all subsets of $\mathbb{R}$ and *invariant under translations*.

6.2 Means

Vitali's theorem shows that for a map $\mu : \mathcal{P}(\mathbb{R}) \to [0, +\infty]$ such that $\mu([0, 1]) = 1$, we cannot impose *simultaneously* invariance under translations (*a fortiori*, under isometries) and σ-additivity. It is thus natural to relax this latter condition a bit, which leads to the following definitions: given a group G (whose law will be written multiplicatively) operating on a non-empty set E, a *finitely additive measure on E* is a map $\mu : \mathcal{P}(E) \to [0, +\infty]$ satisfying

$$\mu\left(\bigcup_{i=1}^{n} A_i\right) = \sum_{i=1}^{n} \mu(A_i),$$

for all $A_1, \ldots, A_n \subset E$ pairwise disjoint. Such a measure is said to be *G-invariant* if it additionally satisfies

$$\mu(gA) = \mu(A) \text{ for } g \in G \text{ and } A \subset E.$$

Finally, a *mean on E* is a finitely additive measure μ on E normalised by the condition $\mu(E) = 1$.

6.2.1 Interpretation in terms of linear functionals

We aim to establish some existence theorems on invariant means. To achieve this, first we interpret the means on E *in terms of positive linear functionals*, which will allow us to profit from the powerful artillery developed by Banach and his successors: functional analysis and specifically, duality.

Let $(\ell^\infty(E), \|\cdot\|_\infty)$ denote the Banach algebra of bounded functions from E to $\mathbb{R}$. Let φ be a linear functional on $\ell^\infty(E)$, positive (in the sense that $\varphi(f) \geqslant 0$ if $f \geqslant 0$) and normalised by $\varphi(1) = 1$, where 1 here is the function everywhere equal to 1. Such a linear functional is automatically continuous: indeed, if $f \in \ell^\infty(E)$, $\|f\|_\infty \pm f \geqslant 0$, thus

$$|\varphi(f)| \leqslant \varphi(1)\|f\|_\infty = \|f\|_\infty. \tag{6.1}$$

Moreover, the map

$$\mu_\varphi : \mathcal{P}(E) \to [0, 1], \quad A \mapsto \varphi(\mathbf{1}_A),$$

where $\mathbf{1}_A$ denotes the indicator function of A, defined by

$$\mathbf{1}_A(x) = \begin{cases} 1 & \text{if } x \in A, \\ 0 & \text{otherwise,} \end{cases}$$

is clearly a mean on E.

6.2.2 Theorem   *The map $\varphi \mapsto \mu_\varphi$ is a bijection from the set of positive normalised linear functionals on $\ell^\infty(E)$ onto the set of means on E.*

Proof   First we prove a density lemma to be used later to extend the linear functionals.

6.2.3 Lemma   *The set of step functions, that is to say, the elements of $\ell^\infty(E)$ that only take on a finite number of values, is dense in $\ell^\infty(E)$.*

Proof   Let $f \in \ell^\infty(E)$. Multiplying by a scalar if necessary, we can suppose that $\|f\|_\infty \leqslant 1$. For $n \geqslant 1$, set

$$f_n = \sum_{k=-n}^{n-1} \frac{k}{n} \mathbf{1}_{f^{-1}([k/n,(k+1)/n[)} + \mathbf{1}_{f^{-1}(\{1\})}.$$

It is clear that f_n is a step function, and that $\|f - f_n\|_\infty \leqslant \dfrac{1}{n}$, hence the result.   □

Now let μ be a mean on E, and $f : E \to \mathbb{R}$ a step function. Denote by $x_1, \ldots, x_r$ the values (pairwise distinct) taken on by f, and set

$$\varphi_\mu(f) = \sum_{i=1}^{r} x_i \mu\left(f^{-1}(\{x_i\})\right).$$

It is easy, though not entirely evident, to verify that φ_μ is a linear functional on the linear space of step functions from E to $\mathbb{R}$, positive – hence continuous – and such that $\varphi_\mu(1) = 1$. By Lemma 6.2.3, φ_μ extends uniquely to a continuous linear functional on $\ell^\infty(E)$ (still denoted by φ_μ), positive, and satisfying $\varphi_\mu(1) = 1$. If $f \in \ell^\infty(E)$, the real number $\varphi_\mu(f)$ can also be denoted by[2]

$$\int_E f \, d\mu = \int_E f(x) d\mu(x). \tag{6.2}$$

The two correspondences $M : \varphi \mapsto \mu_\varphi$ and $\Phi : \mu \mapsto \varphi_\mu$ that we have just described are inverses for one another. Indeed, for all $A \subset E$, we have

$$(\Phi \circ M)(\varphi)(\mathbf{1}_A) = \mu_\varphi(A) = \varphi(\mathbf{1}_A),$$

thus $\Phi \circ M = \mathrm{id}$, by linearity and Lemma 6.2.3. The computation of $M \circ \Phi$ is similar (even simpler). $\qquad\square$

We will now see how the eventual G-invariance of a mean μ affects the linear functional φ_μ. For this, we need the following useful notation: if $f \in \ell^\infty(E)$ and $g \in G$, we call $_g f$ the *translate* of f, defined by:[3]

$$_g f : E \to \mathbb{R}, x \mapsto f(g^{-1}x).$$

A continuous linear functional φ on $\ell^\infty(E)$ is said to be *G-invariant* if

$$\varphi(_g f) = \varphi(f) \text{ for } f \in \ell^\infty(E) \text{ and } g \in G.$$

Given the linearity and the continuity of φ, and Lemma 6.2.3, this is equivalent to

$$\varphi\big(_g(\mathbf{1}_A)\big) = \varphi(\mathbf{1}_A) \text{ for } A \subset E \text{ and } g \in G,$$

or again to

$$\varphi(\mathbf{1}_{gA}) = \varphi(\mathbf{1}_A) \text{ for } A \subset E \text{ and } g \in G,$$

---

[2] We are dealing here with a *finite* theory of integration, the difference from the usual theory being the absence of limit theorems (dominated convergence, etc.), which require the σ-additivity of measures.

[3] The exponent -1 is there in order to have $_h(_g f) = _{hg} f$.

hence finally

$$\mu_\varphi(gA) = \mu_\varphi(A) \text{ for } A \subset E \text{ and } g \in G.$$

Thus we obtain the following statement.

6.2.4 Proposition   *If μ is a mean on E, μ is G-invariant if and only if φ_μ is too.*

6.2.5 Remark   In general, the G-invariance of a mean μ on E can also be written

$$\int_E f(gx)d\mu(x) = \int_E f(x)d\mu(x), \tag{6.3}$$

for $g \in G$ and $f \in \ell^\infty(E)$.

6.2.6 Amenable groups

The illustrious American mathematician John von Neumann deserves recognition as having understood that the existence or not of invariant means depends much more on the properties of the group than those of the set on which it acts. For this, he built on ideas dating back to the *Erlangen program* of Félix Klein. The strategy that we will adopt is a good illustration of this point of view:[4] first construct invariant means *on the group G*, and then transport them to the set E on which G acts. More precisely, a group G is called *amenable* if, acting on itself by left translations, it admits a G-invariant mean.

6.2.7 Example   A finite group is amenable: it suffices to set, for all $A \subset G$,

$$\mu(A) = \frac{|A|}{|G|}.$$

The *transport of invariant means* goes smoothly, because of the following simple fact.

6.2.8 Proposition   *Let G be a group acting on a set E, and $f : G \to E$ a map such that*

$$f(gh) = gf(h) \text{ for } (g, h) \in G^2. \tag{6.4}$$

---

[4] To which we will often return in Section 6.3.

Then, if μ is a G-invariant mean on G, the pull-back *mean defined by*

$$\mu_f(X) = \mu\big(f^{-1}(X)\big) \, \text{for all } X \subset E$$

is a G-invariant mean on E.

Proof It is easy to show that μ_f is a mean on E. Let us verify its G-invariance: if $g \in G$ and $X \subset E$, we have

$$f^{-1}(gX) = \{h \in G/f(h) \in gX\} \overset{(6.4)}{=} \{h \in G/f(g^{-1}h) \in X\}$$
$$= \{h \in G/g^{-1}h \in f^{-1}(X)\} = gf^{-1}(X). \qquad \square$$

But then,

$$\mu_f(gX) = \mu\big(f^{-1}(gX)\big) = \mu\big(gf^{-1}(X)\big) = \mu\big(f^{-1}(X)\big) = \mu_f(X).$$

Here is, among many others, an application of this result. Let H be a normal subgroup of G. The group G acts on the quotient G/H by left translations ($g\,\overline{h} := \overline{gh}$), and the canonical surjection $p : G \to G/H$ satisfies $p(gh) = gp(h)$ for $(g, h) \in G^2$. The following result is thus evident.

6.2.9 Proposition *Any quotient of an amenable group is amenable.*

As for the following two results,[5] we refer to Exercises 6.7 and 6.8.

6.2.10 Proposition *Any subgroup of an amenable group is amenable.*

6.2.11 Proposition *Let G be a group and N a normal subgroup of G. If N and G/N are amenable, then so is G.*

These three propositions show that amenability is an *algebraically stable property*. Here is a more difficult result.

6.2.12 Theorem *An abelian group is amenable.*

Proof We use the following two lemmas, of interest by themselves; we have no choice but to state them in the context of topologies that are not necessarily metrisable.

A brief appendix on topological vector spaces has been added to this chapter (see Section 6.5) in order to make the following developments easier to understand. For a few succinct reminders on the weak-* topology, we refer to Chapter 11 of this book, or to [31] for more detailed explanations.

---

[5] The first makes use of the axiom of choice.

6.2.13 Lemma [Kakutani fixed point theorem]   *Let K be a non-empty convex compact subset of a topological vector space[6] X, and $T : K \to K$ an affine and continuous map. Then, T admits a fixed point.*

Proof Let us fix $a \in K$, and set

$$u_n = \frac{1}{n} \sum_{k=0}^{n-1} T^k(a) \text{ for } n \geqslant 1.$$

As K is convex and $T : K \to K$, $u_n \in K$. Moreover,

$$T(u_n) - u_n = \frac{1}{n} \left(T^n(a) - a \right),$$

with $T^n(a) - a \in K - K$, $K - K$ being a compact subset of X. If we fix a neighbourhood V of 0, by Proposition 6.5.1 of Section 6.5, *there exists an integer $N \geqslant 0$ such that $T(u_n) - u_n \in V$ for all $n \geqslant N$.*

Moreover, as K is compact, the cluster set of $(u_n)_{n \geqslant 0}$ is non-empty:

$$\bigcap_{n \geqslant 0} \overline{\{u_k, k \geqslant n\}} \neq \varnothing.$$

This means that if we take a neighbourhood W of ℓ, the set of integer n such that $u_n \in W$ is infinite. As T is continuous, for every neighbourhood W' of $T(\ell) - \ell$, *the set of integers n such that $T(u_n) - u_n \in W'$ is infinite.*

As X is Hausdorff, the two assertions in italics give $T(\ell) - \ell = 0$. $\square$

6.2.14 Lemma   *We keep the same notations. Let $\mathcal{A}$ be a collection of continuous affine maps from K to itself, pairwise commuting. Then the elements of $\mathcal{A}$ have a common fixed point.*

Proof If $T \in \mathcal{A}$, call F_T the set of fixed points of T. The set F_T is non-empty by Lemma 6.2.13, convex as T is affine, and compact because closed in K. Moreover, as $S \circ T = T \circ S$ for all $S \in \mathcal{A}$, the set F_T is stable by all the elements of $\mathcal{A}$. Finally, by induction and thanks to Lemma 6.2.13, the F_T, for $T \in \mathcal{A}$, have the finite intersection property.[7] By the compactness of K,

$$\bigcap_{T \in \mathcal{A}} F_T \neq \varnothing.$$ $\square$

We return to the proof of Theorem 6.2.12, and denote by $\mathcal{K}$ the set of positive linear functionals φ on $\ell^\infty(G)$ satisfying $\varphi(1) = 1$. By the inequality (6.1), for all $f \in \ell^\infty(G)$, we have

$$|\varphi(f)| \leqslant \|f\|_\infty.$$

[6] Not necessarily locally convex.

[7] Which means that every *finite* collection of sets F_T, $T \in \mathcal{A}$, has a non-empty intersection.

The set $\mathcal{K}$ is thus contained in the closed unit ball of the (strong) dual space $\ell^\infty(G)'$ of $\ell^\infty(G)$, and is weak-* closed. By Alaoglu's theorem, the set $\mathcal{K}$, equipped with the topology induced by the weak-* topology on $\ell^\infty(G)'$, is *compact*. Moreover, $\mathcal{K}$ is *non-empty* (because the evaluation maps are in $\mathcal{K}$) and evidently *convex*.

Now fix $g \in G$. For $\varphi \in \mathcal{K}$ and $f \in \ell^\infty(G)$, define

$$T_g\varphi : \ell^\infty(G) \to \mathbb{R}, \quad f \mapsto \varphi(f_g),$$

where[8] f_g denotes the function $G \to G, h \mapsto f(gh)$. It is clear that $T_g\varphi \in \mathcal{K}$. The map

$$T_g : \mathcal{K} \to \mathcal{K}, \quad \varphi \mapsto T_g\varphi$$

is thus available.

Taking into account Theorem 6.2.2 and Proposition 6.2.4, *showing the existence of a G-invariant mean on G is exactly the same as showing the existence of a common fixed point of the T_g, for g in G*. However, each T_g is affine and continuous. Let us justify this latter point: by the properties of the weak-* topology, it suffices to show that, if $f \in \ell^\infty(G)$ is fixed, the map

$$\mathcal{K} \to \mathbb{R}, \quad \varphi \mapsto T_g(\varphi)(f) = \varphi(_g f)$$

is continuous. But this is true, simply by the definition of the weak-* topology.

Moreover, as G is abelian, the T_g ($g \in G$) commute pairwise. By Lemma 6.2.14, the intersection of the F_{T_g}, $g \in G$, is non-empty, which completes the proof. □

6.2.15 Remark We call a *semigroup* any non-empty set equipped with an associative binary operation (also called a *law*) and a neutral element called a *unit*. A familiar example of a semigroup is $(\mathbb{N}, +)$. The proof of Theorem 6.2.12 can be extended word-for-word to the case of semigroups; it shows the existence, for every semigroup S *whose law is commutative*, of a positive linear functional L on $\ell^\infty(S)$ such that $L(1) = 1$ and S-invariant in the sense that

$$L(f_g) = L(f) \text{ for all } g \in S \text{ and } f \in \ell^\infty(S).$$

We will study an application of Theorem 6.2.12 to a semigroup in Exercise 6.12.

---

[8] We use f_g rather than $_g f$ to avoid the appearance of inverses, because we aim for a generalisation of Theorem 6.2.12 to semigroups (see Remark 6.2.15).

Finally, recall that a group G is said to be *solvable* if there exists a finite sequence

$$\{1\} = G_0 \subset G_1 \subset \cdots \subset G_{p-1} \subset G_p = G$$

of subgroups G_i of G, where each G_i is normal in G_{i+1}, such that all the quotients G_{i+1}/G_i are abelian. As G_0 is evidently amenable, Theorem 6.2.12 and Proposition 6.2.11 give the following corollary immediately.

6.2.16 Corollary   *Any solvable group is amenable.*

6.2.17 The problem of a complete extension of Lebesgue measure

We now tackle a problem close to the one we treated at the beginning of this chapter: the existence of means invariant under isometries (or by a subgroup of the group of isometries), and *defined on all subsets of* $\mathbb{R}^d$. For the smallest dimensions ($d = 1$ or 2), the solution to this problem follows immediately from the preceding study.

6.2.18 Theorem   *For $d \in \{1, 2\}$, the group $Is(\mathbb{R}^d)$ of affine isometries of $\mathbb{R}^d$ is solvable, thus amenable. Consequently, there exists a mean $Is(\mathbb{R}^d)$-invariant defined on $\mathcal{P}(\mathbb{R}^d)$. More generally, given an arbitrary $d \geqslant 1$, if G is an amenable subgroup of $Is(\mathbb{R}^d)$, there exists a G-invariant mean on $\mathbb{R}^d$.*

Proof   We consider the sequence

$$\{\mathrm{id}_{\mathbb{R}^d}\} \subset \mathcal{T}_d \subset \mathrm{Is}^+(\mathbb{R}^d) \subset \mathrm{Is}(\mathbb{R}^d),$$

where $\mathcal{T}_d$ denotes the group of translations of $\mathbb{R}^d$ and $\mathrm{Is}^+(\mathbb{R}^d)$ that of direct isometries of $\mathbb{R}^d$. Moreover, $\mathrm{Is}(\mathbb{R}^d)/\mathrm{Is}^+(\mathbb{R}^d)$ is isomorphic to $\mathbb{Z}/2\mathbb{Z}$, $\mathrm{Is}^+(\mathbb{R}^d)/\mathcal{T}_d$ to $SO(\mathbb{R}^d)$, an abelian group because $d \leqslant 2$, and finally $\mathcal{T}_d$ is abelian. The group $\mathrm{Is}(\mathbb{R}^d)$ is hence solvable, thus amenable (Corollary 6.2.16). By Proposition 6.2.8 (applied to $f : \mathrm{Is}(\mathbb{R}^d) \to \mathbb{R}^d, u \mapsto u(e)$, e a fixed vector of $\mathbb{R}^d$), there exists a mean that is $\mathrm{Is}(\mathbb{R}^d)$-invariant on $\mathbb{R}^d$. The last assertion of the theorem can be established similarly.   □

Clearly, the means that we have just constructed have nothing to do with Lebesgue measure: they have total mass 1. This leads to another question: is it possible to *extend* the Lebesgue measure to a *finitely* additive measure, invariant under isometries, and defined on all subsets of $\mathbb{R}^d$?

Our previous work shows that in reality this equates to a problem of *extension of linear functionals invariant under the action of a group*.

The appropriate tool is thus the Hahn–Banach theorem, which we will give in a G-invariant version.

Before the statement, here is a bit of terminology: let G be a group, and V a linear space. An *action* of G over V is by definition a group homomorphism from G to the group $\mathfrak{S}(V)$ of permutations of V. A *linear action* of G over V, also called a *representation of G*, is an action of G over V, such that, for each $g \in G$, the map

$$V \to V, \ x \mapsto gx$$

is linear; or, if you prefer, a group homomorphism from G to the group of linear automorphisms of V.

6.2.19 Theorem *Let G be an amenable group, acting linearly on a real linear space V, φ_0 a G-invariant linear functional[9] defined on a G-invariant linear subspace[10] V_0 of V, and $p : V \to \mathbb{R}$ a function satisfying the following conditions:*

- $p(x + y) \leqslant p(x) + p(y)$ *for* $(x, y) \in V^2$;
- $p(\lambda x) = \lambda p(x)$ *for* $\lambda \in \mathbb{R}_+$ *and* $x \in V$;
- $p(gx) = p(x)$ *for* $(g, x) \in G \times V$;
- $\varphi_0(x) \leqslant p(x)$ *for* $x \in V_0$.

Then φ_0 admits a linear extension $\varphi : V \to \mathbb{R}$, G-invariant and satisfying $\varphi(x) \leqslant p(x)$ for all $x \in V$.

Proof Fix μ a G-invariant mean on G. The Hahn–Banach theorem painlessly provides a linear extension $\Phi : V \to \mathbb{R}$ of φ_0 such that $\Phi(x) \leqslant p(x)$ for all $x \in V$. The problem is that this extension has no reason to be G-invariant! To fix this, we use a well-known method:[11] *we average the translates of Φ over G.* More precisely, set $v \in V$. For $h \in G$, we have

$$(_h\Phi)(v) = \Phi(h^{-1}v) \leqslant p(h^{-1}v) = p(v)$$

and

$$-(_h\Phi(v)) = -\Phi(h^{-1}v) = \Phi(h^{-1}(-v)) \leqslant p(-v),$$

---

[9] That is to say, such that $\varphi_0(gx) = \varphi_0(x)$ for $(g, x) \in G \times V_0$.

[10] That is to say, such that $gx \in V_0$ for $(g, x) \in G \times V_0$.

[11] Called *von Neumann's unitarian trick*.

since the action of G over V is linear. The function $h \mapsto (_h\Phi)(v)$ is thus an element of $\ell^\infty(G)$, which allows us to define:[12]

$$\varphi(v) = \int_G (_h\Phi)(v)d\mu(h) = \int_G \Phi(h^{-1}v)d\mu(h).$$

It is clear that φ is a linear functional on V. Let us verify that φ still extends φ_0: if $v \in V_0$, we have

$$\varphi(v) = \int_G \Phi(h^{-1}v)d\mu(h) = \int_G \varphi_0(h^{-1}v)d\mu(h) = \int_G \varphi_0(v)d\mu(h) = \varphi_0(v),$$

since V_0 and φ_0 are G-invariant. Moreover, for $g \in G$ and $v \in V$,

$$\begin{aligned}
_g\varphi(v) &= \varphi(g^{-1}v) \\
&= \int_G \Phi(h^{-1}g^{-1}v)d\mu(h) \\
&= \int_G \Phi((gh)^{-1}v)d\mu(h) \\
&= \int_G \Phi(h^{-1}v)d\mu(h) \text{ since } \mu \text{ is } G\text{-invariant (see (6.3))} \\
&= \varphi(v),
\end{aligned}$$

which proves that φ is G-invariant. Finally, for $v \in V$, we have

$$\varphi(v) \leqslant \int_G p(h^{-1}v)d\mu(h) = \int_G p(v)d\mu(h) = p(v),$$

which completes the proof. $\square$

We can now state the following fundamental theorem.

6.2.20 Theorem *Let G be an amenable subgroup of the group of isometries of $\mathbb{R}^d$. Then the Lebesgue measure admits a finitely additive and G-invariant extension to the set of all subsets of $\mathbb{R}^d$.*

Proof Let us denote by $V_0 = L^1(\mathbb{R}^d)$ the linear space of Lebesgue-integrable functions $\mathbb{R}^d \to \mathbb{R}$ and by V the linear space of functions $f : \mathbb{R}^d \to \mathbb{R}$ such that there exists a $u \in V_0$ with $|f| \leqslant u$. Let us define

$$\varphi_0(f) = \int_{\mathbb{R}^d} f(x)\, dx \text{ for } f \in V_0,$$

and

$$p(f) = \inf\left\{ \int_{\mathbb{R}^d} u(x)\, dx, u \in V_0, f \leqslant u \right\} \text{ for } f \in V.$$

[12] See p. 153 for the definition.

The group G acts linearly on V: setting

$$gf := {}_g f : \mathbb{R}^d \to \mathbb{R}, x \mapsto f(g^{-1}x) \text{ for } g \in G \text{ and } f \in V,$$

we can immediately verify that the hypotheses of Theorem 6.2.19 are all satisfied. Thus the linear functional φ_0 admits a linear extension φ to V, G-invariant and dominated by p. Then, let us define, for $A \subset \mathbb{R}^d$:

$$\mu(A) = \begin{cases} \varphi(\mathbf{1}_A) & \text{if } \mathbf{1}_A \in V, \\ +\infty & \text{otherwise.} \end{cases}$$

It is clear that μ extends the Lebesgue measure. Moreover, if the subsets $A_1, \ldots, A_n \subset \mathbb{R}^d$ are pairwise disjoint, and if $A := \bigcup_{k=1}^{n} A_k$, then

$$\mathbf{1}_A \in V \text{ if and only if } \mathbf{1}_{A_k} \in V \text{ for } 1 \leqslant k \leqslant n,$$

because

$$\mathbf{1}_A = \sum_{k=1}^{n} \mathbf{1}_{A_k} \text{ and } \mathbf{1}_{A_k} \leqslant \mathbf{1}_A \text{ for } 1 \leqslant k \leqslant n.$$

In any case, we thus have

$$\mu\left(\bigcup_{k=1}^{n} A_k\right) = \sum_{k=1}^{n} \mu(A_k),$$

which proves that μ is finitely additive.

Let us verify that μ is positive: let $A \subset \mathbb{R}^d$ satisfying $\mathbf{1}_A \in V$ (otherwise the result is evident). As $-\mathbf{1}_A \leqslant 0$ and $0 \in V_0$, we have

$$-\mu(A) = -\varphi(\mathbf{1}_A) = \varphi(-\mathbf{1}_A) \leqslant p(-\mathbf{1}_A) \leqslant \int_{\mathbb{R}^d} 0 \, dx = 0,$$

so that $\mu(A) \geqslant 0$. Finally, μ is G-invariant, mainly because so are V and φ, and $\mathbf{1}_{gA} = {}_g\mathbf{1}_A$. $\qquad\square$

6.2.21 Corollary *For $d \in \{1, 2\}$, the Lebesgue measure on $\mathbb{R}^d$ admits an extension to all subsets of $\mathbb{R}^d$ that is finitely additive and invariant under all isometries.*

Proof We have already shown that $\mathrm{Is}(\mathbb{R}^d)$ is amenable $\qquad\square$

6.2.22 Corollary *Let $d \geqslant 1$. If G is an amenable subgroup of $\mathrm{Is}(\mathbb{R}^d)$, the Lebesgue measure on $\mathbb{R}^d$ admits an extension to all subsets of $\mathbb{R}^d$ that is finitely additive and G-invariant. This is the case, for example, if G is the group of translations.*

Proof The group of translations of $\mathbb{R}^d$ is abelian, hence amenable. $\qquad\square$

6.3 Paradoxes

In this section, we are going to explain two paradoxical constructions to which we alluded in Section 6.1. The first [88], dating back to 1914, is due to Hausdorff, and is commonly known as the "duplication of the sphere". The second [14], quite spectacular, constitutes the Banach–Tarski paradox (1924). These paradoxes allow us to answer the question of the existence of an extension of the Lebesgue measure to all subsets of $\mathbb{R}^d$ for $d \geqslant 3$, that is, finitely additive and invariant under isometries.

To understand these paradoxes, we introduce the key notion of *equidecomposability*. In all that follows, G will denote a group (most of the time we could think of it as the group of isometries of $\mathbb{R}^d$) acting on a non-empty set E. The symbol $\sqcup$ will represent a union of *pairwise disjoint* subsets. Two subsets A and B of E are said to be G-congruent (or congruent) if there exists a $g \in G$ such that $B = gA = \{ga, \, a \in A\}$. We also write

$$A \underset{G}{\equiv} B.$$

Of course, G-congruence is an equivalence relation. The subsets A and B are called *equidecomposable* if they are *piecewise congruent*, in other words, if there exist finite decompositions[13]

$$A = \bigsqcup_{i=1}^{n} A_i \text{ and } B = \bigsqcup_{i=1}^{n} B_i,$$

with $A_i, B_i \subset E$ and $A_i \underset{G}{\equiv} B_i$ for $1 \leqslant i \leqslant n$. In this case, we write

$$A \underset{G}{\sim} B.$$

We leave it to the reader to verify that equidecomposability is an equivalence relation on the set of all subsets of E.

6.3.1 Example The group $\mathbb{T}$ of complex numbers with modulus 1 acts on itself by left translations. Let us show that

$$\mathbb{T} \setminus \{1\} \underset{\mathbb{T}}{\sim} \mathbb{T}.$$

For this, let us fix $z \in \mathbb{T}$, of infinite order,[14] and set $D = \{z^n, n \geqslant 0\}$. Clearly, $zD = D \setminus \{1\}$. Hence,

$$\mathbb{T} = (\mathbb{T} \setminus D) \sqcup D \underset{\mathbb{T}}{\sim} (\mathbb{T} \setminus D) \sqcup zD = (\mathbb{T} \setminus D) \sqcup (D \setminus \{1\}) = \mathbb{T} \setminus \{1\}.$$

---

[13] As the use of the notation $\sqcup$ suggests, the A_i are pairwise disjoint, as are the B_i.

[14] That is, $z^n \neq 1$ for all $n \geqslant 1$.

We will reuse the Hilbert hotel-type idea (whereby we "shift over in order to make room") in the explanation of the Banach–Tarski paradox.

A subset A of E is called *G-paradoxical* if there exist two *disjoint* subsets A_1 and A_2 of A such that $A \underset{G}{\sim} A_1$ and $A \underset{G}{\sim} A_2$. In a way, it means that A contains two disjoint copies of A. A special case will be very useful: a group G is said to be *paradoxical* if it is so when acting on itself by left translations. The following observations are very simple, but crucial.

- If there exists a finitely additive and G-invariant measure μ on E, and if $A \underset{G}{\sim} B$, then $\mu(A) = \mu(B)$.
- If X is a G-paradoxical subset of E, there does not exist any finitely additive and G-invariant measure μ on E such that $\mu(X) = 1$.

It is remarkable that the converse is true. This is a difficult result of Tarski showing that, in a way, *the existence of invariant measures and the absence of paradoxes are consubstantial.*

6.3.2 Theorem [Tarski's alternative] *Let G be a group acting on a set E, and X a subset of E. Then:*

- *either X is G-paradoxical,*
- *or there exists a finitely additive and G-invariant measure μ on E such that $\mu(X) = 1$.*

For the proof of the hard part of this theorem, that is, the existence of a finitely additive and G-invariant measure μ on the set E such that $\mu(X) = 1$ when X is not G-paradoxical, see [163, 183].

- The existence of a finitely additive and G-invariant measure on the set E does not at all exclude the existence of paradoxical subsets of E (see Exercise 6.6). Simply, the measure of a paradoxical subset will then be zero or infinity.

6.3.3 The paradox of the sphere

We denote by $SO(3)$ the group of linear rotations of $\mathbb{R}^3$, in other words, the set of linear isometries with determinant 1 of the Euclidean space $\mathbb{R}^3$. Its unit, the identity of $\mathbb{R}^3$, will be denoted by 1.

6.3.4 Theorem [Hausdorff, 1914 [88]] *The unit sphere of $\mathbb{R}^3$ is $SO(3)$-paradoxical.*

Following the method introduced by Klein and von Neumann, the proof will consist of constructing a paradoxical subset of the group $SO(3)$, which we will then transport onto the sphere.

We start with a few rudiments on free groups. Let a and b be two elements of a group G. We call an *alphabet* the set

$$\mathcal{A} = \{a, b, a^{-1}, b^{-1}\}.$$

Given $n \geqslant 1$, we call a *reduced word of length n built on the alphabet $\mathcal{A}$* every element of G of the form $a_1 \cdots a_n$, with $a_i \in \mathcal{A}$ such that $a_i a_{i+1} \neq 1$ for $1 \leqslant i \leqslant n - 1$ (in other words, there is no evident simplification within the word). By convention, the unit 1 of G is the reduced word with length zero, called the *empty word*. It is clear that the elements of the subgroup $\langle a, b \rangle$ of G generated by a and b are exactly the reduced words built on the alphabet $\mathcal{A}$. The subgroup of G generated by a and b is said to be *free of rank 2* if the following two conditions are satisfied:

 (i) the alphabet $\mathcal{A}$ has four distinct elements;
(ii) every reduced word of length $n \geqslant 1$ built on $\mathcal{A}$ is distinct from 1.

Basically, it means that there is no non-trivial relation between a and b, and it implies that the subgroup $\langle a, b \rangle$ is *as little commutative as possible*.

6.3.5 Proposition *A free group of rank 2 is paradoxical.*

Proof We keep the preceding notations. For any $g \in \mathcal{A}$, let $I(g)$ be the set of reduced words that begin with the letter g. The free nature of $\langle a, b \rangle$ immediately implies the following two facts:

• if $g \neq h$, then $I(g) \cap I(h) = \varnothing$;
• $\langle a, b \rangle = \{1\} \sqcup I(a) \sqcup I(a^{-1}) \sqcup I(b) \sqcup I(b^{-1})$.

Moreover, the decomposition

$$I(a) \sqcup I(a^{-1}) = a\{1\} \sqcup aI(a) \sqcup aI(b) \sqcup aI(b^{-1}) \sqcup I(a^{-1})$$

proves that

$$I(a) \sqcup I(a^{-1}) \underset{\langle a,b \rangle}{\sim} \langle a, b \rangle.$$

Similarly, $I(b) \sqcup I(b^{-1}) \underset{\langle a,b \rangle}{\sim} \langle a, b \rangle$. Thus we have identified two disjoint subsets of $\langle a, b \rangle$, both equidecomposable to $\langle a, b \rangle$. □

We can now state the following fundamental technical result.

6.3.6 Theorem *The group $SO(3)$ contains a free subgroup of rank 2, and thus is paradoxical.*

6.3.7 Remark One can show [183] that when $SO(3)$ is equipped with its usual topology, the set of couples $(a, b) \in SO(3)^2$ that generate a free subgroup is a residual (hence dense) subset of $SO(3)$.

Proof It is possible [183] to give a direct construction of such a subgroup, but with the inconvenience of involving complicated calculations. We will use an indirect, but simpler, method: it consists of judiciously constructing a subgroup of $SO(3)$ that is not free, but in which we can easily find the desired subgroup. Hausdorff's original argument, much in the same line of thought, used two rotations of angles $\frac{2\pi}{3}$ and π, with the angle θ between their axes chosen in such a way that $\cos(2\theta)$ is transcendental.

More recently, Osofsky and Adams [140] gave a simpler example in which $\theta = \frac{\pi}{4}$: let (e_1, e_2, e_3) be the canonical basis of $\mathbb{R}^3$, and consider the two rotations with angles π and $\frac{2\pi}{3}$, respectively, and axes directed by $\frac{e_1 + e_3}{\sqrt{2}}$ and e_3, respectively:

$$\varphi = \begin{bmatrix} 0 & 0 & 1 \\ 0 & -1 & 0 \\ 1 & 0 & 0 \end{bmatrix} \text{ and } \psi = \begin{bmatrix} -1/2 & -\sqrt{3}/2 & 0 \\ \sqrt{3}/2 & -1/2 & 0 \\ 0 & 0 & 1 \end{bmatrix}.$$

Of course, $\varphi^2 = \psi^3 = 1$: these non-trivial relations show that the subgroup $\langle \varphi, \psi \rangle$ of $SO(3)$ generated by φ and ψ is not free. We will nonetheless show that it is *almost free*, meaning that φ and ψ satisfy within $\langle \varphi, \psi \rangle$ *only* the relations that they satisfy separately, that is, $\varphi^2 = \psi^3 = 1$. More precisely, we can make the following statement.

6.3.8 Proposition *If $p \geqslant 1$, $\varepsilon \in \{0, 1\}$, $\varepsilon' \in \{-1, 0, 1\}$ and $\varepsilon_i = \pm 1$, then*

$$\varphi^\varepsilon \psi^{\varepsilon_1} \varphi \psi^{\varepsilon_2} \varphi \cdots \psi^{\varepsilon_p} \varphi \psi^{\varepsilon'} \neq 1. \tag{6.5}$$

Proof Consider a word w of type (6.5), and let us show that $w \neq 1$. Conjugating, if necessary, w by ψ, we can suppose that w ends with φ. We thus have two types of word to investigate:

$$w = \psi^{\varepsilon_1} \varphi \psi^{\varepsilon_2} \varphi \cdots \psi^{\varepsilon_p} \varphi \tag{6.6}$$

and

$$w = \varphi \psi^{\varepsilon_1} \varphi \psi^{\varepsilon_2} \varphi \cdots \psi^{\varepsilon_p} \varphi, \tag{6.7}$$

where $\varepsilon_1, \ldots, \varepsilon_p \in \{-1, 1\}$. We start by examining the form of a word of type (6.6). First of all, a simple calculation shows that

$$
\psi^\varepsilon \varphi = \begin{bmatrix} 0 & \varepsilon\sqrt{3}/2 & -1/2 \\ 0 & 1/2 & \varepsilon\sqrt{3}/2 \\ 1 & 0 & 0 \end{bmatrix} \tag{6.8}
$$

for $\varepsilon = \pm 1$. We then show by induction on $p \geqslant 1$ that

$$
\psi^{\varepsilon_1}\varphi\psi^{\varepsilon_2}\varphi \cdots \psi^{\varepsilon_p}\varphi = \frac{1}{2^p} \begin{bmatrix} p_1 & i_1\sqrt{3} & i_2 \\ p_2\sqrt{3} & i_3 & i_4\sqrt{3} \\ p_3 & p_4\sqrt{3} & p_5 \end{bmatrix}, \tag{6.9}
$$

where $i_1, \ldots, i_4$ (resp. $p_1, \ldots, p_5$) are odd integers (resp. even). This is evident if $p = 1$; suppose the result is true for rank $p \geqslant 1$. Then, using the equalities (6.8) and (6.9), we obtain:

$$
\psi^{\varepsilon_1}\varphi\psi^{\varepsilon_2}\varphi \cdots \psi^{\varepsilon_p}\varphi\psi^{\varepsilon_{p+1}}\varphi
$$
$$
= \frac{1}{2^{p+1}} \begin{bmatrix} 2i_2 & (\varepsilon_{p+1}p_1 + i_1)\sqrt{3} & -p_1 + 3\varepsilon_{p+1}i_1 \\ 2i_4\sqrt{3} & 3\varepsilon_{p+1}p_2 + i_3 & (-p_2 + \varepsilon_{p+1}i_3)\sqrt{3} \\ 2p_5 & (\varepsilon_{p+1}p_3 + p_4)\sqrt{3} & -p_3 + 3\varepsilon_{p+1}p_4 \end{bmatrix},
$$

which obviously proves the induction step. Equation (6.9) shows then that the word w defined by (6.6) is different from 1. But (6.9) shows that this same word is also different from φ; the word defined by (6.7) is thus different from 1, which completes the proof of Proposition 6.3.8. $\quad\square$

To finish the proof of Theorem 6.3.6, we set

$$
a = \psi\varphi\psi \text{ and } b = \varphi\psi\varphi\psi\varphi.
$$

Proposition 6.3.8 shows the free nature of the group $\langle a, b \rangle$. Indeed, the alphabet $\{a, a^{-1}, b, b^{-1}\}$ has four distinct elements, and every non-empty reduced word built on this alphabet is of the form (6.5), hence different from 1. $\quad\square$

It remains to carry the paradox that we have just shown in $SO(3)$ to the unit sphere of $\mathbb{R}^3$. For this, we use the following lemma.

6.3.9 Lemma *Let G be a paradoxical group acting freely on a set E. Then E is G-paradoxical.*

Proof By a *free action* we mean an action without a fixed point: if $g \in G$ is different from 1 and $x \in E$, then $gx \neq x$. An action of G on E can be associated with the orbital equivalence relation on E defined as

$$
x \sim y \Leftrightarrow \text{there exists } g \in G \text{ such that } y = gx.
$$

Using the axiom of choice, let us select a subset M of E that intersects each equivalence class in a single element. As G acts freely over A, if A and B are two disjoint subsets of G, then AM and BM are disjoint subsets of E. Moreover, as G is paradoxical, there exist two disjoint subsets H and K of G, both equidecomposable to G. We can thus write

$$G = \bigsqcup_{i=1}^{n} G_i \text{ and } H = \bigsqcup_{i=1}^{n} g_i G_i,$$

with $G_i \subset G$ and $g_i \in G$. Setting $A_i = G_i M$, we have

$$E = GM = \bigsqcup_{i=1}^{n} G_i M = \bigsqcup_{i=1}^{n} A_i \text{ and } HM = \bigsqcup_{i=1}^{n} g_i G_i M = \bigsqcup_{i=1}^{n} g_i A_i.$$

Consequently, $E \underset{G}{\sim} HM$ and similarly $E \underset{G}{\sim} KM$, with $HM \cap KM = \varnothing$. $\qquad\square$

Unfortunately, the paradoxical group $\langle a, b \rangle$ encountered in the proof of Theorem 6.3.6 does not act freely on the unit sphere S^2 of $\mathbb{R}^3$, as every element of this group other than 1 has exactly two fixed points in S^2. As a first step, we will be satisfied with a somewhat weaker statement.

6.3.10 Proposition *There exists a countable subset D of the unit sphere S^2 of $\mathbb{R}^3$ such that $S^2 \setminus D$ is $SO(3)$-paradoxical.*

Proof Let G be a free subgroup of rank 2 of $SO(3)$. It is certainly countable. Let us denote by D the (countable) set of the fixed points in S^2 of the elements of $G \setminus \{1\}$. The group G acts on D: indeed, if g is an element of G different from 1 and if $x \in D$ is a fixed point of $h \in G$, then $gx := g(x) \in S^2$ is a fixed point of the rotation $ghg^{-1} \in G$. Consequently, G acts on $S^2 \setminus D$ and this action is obviously free. We conclude using Lemma 6.3.9. $\qquad\square$

We seek to improve this result by getting rid of the subset D.

6.3.11 Proposition *The subset $S^2 \setminus D$ is $SO(3)$-equidecomposable to S^2. Consequently, S^2 is $SO(3)$-paradoxical.*

Proof Let us fix $a \in S^2 \setminus D$, and denote by E the set of rotations $r \in SO(3)$ with axis $\mathbb{R}a$ for which there exists $n \geqslant 1$ and $x, y \in D$ such that $r^n(x) = y$. Clearly E is countable, and if $r \in SO(3) \setminus E$, then $r^n(D) \cap D = \varnothing$ for all $n \geqslant 1$. Therefore, the $r^n(D)$, $n \in \mathbb{N}$, are pairwise disjoint. We then set $D' = \bigsqcup_{n \in \mathbb{N}} r^n(D)$. We have $r(D') = D' \setminus D$, so that

$$S^2 = (S^2 \setminus D') \sqcup D' \underset{SO(3)}{\sim} (S^2 \setminus D') \sqcup (D' \setminus D) = S^2 \setminus D.$$

We conclude thanks to Proposition 6.3.10 and the following lemma.

6.3.12 Lemma   *If a group G acts on a set E, two equidecomposable subsets of E are simultaneously G-paradoxical.*

Proof   Let A and B be two equidecomposable subsets of E, A being paradoxical. We can fix two subsets C and D of A, disjoint and both equidecomposable to A. Let us write

$$A = \bigsqcup_{i=1}^{n} A_i \text{ and } B = \bigsqcup_{i=1}^{n} g_i A_i,$$

with $A_i \subset E$ and $g_i \in G$, and set

$$C' = \bigsqcup_{i=1}^{n} g_i(C \cap A_i) \text{ and } D' = \bigsqcup_{i=1}^{n} g_i(D \cap A_i).$$

The sets C' and D' are disjoint subsets of B, and, for example,

$$C' \underset{G}{\sim} \bigsqcup_{i=1}^{n}(C \cap A_i) = C \underset{G}{\sim} A \underset{G}{\sim} B.$$

Similarly, $D' \underset{G}{\sim} B$, which proves the result. $\square$

6.3.13 Remark   Lemma 6.3.12 allows us to "iterate" the definition of the paradoxical nature. Indeed, it follows immediately that if $A \subset E$ is G-paradoxical, for all $n \geqslant 2$ there exist subsets $A_1, \ldots, A_n$ of E, pairwise disjoint and all equidecomposable to A.

6.3.14 Proposition   *The Euclidean closed unit ball B of $\mathbb{R}^3$ is $\mathrm{Is}^+(\mathbb{R}^3)$-paradoxical.*

Proof   We first show that $B \setminus \{0\}$ is $SO(3)$-paradoxical. For this, let C and D be two disjoint subsets of S^2, both equidecomposable to S^2 under the action of the group $SO(3)$. We can then write

$$S^2 = \bigsqcup_{i=1}^{n} A_i \text{ and } C = \bigsqcup_{i=1}^{n} g_i A_i,$$

with $A_i \subset S^2$ and $g_i \in SO(3)$. For any subset P of S^2, set

$$P^* = \{tx, x \in P, 0 < t \leqslant 1\}$$

(solid angle generated by P). Clearly, if P and Q are two disjoint subsets of S^2, then the subsets P^* and Q^* are two disjoint subsets of $B \setminus \{0\}$. Therefore, by using the linearity of the g_i:

$$C^* = \bigsqcup_{i=1}^{n} g_i A_i^* \underset{SO(3)}{\sim} \bigsqcup_{i=1}^{n} A_i^* = (S^2)^* = B \setminus \{0\}.$$

Similarly, $D^* \sim B \setminus \{0\}$, which shows that $B \setminus \{0\}$ is $SO(3)$-paradoxical, hence also $\mathrm{Is}^+(\mathbb{R}^3)$-paradoxical.

Let us now show that B is $\mathrm{Is}^+(\mathbb{R}^3)$-paradoxical. By Lemma 6.3.12, it is sufficient to prove that

$$B \setminus \{0\} \underset{\mathrm{Is}^+(\mathbb{R}^3)}{\sim} B.$$

We will do this by using again the "shifting" technique seen in Example 6.3.1. Let $r \in \mathrm{Is}^+(\mathbb{R}^3)$ be an affine rotation of angle incommensurable with π, whose axis contains the point $\left(0, 0, \dfrac{1}{2}\right)$ but not the origin. Setting $D = \{r^n(0), n \geqslant 0\} \subset B$, we have $r(D) = D \setminus \{0\}$, so that

$$B = (B \setminus D) \sqcup D \underset{\mathrm{Is}^+(\mathbb{R}^3)}{\sim} (B \setminus D) \sqcup r(D) = (B \setminus D) \sqcup (D \setminus \{0\}) = B \setminus \{0\}. \qquad \square$$

6.3.15 The Banach–Tarski paradox in $\mathbb{R}^3$

We now come to the most spectacular form of the Banach–Tarski paradox.

6.3.16 Theorem [Banach–Tarski [14]] *Two bounded subsets $\mathbb{R}^3$ with non-empty interior are $\mathrm{Is}^+(\mathbb{R}^3)$-equidecomposable.*

This statement is paradoxical in that it seems to contradict the conservation of volume by isometry: for example, two balls with different radii, or even one ball and n copies of this ball, are equidecomposable. In reality, of course, this is not a problem: *at least* one[15] of the pieces of the decomposition is certainly non-measurable in the sense of Lebesgue.

The proof consists of using the $\mathrm{Is}^+(\mathbb{R}^3)$-paradoxical nature of the unit ball of $\mathbb{R}^3$ to replicate it, as well as an additional ingredient: a low-cost adaptation of the set-theoretic theorem of Cantor–Bernstein, stated and proved below.

6.3.17 Theorem [Cantor–Bernstein] *Let E and F be two sets, such that there exist two injections $f : E \to F$ and $g : F \to E$. Then, there exists a bijection from E onto F.*

---

[15] Even two!

Proof It is sufficient[16] to find a subset A of E such that

$$g\big(F \setminus f(A)\big) = E \setminus A,$$

or again

$$E \setminus g\big(F \setminus f(A)\big) = A.$$

Seen from this point of view, we appear to have a fixed point theorem. Now, the map

$$\Phi : \mathcal{P}(E) \to \mathcal{P}(E), \, A \mapsto E \setminus g\big(F \setminus f(A)\big)$$

is non-decreasing in the sense of inclusion, and *this is sufficient to ensure the existence of a fixed point of* Φ.[17] Indeed, let us set

$$A = \bigcup_{X \subset E, X \subset \Phi(X)} X.$$

If $X \subset \Phi(X)$, then $X \subset A$, hence $X \subset \Phi(X) \subset \Phi(A)$ since Φ is non-decreasing, so that $A \subset \Phi(A)$. But then, $\Phi(A) \subset \Phi\big(\Phi(A)\big)$, so that $\Phi(A) \subset A$ by definition of A. Finally, A is indeed a fixed point of Φ.   □

We now consider a group G acting on the set E, and A, B two subsets of E. We write

$$A \underset{G}{\preceq} B$$

to mean that A is equidecomposable to some subset of B. We thus define a binary relation on the set of all subsets of E, which is reflexive and transitive. The Banach–Cantor–Bernstein theorem states that, modulo equidecomposability, this is an order relation.

6.3.18 Theorem [Banach–Cantor–Bernstein [13]] *Let G be a group acting on the set E, and A, B two subsets of E such that $A \underset{G}{\preceq} B$ and $B \underset{G}{\preceq} A$. Then $A \underset{G}{\sim} B$.*

Proof We will follow the proof of Theorem 6.3.17. By hypothesis, there exist $A' \subset B$ and $B' \subset A$ such that we can write

$$A = \bigsqcup_{i=1}^{m} A_i, \, A' = \bigsqcup_{i=1}^{m} g_i A_i, \, B = \bigsqcup_{j=1}^{n} B_j \text{ and } B' = \bigsqcup_{j=1}^{n} h_j B_j.$$

---

[16] Indeed, the function $h : E \to F$, $x \mapsto \begin{cases} f(x) & \text{if } x \in A \\ g^{-1}(x) & \text{otherwise} \end{cases}$ is then bijective.

[17] In the same way that if $\varphi : [0, 1] \to [0, 1]$ is a non-decreasing function, $\sup\{x \in [0, 1] / \varphi(x) \geqslant x\}$ is a fixed point of φ.

This allows us to define, piecewise, two injections

$$f : A \to B, x \mapsto g_i x \text{ if } x \in A_i$$

and

$$g : B \to A, x \mapsto h_i x \text{ if } x \in B_i.$$

These injections have a simple but very useful property:

$$\text{if } X \subset A(\text{resp. } X \subset B), \text{ then } X \underset{G}{\sim} f(X) \ (\text{resp.} X \underset{G}{\sim} g(X)).$$

Indeed, it is sufficient to write $X = \bigsqcup_{i=1}^{m}(X \cap A_i)$, so that

$$f(X) = \bigsqcup_{i=1}^{m} g_i(X \cap A_i) \underset{G}{\sim} \bigsqcup_{i=1}^{m}(X \cap A_i) = X.$$

But then, the Cantor–Bernstein theorem provides a subset X of A such that

$$g(B \setminus f(X)) = A \setminus X.$$

As we have seen above, we thus automatically have

$$X \underset{G}{\sim} f(X) \text{ and } B \setminus f(X) \underset{G}{\sim} A \setminus X,$$

hence

$$A = X \sqcup (A \setminus X) \underset{G}{\sim} f(X) \sqcup \big(B \setminus f(X)\big) = B. \qquad \square$$

We can now proceed to the proof of Theorem 6.3.16.

Proof Let A, A' be two bounded subsets of $\mathbb{R}^3$ with non-empty interior, and B, B' two closed balls of $\mathbb{R}^3$ such that $B \subset A$ and $A' \subset B'$. We can cover B' with a finite number $B_1, \dots, B_n$ of translates of B; then there exist subsets $C_1, \dots, C_n$ of $B_1, \dots, B_n$, respectively, pairwise disjoint, such that

$$A' = \bigsqcup_{i=1}^{n} C_i.$$

Moreover, as B is $\mathrm{Is}^+(\mathbb{R}^3)$-paradoxical, by Remark 6.3.13, there exist some subsets $D_1, \dots, D_n$ of B pairwise disjoint and all equidecomposable to B. Finally, let $B'_1, \dots, B'_n$ be pairwise disjoint translates of B. We thus have

$$A' = C_1 \sqcup \cdots \sqcup C_n \underset{\mathrm{Is}^+(\mathbb{R}^3)}{\preceq} B'_1 \sqcup \cdots \sqcup B'_n \underset{\mathrm{Is}^+(\mathbb{R}^3)}{\sim} D_1 \sqcup \cdots \sqcup D_n \subset B \subset A,$$

hence $A' \underset{\mathrm{Is}^+(\mathbb{R}^3)}{\preceq} A$, and also $A \underset{\mathrm{Is}^+(\mathbb{R}^3)}{\preceq} A'$ by switching the roles of A and A'.
We conclude thanks to the Banach–Cantor–Bernstein theorem. $\qquad\square$

What are the impacts of the Hausdorff–Banach–Tarski paradox?

- There does not exist a mean that is $\mathrm{Is}^+(\mathbb{R}^3)$-invariant on the sphere, or on the closed unit ball of $\mathbb{R}^3$.

- There does not exist a finitely additive measure μ on $\mathbb{R}^3$, invariant under isometries (or even only under direct isometries) and such that $\mu([0, 1]^3) = 1$. Indeed, as $[0, 1]^3$ and $[0, 1]^3 \cup [2, 3]^3$ are $\mathrm{Is}^+(\mathbb{R}^3)$-equidecomposable according to Theorem 6.3.16, we would then have

$$2 = \mu([0, 1]^3 \cup [2, 3]^3) = \mu([0, 1]^3) = 1.$$

In particular, the Lebesgue measure does not admit a finitely additive extension to all the subsets of $\mathbb{R}^3$ that is invariant under isometries, and this result can easily be extended to $\mathbb{R}^d$, $d \geqslant 3$. The situation is thus radically different from what happens in dimensions 1 and 2 (cf. Corollary 6.2.21): we can replicate balls, but not disks.

6.3.19 Remarks We conclude with a few remarks concerning the axiom of choice, which plays a crucial role in the proof of the Hausdorff and Banach–Tarski paradoxes.

(1) Consider the following assertions:
- The axiom of choice (AC).
- Every subset of $\mathbb{R}$ is Lebesgue-measurable (LM).

Denote by ZF the Zermelo–Fraenkel axioms of set theory, if necessary extended with the axiom of choice (ZFC). In 1964, by supposing the consistency[18] of ZFC and the existence of an "inaccessible cardinal" (CI), Solovay proved the consistency of the set ZF+CAC+LM, where CAC denotes the countable form of the axiom of choice, weaker than AC and essentially allowing the construction of recursive sequences. In other words, if we suppose the consistency of ZFC+CI, a theory without the Banach–Tarski paradox is possible. See [183] for more details on this subject.

(2) The existence of an inaccessible cardinal implies the consistency of ZF, since it provides a model for ZF. In particular, according to Gödel's theorem, this implies that we cannot establish this existence within ZF.

---

[18] It means for a system of axioms to be free of contradictions.

(3) ZF and ZFC are equiconsistent, meaning that if there is a contradiction in ZFC then there is also one in ZF. It is the same for ZF and (ZFC + there does not exist an inaccessible cardinal). However, ZF and (ZFC + there exists an inaccessible cardinal) are not equiconsistent; there is a risk-taking with respect to ZF.

6.4 Superamenability

We have previously seen that the group $\text{Is}(\mathbb{R}^2)$ of plane affine isometries is amenable, because it is "reasonably" commutative (solvable). This prevents $\mathbb{R}^2$ from being $\text{Is}(\mathbb{R}^2)$-paradoxical (Proposition 6.2.8), but does not preclude the existence of $\text{Is}(\mathbb{R}^2)$-paradoxical subsets of $\mathbb{R}^2$ (Exercise 6.6). Nonetheless, because of Corollary 6.2.21, a bounded subset of $\mathbb{R}^2$ with non-empty interior cannot be $\text{Is}(\mathbb{R}^2)$-paradoxical.

This is a motivation for the following definition: a group G is said to be *superamenable* if for every non-empty subset A of G, there exists a finitely additive and G-invariant measure μ on G satisfying $\mu(A) = 1$. In this case, if G acts on E, E *does not admit any non-empty G-paradoxical subset*.[19] In particular, the group $\text{Is}(\mathbb{R}^2)$ is not superamenable. For $d \geqslant 3$, $\text{Is}(\mathbb{R}^d)$ is not amenable, and hence not superamenable. What happens for $\text{Is}(\mathbb{R})$? To answer this question, we will start by establishing the superamenability of certain groups, those of *sub-exponential growth*.

In what follows, S is a finite non-empty subset of G, and we set

$$S^{-1} = \{g^{-1}, g \in S\}.$$

For every $n \in \mathbb{N}$, we denote by $\gamma_S(n)$ the number of reduced words of length $\leqslant n$ that can be written on the alphabet $S \cup S^{-1}$. The following properties are easy to obtain:

- the function γ_S is non-decreasing;
- $\gamma_S(0) = 1$ (1 is the unique empty word);
- $\gamma_S(m + n) \leqslant \gamma_S(m)\gamma_S(n)$ for $(m, n) \in \mathbb{N}^2$.

We can thus conclude (Exercise 6.9) that the sequence $\left(\gamma_S(n)^{1/n}\right)_{n \geqslant 1}$ is convergent.

Moreover, note that for $n \geqslant 1$,

$$\gamma_S(n) \leqslant \gamma_S(1)^n = |S \cup S^{-1}|^n,$$

---

[19] Indeed, if $\varnothing \neq X \subset E$, if $f : G \to E$, $g \mapsto gx$ (with $x \in X$ fixed) and if μ is a G-invariant mean on $f^{-1}(X)$, then the formula $\tilde{\mu}(Y) = \mu(f^{-1}(Y))$ defines a G-invariant mean on X.

hence

$$\lim_{n \to +\infty} \gamma_S(n)^{1/n} \leqslant |S \cup S^{-1}|.$$

The function γ_S is thus, at worst, a function of exponential growth. The group G is said to be *of sub-exponential growth* if, for every finite non-empty subset S of G,

$$\gamma_S(n)^{1/n} \to 1 \text{ as } n \to +\infty.$$

Let us examine two special cases.

(1) If a and b are two elements of G that generate a free subgroup, and if $S = \{a, b\}$, it is easy to show that for every $n \in \mathbb{N}$, $\gamma_S(n) \geqslant 2^n$. In this case, γ_S is effectively of exponential growth.

(2) If G is abelian, by writing $S = \{g_1, \dots, g_r\}$, we see that every reduced word of length $\leqslant n$ built on $S \cup S^{-1}$ can be written $g_1^{\alpha_1} \cdots g_r^{\alpha_r}$, with $\sum_{i=1}^{r} |\alpha_i| \leqslant n$. As a result, $\gamma_S(n) \leqslant (2n + 1)^r$. In this case, the function γ_S is of polynomial growth, and G is of sub-exponential growth.

For more details on this topic, see [3].

Let us proceed to the fundamental theorem of this section.

6.4.1 Theorem   *Every group of sub-exponential growth is superamenable.*

In particular, every abelian group is superamenable. Before proving this theorem, we show a radical consequence, which profits from the "almost commutativity" of $\mathrm{Is}(\mathbb{R})$ and highlights the fundamental difference between dimensions 1 and 2.

6.4.2 Corollary   *The group $\mathrm{Is}(\mathbb{R})$ is superamenable. In particular, $\mathbb{R}$ does not admit any non-empty $\mathrm{Is}(\mathbb{R})$-paradoxical subset.*

Proof   Every isometry f of $\mathbb{R}$ can be written uniquely in the form $u_f t_f$, where $u_f = \pm \mathrm{id}_{\mathbb{R}}$ and t_f is a translation. Moreover, the composition of isometries can be expressed as follows:

$$(u_f t_f) \circ (u_g t_g) = u_f u_g (\underbrace{u_g^{-1} t_f u_g t_g}_{\text{translation}}).$$

This equality shows the fact that $\mathrm{Is}(\mathbb{R})$ is the semi-direct product of the abelian group $O(\mathbb{R}) = \{\pm \mathrm{id}_{\mathbb{R}}\}$ of linear isometries of $\mathbb{R}$ with the normal subgroup of translations. Then, let S be a finite non-empty subset of $\mathrm{Is}(\mathbb{R})$,

$S' := \{t_f, f \in S\}$ and $S'' := S' \cup \{-\mathrm{id}_{\mathbb{R}}\}$. Any reduced word of length $\leqslant n$ built on the alphabet $S \cup S^{-1}$ is a reduced word of length $\leqslant 2n$ built on the alphabet $S'' \cup S''^{-1}$. Moreover, for all $g \in S'$,

$$g \circ (-\mathrm{id}_{\mathbb{R}}) = (-\mathrm{id}_{\mathbb{R}}) \circ g^{-1}.$$

Therefore, every reduced word of length $\leqslant n$ built on the alphabet $S \cup S^{-1}$ can be written $(-1)^{\varepsilon} \circ g$, where $\varepsilon \in \{0, 1\}$ and g is a reduced word of length $\leqslant n$ built on the alphabet $S' \cup S'^{-1}$ (we "bring back" the $-\mathrm{id}_{\mathbb{R}}$ to the left). Consequently,

$$\gamma_S(n) \leqslant 2\gamma_{S'}(n).$$

As the elements of S' commute pairwise, the function $\gamma_{S'}$ is of polynomial growth. It is hence the same for the function γ_S. $\qquad\square$

We finish with the proof of Theorem 6.4.1.

Proof By Tarski's alternative (Theorem 6.3.2), it is sufficient to show that if $A \subset G$ is non-empty, A is not G-paradoxical. We suppose the contrary. Then there exist two disjoint subsets B and C of A such that

$$A = \bigsqcup_{i=1}^{p} B_i = \bigsqcup_{j=1}^{q} C_j, B = \bigsqcup_{i=1}^{p} g_i B_i \text{ and } C = \bigsqcup_{j=1}^{q} h_j C_j.$$

We define the injections

$$f_B : A \to A, x \mapsto g_i x \text{ if } x \in B_i$$

and

$$f_C : A \to A, x \mapsto h_i x \text{ if } x \in C_i,$$

and set $S = \{g_1, \ldots, g_p, h_1, \ldots, h_q\}$. Finally, we consider the set M_n of words of length n written with the letters f_B and f_C, that is, the maps from A to A of the form $\varphi_1 \circ \cdots \circ \varphi_n$ with $\varphi_i \in \{f_B, f_C\}$. *Two such words, coming from distinct n-tuples* $(\varphi_1, \ldots, \varphi_n)$, *are certainly distinct*. Indeed, suppose that

$$\varphi_1 \circ \cdots \circ \varphi_n = \psi_1 \circ \cdots \circ \psi_n,$$

where $(\varphi_1, \ldots, \varphi_n)$ and $(\psi_1, \ldots, \psi_n)$ are distinct n-tuples of elements of $\{f_B, f_C\}$. Let p be the smallest element of $[\![1, n]\!]$ such that $\varphi_p \neq \psi_p$. As f_B and f_C are injective, we thus have

$$\varphi_p \circ \cdots \circ \varphi_n = \psi_p \circ \cdots \circ \psi_n,$$

with, for example, $\varphi_p = f_B$ and $\psi_p = f_C$. But this is not possible, as the image of f_B (resp. f_C) is contained in B (resp. C). In reality, the argument that we have just given shows even more:

$$\varphi_1 \circ \cdots \circ \varphi_n(x) \neq \psi_1 \circ \cdots \circ \psi_n(x) \text{ for all } x \in A.$$

Consequently, for each $x \in A$, the set $\{w(x), w \in M_n\}$ is of cardinal 2^n. However, each $w(x)$ is the value at x of a word of length n in the g_i and the h_j. As a result, $\gamma_S(n) \geqslant 2^n$, which contradicts the hypothesis that G is of sub-exponential growth. $\qquad\qquad\qquad\qquad\qquad\qquad\qquad\qquad\qquad\qquad\quad\square$

6.5 Appendix: Topological vector spaces

We limit ourselves here to the strict minimum required, refering to [161] for more details.

A topological vector space is a linear space E on the field $\mathbb{K}$ of real or complex numbers, equipped with a Hausdorff topology that makes the maps

$$E \times E \to E, \ (x, y) \mapsto x + y \qquad\qquad (6.10)$$

and

$$\mathbb{K} \times E \to E, \ (\lambda, x) \mapsto \lambda x \qquad\qquad (6.11)$$

continuous, where $E \times E$ and $\mathbb{K} \times E$ are equipped with the product topologies.

In this case, the translations $\tau_a : E \to E, x \mapsto x + a$ (for all a in E) are homeomorphisms. Consequently, if $a \in E$, the neighbourhoods of a are exactly the $a + V$, where V is a neighbourhood of 0: the topology of E is completely known as soon as a fundamental system of neighbourhoods of 0 is at our disposal.

A subset X of E is said to be *bounded* if it is absorbed by all neighbourhoods of 0, that is, for any neighbourhood V of 0, there exists a strictly positive real number δ such that $\lambda X \subset V$ for all $\lambda \in \mathbb{K}$ with $|\lambda| \leqslant \delta$.

6.5.1 Proposition *Every compact subset of E is bounded.*

Proof Let V be a neighbourhood of 0. By continuity of (6.10), there exists a neighbourhood W_1 of 0 such that $W_1 + W_1 \subset V$. Then, by continuity of (6.11), there exists a $\delta > 0$ and a neighbourhood W_2 of 0 such that $\lambda W_2 \subset W_1$ for $|\lambda| \leqslant \delta$. As K is compact, we can find $x_1, \ldots, x_p \in K$ such that

$$K \subset \bigcup_{k=1}^{p} (x_k + W_2).$$

By decreasing δ if necessary, we can also suppose that $\lambda x_k \in W_1$ for $|\lambda| \leqslant \delta$ and $1 \leqslant k \leqslant p$. Hence, if $|\lambda| \leqslant \delta$, we have

$$\lambda K \subset \bigcup_{k=1}^{p} (\lambda x_k + \lambda W_2) \subset W_1 + W_1 \subset V. \qquad \square$$

Exercises

6.1. The Steinhaus theorem. Let λ be the Lebesgue measure on $\mathbb{R}^n$.

(a) Show that if $f \in L^1(\mathbb{R}^n)$ and $g \in L^\infty(\mathbb{R}^n)$, the convolution $f * g$, defined by

$$(f * g)(x) = \int f(x - t)g(t)\,dt$$

(the integral being taken over $\mathbb{R}^n$) is continuous on $\mathbb{R}^n$.

(b) Let A and B be two measurable subsets of $\mathbb{R}^n$ such that $0 < \lambda(A) < +\infty$ and $0 < \lambda(B) < +\infty$. Show that $\{x \in \mathbb{R}^n / 1_A * 1_B(x) \neq 0\}$ is a non-empty open subset of $\mathbb{R}^n$ contained in $A + B$. Show that the fact that $A + B$ has non-empty interior remains true if A or B has infinite measure.

(c) In particular, if A is a measurable subset of $\mathbb{R}^n$ such that $\lambda(A) > 0$, $A - A$ contains a neighbourhood of 0 (Steinhaus theorem).

6.2. Let $f : \mathbb{R}^n \to \mathbb{R}$ be a measurable and "midconvex" function, that is,

$$f\left(\frac{x + y}{2}\right) \leqslant \frac{f(x) + f(y)}{2} \text{ for } x, y \in \mathbb{R}^n.$$

(a) Show that f is bounded on a neighbourhood of 0.

(b) Show that f is convex.

6.3. Let λ be the Lebesgue measure on $\mathbb{R}$, and A be a measurable subset of $\mathbb{R}$ such that $\lambda(A) > 0$. We intend to show that A contains a non-measurable subset. For this, we consider the equivalence relation on $\mathbb{R}$ (congruence modulo $\mathbb{Q}$) defined as follows:

$$x \sim y \Leftrightarrow x - y \in \mathbb{Q},$$

and we select (axiom of choice) a subset M of $\mathbb{R}$ that intersects each equivalence class in exactly one point. For every rational number r, we set

$$A_r = A \cap (r + M).$$

(a) Show that if all the A_r are measurable, then at least one of them has strictly positive measure.

(b) Conclude by using the Steinhaus theorem (Exercise 6.1).

6.4. The aim of this exercise is to show that the set of all Borel sets of $\mathbb{R}$ has the same cardinality as $\mathbb{R}$. We call a *Polish space* any separable topological space whose topology can be defined by a complete metric.

(a) Let $X_0 = (\mathbb{N}^*)^{\mathbb{N}^*}$ be the set of sequences of integers $\geqslant 1$. Show that X_0 equipped with the product topology is a Polish space.

(b) Show that any Polish space is a continuous image of X_0.

(c) Let $\mathcal{A}$ be a class of subsets of $\mathbb{R}$, containing the closed sets and stable under countable unions and intersections. Show that $\mathcal{A}$ contains the Borel σ-algebra of $\mathbb{R}$.

(d) Let $\mathcal{A}$ be the class of subsets of $\mathbb{R}$ formed by the empty set and the continuous images of X_0. Show that $\mathcal{A}$ satisfies the hypothesis of part (c).

(e) Show that any Borel set of $\mathbb{R}$ is a continuous image of X_0.

(f) Conclude.

6.5. Show that the Cantor middle third set contains transcendental numbers and non-Borel subsets, but no non-Lebesgue-measurable subset.

6.6. The Sierpinski–Mazurkiewicz paradox. Let $u \in \mathbb{C}$ be a transcendental number with modulus 1. Set

$$\mathbb{N}[u] = \{P(u), \ P \in \mathbb{N}[X]\}.$$

Show that $\mathbb{N}[u]$ is paradoxical under the action of the group of direct isometries, and more precisely under the action of the group generated by the rotation $r : z \mapsto uz$ and the translation $\tau : z \mapsto z + 1$.

6.7. Let G be an amenable group and H a subgroup of G. We intend to show that H is amenable. For this, we define an equivalence relation on G as follows:

$$x \sim y \ \Leftrightarrow \ yx^{-1} \in H.$$

The equivalence class of x is thus Hx. Select M (axiom of choice), a subset of G intersecting each class in exactly one element. For any subset A of H, set

$$v(A) = \mu(AM),$$

where $AM = \{am, (a, m) \in A \times M\}$. Show that v is an H-invariant mean on H.

6.8. Let G be an amenable subgroup and H a normal subgroup of G. We suppose that the groups H and G/H are amenable, and equipped with means μ

and $\overline{\mu}$, respectively H and G/H-invariant. We intend to construct a G-invariant mean on G, which will prove Theorem 6.2.11. For any subset A of G, we set $\mu_H(A) = \mu(A \cap H)$; clearly μ_H is a mean on G, but it has no reason to be G-invariant.

(a) Show that if $g \in G$, $\mu_H(g^{-1}A)$ depends only on the right class of g modulo H. This allows us to set

$$\mu_H^*(\overline{g}^{-1}A) = \mu_H(g^{-1}A) \text{ for } \overline{g} \in G/H.$$

(b) In order to obtain a G-invariant mean on G, we are brought to averaging the $\mu_H^*(\overline{g}^{-1}A)$, with $\overline{g}$ taken over G/H. We thus set

$$\nu(A) = \int_{G/H} \mu_H^*(x^{-1}A)d\overline{\mu}(x).$$

Show that ν is a G-invariant mean on G.

6.9. Let $(u_n)_{n \geqslant 0}$ be a sequence of non-negative real numbers such that

$$u_{m+n} \leqslant u_m + u_n \text{ for } m, n \geqslant 0.$$

(a) Show that $u_{qb+r} \leqslant qu_b + u_r$ for $(q, b, r) \in \mathbb{N}^3$.
(b) Show that the sequence $\left(\dfrac{u_n}{n}\right)_{n \geqslant 1}$ converges to $\inf\limits_{n \geqslant 1} \dfrac{u_n}{n}$.

6.10. The invariant Hahn–Banach theorem, according to Agnew–Morse (variant of Theorem 6.2.19). Given:

• X a real normed space;
• $\mathcal{A}$ a collection of continuous linear maps from X to itself, pairwise commuting;
• $p : X \rightarrow \mathbb{R}$ a positively homogeneous map,[20] sub-additive and $\mathcal{A}$-invariant;[21]
• Y a linear subspace of X stable by all the elements of $\mathcal{A}$;
• ℓ a linear functional on Y, $\mathcal{A}$-invariant and such that $\ell(x) \leqslant p(x)$ for $x \in Y$.

(a) Let $\mathcal{C}$ be the convex hull of the semigroup generated[22] by $\mathcal{A}$. For $x \in X$, we set

$$q(x) = \inf\{p(ux), u \in \mathcal{C}\}.$$

Justify this definition.

---

[20] That is: $p(\lambda x) = \lambda p(x)$ if $x \in X$ and $\lambda \in \mathbb{R}_+$.
[21] That is: $p(ax) = p(x)$ if $a \in \mathcal{A}$ and $x \in X$.
[22] This semigroup is made up of the identity and of finite products of elements of $\mathcal{A}$.

(b) Verify that $q \leqslant p$, that q is sub-additive[23] and positively homogeneous, and that $\ell \leqslant q$. By the Hahn–Banach theorem, ℓ thus admits a linear extension to X, still called ℓ, such that $\ell(x) \leqslant q(x)$ for all $x \in X$ (hence also $\ell \leqslant p$).

(c) Let $A \in \mathcal{A}$. Show that, for all $x \in X$, we have $q(x - Ax) \leqslant 0$.

(d) Finally, show that ℓ is $\mathcal{A}$-invariant.

6.11. The Dixmier stability theorem [49]. Let H be a Hilbert space. We denote by $\mathcal{L}(H)$ the set of continuous linear maps from H to itself.

(a) Let G be an amenable group and $\pi : G \to \mathcal{L}(H)$ a representation in the sense of semigroups, that is, a map $\pi : G \to \mathcal{L}(H)$ satisfying

$$\pi(1) = \mathrm{id}_H \text{ and } \pi(st) = \pi(s)\pi(t) \text{ for } s, t \in G.$$

Moreover, we suppose that π is uniformly bounded:

$$\|\pi(t)\| \leqslant C \text{ for all } t \in G.$$

Show that π is conjugate to a unitary representation, that is, there exists an invertible element A of $\mathcal{L}(H)$ such that $A\pi(t)A^{-1}$ is a unitary operator[24] for all $t \in G$.

(b) Let $T \in \mathcal{L}(H)$, invertible and such that $\|T^n\| \leqslant C$ for all $n \in \mathbb{Z}$. Show that T is conjugate to a unitary operator.

(c) Under the assumptions of part (b), show the existence of a constant $C' > 0$ such that

$$\left\| \sum_{|n| \leqslant N} a_n T^n \right\| \leqslant C' \left\| \sum_{|n| \leqslant N} a_n e^{int} \right\|_\infty$$

for every $N \geqslant 0$ and every finite sequence $(a_n)_{|n| \leqslant N}$ of complex numbers (two-sided von Neumann inequality).

6.12. The Nagy similarity theorem

(a) Let H be a Hilbert space, G an amenable semigroup and $\pi : G \to \mathcal{L}(H)$ a representation.[25] Suppose in addition that there exist constants $a, b > 0$ such that

$$a\|x\| \leqslant \|\pi(t)x\| \leqslant b\|x\| \text{ for } t \in G \text{ and } x \in H.$$

---

[23] It is here that we use the commutativity of the elements of $\mathcal{A}$.

[24] An operator T of H is called unitary if it satisfies $TT^* = T^*T = \mathrm{id}_H$.

[25] See the definition in Exercise 6.11.

Show that there exists an invertible element A of $\mathcal{L}(H)$ such that $A\pi(t)A^{-1}$ is an isometry[26] for all $t \in G$.

(b) Let T_1, T_2 be two commuting elements of $\mathcal{L}(H)$, such that there exist constants $a, b > 0$ such that

$$a\|x\| \leqslant \|T_1^m T_2^n x\| \leqslant b\|x\| \text{ for } m, n \in \mathbb{N} \text{ and } x \in H.$$

Show that T_1 and T_2 are simultaneously conjugate to isometries that commute: there exist an invertible element A of $\mathcal{L}(H)$ and isometries S_1 and S_2 such that $AT_j A^{-1} = S_j$ for $j \in \{1, 2\}$.

---

[26] An operator T of H is said to be isometric if it preserves the inner product, that is, if $T^*T = \mathrm{id}_H$. The invertibility of T is not required.

7
Riemann's "other" function

7.1 Introduction

Is it exceptional for a continuous function to be nowhere differentiable? Current practice is misleading on this point, and Baire's theory shows that, in a generic fashion, a continuous function is not differentiable at any point [24]. More explicitly, the first example of a continuous but nowhere differentiable function was apparently given by Bolzano in 1834. As for Riemann, according to an oral tradition [26], he supposedly introduced, around 1860, the function R (which from now on, we will call the *Riemann function*) defined on $\mathbb{R}$ by

$$R(x) = \sum_{n=1}^{\infty} \frac{\sin(n^2 \pi x)}{n^2}.$$

Figure 7.1 shows the shape of the graph of a partial sum of large index of the function R. For the time being, let us simply indicate that the observation of what happens in the neighbourhoods of the points 0 and 1 is already instructive, and gives a first idea of the results to be obtained later in this chapter.

The next year, in 1861, Weierstrass – who had not been able to establish or refute Riemann's conjecture about the function R – showed that if a and b are two real numbers such that $0 < a < 1$ and $ab > 1 + \frac{3\pi}{2}$, the continuous function W defined by

$$W(x) = \sum_{n=0}^{\infty} a^n \cos(b^n x)$$

is not differentiable at any point. The conditions of Weierstrass on a and b, clearly artificial, were later improved, and the "optimal" result was obtained by Hardy in 1916 (see Exercise 7.2).

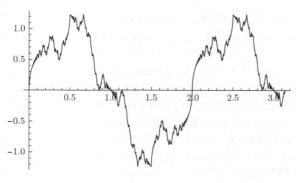

Figure 7.1

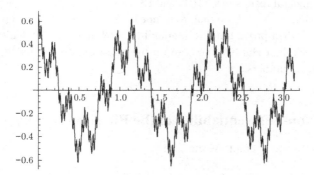

Figure 7.2

7.1.1 Theorem [Hardy [74]]   *If* $0 < a < 1$ *and* $ab \geqslant 1$*, the function W is not differentiable at any point.*

Figure 7.2 shows the shape of a partial sum of large index of the function W.

In the article [74], Hardy also studied Riemann's function and showed a first important theorem based on his joint work with Littlewood [81], which however still did not refute Riemann's conjecture.

7.1.2 Theorem [Hardy [74]]   *Riemann's function is not differentiable at any irrational point, nor at any rational point of the form* $\dfrac{2p + 1}{2q}$ *or* $\dfrac{2p}{4q + 1}$.

The question was finally answered completely (in the negative) in 1968 by Joseph Gerver, a student of Serge Lang.[1]

---

[1] Who had the habit of proposing this problem to his *undergraduate* students.

7.1.3 Theorem [Gerver [68]] *Riemann's function is differentiable at a rational point r if and only if r can be written as the ratio of two odd integers.*

Gerver's proof [68] is completely elementary, and is a bravura piece, but we must admit that it is not easily understandable. In 1981, Itatsu, thanks to a new method[2] [95], simplified Gerver's proof considerably, at the same time obtaining more precise estimations of R in the neighbourhood of the rational points. What makes Riemann's example much more delicate than Weierstrass' one is the slow growth of the sequence $(n^2)_{n \geqslant 1}$ compared with that of a geometric sequence $(b^n)_{n \geqslant 0}$ for $b > 1$. Consequently, Riemann's function does not belong to the category of Fourier series known as *lacunary*, for which we have numerous general results (see [104] and Exercise 7.2, as well as [150]).

In this chapter, we show that R is not differentiable at 0 using elementary methods, then prove the differentiability of R at 1 using Itatsu's method, and finally explain Hardy's theorem on the non-differentiability of R at the irrational real numbers.

7.2 Non-differentiability of the Riemann function at 0

We fix $x > 0$, and decompose the sum

$$R(x) = \sum_{n=1}^{\infty} \frac{\sin(n^2 \pi x)}{n^2}$$

into three parts:

$$R(x) = \sum_{n^2 x \leqslant \frac{1}{2}} \frac{\sin(n^2 \pi x)}{n^2} + \sum_{\frac{1}{2} < n^2 x \leqslant 1} \frac{\sin(n^2 \pi x)}{n^2} + \sum_{n^2 x > 1} \frac{\sin(n^2 \pi x)}{n^2}.$$

We bound the first sum from below using the inequality $\sin u \geqslant \frac{2}{\pi} u$, valid on $\left[0, \frac{\pi}{2}\right]$, the second by 0 and the third by $-\sum_{n^2 x > 1} \frac{1}{n^2}$. As the first sum contains $[(2x)^{-1/2}]$ terms, we obtain

$$R(x) \geqslant [(2x)^{-1/2}] 2x - \sum_{n > [x^{-1/2}]} \frac{1}{n^2}.$$

---

[2] But coming back to basics, in a sense, with the use of the Poisson summation formula (see Section 7.4 and the remarks at the end of this chapter).

Moreover, for $N \geqslant 1$,

$$\sum_{p=N+1}^{\infty} \frac{1}{p^2} \leqslant \int_{N}^{+\infty} \frac{dt}{t^2} = \frac{1}{N},$$

so that

$$R(x) \geqslant [(2x)^{-1/2}]2x - \frac{1}{[x^{-1/2}]}.$$

The minorant function is equivalent to $(\sqrt{2} - 1)\sqrt{x}$ as x tends to 0 by positive values, which proves that

$$\frac{R(x)}{x} \to +\infty \text{ as } x \searrow 0.$$

The function R is therefore not differentiable at 0. In fact, we have shown the following more precise result.

7.2.1 Proposition *There exist positive real numbers c, δ such that*

$$R(x) \geqslant c\sqrt{x} \text{ for } x \in [0, \delta].$$

7.3 Itatsu's method

The purpose of this section is to prove the following statement.

7.3.1 Theorem *Riemann's function is differentiable at 1, and*

$$R'(1) = -\frac{\pi}{2}.$$

This differentiability is delicate to prove. To achieve it, and to obtain even more results, it is in our interest to consider the complex-valued function F, defined by

$$F(x) = \sum_{n=1}^{\infty} \frac{e^{in^2\pi x}}{n^2} \text{ for } x \in \mathbb{R}.$$

We evidently have

$$R(x) = \text{Im } F(x).$$

Symmetrically, we set

$$C(x) = \text{Re } F(x) = \sum_{n=1}^{\infty} \frac{\cos(n^2\pi x)}{n^2} \text{ for } x \in \mathbb{R}.$$

The result that we aim for will follow from the following two facts.

- *Fact 1.* A "bad" behaviour of F at 0, $\frac{1}{2}$-Hölderian, but that can be described very precisely and, as we will see, follows from a *first* functional equation (behaviour of F with respect to the transformation $\sigma : x \mapsto -\frac{1}{x}$).

7.3.2 Proposition   *As $x \searrow 0$,*

$$F(x) = F(0) + i\pi e^{i\pi/4}\sqrt{x} - \frac{i\pi}{2}x + O(x^{3/2}).$$

- *Fact 2.* A *second* functional equation. This one is easy to show (behaviour of F with respect to the transformation $\tau : x \mapsto x + 1$).

7.3.3 Proposition   *For all $x \in \mathbb{R}$,*

$$F(1 + x) = \frac{1}{2}F(4x) - F(x).$$

Proof   We have

$$F(1+x) = \sum_{n=1}^{\infty} \frac{e^{in^2\pi(x+1)}}{n^2}$$

$$= \sum_{n=1}^{\infty} (-1)^n \frac{e^{in^2\pi x}}{n^2} \quad \text{as } (-1)^{n^2} = (-1)^n \, !$$

$$= \sum_{n=1}^{\infty} \frac{e^{i4n^2\pi x}}{4n^2} - \sum_{n=0}^{\infty} \frac{e^{i(2n+1)^2\pi x}}{(2n+1)^2}$$

$$= \frac{1}{4}F(4x) - \left(F(x) - \sum_{n=1}^{\infty} \frac{e^{i4n^2\pi x}}{4n^2}\right)$$

$$= \frac{1}{2}F(4x) - F(x). \qquad \qquad \square$$

If we temporarily admit Proposition 7.3.2, we then have at hand the following two pieces of information:

$$F(x) = F(0) + a\sqrt{x} + bx + O(x^{3/2}) \text{ as } x \searrow 0,$$

with $a = i\pi e^{i\pi/4}$ and $b = -i\frac{\pi}{2}$, and

$$F(1 + x) = \frac{1}{2}F(4x) - F(x).$$

We can immediately see that, as $x \searrow 0$:

$$F(1 + x) = -\frac{1}{2}F(0) + a\left(\frac{1}{2}\sqrt{4} - 1\right)\sqrt{x} + b\left(\frac{1}{2}4 - 1\right)x + O(x^{3/2})$$

$$= F(1) + bx + O(x^{3/2}),$$

thanks to the amazing equality

$$\frac{1}{2}\sqrt{4} - 1 = 0 \, !$$

We can immediately deduce that

$$R(1 + x) = -\frac{\pi}{2}x + O(x^{3/2}) \text{ as } x \searrow 0,$$

which completes the proof of Theorem 7.3.1, thanks to the oddness of $R(1+x)$.

Thus, Proposition 7.3.2 constitutes the core of the proof; we aim to obtain it in the next section.

7.3.4 Remark The estimation

$$F(1 + x) = F(1) + bx + O(x^{3/2}) \text{ when } x \searrow 0$$

proves the differentiability of F at 1, since $F(1 - x) = \overline{F(1 + x)}$ and $\overline{b} = -b$.

7.3.5 The link between F and the Jacobi θ_0 function

Let H be the open upper half-plane of complex numbers with positive imaginary part and let $\overline{H} = \{z \in \mathbb{C} / \operatorname{Im} z \geqslant 0\}$ be its closure. The function F previously defined admits a natural extension to $\overline{H}$, defined by

$$F(z) = \sum_{n=1}^{\infty} \frac{e^{in^2 \pi z}}{n^2} \text{ for } z \in \overline{H},$$

which has the following properties:

- F is continuous and bounded on $\overline{H}$,

$$|F(z)| \leqslant \sum_{n=1}^{\infty} \frac{1}{n^2} = \frac{\pi^2}{6} \text{ for } z \in \overline{H} \, ; \tag{7.1}$$

- F is holomorphic on H, and

$$F'(z) = i\pi \sum_{n=1}^{\infty} e^{in^2 \pi z} \text{ for } z \in H.$$

Thus, F is (almost) a primitive of the *Jacobi θ_0 function* defined by

$$\theta_0(z) = \sum_{n \in \mathbb{Z}} e^{in^2 \pi z} = \sum_{n \in \mathbb{Z}} q^{n^2} \text{ for } z \in H \text{ and } q = e^{i\pi z}.$$

More precisely, we have

$$F'(z) = \frac{i\pi}{2}\left(\theta_0(z) - 1\right) \text{ for } z \in H.$$

The function θ_0 will play a central role in this chapter, essentially because it has some remarkable properties with respect to the action of the modular group **PSL**$(2, \mathbb{Z})$ on H. The most fundamental of these is contained in Theorem 7.3.6 below.

7.3.6 Theorem   *For all* $z \in H$,

$$\theta_0(z) = \frac{e^{i\pi/4}}{\sqrt{z}}\theta_0\left(-\frac{1}{z}\right), \tag{7.2}$$

where $\sqrt{}$ *denotes the principal branch[3] of the square root function on* $\mathbb{C} \setminus \mathbb{R}_-$.

Proof   This functional equation has two ingredients.

(1) *The Poisson summation formula.* Let $\mathcal{S}$ be the Schwartz space of smooth, rapidly decreasing functions:

$$\mathcal{S} = \left\{ f \in C^\infty(\mathbb{R}, \mathbb{C}) \;\middle|\; x^p f^{(q)}(x) \xrightarrow[x \to \pm\infty]{} 0 \text{ for all } p, q \geqslant 0 \right\}.$$

For any $f \in \mathcal{S}$, define the Fourier transform of f by[4]

$$\widehat{f}(x) = \int e^{-2i\pi xt} f(t)\, dt \text{ for } x \in \mathbb{R}.$$

Then,

$$\sum_{n\in\mathbb{Z}} \widehat{f}(n) = \sum_{n\in\mathbb{Z}} f(n). \tag{7.3}$$

To show this formula, it suffices to remark that the 1-periodic and C^∞ function $t \mapsto \sum_{n\in\mathbb{Z}} f(t + n)$ is the sum of its Fourier series, and that its Fourier coefficients are none other than the $\widehat{f}(n)$.

(2) *The computation of the Fourier transform of a Gaussian.* First of all, let us note that

$$\int e^{-\pi(t+a)^2}\, dt = 1 \text{ for any } a \in \mathbb{R}, \tag{7.4}$$

and that the left-hand side of (7.4) is an entire function of a. By analytic extension, (7.4) is thus valid for $a \in \mathbb{C}$. By choosing $a = iy$, $y \in \mathbb{R}$, we deduce that

$$\int e^{-\pi t^2} e^{-2i\pi t y}\, dt = e^{-\pi y^2}.$$

---

[3] In other words: $\sqrt{z} = \sqrt{r}\, e^{i\varphi/2}$ if $z = re^{i\varphi}, r > 0, -\pi < \varphi < \pi$. Recall that the function $z \mapsto \sqrt{z}$ is holomorphic in $\mathbb{C} \setminus \mathbb{R}_-$.

[4] Here, as usual, an integral without bounds is taken over $\mathbb{R}$.

Hence, if x is a positive real number, we have

$$\int e^{-\pi t^2 x} e^{-2i\pi t y} dt = \frac{1}{\sqrt{x}} e^{-\pi y^2/x},$$

that is to say,

$$\int e^{i\pi t^2(ix)} e^{-2i\pi t y} dt = \frac{e^{i\pi/4}}{\sqrt{ix}} e^{-i\pi y^2/(ix)}.$$

By analytic extension, we thus have

$$\int e^{i\pi t^2 z} e^{-2i\pi t y} dt = \frac{e^{i\pi/4}}{\sqrt{z}} e^{-i\pi y^2/z} \text{ for } z \in H.$$

The Poisson formula then gives

$$\sum_{n \in \mathbb{Z}} e^{i\pi n^2 z} = \frac{e^{i\pi/4}}{\sqrt{z}} \sum_{n \in \mathbb{Z}} e^{-i\pi n^2/z},$$

and hence (7.2). $\qquad\qquad\qquad\qquad\qquad\qquad\qquad\qquad\qquad\square$

7.3.7 Application to an estimation of F in a neighbourhood of 0

We are going to *exploit the equality (7.2) in order to deduce a functional equation satisfied by F*. Ideally, we would like to write

$$F(x) = F(0) + \int_0^x F'(t) \, dt = F(0) + \frac{i\pi}{2} \int_0^x (\theta_0(t) - 1) \, dt.$$

More correctly, let us fix $x > 0$. For $y \in \,]0, 1]$, we have

$$\int_0^x \theta_0(t+iy)dt = \int_0^x \left(1 + \frac{2}{i\pi} F'(t+iy)\right)dt = x + \frac{2}{i\pi}\left(F(x+iy) - F(iy)\right). \quad (7.5)$$

As the function F is continuous on $\overline{H}$, the right-hand side of (7.5) tends, as $y \searrow 0$, to

$$x + \frac{2}{i\pi}\left(F(x) - F(0)\right). \quad (7.6)$$

We now focus our interest on the left-hand side of (7.5). We have

$$\int_0^x \theta_0(t+iy)\,dt = e^{i\pi/4} \int_0^x \frac{1}{\sqrt{t+iy}} \theta_0\left(-\frac{1}{t+iy}\right) dt \quad \text{thanks to (7.2)}$$

$$= e^{i\pi/4} \int_0^x \frac{dt}{\sqrt{t+iy}} + \frac{2e^{i\pi/4}}{i\pi} \int_0^x \frac{1}{\sqrt{t+iy}} F'\left(-\frac{1}{t+iy}\right) dt$$

$$= 2e^{i\pi/4}\left(\sqrt{x+iy} - \sqrt{iy}\right)$$

$$+ \frac{2e^{i\pi/4}}{i\pi} \int_0^x (t+iy)^{3/2} \frac{d}{dt}\left(F\left(-\frac{1}{t+iy}\right)\right) dt.$$

An integration by parts then gives

$$\int_0^x (t + iy)^{3/2} \frac{d}{dt}\left(F\left(-\frac{1}{t+iy}\right)\right) dt$$

$$= (x + iy)^{3/2} F\left(-\frac{1}{x+iy}\right) - (iy)^{3/2} F\left(-\frac{1}{iy}\right)$$

$$- \frac{3}{2}\int_0^x \sqrt{t+iy}\, F\left(-\frac{1}{t+iy}\right) dt,$$

so that

$$\int_0^x \theta_0(t + iy)\, dt = 2e^{i\pi/4}\left(\sqrt{x+iy} - \sqrt{iy}\right)$$

$$+ \frac{2e^{i\pi/4}}{i\pi}\left((x + iy)^{3/2} F\left(-\frac{1}{x+iy}\right) - (iy)^{3/2} F\left(-\frac{1}{iy}\right)\right)$$

$$- \frac{3e^{i\pi/4}}{i\pi}\int_0^x \sqrt{t+iy}\, F\left(-\frac{1}{t+iy}\right) dt. \tag{7.7}$$

We now let y tend to 0 in the identity (7.7), while observing that, thanks to (7.1):

$$\left|\sqrt{t+iy}\, F\left(-\frac{1}{t+iy}\right)\right| \leqslant \frac{\pi^2}{6}(t^2 + y^2)^{1/4} \leqslant \frac{\pi^2}{6}(t^2 + 1)^{1/4} \in L^1([0, x]).$$

By the dominated convergence theorem,

$$\int_0^x \sqrt{t+iy}\, F\left(-\frac{1}{t+iy}\right) dt \to \int_0^x \sqrt{t}\, F\left(-\frac{1}{t}\right) dt \text{ as } y \searrow 0.$$

Finally, as $y \searrow 0$, the left-hand side of (7.5) tends to

$$2e^{i\pi/4}\sqrt{x} + \frac{2e^{i\pi/4}}{i\pi} x^{3/2} F\left(-\frac{1}{x}\right) - \frac{3e^{i\pi/4}}{i\pi}\int_0^x \sqrt{t}\, F\left(-\frac{1}{t}\right) dt. \tag{7.8}$$

Ultimately, (7.5), (7.6) and (7.8) provide the equality

$$x + \frac{2}{i\pi}\left(F(x) - F(0)\right) = 2e^{i\pi/4}\sqrt{x} + \frac{2e^{i\pi/4}}{i\pi} x^{3/2} F\left(-\frac{1}{x}\right)$$

$$- \frac{3e^{i\pi/4}}{i\pi}\int_0^x \sqrt{t}\, F\left(-\frac{1}{t}\right) dt,$$

from which we immediately deduce the following theorem.

7.3.8 Theorem   *For all $x > 0$,*

$$F(x) = F(0) + i\pi e^{i\pi/4}\sqrt{x} - \frac{i\pi}{2}x + e^{i\pi/4}x^{3/2} F\left(-\frac{1}{x}\right) - \frac{3}{2}e^{i\pi/4}\int_0^x \sqrt{t}\, F\left(-\frac{1}{t}\right) dt. \tag{7.9}$$

To complete the proof of Proposition 7.3.2, we only need to recall that F is bounded on the real axis.

7.3.9 Other rational points

As the function F is 2-periodic and satisfies $F(-x) = \overline{F(x)}$, if it is differentiable at $x \in \mathbb{R}$, then it is also differentiable at $-x$ and $x + 2$. Moreover, the equality (7.9) can also be written in the form

$$F(x) = F(0) + a\sqrt{x} + bx + \varphi(x)F\left(-\frac{1}{x}\right) + \psi(x) \text{ for } x > 0,$$

with $a = i\pi e^{i\pi/4}$ and $b = -\dfrac{i\pi}{2}$, the functions φ and ψ being defined by

$$\varphi(x) = -\frac{a}{2b}x^{3/2} \text{ and } \psi(x) = -\int_0^x \varphi'(t)F\left(-\frac{1}{t}\right)dt \text{ for } x > 0.$$

We observe that the functions φ and ψ are C^1 on $\mathbb{R}_+^*$, and that φ is never zero on this interval. Consequently, if F is differentiable at $x > 0$, then it is also differentiable at $-\dfrac{1}{x}$. More precisely, a simple calculation shows that

$$\text{if } F'(x) = b, \text{ then } F'\left(-\frac{1}{x}\right) = b.$$

By using Exercise 7.7, which describes the orbits of the action of the θ_0-modular group on the rational points, we can deduce from what we have already established, that is, the non-differentiability at 0 and the differentiability at 1 of the function F, the following two facts:

- the differentiability of F, *hence also of R and C*, at all the rational points that can be written with both numerator and denominator odd, the derivative of R (resp. C) at each of these points being $-\dfrac{\pi}{2}$ (resp. 0);
- the non-differentiability of F at all other rational numbers.

Note that this latter point implies only that if x is a rational point with numerator and denominator of different parity, *at least one* of the functions R and C is not differentiable at x. We will be more precise in Exercise 7.11, showing that in reality *neither R nor C is differentiable at x*. If we can wait until then, we can say that the proof of Gerver's Theorem 7.1.3 is complete.

The case of irrational points will prove to be much tougher.

7.4 Non-differentiability at the irrational points

This section is devoted to the proof of the non-differentiability of R at the irrational points (Hardy's Theorem 7.1.2). There are now numerous proofs of this result, more or less accessible to a non-specialist. We cite, for example, Duistermaat [52], Jaffard *et al.* [97], Jaffard [96] and Holschneider–Tchamitchian [91].

Duistermaat's proof is probably the most complete and accessible, and provides additional information on the behaviour of R in the neighbourhood of an irrational point x_0, depending on the degree of irrationality of x_0. Nonetheless, as this book has an historical flavour and gives pride of place to the works of Hardy and Littlewood, it is their proof that we will present,[5] while comparing it with that of Duistermaat. This proof has not aged very much, and its general principle is very clear, even if the technical details are sometimes difficult to follow.[6]

Before tackling the heart of the subject, we introduce a few definitions.

With $0 < \alpha \leqslant 1$ and a function $g : \mathbb{R} \to \mathbb{C}$, we say that g is α-*Hölderian* if there exists a constant $C > 0$ such that

$$|g(x) - g(y)| \leqslant C|x - y|^{\alpha} \text{ for } x, y \in \mathbb{R}.$$

We say that g is α-*Hölderian* (resp. *strongly α-Hölderian*) *at a given* x_0 if

$$g(x) - g(x_0) = O(|x - x_0|^{\alpha}) \text{ (resp. } o(|x - x_0|^{\alpha}) \text{ when } x \to x_0.$$

For example, if $g'(x_0)$ exists, g is 1-Hölderian at x_0; if, additionally, $g'(x_0) = 0$, then g is strongly 1-Hölderian at x_0.

We denote the three preceding conditions by

$$g \in \text{Lip}_{\alpha} \,;\; g \in \text{Lip}_{\alpha}(x_0) \,;\; g \in \text{lip}_{\alpha}(x_0)$$

respectively. We denote by $\alpha(x_0)$ the least upper bound of the $\alpha \geqslant 0$ such that $g \in \text{Lip}_{\alpha}(x_0)$. This bound is called the *Hölder exponent of g at x_0*.

The *modular group* Γ is the group of linear fractional maps

$$z \mapsto \frac{az + b}{cz + d}$$

of the complex projective line $\mathbb{P}^1(\mathbb{C})$ with integer coefficients ($a, b, c, d \in \mathbb{Z}$) and with determinant 1: $ad - bc = 1$. The group Γ is isomorphic to the quotient of $\mathbf{SL}(2, \mathbb{Z})$ by $\{-I, I\}$ (I being the identity matrix of order 2), which will allow us to identify the linear fractional map $z \mapsto \dfrac{az + b}{cz + d}$ and the matrix $\begin{bmatrix} a & b \\ c & d \end{bmatrix}$ – the quadruplet (a, b, c, d) being defined up to a change of sign –.

The composition of linear fractional maps corresponds to the matrix product. The group Γ is generated by the two linear fractional maps σ and τ, defined by

$$\sigma(z) = -\frac{1}{z} \text{ and } \tau(z) = z + 1$$

---

[5] It is, moreover, in the spirit of the approximate functional equation of the function θ_0 (see Chapter 9), due to Hardy and Littlewood.

[6] Especially because Hardy and Littlewood gave very few details!

(see Exercise 7.6), with the relations

$$\sigma^2 = I \, ; \quad \tau \text{ is of infinite order}; \quad (\sigma\tau)^3 = I.$$

The group Γ acts on the upper half-plane $H = \{z \in \mathbb{C}/\operatorname{Im} z > 0\}$, because if $g(z) = \dfrac{az+b}{cz+d}$, then

$$\operatorname{Im} g(z) = \frac{(ad-bc)\operatorname{Im} z}{|cz+d|^2} = \frac{\operatorname{Im} z}{|cz+d|^2},$$

hence $g(z) \in H$ if $z \in H$.

The Jacobi function, which we have already met in Section 7.3, is the function $\theta_0 : H \to \mathbb{C}$ defined by

$$\theta_0(z) = \sum_{n \in \mathbb{Z}} e^{i\pi n^2 z} = \sum_{n \in \mathbb{Z}} q^{n^2},$$

with $q = e^{i\pi z}$, $|q| = e^{-\pi \operatorname{Im} z} < 1$. Here we will also need the functions $\theta_1, \theta_2 : H \to \mathbb{C}$, defined by

$$\theta_1(z) = \sum_{n \in \mathbb{Z}} (-1)^n e^{i\pi n^2 z} \text{ and } \theta_2(z) = \sum_{n \in \mathbb{Z}} e^{i\pi (n-1/2)^2 z}.$$

We also set

$$\theta(z, a) = \sum_{n \in \mathbb{Z}} e^{i\pi n^2 z} e^{2i\pi na} \text{ for } z \in H \text{ and } a \in \mathbb{C},$$

so that $\theta_1(z) = \theta\left(z, \dfrac{1}{2}\right)$.

Finally, recall that

$$F(x) = \sum_{n=1}^{\infty} \frac{e^{i\pi n^2 x}}{n^2} = \sum_{n=1}^{\infty} \frac{\cos(n^2 \pi x)}{n^2} + i \sum_{n=1}^{\infty} \frac{\sin(n^2 \pi x)}{n^2} = C(x) + i R(x),$$

where R is the Riemann function that interested us from the beginning.

Here is a first simple result on the behaviour of R.

7.4.1 Theorem *The functions R and F satisfy:*

(1) $F \in \text{Lip}_{1/2}$, in particular, $R \in \text{Lip}_{1/2}$;

(2) $R \notin \text{lip}_{1/2}(0)$, its pointwise Hölder exponent α thus satisfying

$$\alpha(0) = \frac{1}{2}.$$

Proof (1) See Exercise 7.1. Another way to prove the result would be the following: let $E_n(F)$ be the distance from F to the space of trigonometric polynomials of degree $\leqslant n$ for the L^∞-norm. Then [126], for $0 < \alpha < 1$, we have

$$F \in \text{Lip}_\alpha \Leftrightarrow E_n(F) = O(n^{-\alpha}).$$

Here, approximating F by the partial sums of its Fourier series, we see that $E_n(F) = O(n^{-1/2})$, hence $F \in \text{Lip}_{1/2}$.
(2) See Section 7.2.   □

The fact that R is in $\text{Lip}_{1/2}$ (and not any better) says nothing *a priori* about its properties of differentiability; if $g \in \text{Lip}_1$, g is of bounded variation, hence differentiable almost everywhere, but for $\alpha < 1$, there exist certain $g \in \text{Lip}_\alpha$ that are everywhere non-differentiable, for example $g(x) = \sum_{n=1}^\infty \dfrac{\sin(2^n x)}{2^{n\alpha}}$. The fact that we are able to decide if R is differentiable at the different points of $\mathbb{R}$ is all the more difficult, and all the more remarkable.

We now proceed to Hardy's Theorem 7.1.2, which reasonably should also be attributed to Littlewood. With the above preparations, we are going to establish it in the following form.

7.4.2 Theorem [Hardy and Littlewood [74, 81]] *If x_0 is an irrational real number, the functions F, R and C do not belong to $\text{lip}_{3/4}(x_0)$. In particular, they are not differentiable at the point x_0.*

The Hardy–Littlewood proof has three ingredients.

(1) An *analysis using the Poisson kernel* (or rather its derivative). This provides a necessary condition for a function h to be in $\text{lip}_\alpha(x_0)$ in terms of the behaviour of a generating series associated with h, which for $h = R$ is none other than the real part of the Jacobi function θ_0 at certain points of H. Jaffard *et al.* and Holschneider–Tchamitchian replace this step by an analysis using wavelets, and Duistermaat does not need it at all.
(2) *The invariance of the trio* $(\theta_0, \theta_1, \theta_2)$ *composed of the Jacobi function θ_0 and its twin sisters θ_1 and θ_2 under the action of the modular group* Γ. This allows the determination of lower bounds of the modulus of $\theta_0(z)$, or even of the modulus of its real part at certain points of H. Here, Duistermaat's approach is different: he works only with a subgroup Γ_0 of the modular group, called the θ_0-*modular group*, consisting of the linear fractional maps $g : z \mapsto \dfrac{az + b}{cz + d}$ such that the reduction modulo 2 of the matrix

$\begin{bmatrix} a & b \\ c & d \end{bmatrix}$ is either I_2 or $\begin{bmatrix} 0 & 1 \\ 1 & 0 \end{bmatrix}$, which leaves θ_0 invariant. This allows him to avoid introducing the functions θ_1 and θ_2. In return, it requires a thorough study of the behaviour of R in a neighbourhood of a rational point $\frac{p}{q}$, depending on the respective parities of p and q. The method of Hardy and Littlewood avoids this stage.

(3) *An enhancement of the second ingredient with the help of the contin-ued fractions expansion of x_0*, and more precisely of the element g of Γ defined by

$$g(z) = (-1)^n \frac{q_{n-1}z - p_{n-1}}{q_n z - p_n},$$

where $\frac{p_n}{q_n}$ is a convergent of x_0. This approach requires the control of an infinity of pairs of consecutive convergents $\left(\frac{p_{n-1}}{q_{n-1}}, \frac{p_n}{q_n} \right)$, and thus forces the use of some combinatorics in the expansion of x_0, but the latter remains elementary and is interesting on its own. Here, Duistermaat needs only to control an infinity of convergents $\frac{p_n}{q_n}$, but in return uses the control of R in a neighbourhood of the points $\frac{p_n}{q_n}$. The superiority of his method is that it gives both upper and lower bounds, whereas the method of Hardy and Littlewood gives only lower bounds.

It should be noted that, in terms of the results, Duistermaat goes further than Hardy and Littlewood: he brings the following information concerning the Hölder exponent $\alpha(x_0)$, when x_0 is irrational (without speaking of the case of rational points). Let us suppose that there exist $\delta > 0$ and $d \geqslant 2$ such that

$$\left| x_0 - \frac{p_n}{q_n} \right| \geqslant \delta q_n^{-d} \text{ for all } n \geqslant 1,$$

where $\frac{p_n}{q_n}$ denotes the nth convergent of x_0. Then, we have

$$R \in \text{lip}_\alpha(x_0), \text{ where } \alpha = \frac{3}{4}\left(1 - d(d-2)\right).$$

In particular, we can affirm that

- if x_0 has bounded partial quotients, then $\alpha(x_0) = \frac{3}{4}$;
- $\alpha(x_0) = \frac{3}{4}$ for almost all irrational numbers x_0.

A very nice result of Jaffard [96] refines that of Duistermaat: if we denote by $\tau(x_0)$ the least upper bound of the set of real numbers τ for which there exists an infinity of n such that

- p_n and q_n are not both odd,
- $\left| x_0 - \dfrac{p_n}{q_n} \right| \leqslant \dfrac{1}{q_n^{\tau}}$,

then we have exactly

$$\alpha(x_0) = \frac{1}{2} + \frac{1}{2\tau(x_0)}.$$

In this chapter, we limit ourselves to the proof of Theorem 7.4.2.

7.4.3 First ingredient

7.4.4 Lemma *Let $f : \mathbb{R} \to \mathbb{C}$ be a continuous 2π-periodic function, $x_0 \in \mathbb{R}$, and $0 < \alpha < 1$. For $0 < r < 1$ and $x \in \mathbb{R}$, set*

$$f_r(x) = \sum_{n \in \mathbb{Z}} \widehat{f}(n) r^{|n|} e^{inx}.$$

Then, if $f \in \mathrm{lip}_\alpha(x_0)$, we have

$$f_r'(x_0) = \sum_{n \in \mathbb{Z}} in\, \widehat{f}(n) r^{|n|} e^{inx_0} = o\left((1-r)^{\alpha-1} \right) \text{ as } r \nearrow 1. \tag{7.10}$$

Proof Let

$$P_r(t) = \sum_{n \in \mathbb{Z}} r^{|n|} e^{int} = \frac{1 - r^2}{1 - 2r \cos t + r^2}$$

be the Poisson kernel. We have $f_r = f * P_r$, so that $f_r'(x_0) = f * P_r'(x_0)$ since differentiation and convolution commute, or again

$$f_r'(x_0) = \frac{1}{2\pi} \int_{-\pi}^{\pi} f(x_0 - t) P_r'(t)\, dt.$$

As P_r' has mean zero on $[-\pi, \pi]$, this can also be written as

$$f_r'(x_0) = \frac{1}{2\pi} \int_{-\pi}^{\pi} \left(f(x_0 - t) - f(x_0) \right) P_r'(t)\, dt. \tag{7.11}$$

Moreover, the hypothesis on f allows us to write

$$|f(x_0 - t) - f(x_0)| \leqslant \varepsilon(t) |t|^\alpha, \tag{7.12}$$

where $\varepsilon : \mathbb{R} \to \mathbb{R}_+$ is a bounded and even function that tends to 0 at 0. Moreover,

$$P_r'(t) = \frac{-2r(1 - r^2) \sin t}{(1 - 2r \cos t + r^2)^2},$$

with

$$1 - 2r\cos t + r^2 = (1 - r)^2 + 4r\sin^2\frac{t}{2}$$

$$\geqslant \delta\left((1-r)^2 + t^2\right) \text{ for } |t| \leqslant \pi \text{ and } \frac{1}{2} \leqslant r < 1,$$

where δ is a positive constant, because

$$|\sin u| \geqslant \frac{2}{\pi}|u| \text{ for } |u| \leqslant \frac{\pi}{2}.$$

As a result, for $|t| \leqslant \pi$ and $\frac{1}{2} \leqslant r < 1$, we have

$$|P_r'(t)| \leqslant \frac{C(1-r)|t|}{\left((1-r)^2 + t^2\right)^2},$$

where $C > 0$ is a new numerical constant. By inserting this inequality in (7.11) and taking account of (7.12), we obtain

$$|f_r'(x_0)| \leqslant \frac{C}{\pi}\int_0^\pi \frac{(1-r)t^{\alpha+1}\varepsilon(t)}{\left((1-r)^2 + t^2\right)^2}\,dt,$$

or again, by making the change of variable $t = (1-r)u$ and setting now that $\chi_r(u) = \varepsilon\left((1-r)u\right)\mathbf{1}_{0\leqslant u\leqslant\frac{\pi}{1-r}}$:

$$|f_r'(x_0)| \leqslant \frac{C}{\pi}\int_0^{+\infty} \frac{(1-r)^{\alpha+3}}{(1-r)^4}\frac{u^{\alpha+1}}{(1+u^2)^2}\chi_r(u)\,du$$

$$= \frac{C}{\pi}(1-r)^{\alpha-1}\int_0^{+\infty}\frac{u^{\alpha+1}}{(1+u^2)^2}\chi_r(u)\,du.$$

This last integral tends to zero as $r \nearrow 1$, by the hypothesis on f (which implies that $\chi_r(u) \longrightarrow 0$ when $r \nearrow 1$) and by the dominated convergence theorem. This completes the proof of Lemma 7.4.4. □

Now, if we apply Lemma 7.4.4 to the function $f : x \mapsto R\left(\frac{x}{\pi}\right)$ at the point $x_0\pi$, with x_0 an irrational real number, then

$$f_r'(x_0\pi) = \sum_{n=1}^\infty r^{n^2}\cos(\pi n^2 x_0),$$

and if we set $r = e^{-\pi y}$ with $y > 0$, we recognise, within $f_r'(x_0\pi)$, the value

$$\operatorname{Re}\sum_{n=1}^\infty e^{i\pi n^2(x_0+iy)} = \frac{1}{2}\operatorname{Re}\left(\theta_0(x_0 + iy)\right) - \frac{1}{2},$$

and the conclusion (7.10) can also be written

$$\text{if } R \in \text{lip}_\alpha(x_0), \text{ then } \text{Re}\left(\theta_0(x_0 + iy)\right) = o(y^{\alpha-1}) \text{ as } y \searrow 0.$$

Indeed, if $R \in \text{lip}_\alpha(x_0)$, then $f \in \text{lip}_\alpha(x_0\pi)$, and

$$1 - r \sim \pi y \text{ when } y \searrow 0.$$

The strategy of the proof of Hardy–Littlewood (with $\alpha = 3/4$) will thus consist of showing (where the notation Ω denotes the negation of o) that

$$x_0 \text{ irrational} \Rightarrow \text{Re}\left(\theta_0(x_0 + iy)\right) = \Omega(y^{-1/4}) \text{ when } y \searrow 0. \qquad (7.13)$$

The following ingredients are totally dedicated to the proof of (7.13).

7.4.5 Second ingredient

This section is the longest in the approach of Hardy and Littlewood, but at the same time consists of well-known results. We limit ourselves to a few naive rudiments about elliptic functions, entirely expressed in the Poisson summation formula (7.14) that follows. The alternative is to read the four monumental volumes of Tannery and Molk [172]! As remarkable as these works are, reading them is not indispensable here, if we only wish to understand the proof of Theorem 7.4.2 in detail.

We thus generalise the identity seen before for the Jacobi function to the function $\theta(z, a)$, as follows:

$$\theta(z, a) = \frac{e^{i\pi/4}}{\sqrt{z}} \sum_{n \in \mathbb{Z}} e^{-i\pi(n-a)^2/z} \text{ for } z \in H \text{ and } a \in \mathbb{R}. \qquad (7.14)$$

The functions φ and ψ defined by

$$\varphi(x) = e^{i\pi z x^2} \text{ and } \psi(x) = \varphi(x)e^{2i\pi ax} \text{ for } x \in \mathbb{R}$$

are in the Schwartz class $\mathcal{S}$, and $\widehat{\psi}(x) = \widehat{\varphi}(x - a)$. But we have already seen that $\widehat{\varphi}(x) = \dfrac{e^{i\pi/4}}{\sqrt{z}} e^{-i\pi x^2/z}$. The Poisson summation formula (7.3), applied to ψ, then gives

$$\theta(z, a) = \sum_{n \in \mathbb{Z}} \psi(n) = \sum_{n \in \mathbb{Z}} \widehat{\psi}(n) = \sum_{n \in \mathbb{Z}} \widehat{\varphi}(n - a),$$

which proves (7.14), still true for any $a \in \mathbb{C}$ (although we will not use this fact). This formula will enable us to describe completely the action of the generators σ and τ of the group Γ on the functions θ_0, θ_1 and θ_2, in the form of the following lemma.

7.4.6 Lemma *We have the following three pairs of relations, valid for* $z \in H$:

(1) $\theta_0(z) = \dfrac{e^{i\pi/4}}{\sqrt{z}} \theta_0(\sigma z)$; $\theta_0(z) = \theta_1(\tau z)$,

(2) $\theta_1(z) = \dfrac{e^{i\pi/4}}{\sqrt{z}} \theta_2(\sigma z)$; $\theta_1(z) = \theta_0(\tau z)$,

(3) $\theta_2(z) = \dfrac{e^{i\pi/4}}{\sqrt{z}} \theta_1(\sigma z)$; $\theta_2(z) = e^{-i\pi/4}\theta_2(\tau z)$.

Proof (1) The first relation has already been seen $\big($in (7.2)$\big)$; anyway, $\theta_0(z) = \theta(z, 0)$ so it is obtained again with (7.14). Moreover, changing z to $z + 1$ multiplies $e^{i\pi n^2 z}$ by $(-1)^n$, hence

$$\theta_1(\tau z) = \sum_{n \in \mathbb{Z}} (-1)^n (-1)^n e^{i\pi n^2 z} = \theta_0(z),$$

and similarly for $\theta_0(\tau z) = \theta_1(z)$.

(2) We have $\theta_1(z) = \theta\big(z, \frac{1}{2}\big)$ and so can apply (7.14).

(3) As σ is involutive, changing z to σz in part (2) gives us

$$\theta_2(z) = e^{-i\pi/4}\sqrt{\sigma z}\,\theta_1(\sigma z).$$

Moreover, if $z = re^{i\theta}$ with $r > 0$ and $0 < \theta < \pi$, we have $\sigma z = r^{-1}e^{i(\pi-\theta)}$, so that $\sqrt{\sigma z} = r^{-1/2}e^{i\left(\frac{\pi}{2} - \frac{\theta}{2}\right)} = \dfrac{i}{\sqrt{z}}$ and $\theta_2(z) = \dfrac{e^{i\pi/4}}{\sqrt{z}} \theta_1(\sigma z)$. Finally, changing z to τz multiplies $e^{i\pi\left(n - \frac{1}{2}\right)^2 z}$ by $e^{i\pi\left(n - \frac{1}{2}\right)^2} = e^{i\pi/4}$, as $n^2 - n$ is even; this proves the final relation of part (3). $\qquad\square$

In what follows, ε will be the unique ring homomorphism from $\mathbb{Z}$ onto $\mathbb{F}_2$ (the reduction modulo 2 homomorphism), where $\mathbb{F}_2 = \mathbb{Z}/2\mathbb{Z}$ is the field with two elements; if $n \in \mathbb{Z}$,

$$\varepsilon(n) = \begin{cases} 0 & \text{if } n \text{ is even,} \\ 1 & \text{if } n \text{ is odd.} \end{cases}$$

By extension, if $g = \begin{bmatrix} a & b \\ c & d \end{bmatrix} \in \Gamma$, we set

$$\varepsilon(g) = \begin{bmatrix} \varepsilon(a) & \varepsilon(b) \\ \varepsilon(c) & \varepsilon(d) \end{bmatrix},$$

and ε becomes a group homomorphism from the modular group Γ onto the group $\mathbf{GL}(2, \mathbb{F}_2)$. We note that $\varepsilon(g)$ is well-defined even though a, b, c, d are defined only up to a change of sign, since $\varepsilon(n) = \varepsilon(-n)$. The matrix $\varepsilon(g)$ can only take on one of the following *six* values, where I is the identity element of Γ.

7.4.7 Lemma *Let* $g = \begin{bmatrix} a & b \\ c & d \end{bmatrix} \in \Gamma$. *Then,* $\varepsilon(g)$ *is of the six types below.*

(1) $\varepsilon(g) = \begin{bmatrix} 1 & 0 \\ 0 & 1 \end{bmatrix} = \varepsilon(\sigma^2) = \varepsilon(I)$,

(2) $\varepsilon(g) = \begin{bmatrix} 1 & 1 \\ 0 & 1 \end{bmatrix} = \varepsilon(\tau)$,

(3) $\varepsilon(g) = \begin{bmatrix} 1 & 0 \\ 1 & 1 \end{bmatrix} = \varepsilon(\sigma\tau\sigma) = \varepsilon(\tau\sigma\tau)$,

(4) $\varepsilon(g) = \begin{bmatrix} 0 & 1 \\ 1 & 1 \end{bmatrix} = \varepsilon(\sigma\tau)$,

(5) $\varepsilon(g) = \begin{bmatrix} 0 & 1 \\ 1 & 0 \end{bmatrix} = \varepsilon(\sigma)$,

(6) $\varepsilon(g) = \begin{bmatrix} 1 & 1 \\ 1 & 0 \end{bmatrix} = \varepsilon(\tau\sigma)$.

In what follows, we say that g is "of type i" if $\varepsilon(g)$ *is of type i in Lemma 7.4.7.*

Proof First of all, if $g \in \Gamma$, then $\varepsilon(g) \in \mathbf{GL}(2, \mathbb{F}_2)$ and the group $\mathbf{GL}(2, \mathbb{F}_2)$ is a group of order 6 (moreover, isomorphic to the symmetric group $\mathbf{S}_3$). The rest of the proof is reduced to simple matrix calculations; for example, for type 3, if we choose $\sigma = \begin{bmatrix} 0 & -1 \\ 1 & 0 \end{bmatrix}$ and $\tau = \begin{bmatrix} 1 & 1 \\ 0 & 1 \end{bmatrix}$, we have $\sigma\tau\sigma = \begin{bmatrix} -1 & 0 \\ 1 & -1 \end{bmatrix}$, so that $\varepsilon(\sigma\tau\sigma) = \begin{bmatrix} 1 & 0 \\ 1 & 1 \end{bmatrix}$. Similarly, $\tau\sigma\tau = \begin{bmatrix} 1 & 0 \\ 1 & 1 \end{bmatrix}$, so that $\varepsilon(\tau\sigma\tau) = \begin{bmatrix} 1 & 0 \\ 1 & 1 \end{bmatrix}$. $\quad\square$

The two preceding lemmas will be exploited in the following way.

7.4.8 Lemma *Let* $T = \begin{bmatrix} a & b \\ c & d \end{bmatrix} \in \Gamma$, $j \in \{0, 1, 2\}$ *then there exists* $\ell \in \{0, 1, 2\}$ *such that*

$$\theta_j(z) = \frac{u}{\sqrt{cz+d}} \theta_\ell(Tz) \text{ for any } z \in H, \qquad (7.15)$$

where u is an 8th root of unity depending on T and j, but not on z, and where $\sqrt{cz+d}$ *denotes a holomorphic branch of the square root of* $cz + d$ *on the open set* $\mathbb{C} \backslash \mathbb{R}_-$. *For* $j = 0$, ℓ *is called the index of T and is written* $i(T)$. *The index* $i(T)$ *depends only on* $\varepsilon(T)$, *and can take the following values:*

(1) $i(T) = 0$ *if T is of type 1 or 5;*
(2) $i(T) = 1$ *if T is of type 2 or 6;*
(3) $i(T) = 2$ *if T is of type 3 or 4.*

Proof The equality (7.15) can be shown by induction on the length of the expression of T using the generators σ and τ. It is true if $T = I$; it suffices then to show that if it is true for T, it is also true for σT, τT and $\tau^{-1} T$. Now, if $T = \begin{bmatrix} a & b \\ c & d \end{bmatrix}$, we have

$$\sigma T = \begin{bmatrix} 0 & -1 \\ 1 & 0 \end{bmatrix} \begin{bmatrix} a & b \\ c & d \end{bmatrix} = \begin{bmatrix} -c & -d \\ a & b \end{bmatrix},$$

so that, via Lemma 7.4.6:

- if $\ell = 0$, then

$$\theta_j(z) = \frac{u}{\sqrt{cz+d}} \theta_0(Tz) = \frac{u}{\sqrt{cz+d}} \frac{e^{i\pi/4}}{\sqrt{Tz}} \theta_0(\sigma Tz) = \frac{v}{\sqrt{az+b}} \theta_0(\sigma Tz),$$

with $v^8 = 1$ and $\sigma Tz = \dfrac{-cz-d}{az+b}$, because if φ (resp. ψ) is a holomorphic branch of the square root of F (resp. G), then $\varphi\psi$ is a holomorphic branch of the square root of FG;

- if $\ell = 1$,

$$\theta_j(z) = \frac{u}{\sqrt{cz+d}} \theta_1(Tz) = \frac{u}{\sqrt{cz+d}} \frac{e^{i\pi/4}}{\sqrt{Tz}} \theta_2(\sigma Tz) = \frac{v}{\sqrt{az+b}} \theta_2(\sigma Tz) ;$$

- finally, if $\ell = 2$,

$$\theta_j(z) = \frac{u}{\sqrt{cz+d}} \theta_2(Tz) = \frac{u}{\sqrt{cz+d}} \frac{e^{i\pi/4}}{\sqrt{Tz}} \theta_1(\sigma Tz) = \frac{v}{\sqrt{az+b}} \theta_1(\sigma Tz).$$

Similarly, we have $\tau T = \begin{bmatrix} 1 & 1 \\ 0 & 1 \end{bmatrix} \begin{bmatrix} a & b \\ c & d \end{bmatrix} = \begin{bmatrix} a+c & b+d \\ c & d \end{bmatrix}$; hence, via Lemma 7.4.6:

- if $\ell = 0$,

$$\theta_j(z) = \frac{u}{\sqrt{cz+d}} \theta_0(Tz) = \frac{u}{\sqrt{cz+d}} \theta_1(\tau Tz) ;$$

- if $\ell = 1$,

$$\theta_j(z) = \frac{u}{\sqrt{cz+d}} \theta_1(Tz) = \frac{u}{\sqrt{cz+d}} \theta_0(\tau Tz) ;$$

- if $\ell = 2$,

$$\theta_j(z) = \frac{u}{\sqrt{cz+d}} \theta_2(Tz) = \frac{u}{\sqrt{cz+d}} e^{-i\pi/4} \theta_2(\tau Tz) = \frac{v}{\sqrt{cz+d}} \theta_2(\tau Tz).$$

The case of $\tau^{-1}T = \begin{bmatrix} a-c & b-d \\ c & d \end{bmatrix}$ can be handled similarly, which completes the proof of (7.15).

We now justify the fact that $i(T)$ depends only on $\varepsilon(T)$. If ever $\varepsilon(T_1) = \varepsilon(T_2)$, then, by setting $T := T_1^{-1}T_2$, T belongs to the kernel of the homomorphism ε, hence to the θ_0-modular group Γ_0 defined and studied in Exercise 7.7, which is none other than the subgroup of Γ generated by σ and τ^2. As each of the transformations σ and τ^2 is of index zero, and $T_2 = T_1 T$, we immediately find that $i(T_2) = i(T_1)$.

Finally, if we use the notation

$$j \xrightarrow{\ T\ } \ell$$

(that is, j leads to ℓ via T) to describe the relation (7.15), Lemma 7.4.6 says that

$$0 \xrightarrow{\ \sigma\ } 0, \quad 1 \xrightarrow{\ \sigma\ } 2, \quad 2 \xrightarrow{\ \sigma\ } 1,$$
$$0 \xrightarrow{\ \tau\ } 1, \quad 1 \xrightarrow{\ \tau\ } 0, \quad 2 \xrightarrow{\ \tau\ } 2.$$

Therefore:

- if T is of type 1, $i(T) = i(I) = 0$;
- if T is of type 2, $i(T) = i(\tau) = 1$;
- if T is of type 3, $i(T) = i(\sigma\tau\sigma) = 2$ as $0 \xrightarrow{\ \sigma\ } 0 \xrightarrow{\ \tau\ } 1 \xrightarrow{\ \sigma\ } 2$;
- if T is of type 4, $i(T) = i(\sigma\tau) = 2$ as $0 \xrightarrow{\ \tau\ } 1 \xrightarrow{\ \sigma\ } 2$;
- if T is of type 5, $i(T) = i(\sigma) = 0$;
- if T is of type 6, $i(T) = i(\tau\sigma) = 1$ as $0 \xrightarrow{\ \sigma\ } 0 \xrightarrow{\ \tau\ } 1$.

This completes the proof of Lemma 7.4.8. $\square$

7.4.9 Remark   For technical reasons that will be explained in Section 7.4.11, the transformations of type 1, 2, 5 and 6, that lead to $i(T) = 0$ or 1, will be considered "good" and those of type 3 or 4, that lead to $i(T) = 2$, will be considered "bad".

7.4.10 Lemma   *Suppose that* $T = \begin{bmatrix} a & b \\ c & d \end{bmatrix} \in \Gamma$ *is of type 1, 2, 5 or 6 and that, for a* $z \in H$ *and* $Q = e^{i\pi Tz}$, *we have* $|Q| \leqslant \rho < 1$. *Set*

$$\lambda = 2\Big(\rho + \frac{\rho^4}{1-\rho}\Big).$$

Then, we have the following two inequalities:

$$|\theta_0(z)| \geqslant |cz+d|^{-1/2}(1-\lambda) \tag{7.16}$$

and

$$|\arg \theta_\ell(Tz)| \leqslant \frac{\pi}{12} \ \textit{if}\ \lambda < 0.211,\ \textit{where}\ \ell = i(T).\qquad (7.17)$$

Proof By Lemma 7.4.8 we have:

$$\theta_0(z) = \frac{u}{\sqrt{cz+d}}\,\theta_\ell(Tz) = \frac{u}{\sqrt{cz+d}}\,(1 \pm 2Q \pm 2Q^4 \pm 2Q^9 \pm \ldots),$$

so that

$$|\theta_0(z)| = |cz+d|^{-1/2}|1 \pm 2Q \pm 2Q^4 \pm 2Q^9 \pm \ldots|$$
$$\geqslant |cz+d|^{-1/2}\left(1 - 2(\rho + \rho^4 + \rho^9 + \ldots)\right)$$
$$\geqslant |cz+d|^{-1/2}\left(1 - 2\rho - \frac{2\rho^4}{1-\rho}\right),$$

which gives (7.16).

  More precisely, we have

$$|\theta_\ell(Tz) - 1| \leqslant \lambda,$$

so that $\theta_\ell(Tz)$ belongs to the cone of vertex O and tangent to the circle of centre 1 and radius λ, whose half-angle at the vertex is

$$\alpha = \arctan \frac{\lambda}{\sqrt{1-\lambda^2}} \leqslant \arctan \frac{\lambda}{1-\lambda}$$

if $\lambda < 1$. See Figure 7.3.

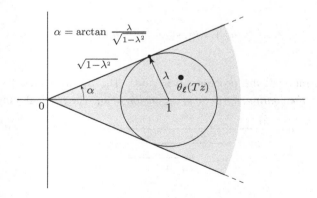

Figure 7.3

We thus have $|\arg \theta_\ell(Tz)| \leqslant \dfrac{\pi}{12}$ if

$$\arctan \frac{\lambda}{1-\lambda} \leqslant \frac{\pi}{12},$$

or, equivalently,

$$\frac{\lambda}{1-\lambda} \leqslant \tan \frac{\pi}{12} = 2 - \sqrt{3},$$

or again,

$$\lambda \leqslant \frac{2-\sqrt{3}}{3-\sqrt{3}} = \frac{3-\sqrt{3}}{6}.$$

Now, we have

$$\frac{3-\sqrt{3}}{6} > \frac{1.267}{6} > \frac{1.266}{6} = 0.211,$$

hence the inequality (7.17), which we will use later to find a lower bound for the real part of $\theta_0(z)$, and not just its modulus. □

7.4.11 A brief reminder on continued fractions

We limit ourselves here to a few basic notions. For more details, see [20, 87, 138]. The method essentially consists of extending Euclid's algorithm to the case of irrational numbers. Given an irrational number x_0, we write

$$\begin{cases}
x_0 & = a_0 + \omega_0, & \text{with } a_0 = [x_0] \in \mathbb{Z},\ 0 < \omega_0 < 1 \\
\dfrac{1}{\omega_0} & = a_1' = a_1 + \omega_1, & \text{with } a_1 \in \mathbb{N}^*,\ 0 < \omega_1 < 1 \\
\dfrac{1}{\omega_1} & = a_2' = a_2 + \omega_2, & \text{with } a_2 \in \mathbb{N}^*,\ 0 < \omega_2 < 1 \\
\ \vdots & & \\
\dfrac{1}{\omega_n} & = a_{n+1}' = a_{n+1} + \omega_{n+1}, & \text{with } a_{n+1} \in \mathbb{N}^*,\ 0 < \omega_{n+1} < 1 \\
\ \vdots & &
\end{cases}$$

As x_0 is irrational, the process never stops. The a_p are called the *partial quotients of the continued fractions expansion of x_0*. The rational number

$$a_0 + \cfrac{1}{a_1 + \cfrac{1}{a_2 + \cfrac{1}{a_3 + \cfrac{\ddots}{a_{n-1} + \cfrac{1}{a_n}}}}} \tag{7.18}$$

is denoted by $[a_0; a_1, \ldots, a_n]$ and is called the nth *convergent* of the expansion of x_0. The a'_n are called the *complete quotients* of the expansion of x_0. To obtain x_0 exactly, it suffices to replace a_n by a'_n in (7.18).

Moreover, we define the following two sequences of integers $(p_n)_{n \geqslant 0}$ and $(q_n)_{n \geqslant 0}$ by

$$p_{-1} = 1, \; p_0 = a_0 \text{ and } p_{n+1} = a_{n+1}p_n + p_{n-1} \text{ for } n \geqslant 0,$$
$$q_{-1} = 0, \; q_0 = 1 \text{ and } q_{n+1} = a_{n+1}q_n + q_{n-1} \text{ for } n \geqslant 0.$$

Then, induction on n shows that

$$[a_0; a_1, \ldots, a_n] = \frac{p_n}{q_n},$$

$$h_n := p_{n-1}q_n - p_nq_{n-1} = (-1)^n \text{ for } n \geqslant 0 \qquad (7.19)$$

$$\text{and } x_0 = \frac{a'_{n+1}p_n + p_{n-1}}{a'_{n+1}q_n + q_{n-1}} \text{ for } n \geqslant 0.$$

The relation (7.19) shows in particular that p_n and q_n are relatively prime. The fraction $\frac{p_n}{q_n}$ provides a good rational approximation of x_0, because

$$q_nx_0 - p_n = \frac{h_n}{q'_{n+1}} \text{ and } |q_nx_0 - p_n| = \frac{1}{q'_{n+1}},$$
$$\text{where } q'_{n+1} = a'_{n+1}q_n + q_{n-1} > q_{n+1}. \qquad (7.20)$$

Indeed,

$$q_nx_0 - p_n = \frac{q_n(a'_{n+1}p_n + p_{n-1}) - p_n(a'_{n+1}q_n + q_{n-1})}{q'_{n+1}}$$
$$= \frac{p_{n-1}q_n - p_nq_{n-1}}{q'_{n+1}} = \frac{h_n}{q'_{n+1}}.$$

In particular, as the sequence of integers $(q_n)_{n \geqslant 1}$ is increasing,

$$\frac{p_n}{q_n} \to x_0,$$

which justifies the usual notation

$$x_0 = [a_0; a_1, a_2, \ldots].$$

The link with the second ingredient is the following: let n be an integer $\geqslant 1$ and T_n the element of Γ defined by

$$T_nz = h_n\frac{q_{n-1}z - p_{n-1}}{q_nz - p_n},$$

that is,

$$T_n = \begin{bmatrix} h_n q_{n-1} & -h_n p_{n-1} \\ q_n & -p_n \end{bmatrix} = \begin{bmatrix} a & b \\ c & d \end{bmatrix}.$$

We obtain $T_n \in \Gamma$, since the coefficients of T_n are integers, and in addition

$$\det T_n = h_n(p_{n-1}q_n - p_n q_{n-1}) = h_n^2 = 1.$$

Moreover,

$$\varepsilon(T_n) = \begin{bmatrix} \varepsilon(q_{n-1}) & \varepsilon(p_{n-1}) \\ \varepsilon(q_n) & \varepsilon(p_n) \end{bmatrix}.$$

Let us attempt to explain how the intervention of the linear fractional map T_n comes naturally, thanks to two observations:

- the linear fractional map T_n sends the rational number $\dfrac{p_n}{q_n}$ to infinity;[7]
- if $z \in H$ and $q = e^{i\pi z}$, as

$$\theta_0(z) = 1 + 2\sum_{n=1}^{\infty} q^{n^2} \text{ and } \theta_1(z) = 1 + 2\sum_{n=1}^{\infty}(-1)^n q^{n^2},$$

if $|q|$ is small, $\theta_0(z)$ and $\theta_1(z)$ are close to 1; this is not the case for

$$\theta_2(z) = \sum_{n \in \mathbb{Z}} e^{i\pi(n-\frac{1}{2})^2 z},$$

which would be close to 0.

We can conclude from these two observations that *if T_n is "good"* (i.e., $i(T_n) = 0$ or 1) and if z is close to $\dfrac{p_n}{q_n}$, $\theta_{i(T_n)}(T_n z)$ will be close to 1; under these conditions, by Lemma 7.4.8, $\theta_0(z)$ will be close to $\dfrac{u}{\sqrt{q_n z - p_n}}$. This local approximation of the function θ_0 by a very simple function gives us a chance to find a lower bound for the modulus, or even the modulus of the real part, of $\theta_0(z)$ for $z \in H$ close to x_0.

We conclude this section with two definitions that will be useful later. Remember that the transformation T_n is called *good* if T_n is of one of the four types 1, 2, 5 or 6 of Lemma 7.4.7, in other words, if $i(T_n) = 0$ or 1; it is said to be *very good* if it is good and if in addition:

$$\text{either } a_{n+1} \geqslant 2, \text{ or } a_{n+1} = a_{n+2} = 1.$$

---

[7] However, as x_0 is irrational, there is no element of Γ that sends x_0 to infinity: the rational approximation of x_0 thus appears essential.

7.4.12 Third ingredient

The following crucial lemma gives the combinatorics of the continued fractions expansion of x_0 necessary to conclude.

7.4.13 Lemma (*1*) *The transformation T_n is good for an infinity of n.*
(*2*) *The transformation T_n is very good for an infinity of n.*
(*3*) *If T_n is very good, then we have $q'_{n+1} \geqslant \dfrac{3}{2} q_n$.*

Proof (1) The first point is easy: if T_n is bad, the second line of $\varepsilon(T_n)$ is made up of 1. But then, the first line of $\varepsilon(T_{n+1})$ is made up of 1, so T_{n+1} is of type 2 or 6 and in consequence is good.
(2) The proof of the second point is more difficult, because we are more demanding. We will show that *if T_n ($n \geqslant 1$) is good, but not very good, then one of the linear fractional maps T_{n+1}, T_{n+2}, T_{n+3} is very good.* This, with part (1), will allow us to conclude. Choose then $n \geqslant 1$ such that T_n is good but not very good. Then, T_n is of type 1, 2, 5 or 6, and additionally $a_{n+1} = 1$ and $a_{n+2} \geqslant 2$. We distinguish four cases, beginning with the easiest.

- *Case 1. T_n is of type 2.* Then, $\varepsilon(T_n) = \begin{bmatrix} 1 & 1 \\ 0 & 1 \end{bmatrix}$. Moreover, as $a_{n+1} = 1$,

$$\begin{cases} p_{n+1} = p_n + p_{n-1}, \\ q_{n+1} = q_n + q_{n-1}, \end{cases}$$

so that

$$\begin{cases} \varepsilon(p_{n+1}) & = \varepsilon(p_n) + \varepsilon(p_{n-1}) & = 1 + 1 & = 0, \\ \varepsilon(q_{n+1}) & = \varepsilon(q_n) + \varepsilon(q_{n-1}) & = 0 + 1 & = 1, \end{cases}$$

and hence $\varepsilon(T_{n+1}) = \begin{bmatrix} 0 & 1 \\ 1 & 0 \end{bmatrix}$. The linear fractional map T_{n+1} is thus of type 5, and as $a_{n+2} \geqslant 2$, T_{n+1} is very good.

- *Case 2. T_n is of type 6.* We can verify, exactly as in case 1, that T_{n+1} is of type 1. As $a_{n+2} \geqslant 2$, T_{n+1} is very good.

- *Case 3. T_n is of type 1.* Then $\varepsilon(T_n) = \begin{bmatrix} 1 & 0 \\ 0 & 1 \end{bmatrix}$ and, as $a_{n+1} = 1$, $\varepsilon(T_{n+1}) = \begin{bmatrix} 0 & 1 \\ 1 & 1 \end{bmatrix}$ (same calculation as in case 1): T_{n+1} is of type 4. Moreover,

$$\begin{aligned} 1 &= \varepsilon(q_{n+1}p_{n+2} - p_{n+1}q_{n+2}) \\ &= \varepsilon(q_{n+1})\varepsilon(p_{n+2}) - \varepsilon(p_{n+1})\varepsilon(q_{n+2}) \\ &= \varepsilon(p_{n+2}) - \varepsilon(q_{n+2}). \end{aligned}$$

The integers p_{n+2} and q_{n+2} are thus of different parity, which leads us to examine two sub-cases.

- *Case 3'.* $\varepsilon(p_{n+2}) = 1$ and $\varepsilon(q_{n+2}) = 0$. Then, $\varepsilon(T_{n+2}) = \begin{bmatrix} 1 & 1 \\ 0 & 1 \end{bmatrix}$. The linear fractional map T_{n+2} is thus of type 2, and T_{n+2} is very good unless $a_{n+3} = 1$ and $a_{n+4} \geqslant 2$. But in this case, $\varepsilon(T_{n+3}) = \begin{bmatrix} 0 & 1 \\ 1 & 0 \end{bmatrix}$ (type 5) and $a_{n+4} \geqslant 2$, so T_{n+3} is very good.

- *Case 3''.* $\varepsilon(p_{n+2}) = 0$ and $\varepsilon(q_{n+2}) = 1$. In this case, $\varepsilon(T_{n+2}) = \begin{bmatrix} 1 & 1 \\ 1 & 0 \end{bmatrix}$ (type 6). It follows that T_{n+2} is very good, except if $a_{n+3} = 1$ and $a_{n+4} \geqslant 2$. But then, $\varepsilon(T_{n+3}) = \begin{bmatrix} 1 & 0 \\ 0 & 1 \end{bmatrix}$ (type 1) and $a_{n+4} \geqslant 2$, so T_{n+3} is very good.

- *Case 4.* T_n is of type 5. An argument strictly identical to that developed for case 3 shows that T_{n+2} or T_{n+3} is very good.

(3) Let $n \geqslant 1$ be such that T_n is very good. Then, thanks to (7.20), we have

$$q'_{n+1} = a'_{n+1}q_n + q_{n-1} = (a_{n+1} + \omega_{n+1})q_n + q_{n-1}.$$

As $q_{n-1} \geqslant 0$, if $a_{n+1} \geqslant 2$, then $q'_{n+1} \geqslant 2q_n$. If, instead, $a_{n+1} = a_{n+2} = 1$, then

$$\begin{aligned} q'_{n+1} &= (a_{n+1} + \omega_{n+1})q_n + q_{n-1} \\ &= \left(a_{n+1} + \frac{1}{a_{n+2} + \omega_{n+2}}\right)q_n + q_{n-1} \\ &\geqslant \left(1 + \frac{1}{1 + \omega_{n+2}}\right)q_n \\ &\geqslant \frac{3}{2}q_n \text{ since } 0 < \omega_{n+2} < 1, \end{aligned}$$

which completes the proof of Lemma 7.4.13.   □

7.4.14 End of the proof of the Hardy–Littlewood theorem

Let x_0 be an irrational real number. As we have already said, to show that $R \notin \mathrm{lip}_{3/4}(x_0)$, we need to show (7.13), by exhibiting a sequence $(y_n)_{n \geqslant 1}$ of positive real numbers with limit zero such that

$$|\mathrm{Re}\,(\theta_0(x_0 + iy_n))| \geqslant \delta y_n^{-1/4}, \tag{7.21}$$

where δ is a positive numerical constant.

We start with something easier, enabling us to show that $F \notin \mathrm{lip}_{3/4}(x_0)$, where

$$F(x) = \sum_{n=1}^{\infty} \frac{e^{i\pi n^2 x}}{n^2},$$

in fact with the estimate

$$|\theta_0(x_0 + iy_n)| \geqslant \delta y_n^{-1/4}. \tag{7.22}$$

For this, let us fix an integer $n \geqslant 1$ and $y > 0$, and set $z = x_0 + iy$ and $Q = e^{i\pi T_n z}$. We have

$$\mathrm{Im}\, T_n z = \frac{y}{|q_n z - p_n|^2} = \frac{y}{(q_n x_0 - p_n)^2 + q_n^2 y^2} = \frac{y}{q_{n+1}^{\prime -2} + q_n^2 y^2}$$

by (7.20). To y, we assign the value that maximises this quotient, that is to say, $y = \dfrac{1}{q_n q_{n+1}'}$. We then set

$$y_n = \frac{1}{q_n q_{n+1}'}, \quad z_n = x_0 + iy_n \quad \text{and} \quad Q_n = e^{i\pi T_n z_n}.$$

Therefore,

$$\mathrm{Im}\, T_n z_n = \frac{q_{n+1}'}{2q_n}.$$

We always have $\mathrm{Im}\, T_n z_n \geqslant \dfrac{1}{2}$, because $q_{n+1}' \geqslant q_n$ for all $n \geqslant 1$. Moreover,

$$|Q_n| = e^{-\pi\, \mathrm{Im}(T_n z_n)} \leqslant e^{-\pi/2} < 0.2$$

so that, if we set $\rho = 0.21$,

$$\lambda = 2\left(\rho + \frac{\rho^4}{1 - \rho}\right) = 2\left(0.21 + \frac{(0.21)^4}{0.79}\right) < \frac{1}{2}.$$

Finally,

$$|q_n z_n - p_n|^2 = q_{n+1}'^{-2} + q_n^2 y_n^2 = \frac{2}{q_{n+1}'^2} < 2y_n.$$

The lower bound (7.16) in Lemma 7.4.10 then gives us, if T_n is good,

$$|\theta_0(x_0 + iy_n)| \geqslant |q_n z_n - p_n|^{-1/2}(1 - \lambda) \geqslant \frac{1}{2}(2y_n)^{-1/4} = \delta y_n^{-1/4}.$$

This proves (7.22) whenever the linear fractional map T_n is good, which is the case for an infinity of n by Lemma 7.4.13, and, combined with Lemma 7.4.4, shows the following highly non-trivial result:

$$\text{if } x_0 \in \mathbb{R} \setminus \mathbb{Q}, \text{ then } F \notin \mathrm{lip}_{3/4}(x_0).$$

But we want more, namely the same property[8] for $R = \mathrm{Im}\, F$, and for this we need to track not only the modulus of $\theta_0(x_0 + iy_n)$, but also its argument (this is the technically delicate point of the reasoning of Hardy and Littlewood). For this, the upper bound $|Q_n| \leqslant e^{-\pi/2} < 0.21$ is no longer sufficient. In Lemma 7.4.13, we go to a lot of trouble to improve this upper bound in the following way. Let I be the (infinite) set of integers $n \geqslant 1$ for which T_n is very good. If $n \in I$, we have, by setting $\ell = i(T_n)$ and $Q_n = e^{i\pi T_n z_n}$,

$$|Q_n| \leqslant e^{-3\pi/4} < 0.1 \text{ and } |\arg \theta_\ell(T_n z_n)| \leqslant \frac{\pi}{12}. \tag{7.23}$$

Indeed, if $n \in I$,

$$|Q_n| = e^{-\pi \,\mathrm{Im}\, T_n z_n} = \exp\left(-\pi \frac{q'_{n+1}}{2q_n}\right) \leqslant e^{-3\pi/4} < 0.1$$

by part (3) of Lemma 7.4.13. Then, if

$$\rho_n = 0.1 \text{ and } \lambda_n = 2\left(\rho_n + \frac{\rho_n^4}{1 - \rho_n}\right),$$

we obtain

$$\lambda_n = 2\left(0.1 + \frac{10}{9}(0.1)^4\right) < 2\left(0.1 + (0.1)^3\right) < 0.211$$

and the inequality (7.17) then gives us (7.23). Hence, (7.15) gives

$$\theta_0(z_n) = \frac{u_n}{\sqrt{q_n z_n - p_n}} \theta_\ell(T_n z_n) \text{ with } u_n = e^{ik\pi/4} \ (0 \leqslant k \leqslant 7). \tag{7.24}$$

Moreover,

$$q_n z_n - p_n = (q_n x_0 - p_n) + i q_n y_n = \frac{h_n + i}{q'_{n+1}}$$

by (7.20), hence an argument of $q_n z_n - p_n$ is $\dfrac{\pi}{4}$ if $h_n = 1$ and $\dfrac{3\pi}{4}$ if $h_n = -1$. If α is an argument of $\theta_\ell(T_n z_n)$ such that $|\alpha| \leqslant \dfrac{\pi}{12}$ (such an argument exists by (7.23)), (7.24) shows that an argument of $\theta_0(z_n)$ is

$$\arg \theta_0(z_n) = k\frac{\pi}{4} - \varepsilon\frac{\pi}{8} + \alpha,$$

with $k \in \{0, \ldots, 7\}$, $\varepsilon \in \{1, 3\}$ and $|\alpha| \leqslant \dfrac{\pi}{12}$. But then,

$$\arg \theta_0(z_n) = (2k - \varepsilon)\frac{\pi}{8} + \alpha.$$

---

[8] And also for $C = \mathrm{Re}\, F$, but this proof will not be explained here.

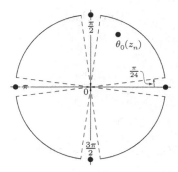

Figure 7.4

Hence, if $k' \in \mathbb{Z}$,

$$\arg \theta_0(z_n) - k' \frac{\pi}{2} = (2k - 4k' - \varepsilon) \frac{\pi}{8} + \alpha.$$

As $2k - 4k' - \varepsilon$ is an *odd* integer and $|\alpha| \leqslant \frac{\pi}{12}$, we obtain

$$\left| \arg \theta_0(z_n) - k' \frac{\pi}{2} \right| \geqslant \frac{\pi}{8} - \frac{\pi}{12} = \frac{\pi}{24}$$

(see Figure 7.4). It follows that

$$| \cos \arg \theta_0(z_n)| \geqslant \cos \left(\frac{\pi}{2} - \frac{\pi}{24} \right) = \sin \frac{\pi}{24},$$

so

$$|\mathrm{Re}\,(\theta_0(z_n))| \geqslant |\theta_0(z_n)| \sin \frac{\pi}{24} \geqslant \delta \sin \frac{\pi}{24} y_n^{-1/4}$$

by (7.22). Modifying if necessary the numerical constant δ, we obtain (7.21) for $n \in I$, which proves (7.13) and completes the proof of the Hardy–Littlewood Theorem 7.4.2.

We conclude with a few historical remarks. First of all, "a good gambler is always lucky", and Riemann had the knack of presenting the function R as a subject of study. The properties of perfect squares (via the Poisson summation formula and the fact that a Gaussian is its own Fourier transform) resonate amazingly.

Next, the study of the function R offers a fine example of the "blindness" of great mathematicians, or even of an entire scientific community. Hardy and Littlewood elaborately and repetitively used the functional equations of the theta functions to show the non-differentiability of R at the irrational points; instead, they could have proved the differentiability of R at 1 much more simply using a *single* functional equation of *the* function θ_0. Gerver obtained this differentiability differently, and the merit goes to Itatsu for his proof using this single functional equation, 70 years after Hardy and Littlewood . . .

Exercises

7.1. Set

$$F(x) = \sum_{n=1}^{\infty} \frac{e^{i\pi n^2 x}}{n^2} \text{ for } x \in \mathbb{R}.$$

(a) Show that if $x, h \in \mathbb{R}$ and $N \geqslant 1$,

$$|F(x+h) - F(x)| \leqslant \pi N |h| + 2N^{-1}.$$

(b) By choosing N as a function of h in an optimal manner, show now that $F \in \text{Lip}_{1/2}$.

7.2. The aim of this exercise is to show the non-differentiability of the Weierstrass function by using a method due to G. Freud. In what follows, a and b will be positive real numbers such that $a < 1$ and $ab \geqslant 1$. We set

$$W(t) = \sum_{n=0}^{\infty} a^n \cos(b^n t) \text{ for } t \in \mathbb{R}.$$

Let[9] ψ be a function in the Schwartz class such that
- $\widehat{\psi}(x) = 0$ if $x \leqslant b^{-1}$ or $x \geqslant b$,
- $\widehat{\psi}(1) = 1$,

with this choice for the definition of the Fourier transform:

$$\widehat{\psi}(x) = \int \psi(t) e^{-ixt} dt.$$

Finally, for $j \in \mathbb{N}$, we define the function $\psi_j : t \mapsto b^j \psi(b^j t)$.

(a) Show that $\int t^k \psi(t) \, dt = 0$ for all $k \in \mathbb{N}$.

(b) Let $f : \mathbb{R} \to \mathbb{C}$ be a measurable and bounded function. Suppose that f is differentiable at $t_0 \in \mathbb{R}$. Show that

$$f * \psi_j(t_0) = o(b^{-j}) \text{ as } j \to +\infty.$$

(c) Show that the function W is not differentiable at any point of $\mathbb{R}$.

(d) Is it the same for the function $\widetilde{W}$ defined by

$$\widetilde{W}(t) = \sum_{n=0}^{\infty} a^n \sin(b^n t) ?$$

---

[9] Remember that the Fourier transform is an isomorphism from the Schwartz class onto itself. In particular, any C^{∞} and compactly supported function is a Fourier transform.

7.3. The *Zygmund class* Λ is the set of continuous functions $f : \mathbb{R} \to \mathbb{R}$ for which there exists a constant $C > 0$ such that

$$|f(x + h) + f(x - h) - 2f(x)| \leqslant C|h| \text{ for } x, h \in \mathbb{R}.$$

Show that the function g defined by

$$g(x) = \sum_{n=1}^{\infty} \frac{\sin(2^n x)}{2^n}$$

is an element of Λ. By Exercise 7.2, the class Λ thus contains nowhere differentiable functions.

7.4. We define the *little Zygmund class* λ as the set of continuous functions $f : \mathbb{R} \to \mathbb{R}$, 2π-periodic and such that

$$\sup_{x \in \mathbb{R}} |f(x + h) + f(x - h) - 2f(x)| = o(h) \text{ as } h \to 0.$$

(a) Show that if $g \in \lambda$ has a local extremum at x, then g is differentiable at x.

(b) Show that in reality, the set of points where g is differentiable is dense in $\mathbb{R}$ and uncountable.

(c) Let $(\varepsilon_n)_{n \geqslant 1}$ be a sequence of real numbers converging to 0, and let φ be the function defined by

$$\varphi(x) = \sum_{n=1}^{\infty} \varepsilon_n \frac{\sin(2^n x)}{2^n}.$$

Show that φ is differentiable at certain points.

7.5. Let $f : \mathbb{R} \to \mathbb{C}$ be a continuous, 2π-periodic function. For $n \geqslant 1$, let $E_n(f)$ be the distance from f to the space of trigonometric polynomials of degree $\leqslant n$, in the sense of the norm $\|\cdot\|_\infty$. Suppose that

$$E_n(f) = O(n^{-1}) \text{ as } n \to +\infty.$$

(a) Show that there exists a sequence $(P_n)_{n \geqslant 0}$ of trigonometric polynomials, with P_n of degree $\leqslant 2^n$, such that

$$f = \sum_{n=0}^{\infty} P_n, \text{ with } \|P_n\|_\infty \leqslant C 2^{-n},$$

where C is a real constant.

(b) By using Bernstein's inequality,[10] show that

$$|P_n(x + h) + P_n(x - h) - 2P_n(x)| \leqslant 4^n h^2 \|P_n\|_\infty.$$

[10] Which states that $\|P'\|_\infty \leqslant d\|P\|_\infty$ if P is a trigonometric polynomial of degree $\leqslant d$ [120].

(c) Show that $f \in \Lambda$ (for the definition of Λ, see Exercise 7.3). One can show [126] that the converse is true: if $f \in \Lambda$, then $E_n(f) = O(n^{-1})$.

(d) State a similar necessary and sufficient condition to belong to λ (for the definition of λ, see Exercise 7.4). Prove the sufficiency of this condition.

7.6. Let G be the subgroup of the modular group Γ generated by σ and τ. The elements of Γ are seen here as bijections from the Riemann sphere $\mathbb{C} \cup \{\infty\}$ onto itself. If $a \in \mathbb{C}$, we denote by $\Omega_G(a)$ the orbit of a under the action of G, that is,

$$\Omega_G(a) = \{\delta(a), \delta \in G\} \subset \mathbb{C} \cup \{\infty\}.$$

(a) Show that $\Omega_G(0) = \Omega_G(\infty) = \mathbb{Q} \cup \{\infty\}$.

(b) Let $\gamma \in \Gamma$. Show that there exists a $\delta \in G$ such that $\delta(\infty) = \gamma(\infty)$.

(c) Show that there exists an $n \in \mathbb{Z}$ such that $\delta^{-1}\gamma = \tau^n$, and then that $\gamma \in G$. Hence, $G = \Gamma$.

7.7. The θ_0-modular group. The θ_0-*modular group* is the subgroup Γ_0 of Γ generated by σ and τ^2. Let B be the set of rational numbers $\frac{p}{q}$ with p and q odd and relatively prime, and let $A = (\mathbb{Q} \setminus B) \cup \{\infty\}$.

(a) Verify that $\Omega_{\Gamma_0}(0) = \Omega_{\Gamma_0}(\infty) \subset A$ and that $\Omega_{\Gamma_0}(1) \subset B$ (this is how we denote the orbits of 0, ∞ and 1 under the action of the group Γ_0).

(b) Show that these inclusions are in fact equalities.

(c) Show that Γ_0 is equal to the set of $g \in \Gamma$ such that

$$\varepsilon(g) = \begin{bmatrix} 0 & 1 \\ 1 & 0 \end{bmatrix} \text{ or } \varepsilon(g) = \begin{bmatrix} 1 & 0 \\ 0 & 1 \end{bmatrix}.$$

7.8. The Gauss sums

(a) Let a and b be two integers such that $0 \leqslant b < a$. Show that

$$\sqrt{\varepsilon}\, S(a, b, \varepsilon) := \sqrt{\varepsilon} \sum_{n \in \mathbb{Z}} e^{-\pi(na+b)^2 \varepsilon} \to \frac{1}{a} \text{ as } \varepsilon \searrow 0.$$

(b) Let $r = \frac{p}{q}$ be a rational number, with $p \in \mathbb{Z}$ and $q \in \mathbb{N}^*$ relatively prime, and of different parity.

(i) Show that

$$\theta_0(r + i\varepsilon) = \sum_{k=0}^{q-1} e^{i\pi k^2 r} S(q, k, \varepsilon),$$

and hence deduce that

$$\sqrt{\varepsilon}\, \theta_0(r + i\varepsilon) \to \frac{1}{\sqrt{q}} S_r \text{ as } \varepsilon \searrow 0, \text{ where } S_r = \frac{1}{\sqrt{q}} \sum_{k=0}^{q-1} e^{i\pi k^2 r}.$$

(ii) By Exercise 7.7, there exists an element g of the θ_0-modular group Γ_0 such that $g(r) = \infty$. Show that the linear fractional map g is "good", and more precisely of type 1 or 5, so that $i(g) = 0$ (with the notations of Lemmas 7.4.7 and 7.4.8). Deduce that, if we choose for $\sqrt{\ }$ the principal branch of the square-root function in $\mathbb{C} \setminus \mathbb{R}_-$, the 8th root of unity referred to in Lemma 7.4.8, associated with $T = g$ and with $j = 0$, is equal to $e^{i\pi/4} S_r$.

(iii) Establish the following discrete functional equations:

$$\begin{cases} S_0 &= 1, \\ S_{r+2} &= S_r, \\ S_{-1/r} &= e^{i\pi\rho(r)/4} S_r & \text{if } r \neq 0, \end{cases}$$

where

$$\rho(r) = \begin{cases} -1 & \text{if } r > 0 \\ 1 & \text{if } r < 0 \end{cases}$$

and note that these equalities provide an algorithm to calculate S_r.

(iv) Verify, for example, that if q is an even positive integer, then

$$S_{1/q} = e^{i\pi/4}.$$

(c) Hereafter, let q be an arbitrary positive integer. Define the Gauss sum

$$G_q = \sum_{k=0}^{q-1} e^{2ik^2\pi/q}.$$

Show that

$$G_q = \begin{cases} (1+i)\sqrt{q} & \text{if } q \equiv 0 \mod 4, \\ \sqrt{q} & \text{if } q \equiv 1 \mod 4, \\ 0 & \text{if } q \equiv 2 \mod 4, \\ i\sqrt{q} & \text{if } q \equiv 3 \mod 4. \end{cases}$$

7.9. An elementary method of Fourier analysis due to Cellérier [38]. Let a be an even integer $\geqslant 12$. Set

$$W(x) = \sum_{k=0}^{\infty} a^{-k} \sin(a^k x) \text{ for } x \in \mathbb{R}.$$

We provide an elementary proof that W is everywhere non-differentiable.

(a) For $x \in \mathbb{R}$ and $h > 0$, show the inequality

$$\left| \frac{\sin(x+h) - \sin x}{h} - \cos x \right| \leqslant \frac{h}{2}.$$

(b) We fix $x \in \mathbb{R}$, $n \in \mathbb{N}^*$ and set $h_n = \dfrac{2\pi}{a^n}$ and $h'_n = \dfrac{\pi}{a^n}$, as well as

$$\Delta_n = \frac{W(x + h_n) - W(x)}{h_n} \text{ and } \Delta'_n = \frac{W(x + h'_n) - W(x)}{h'_n}.$$

Show that

$$\Delta_n = \sum_{k=0}^{n-1} \cos(a^k x) + \varepsilon_n, \text{ with } |\varepsilon_n| \leqslant \frac{\pi}{a - 1},$$

and that

$$\Delta'_n = \sum_{k=0}^{n-1} \cos(a^k x) - \frac{2\sin(a^n x)}{\pi} + \varepsilon'_n, \text{ with } |\varepsilon'_n| \leqslant \frac{\pi}{2(a - 1)}.$$

(c) Suppose that W is differentiable at x. With the notations of the preceding question, verify that

$$\cos(a^n x) = \Delta_{n+1} - \Delta_n + \varepsilon_n - \varepsilon_{n+1} \text{ and } \sin(a^n x) = \frac{\pi}{2}(\Delta_n - \Delta'_n + \varepsilon'_n - \varepsilon_n),$$

and thus deduce the two inequalities

$$|\cos(a^n x)| \leqslant \frac{2\pi}{a - 1} + \delta_n \text{ and } |\sin(a^n x)| \leqslant \frac{3\pi^2}{4(a - 1)} + \delta'_n,$$

where δ_n and δ'_n tend to 0 when $n \to +\infty$.

(d) By using a famous formula of trigonometry, show that there is a contradiction.

7.10. Hardy and Littlewood series. In this exercise, let x be an irrational real number. We define $\Omega(\cdot)$ as the negation of $o(\cdot)$.

(a) Show that if

$$\sum_{k=1}^{n} \cos(\pi k^2 x) = o(\sqrt{n}) \text{ as } n \to +\infty,$$

then

$$\sum_{n=1}^{\infty} \cos(\pi n^2 x) r^{n^2} = o\big((1 - r)^{-1/4}\big) \text{ as } r \nearrow 1.$$

By using the assertion (7.13), by which

$$\mathrm{Re}\,(\theta_0(x + iy)) = \Omega(y^{-1/4}) \text{ as } y \searrow 0,$$

show that this leads to a contradiction.

(b) Prove Kronecker's lemma, which states that if $(a_n)_{n \geqslant 1}$ is a sequence of real numbers, and $(\lambda_n)_{n \geqslant 1}$ a non-decreasing sequence of positive real numbers tending to $+\infty$ such that the series $\sum \dfrac{a_n}{\lambda_n}$ converges, then

$$\frac{1}{\lambda_n} \sum_{k=1}^{n} a_k \to 0.$$

(c) Let α be a real number such that $0 < \alpha \leqslant \dfrac{1}{2}$. Show that the series

$$\sum \frac{\cos(\pi n^2 x)}{n^\alpha}$$

is divergent.[11]

7.11. Gerver's results for the function R. Let I be an open subset of $\mathbb{R}$ and $f : I \to \mathbb{C}$ a continuous function. We say that f is *pseudo-differentiable at* $x \in I$ if the two limits

$$\lim_{h \searrow 0} \frac{f(x+h) - f(x)}{\sqrt{h}} = \delta_+(f, x) \quad \text{and} \quad \lim_{h \searrow 0} \frac{f(x-h) - f(x)}{\sqrt{h}} = \delta_-(f, x)$$

exist.

(a) Show that the pseudo-derivatives follow the usual rules of computation:

$$\begin{cases} \delta_\pm(f + g, x) &= \delta_\pm(f, x) + \delta_\pm(g, x), \\ \delta_\pm(fg, x) &= \delta_\pm(f, x)g(x) + \delta_\pm(g, x)f(x). \end{cases}$$

(b) Let $g : I \to I$ be differentiable and non-decreasing. Show that we have the following chain rule:

$$\delta_\pm(f \circ g, x) = \sqrt{g'(x)}\, \delta_\pm\big(f, g(x)\big).$$

(c) Show that if f is differentiable at x, then $\delta_\pm(f, x) = 0$, and that if $\delta_+(f, x) \neq 0$ or $\delta_-(f, x) \neq 0$, then f is not differentiable at x.

---

[11] On the contrary, if $\alpha > 1/2$, our series is the Fourier series of a 2-periodic function, square-summable by the Riesz–Fischer theorem, and hence converges almost everywhere according to a deep theorem of Carleson. We can be more specific by using some results of the article [81]: while non-trivial, these results are much more elementary than those of Carleson. Indeed, on p. 214 of [81], we find the following estimate:

$$\text{if } a_n(x) = O(n^\rho), \text{ then } s_n(x) = O\big(n^{1/2}(\ln n)^{\rho/2}\big),$$

where the $a_n(x)$ are the partial quotients of the continued fractions expansion of x, and where $s_n(x) = \sum_{k=1}^{n} e^{i\pi k^2 x}$. Now, we will see in Exercise 9.6, in Chapter 9, that $a_n(x) = O(n^2)$ for almost all irrational x. An Abel transformation then shows that for such an x, the series
$$\sum \frac{e^{i\pi n^2 x}}{n^\alpha} \text{ is convergent.}$$

In what follows, we take $I = \mathbb{R}^*$ and $f = F$, with the notations as in the current chapter, in particular those of Section 7.3, p. 185 and of Exercise 7.7.

(d) Show that if F is pseudo-differentiable at $x > 0$, then it is also pseudo-differentiable at $\sigma x = -\dfrac{1}{x}$, with

$$\delta_\pm(F, x) = e^{i\pi/4} \sqrt{x}\, \delta_\pm(F, \sigma x).$$

(e) Show that $\delta_+(F, 0) = i\pi e^{i\pi/4} =: a$ and that $\delta_-(F, 0) = \overline{a}$.

(f) Show that $\delta_+(F, -x) = \overline{\delta_-(F, x)}$ and $\delta_-(F, -x) = \overline{\delta_+(F, x)}$ for any $x \in \mathbb{R}$.

(g) Let x be a rational number belonging to the orbit of 0 under the action of the θ_0-modular group.[12] Show that the arguments of the two complex numbers $\delta_+(F, x)$ and $\delta_-(F, x)$ can be chosen so that

$$\left| \arg \delta_+(F, x) - \arg \delta_-(F, x) \right| = \frac{\pi}{2}.$$

Then, deduce that the numbers $\delta_+(F, x)$ and $\delta_-(F, x)$ cannot both be real numbers, nor both pure imaginary numbers, and hence that *the functions R and C are not differentiable at x*.

(h) Calculate $\delta_+\left(F, 1/2\right)$ and $\delta_-\left(F, 1/2\right)$. Conclude that the function R (resp. C) is not left (resp. right) differentiable at $1/2$. Why is it plausible for R (resp. C) to be left (resp. right) differentiable at $1/2$?[13]

(i) Establish the functional equation

$$F\left(\frac{1}{2} + x\right) = \frac{1-i}{4} F(4x) + i F(x),$$

and use it to study the left and right differentiability of the functions R and C at $1/2$.

---

[12] In other words, a rational number with numerator and denominator of different parity by Exercise 7.7.

[13] The next question will confirm this.

8

Partitio numerorum

8.1 Introduction

Given an integer $n \geqslant 1$, a *partition* of n is a representation of n as a sum of positive integers, without regard to the order of the terms. We denote by $p(n)$ the number of partitions of the integer n, with the convention $p(0) = 1$. For example, $p(5) = 7$ since[1]

$$5 = 4+1 = 3+2 = 3+1+1 = 2+2+1 = 2+1+1+1 = 1+1+1+1+1.$$

We can also see $p(n)$ as the number of sequences[2] $(m_k)_{k \geqslant 1}$ of non-negative integers satisfying

$$n = \sum_{k=1}^{\infty} k m_k,$$

where the integer m_k is then the number of times the "atom" k is repeated in the corresponding partition of n. The study of the arithmetical function $p(n)$ dates back to Euler and, at least until the beginning of the twentieth century, it consisted essentially in the accumulation of algebraic identities, sometimes highly virtuosic (Euler's pentagonal theorem, Rogers–Ramanujan identities, etc.). It was Hardy and Ramanujan who became interested in the behaviour of $p(n)$ as $n \rightarrow +\infty$. Their work on partitions marked simultaneously the beginning and perhaps the apex of their brief collaboration, which lasted from 1914 – the year of Ramanujan's arrival in Cambridge – to early 1919, at the time of Ramanujan's return to India, where he died on 26 April 1920.

---

[1] We repeat, we do not distinguish the sums $3 + 2$ and $2 + 3$.
[2] Of course, such a sequence $(m_k)_{k \geqslant 1}$ is zero beyond a certain rank.

Their starting point is a fundamental idea due to Euler, based on the consideration of the *generating function*[3]

$$f(z) = \sum_{n=0}^{\infty} p(n)z^n$$

rather than the sequence $(p(n))_{n \geqslant 0}$. As we will see, the function f can be expressed in the form of an infinite product:

$$f(z) = \prod_{n=1}^{\infty} \frac{1}{1 - z^n} \text{ for } |z| < 1 \tag{8.1}$$

(identity due to Euler). Even if we remain in the domain of real numbers, this identity alone enables us to obtain non-trivial information about $p(n)$, for example the order of magnitude of $\ln p(n)$, which is $C\sqrt{n}$ (C being a real constant). This already informs us that $p(n)$ increases rapidly to infinity![4]

By using Tauberian methods, Hardy and Ramanujan obtained the following more precise estimate [84]:

$$\ln p(n) \sim K\sqrt{n}, \text{ where } K = \pi\sqrt{\frac{2}{3}}.$$

However, as Hardy remarked [79],[5] an asymptotic formula for the logarithm of an arithmetical function is a very coarse result: it makes no distinction, for example, between the orders of magnitude of $n^{1000}e^{K\sqrt{n}}$ and $n^{-1000}e^{K\sqrt{n}}$!

Having exhausted the resources of real analysis, Hardy and Ramanujan turned towards what at the time was known as the "theory of functions", today called "functions of a complex variable". We can indeed very naturally recover $p(n)$ knowing $f(z)$, via the Cauchy–Fourier formula

$$p(n) = \frac{1}{2i\pi} \int_{|z|=r} \frac{f(z)}{z^{n+1}} dz, \tag{8.2}$$

where r is an arbitrary element of the interval $]0, 1[$ and the circle $|z| = r$ is oriented counter-clockwise. We are thus reduced to the asymptotic study of an integral.

Let us put the function f aside momentarily, and replace it with a more congenial function, for example a rational function g whose poles are all situated on the unit circle. A natural idea to study the behaviour of the integral

---

[3] As Hardy remarks in [75], the power series constitute the "appropriate weapon" to tackle problems in additive number theory, simply because $z^m z^n = z^{m+n}$. The "machinery" adapted to multiplicative problems – for example the repartition of prime numbers – is more to be found in the theory of Dirichlet series (of the form $\sum a_n n^{-s}$), since $m^{-s} n^{-s} = (mn)^{-s}$.

[4] See Remark 8.3.7 and Exercise 8.1.

[5] With a severity that we could perhaps find excessive.

$\int_{|z|=r} \dfrac{g(z)}{z^{n+1}} dz$ as $n \to +\infty$ is to deform the contour of integration in order to *cross over the barrier of the singularities of g*. More precisely, if R is a real number greater than 1, we have

$$\frac{1}{2i\pi} \int_{|z|=R} \frac{g(z)}{z^{n+1}} dz = S_n + \frac{1}{2i\pi} \int_{|z|=r} \frac{g(z)}{z^{n+1}} dz,$$

where S_n is the sum of the residues of $\dfrac{g(z)}{z^{n+1}}$ at its different poles of modulus 1. But then, at least if g tends to 0 at infinity, the integral on the left is likely to be small compared with that on the right, which makes plausible the equivalence

$$\frac{1}{2i\pi} \int_{|z|=r} \frac{g(z)}{z^{n+1}} dz \sim -S_n.$$

For example, in the case of

$$g(z) = \frac{1}{(1-z)(1-z^3)(1-z^5)(1-z^7)} = \sum_{n=0}^{\infty} a_n z^n,$$

we find[6]

$$a_n = \frac{1}{2i\pi} \int_{|z|=r} \frac{g(z)}{z^{n+1}} dz \sim \frac{n^3}{630},$$

the integer a_n being the number of partitions of the integer n using uniquely the "atoms" 1, 3, 5 and 7.

Unfortunately, the function f is very far from being rational.[7] In this regard, Hardy and Ramanujan make an instructive comparison, in the article [85], between the problem of partitions and that of the distribution of prime numbers. Since Euler, we know that the latter is intimately linked to Riemann's zeta function and to certain associated arithmetical functions, such as the Λ von Mangoldt function[8] defined on $\mathbb{N}^*$ by

$$\Lambda(n) = \begin{cases} \ln p \text{ if } n = p^k, p \text{ prime}, k \geqslant 1 \\ 0 \text{ otherwise} \end{cases}$$

(in particular, $\Lambda(1) = 0$). The natural generating function of the sequence $\Lambda(n)$ is the *Dirichlet* series $\sum \Lambda(n) n^{-s}$, which we know how to sum:

$$\sum_{n=1}^{\infty} \Lambda(n) n^{-s} = -\frac{\zeta'(s)}{\zeta(s)} \text{ for } \operatorname{Re} s > 1.$$

---

[6] See Exercise 8.4 for a generalisation.

[7] Further on (see Remark 8.3.7), we will see that the order of magnitude of $f(x)$ is $e^{\frac{\pi^2}{6(1-x)}}$ as $x \nearrow 1^-$.

[8] See Chapter 3 of this book.

The famous prime number theorem can be stated as follows:

$$\psi(x) \sim x,$$

where ψ is the summatory function associated with Λ, defined by

$$\psi(x) = \sum_{n \leqslant x} \Lambda(n).$$

A first step of the proof consists of recovering the summatory function from the generating function, as in (8.2). The Cauchy–Fourier formula is replaced here by an inversion formula due to Perron [176]:

$$\psi(x) = -\frac{1}{2i\pi} \int_{c-i\infty}^{c+i\infty} \frac{\zeta'(s)}{\zeta(s)} \cdot \frac{x^s}{s} \, ds, \qquad (8.3)$$

where c is a real number > 1, x a *non-integer* real number and the integral is to be understood as $\lim_{T \to +\infty} \int_{c-iT}^{c+iT}$. The idea is then to shift the contour of integration to the left, to get it past the singularity $s = 1$. When we do this, the error introduced is equal to the residue at $s = 1$ of $-\dfrac{\zeta'(s)}{\zeta(s)} \cdot \dfrac{x^s}{s}$, which is equal to x. We thus obtain

$$\psi(x) = x - \frac{1}{2i\pi} \int_{c'-i\infty}^{c'+i\infty} \frac{\zeta'(s)}{\zeta(s)} \cdot \frac{x^s}{s} \, ds,$$

where c' is this time a real number < 1. It is reasonable to believe that the last integral above – if indeed it exists– will be negligible compared with that of (8.3) (since the exponent s of the integrand has diminished): this leads to the prime number theorem. Naturally, the argument that we have just given has numerous shady zones. Among others, it requires that the line $\mathrm{Re}\, s = 1$ is contained in a stripe without zeros of the function ζ; moreover, it faces difficulties related to the behaviour at infinity[9] of the function $\dfrac{\zeta'}{\zeta}$.

In summary:

- in the case of the prime number theorem, the generating function does not cause a problem related to its singularities (a unique pole of order 1), the difficulty is rather focused on its behaviour at infinity;
- in the case of the partitions of integers, there is no problem with the behaviour at infinity of the generating function (as it cannot be extended outside the unit disk), but the unit circle is bristling with absolutely unavoidable singularities. Our only hope is to let the radius r of the circle of integration of (8.2) *tend* to 1.

[9] Later developments of the theory, notably the works of Wiener, will show that these difficulties are in fact inessential.

Fortunately, amongst these difficulties, we have some substantial compensations: the function f has nice properties with respect to the modular group Γ of linear fractional maps of the form $z \mapsto \dfrac{az+b}{cz+d}$, where $a, b, c, d \in \mathbb{Z}$ satisfy $ad - bc = 1$ (a group already encountered in the preceding chapter). Concretely, f satisfies a set of functional equations that allow us to *obtain good estimations of f in the neighbourhood of the rational points of the unit circle* (we mean by this the points of the form $e^{2i\pi r}$, where $r \in \mathbb{Q}$). Among these points, 1 is, in a way, the worst singularity of f, as it is a pole of each of the factors in the infinite product of Euler's identity (8.1). By using only an approximation of f in the neighbourhood of 1, Hardy and Ramanujan obtained a first highly non-trivial estimate of $p(n)$:

$$p(n) \sim \frac{1}{4n\sqrt{3}} e^{K\sqrt{n}}, \text{ where } K = \pi\sqrt{\frac{2}{3}}.$$

In fact, they expressed their result in the following precise form, a bit mysterious for the moment, but to be discussed later:[10]

$$p(n) = \frac{1}{2\pi\sqrt{2}} \frac{d}{dn}\left(\frac{e^{K\lambda_n}}{\lambda_n}\right) + O(e^{H\sqrt{n}}),$$

where $\lambda_n = \left(n - \dfrac{1}{24}\right)^{\frac{1}{2}}$ and where H is a constant such that $H < K$.

With the use of not just one singularity of f, but a finite number of these: $1, -1, e^{\pm\frac{2i\pi}{3}}, \pm i$, etc., the same method provides an asymptotic expansion of $p(n)$ with an arbitrary – and fixed – number Q of terms:

$$p(n) = P_1(n) + P_2(n) + \cdots + P_Q(n) + R(n),$$

where

$$P_q(n) = L_q(n)\phi_q(n),$$

---

[10] We will give a detailed proof of this result in Section 8.4 of this chapter. The notation $\dfrac{d}{dn}\left(\dfrac{e^{K\lambda_n}}{\lambda_n}\right)$ might seem strange given that n is an integer. In general, if u is a differentiable function of a single real variable, the notation $\dfrac{d}{dn}u(n)$ simply means $u'(n)$. Hence,

$\dfrac{d}{dn}\left(\dfrac{e^{K\lambda_n}}{\lambda_n}\right)$ denotes the derivative of the function $t \mapsto \dfrac{e^{K\left(t-\frac{1}{24}\right)^{\frac{1}{2}}}}{\left(t - \dfrac{1}{24}\right)^{\frac{1}{2}}}$, calculated at the point n.

with

$$\phi_q(n) = \frac{q^{\frac{1}{2}}}{2\pi\sqrt{2}} \frac{d}{dn} \left(\frac{e^{\frac{K\lambda_n}{q}}}{\lambda_n} \right),$$

$L_q(n)$ being the sum of at most q $24q$th roots of unity whose exact form is of little importance here, and $R(n)$ a negligible term compared with $P_Q(n)$ as $n \to +\infty$.

Our story could have come to an end, had not Major P. A. MacMahon appeared on stage. A retired officer of the British Indian Army, he converted to the study of combinatorial problems, and was a formidable calculator. He provided Hardy and Ramanujan with the values[11] of $p(n)$ for $n \leqslant 200$, which enabled them to test the precision of their results. To their great surprise, $\sum_{q=1}^{8} P_q(200)$ gave an approximation of $p(200)$ accurate to less than 0.004.[12] From there, the grand idea – that Littlewood attributes to Ramanujan in [125] and qualifies as a *very great step* – was to attempt to obtain an *exact* formula for $p(n)$, by making *the integer Q dependent on n*, with this Q roughly on the order of $\sqrt{n}$. The final result of Hardy and Ramanujan is as follows:

$$p(n) = \sum_{q \leqslant \sqrt{n}} P_q(n) + O(n^{-1/4}). \tag{8.4}$$

From this, we can deduce that, for n large enough, $p(n)$ is the integer nearest to $\sum_{q \leqslant \sqrt{n}} P_q(n)$: the formula (8.4) is hence at the same time asymptotic and exact. Even if its practical use is compromised by the non-effectiveness of the error term, it allowed Lehmer to conjecture the value

$$p(721) = 161\,061\,755\,750\,279\,477\,635\,534\,762$$

by calculating a sum of around 20 terms.[13]

From there, it seems clear that Hardy and Ramanujan were musing about an exact formula of the form, for example,

$$p(n) = \sum_{q=1}^{\infty} L_q(n)\phi_q(n),$$

but in 1937 Lehmer proved that this series is divergent. It was Rademacher who obtained that same year a representation of $p(n)$ in the form of a convergent series [155]:

---

[11] For example, $p(200) = 3\,972\,999\,029\,388$.

[12] To see that Hardy's research depended on results equivalent to today's computerised numerical computations contributes to soften a bit the reputation of this paragon of pure mathematics that he himself worked hard to forge [78].

[13] The result was later confirmed by Rademacher.

$$p(n) = \sum_{q=1}^{\infty} L_q(n) \psi_q(n),$$

where

$$\psi_q(n) = \frac{q^{\frac{1}{2}}}{\pi\sqrt{2}} \frac{d}{dn} \left(\frac{\sinh\left(\frac{K\lambda_n}{q}\right)}{\lambda_n} \right).$$

The difference with the Hardy–Ramanujan theorem seems infinitesimal ("a very fortunate formal change", wrote Hardy in [79]): an exponential is replaced by a hyperbolic sine. Actually, this sinh is already present in one of the surprising statements contained in the first letter sent to Hardy by Ramanujan on 16 January 1913:

> The coefficient of x^n in $(1 - 2x + 2x^4 - 2x^9 + 2x^{16} + \ldots)^{-1}$ is the closest integer to
>
> $$\frac{1}{4n} \left(\cosh(\pi\sqrt{n}) - \frac{\sinh(\pi\sqrt{n})}{\pi\sqrt{n}} \right),$$

in other words, $\frac{1}{2} \frac{d}{dn} \left(\frac{\sinh(\pi\sqrt{n})}{\pi\sqrt{n}} \right)$! This led the distinguished number theorist Atle Selberg to write [166]:

> In the work on the partition function, studying the paper it seems clear to me that it must have been, in a way, Hardy who did not fully trust Ramanujan's insight and intuition, when he chose the other form of the terms in their expression, for a purely technical reason, which one analyses as not very relevant. **I think that if Hardy had trusted Ramanujan more, they should have inevitably ended with the Rademacher series. There is little doubt about that.**

In any case, one thing is clear: the merit of Ramanujan's "discovery" should be attributed wholly to the great British analyst, and this is really something.

In the rest of this chapter, we describe the proof of Rademacher's theorem in detail. In a way, this does nothing but simplify the analysis of Hardy and Ramanujan while perhaps better illustrating the value of their main idea: the "circle method", which later became fundamental in additive number theory and a few years later gave Hardy and Littlewood the solution to the Waring problem [82].

8.2 The generating function

The generating function is defined, for all suitable $z \in \mathbb{C}$, as

$$f(z) = \sum_{n=0}^{\infty} p(n)z^n.$$

8.2.1 Proposition [Euler] *For* $|z| < 1$, *we have*

$$f(z) = \prod_{n=1}^{\infty} \frac{1}{1 - z^n}. \tag{8.5}$$

Proof First, let us note that

$$\frac{1}{1 - z^n} = 1 + \frac{z^n}{1 - z^n},$$

where the series $\sum \frac{z^n}{1 - z^n}$ is normally convergent on all compact subsets of the open unit disk. Consequently, the right-hand side of (8.5) is a holomorphic function of the variable z in this same disk. By analytic extension, it thus suffices to establish (8.5) in the case where $z = x \in]0, 1[$. First of all, for each integer $k \geqslant 1$, we have

$$\frac{1}{1 - x^k} = \sum_{n=0}^{\infty} \varepsilon_k(n)x^n,$$

where $\varepsilon_k(n) = 1$ if k divides n, and 0 otherwise. For $N \geqslant 1$, by taking the product of N absolutely convergent series, we can thus infer that

$$\prod_{k=1}^{N} \frac{1}{1 - x^k} = \sum_{n=0}^{\infty} p_N(n)x^n, \tag{8.6}$$

where

$$p_N(n) = \sum_{i_1 + \ldots + i_N = n} \varepsilon_1(i_1) \ldots \varepsilon_N(i_N) = \sum_{i_1' + 2i_2' + \ldots + Ni_N' = n} 1.$$

Here, $p_N(n)$ is the number of partitions of the integer n using only the "atoms" $1, 2, \ldots, N$. To conclude, it remains to let N tend to $+\infty$ in (8.6). Now, for each $n \geqslant 0$, $p_N(n)x^n$ *increases*[14] to $p(n)x^n$ as $N \to +\infty$. By the Beppo Levi theorem, we thus have

---

[14] In fact, $p_N(n) = p(n)$ as soon as $N \geqslant n$.

$$\sum_{n=0}^{\infty} p_N(n)x^n \to \sum_{n=0}^{\infty} p(n)x^n \text{ as } N \to +\infty. \qquad (8.7)$$

The combination of (8.6) and (8.7) gives the result. $\qquad\qquad\qquad\square$

8.3 The Dedekind η function

Let H be the upper half-plane of complex numbers with positive imaginary part. The function η is defined by

$$\eta(z) = e^{\frac{i\pi z}{12}} \prod_{n=1}^{\infty} (1 - e^{2i\pi nz}) \text{ for } z \in H.$$

As, for $\operatorname{Im} z \geqslant a > 0$,

$$|e^{2i\pi nz}| = e^{-2\pi n \operatorname{Im} z} \leqslant e^{-2\pi na}, \text{ with } \sum_{n=1}^{\infty} e^{-2\pi na} < +\infty,$$

the infinite product that defines η converges normally on all half-planes of the form $\operatorname{Im} z \geqslant a$ contained in H, and hence the function η is holomorphic in H. The link between η and the generating function f is evident:

$$f(e^{2i\pi z}) = e^{\frac{i\pi z}{12}} \eta(z)^{-1}. \qquad (8.8)$$

8.3.1 The functional equation of the η function

Let Γ be the modular group of linear fractional maps[15]

$$T : H \to H, z \mapsto \frac{az+b}{cz+d}$$

of the upper half-plane, where $a, b, c, d \in \mathbb{Z}$ satisfy $ad - bc = 1$. It is convenient to write T in the form

$$T = \begin{bmatrix} a & b \\ c & d \end{bmatrix},$$

with the convention that we identify a matrix with its opposite. In this correspondence, the matrix product translates simply into composition of applications. The group Γ is generated[16] by the two linear fractional maps

$$\tau z = z + 1 \text{ and } \sigma z = -\frac{1}{z},$$

---

[15] See Chapter 7.
[16] See Exercise 7.6.

represented by the matrices

$$\begin{bmatrix} 1 & 1 \\ 0 & 1 \end{bmatrix} \text{ and } \begin{bmatrix} 0 & -1 \\ 1 & 0 \end{bmatrix}.$$

8.3.2 Theorem *The function η satisfies the following two functional equations, valid for $z \in H$:*

$$\eta(\tau z) = e^{\frac{i\pi}{12}} \eta(z) \tag{8.9}$$

and

$$\eta(\sigma z) = (-iz)^{\frac{1}{2}} \eta(z), \tag{8.10}$$

where the notation $(\cdot)^{\frac{1}{2}}$ denotes the principal branch of the square root in the slit plane $\mathbb{C} \setminus \mathbb{R}_-$, defined by

$$z^{\frac{1}{2}} = \sqrt{r}\, e^{\frac{i\theta}{2}} \ \text{if } z = re^{i\theta}, r > 0, |\theta| < \pi.$$

The first functional equation is evident, the second highly non-trivial. Given its importance, we provide two proofs, using methods of complex and real analysis, respectively.

The fact that τ and σ generate Γ can be combined with our two functional equations to give the following result.

8.3.3 Theorem *Let a, b, c, d be integers satisfying $ad - bc = 1$, and T the linear fractional map defined by*[17]

$$Tz = \frac{az + b}{cz + d} \ \text{for } z \in H.$$

Then there exists a 24th root of unity $\omega = \omega(a, b, c, d)$ such that

$$\eta(z) = \frac{\omega}{(cz + d)^{\frac{1}{2}}} \eta(Tz) \text{ for } z \in H, \tag{8.11}$$

where $(cz + d)^{\frac{1}{2}}$ again denotes the principal branch of the square root in the slit plane $\mathbb{C} \setminus \mathbb{R}_-$.

Proof The proof is by induction on the length n of the expression of T in the form of a word of length n written with the alphabet $\{\tau, \tau^{-1}, \sigma\}$. The result is clear for the word of length zero $T = I_2$ (identity matrix of order 2). Then it

---

[17] In the case $c = 0$, we take by convention $a = d = 1$, so that $Tz = z + b$ and $(cz + d)^{\frac{1}{2}} = 1$ for $z \in H$.

suffices to show that if (8.11) is true for T, it also holds for τT, $\tau^{-1} T$ and σT. We write

$$T = \begin{bmatrix} a & b \\ c & d \end{bmatrix},$$

and distinguish three cases.

- First of all

$$\tau T = \begin{bmatrix} 1 & 1 \\ 0 & 1 \end{bmatrix} \cdot \begin{bmatrix} a & b \\ c & d \end{bmatrix} = \begin{bmatrix} a+c & b+d \\ c & d \end{bmatrix},$$

so that the denominator of $\tau T(z)$ is $cz + d$. Moreover, by using successively the induction hypothesis and the first functional equation of Theorem 8.3.2, we can write

$$\eta(z) = \frac{\omega}{(cz+d)^{\frac{1}{2}}} \eta(Tz) = \frac{\omega'}{(cz+d)^{\frac{1}{2}}} \eta(\tau Tz),$$

where $\omega' := \omega e^{-\frac{i\pi}{12}}$ is indeed a 24th root of unity.
- The case of $\tau^{-1} T$ is handled similarly.
- Finally, consider the case of σT. We have

$$\sigma T = \begin{bmatrix} 0 & -1 \\ 1 & 0 \end{bmatrix} \cdot \begin{bmatrix} a & b \\ c & d \end{bmatrix} = \begin{bmatrix} -c & -d \\ a & b \end{bmatrix}.$$

This time, the denominator of $\sigma T(z)$ is $az + b$. By using successively the induction hypothesis and the second functional equation of Theorem 8.3.2,[18] we obtain

$$\eta(z) = \frac{\omega}{(cz+d)^{\frac{1}{2}}} \eta(Tz) = \frac{\omega}{(cz+d)^{\frac{1}{2}}} \left(i \frac{cz+d}{az+b} \right)^{\frac{1}{2}} \eta(\sigma Tz)$$

$$= \frac{\omega'}{(az+b)^{\frac{1}{2}}} \eta(\sigma Tz)$$

with $\omega' = \pm \omega e^{\frac{i\pi}{4}}$, hence again $\omega'^{24} = 1$. To obtain the last equality, we used the following fact implicitly: if φ (resp. ψ) is a holomorphic branch of the square root of F (resp. G), then $\varphi\psi$ is a holomorphic branch of the square root of FG. $\qquad\square$

---

[18] In the form

$$\eta(z) = (-i\sigma z)^{\frac{1}{2}} \eta(\sigma z) = \left(\frac{i}{z} \right)^{\frac{1}{2}} \eta(\sigma z).$$

In the special case where $T = \sigma$, we have

$$\eta(z) = \left(\frac{i}{z}\right)^{\frac{1}{2}} \eta(\sigma z) = \frac{e^{\frac{i\pi}{4}}}{z^{\frac{1}{2}}} \eta(\sigma z) \text{ for } z \in H,$$

so that

$$\omega = e^{\frac{i\pi}{4}}.$$

What do the functional equations (8.11) tell us about the generating function f of the partitions? Thanks to (8.8), we immediately obtain the following result.[19]

8.3.4 Theorem *To each element* $T = \begin{bmatrix} a & b \\ c & d \end{bmatrix}$ *of* Γ, *we can associate a* 24*th root of unity* $\omega = \omega(a, b, c, d)$ *such that*

$$f(e^{2i\pi z}) = \omega (cz + d)^{\frac{1}{2}} e^{\frac{i\pi}{12}(z - Tz)} f(e^{2i\pi Tz}) \text{ for } z \in H. \tag{8.12}$$

The root of unity appearing in this theorem is the inverse of that in Theorem 8.3.3. In the special case where $T = \sigma = \begin{bmatrix} 0 & -1 \\ 1 & 0 \end{bmatrix}$, we thus have

$$\omega = e^{-\frac{i\pi}{4}}. \tag{8.13}$$

The identities (8.12) are going to provide good approximations of the generating function f in a neighbourhood of the rational points of the unit circle, that is, points of the form $e^{\frac{2ip\pi}{q}}$, $0 \leqslant p < q$, with p and q relatively prime. We begin by making explicit a linear fractional map of the group Γ that maps the rational $\frac{p}{q}$ to infinity: T is defined by

$$Tz = \frac{-uz - v}{qz - p},$$

where u and v satisfy Bézout's identity $up + vq = 1$. Let ω be the 24th root of unity associated with this T in (8.12), and z a complex number with positive real part. By setting $Z = \frac{p}{q} + \frac{iz}{q} \in H$, we have

---

[19] For the case $c = 0$, the convention is the same as in Theorem 8.3.3.

$$qZ - p = q\left(\frac{p}{q} + \frac{iz}{q}\right) - p = iz,$$

$$T(Z) = \frac{-u\left(\frac{p}{q} + \frac{iz}{q}\right) - v}{q\left(\frac{p}{q} + \frac{iz}{q}\right) - p} = \frac{-\frac{1}{q}(iuz + up + vq)}{iz} = -\frac{u}{q} - \frac{1}{iqz}$$

and

$$Z - T(Z) = \frac{p}{q} + \frac{iz}{q} - \left(-\frac{u}{q} - \frac{1}{iqz}\right) = \frac{iz}{q} + \frac{1}{iqz} + \frac{p+u}{q}.$$

By applying (8.12) to Z instead of z, we find

$$f\left(e^{2i\pi Z}\right) = \omega\,(qZ - p)^{\frac{1}{2}}\,e^{\frac{i\pi}{12}(Z - T(Z))}\,f(e^{2i\pi T(Z)})$$

$$= \omega(iz)^{\frac{1}{2}}\,e^{\frac{i\pi(p+u)}{12q}}\,e^{\frac{\pi}{12qz} - \frac{\pi z}{12q}}\,f\left(e^{-\frac{2i\pi u}{q} - \frac{2\pi}{qz}}\right).$$

By setting

$$\omega_{p,q} = \omega\, i^{\frac{1}{2}}\,e^{\frac{i\pi(p+u)}{12q}} = \omega\, e^{i\frac{\pi}{4}}\,e^{\frac{i\pi(p+u)}{12q}},$$

we finally obtain

$$f\left(e^{\frac{2i\pi p}{q} - \frac{2\pi z}{q}}\right) = \omega_{p,q}\,z^{\frac{1}{2}}\,e^{\frac{\pi}{12qz} - \frac{\pi z}{12q}}\,f\left(e^{-\frac{2i\pi u}{q} - \frac{2\pi}{qz}}\right) \text{ for } \operatorname{Re} z > 0.$$

$$(8.14)$$

This time, $\omega_{p,q}$ is a $24q$th root of unity.

In the special case where $p = 0$ and $q = 1$ (which corresponds to the point 1 of the unit circle), we can choose $(u, v) = (0, 1)$, which gives $T = \sigma$ and (see (8.13)) $\omega = e^{-\frac{i\pi}{4}}$, so that

$$\omega_{0,1} = 1. \tag{8.15}$$

Now, if z is close to 0, $f\left(e^{-\frac{2i\pi u}{q} - \frac{2\pi}{qz}}\right)$ is close to $f(0) = 1$, so that we have at hand the approximation

$$f\left(e^{\frac{2i\pi p}{q} - \frac{2\pi z}{q}}\right) \simeq \omega_{p,q}\,z^{\frac{1}{2}}\,e^{\frac{\pi}{12qz} - \frac{\pi z}{12q}} \tag{8.16}$$

near the rational point $e^{\frac{2i\pi p}{q}}$ of the unit circle. This approximation serves as the basis of the fine study of the integral (8.2) to follow in this chapter.

8.3.5 Proof of the functional equation by complex analysis

Here we present a proof of the second functional equation of Theorem 8.3.2 (the first is evident) using complex variable methods, due to Siegel. Remember that H denotes the upper half-plane of the complex plane, and that

$$\eta(z) = e^{\frac{i\pi z}{12}} \prod_{n=1}^{\infty} (1 - e^{2i\pi nz}) \text{ for } z \in H.$$

We thus need to show that

$$\eta\left(-\frac{1}{z}\right) = (-iz)^{\frac{1}{2}} \eta(z) \text{ for } z \in H. \tag{8.17}$$

By the uniqueness of the analytic extension, it suffices to satisfy (8.17) when $z = ia$ with $a > 0$. Thus we must see that

$$e^{-\frac{\pi}{12a}} \prod_{n=1}^{\infty} (1 - e^{-\frac{2\pi n}{a}}) = \sqrt{a}\, e^{-\frac{\pi a}{12}} \prod_{n=1}^{\infty} (1 - e^{-2\pi na}),$$

or equivalently that

$$-\frac{\pi}{12a} + \sum_{n=1}^{\infty} \log(1 - e^{-\frac{2\pi n}{a}}) = -\frac{\pi a}{12} + \frac{1}{2}\log a + \sum_{n=1}^{\infty} \log(1 - e^{-2\pi na}).$$

In order to get rid of the transcendental function log, we use the following lemma.

8.3.6 Lemma *For $0 \leqslant x < 1$, we have*

$$\sum_{n=1}^{\infty} \log(1 - x^n) = \sum_{m=1}^{\infty} \frac{1}{m} \cdot \frac{x^m}{x^m - 1}. \tag{8.18}$$

The proof is very simple: we expand each log as a series, permute two summations and re-sum a geometric series to obtain (8.18).

8.3.7 Remark This lemma is all we need to obtain an estimation of the generating function of the sequence $(p(n))_{n \geqslant 0}$. For $x \in [0, 1[$, we have

$$\ln f(x) = -\sum_{n=1}^{\infty} \ln(1 - x^n) = \sum_{m=1}^{\infty} \frac{x^m}{m(1 - x^m)}.$$

Using the bounds

$$m(1 - x)x^{m-1} \leqslant 1 - x^m = (1 - x)(1 + x + \ldots + x^{m-1}) \leqslant m(1 - x),$$

we obtain

$$\frac{1}{1-x} \sum_{m=1}^{\infty} \frac{x^m}{m^2} \leqslant \ln f(x) \leqslant \frac{x}{1-x} \sum_{m=1}^{\infty} \frac{1}{m^2}$$

and hence the equivalence

$$\ln f(x) \sim \frac{\pi^2}{6(1-x)} \quad \text{as } x \to 1^-.$$

In what follows, we only use the upper bound

$$\ln f(x) \leqslant \frac{\pi^2}{6\left(x^{-1}-1\right)} \quad \text{for } x \in \left]0, 1\right[. \tag{8.19}$$

Using the lemma for $x = e^{-\frac{2\pi}{a}}$ and $x = e^{-2\pi a}$, we are reduced to showing

$$-\frac{\pi}{12a} + \sum_{m=1}^{\infty} \frac{1}{m} \cdot \frac{e^{-\frac{2\pi m}{a}}}{e^{-\frac{2\pi m}{a}} - 1} = -\frac{\pi}{12} a + \frac{1}{2} \log a + \sum_{m=1}^{\infty} \frac{1}{m} \cdot \frac{e^{-2\pi ma}}{e^{-2\pi ma} - 1},$$

or, rearranging,

$$\sum_{m=1}^{\infty} \frac{1}{m} \cdot \frac{1}{e^{2\pi ma} - 1} - \sum_{m=1}^{\infty} \frac{1}{m} \cdot \frac{1}{e^{\frac{2\pi m}{a}} - 1} + \frac{\pi}{12}(a - a^{-1}) = \frac{1}{2} \log a.$$

Setting

$$\Phi(a) = \sum_{m=1}^{\infty} \frac{1}{m} \cdot \frac{1}{e^{2\pi ma} - 1},$$

the last identity can be written in a more compressed form:

$$\Phi(a) - \Phi\left(\frac{1}{a}\right) + \frac{\pi}{12}(a - a^{-1}) = \frac{1}{2} \log a. \tag{8.20}$$

To validate the form (8.20), we apply the residue theorem to the following meromorphic function:

$$F(z) = \frac{1}{z} \cdot \frac{e^{2\pi z} + 1}{e^{2\pi z} - 1} \cdot \frac{e^{2i\pi \frac{z}{a}} + 1}{e^{2i\pi \frac{z}{a}} - 1}.$$

This function, introduced by Siegel, is well adapted to the proof of (8.20). More generally, in fact, to calculate a sum of the form $\sum_{k \in \mathbb{Z}} f(k)$, where f is a rational function without integer poles, we apply the residue theorem to the function $F(z) = f(z) \cdot \frac{e^{2i\pi z} + 1}{e^{2i\pi z} - 1}$ whose residue at $k \in \mathbb{Z}$ is $f(k)$ up to a constant, and whose modulus is well-controlled. Here, to take into account a and $\frac{1}{a}$ at the same time, Siegel used a somewhat more complicated function.

The new function F has simple poles at all elements of $i\mathbb{Z}^* \cup a\mathbb{Z}^*$ and a triple pole at 0; the values of the residues at the simple poles are given by the "P/Q' rule": for $k \in \mathbb{Z}^*$, we have

$$\text{Res}(F, ik) = \frac{1}{ik} \cdot \frac{2}{2\pi} \cdot \frac{e^{-\frac{2k\pi}{a}} + 1}{e^{-\frac{2k\pi}{a}} - 1} = -\frac{1}{ik\pi} \cdot \frac{e^{\frac{2k\pi}{a}} + 1}{e^{\frac{2k\pi}{a}} - 1}, \qquad (8.21)$$

or again

$$\text{Res}(F, ik) = -\frac{1}{ik\pi} \left(1 + \frac{2}{e^{\frac{2k\pi}{a}} - 1} \right). \qquad (8.22)$$

Expression (8.21) of the residue shows that it is an *even* function of k, and (8.22) links its value to the function Φ that interests us, as this residue is the same as the general term of $\Phi\left(\frac{1}{a}\right)$ up to a constant (dependent on k, and to be simplified later on). Similarly, for $k \in \mathbb{Z}^*$, we have

$$\text{Res}(F, ak) = \frac{1}{ka} \cdot \frac{e^{2k\pi a} + 1}{e^{2k\pi a} - 1} \cdot \frac{2}{2i\pi/a} = \frac{1}{ik\pi} \cdot \frac{e^{2k\pi a} + 1}{e^{2k\pi a} - 1},$$

or

$$\text{Res}(F, ak) = \frac{1}{ik\pi} \left(1 + \frac{2}{e^{2k\pi a} - 1} \right). \qquad (8.23)$$

Here also, the residue is an even function of k, and is the same as the general term of $\Phi(a)$ up to a constant.

We now handle the triple pole of F at the origin. We can easily calculate its residue, using the expansion

$$\coth z = \frac{e^{2z} + 1}{e^{2z} - 1} = \frac{1}{z} + \frac{z}{3} + o(z),$$

which gives

$$F(z) = \frac{1}{z} \left(\frac{1}{\pi z} + \frac{\pi z}{3} + o(z) \right) \left(\frac{a}{i\pi z} + \frac{i\pi z}{3a} + o(z) \right)$$

$$= \frac{1}{z} \left(\frac{a}{i\pi^2 z^2} + \frac{1}{3i} (a - a^{-1}) + o(1) \right),$$

and hence

$$\text{Res}(F, 0) = \frac{1}{3i} (a - a^{-1}).$$

Let n be an integer ≥ 1. Set $N = n + \frac{1}{2}$ (it is a half-integer) and consider the parallelogram γ_N of vertices $\pm Na$ and $\pm Ni$, traversed once counter-clockwise (Figure 8.1).

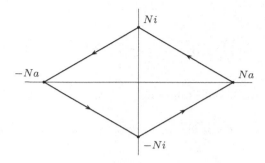

Figure 8.1

The residue theorem gives us, by *parity*, and by the use of (8.22) and (8.23):

$$\int_{\gamma_N} F(w)dw = 2i\pi \left(\frac{1}{3i}(a - a^{-1}) \right.$$

$$\left. + 2\sum_{k=1}^{n} \frac{1}{ik\pi} \left(1 + \frac{2}{e^{2k\pi a} - 1} - 1 - \frac{2}{e^{\frac{2k\pi}{a}} - 1} \right) \right).$$

The 1s cancel themselves out, as announced, so

$$\int_{\gamma_N} F(w)dw = \frac{2\pi}{3}(a - a^{-1}) + 8\sum_{k=1}^{n} \frac{1}{k} \left(\frac{1}{e^{2k\pi a} - 1} - \frac{1}{e^{\frac{2k\pi}{a}} - 1} \right).$$

By letting n tend to infinity, we thus obtain

$$\ell := \lim_{n\to\infty} \int_{\gamma_N} F(w)dw = \frac{2\pi}{3}(a - a^{-1}) + 8\left(\Phi(a) - \Phi\left(\frac{1}{a}\right) \right).$$

We now calculate ℓ directly and show that

$$\ell = \lim_{n\to\infty} \int_{\gamma_N} F(w)dw = 4\log a. \tag{8.24}$$

After division by 8, the comparison of the two values gives the desired result (8.20).

To prove (8.24), we first make the contour fixed and the function variable, according to a general principle of integration theory, using the change of variable $w = Nz$ (recall that $N = n + \frac{1}{2}$). We thus obtain, via $\frac{dw}{w} = \frac{dz}{z}$, the formula

$$\int_{\gamma_N} F(w)dw = \int_{\gamma_1} F_n(z)dz$$

in which we have set

$$F_n(z) = \frac{1}{z} \cdot \frac{e^{2\pi N z} + 1}{e^{2\pi N z} - 1} \cdot \frac{e^{\frac{2i\pi N z}{a}} + 1}{e^{\frac{2i\pi N z}{a}} - 1}. \tag{8.25}$$

To conclude, we need the following two lemmas.

8.3.8 Lemma *Let C_j be the open side of γ_1 situated in the quadrant[20] Q_j.* *Then, we have:*

- *if $z \in C_1 \cup C_3$, $F_n(z) \to -\dfrac{1}{z}$ as $n \to +\infty$;*
- *if $z \in C_2 \cup C_4$, $F_n(z) \to \dfrac{1}{z}$ as $n \to +\infty$.*

8.3.9 Lemma *The sequence $(F_n)_{n \geqslant 1}$ is uniformly bounded on γ_1.*

These two lemmas allow us to finish the proof of (8.24). Indeed, with ℓ the limit we are seeking, the dominated convergence theorem gives

$$\ell = \int_a^i -\frac{dz}{z} + \int_i^{-a} \frac{dz}{z} + \int_{-a}^{-i} -\frac{dz}{z} + \int_{-i}^a \frac{dz}{z}$$

$$= -2 \int_a^i \frac{dz}{z} + 2 \int_{-i}^a \frac{dz}{z}$$

$$= -2 \left(\log i - \log a \right) + 2 \left(\log a - \log(-i) \right)$$

$$= -2 \left(i \frac{\pi}{2} - \log a \right) + 2 \left(\log a + i \frac{\pi}{2} \right)$$

$$= 4 \log a.$$

We have implicitly used the following two facts.

- If f is a holomorphic function in an open set Ω and if γ is a continuous and piecewise C^1 path traced within Ω, with origin u and extremity v, we have

$$\int_\gamma f'(z) dz = f(v) - f(u).$$

Here, Ω is the slit plane $\mathbb{C} \setminus \mathbb{R}_-$, γ a segment and f the principal branch of the logarithm in Ω, which satisfies

$$f(\pm i) = \pm \frac{i\pi}{2} \quad \text{and} \quad f'(z) = \frac{1}{z} \quad \text{for } z \in \Omega.$$

---

[20] The quadrant Q_1 is defined by $\min (\operatorname{Re} z, \operatorname{Im} z) \geqslant 0$, and Q_{j+1} is derived from Q_j by a rotation of angle $\dfrac{\pi}{2}$.

- If $0 \notin [\alpha, \beta] \cup [-\alpha, -\beta]$, then

$$\int_{-\alpha}^{-\beta} \frac{dz}{z} = \int_{\alpha}^{\beta} \frac{dz}{z}.$$

It remains to prove the two lemmas. For the first, let us write $z = x + iy$ and suppose to begin with that $z \in C_1$ (i.e., $x, y > 0$). Then, using equivalents, we obtain

$$z F_n(z) = \frac{e^{2\pi Nx} e^{2i\pi Ny} + 1}{e^{2\pi Nx} e^{2i\pi Ny} - 1} \cdot \frac{e^{-\frac{2\pi Ny}{a}} e^{\frac{2i\pi Nx}{a}} + 1}{e^{-\frac{2\pi Ny}{a}} e^{\frac{2i\pi Nx}{a}} - 1} \to 1 \times -1 = -1. \tag{8.26}$$

The case $z \in C_2$ is handled similarly, and the function $z F_n(z)$ is even.

The proof of Lemma 8.3.9 is a bit more subtle, making use of $N = n + \frac{1}{2}$, which previously only served to avoid the presence of poles on the path of integration, but now implies the important equality

$$e^{2i\pi N} = -1. \tag{8.27}$$

Hence, let $z \in \gamma_1$. First, we suppose that $z \in C_1$. We have

$$z F_n(z) = \frac{1 + e^{-2\pi Nz}}{1 - e^{-2\pi Nz}} \cdot \frac{1 + e^{-\frac{2i\pi Nz}{a}}}{1 - e^{-\frac{2i\pi Nz}{a}}}.$$

As z has positive real and imaginary parts, the two numerators are bounded in modulus by 2. It remains to see that the two denominators do not approach 0, in the sense that they are both bounded below by a positive constant. Let us start by showing this for the first denominator. If this is not the case, there exists a sequence $(z_n)_{n \geqslant 0}$ of points of C_1 such that

$$e^{-2\pi Nz_n} \to 1.$$

By writing $z_n = t_n a + (1 - t_n)i$ with $t_n \in [0, 1]$, and by taking the modulus, we obtain $e^{-2\pi Nt_n a} \to 1$, so that

$$Nt_n \to 0.$$

But then, taking into account (8.27),

$$e^{-2\pi Nz_n} = e^{-2\pi Nt_n a} \cdot e^{-2i\pi N(1-t_n)} = -e^{-2\pi Nt_n a} \cdot e^{2\pi Nt_n i} \to -1,$$

which is absurd. Similarly, if ever

$$e^{-\frac{2i\pi Nz_n'}{a}} \to 1,$$

with this time $z'_n = (1 - t'_n)a + t'_n i$, then $N t'_n \to 0$ by taking the modulus. But then,

$$e^{-\frac{2i\pi N z'_n}{a}} = e^{-2i\pi N(1-t'_n)} e^{\frac{2\pi N t'_n}{a}} = -e^{2i\pi N t'_n} e^{\frac{2\pi N t'_n}{a}} \to -1,$$

which is just as absurd. On C_2, we proceed similarly, and profit from the parity of the function $z F_n(z)$ to complete the proof of Lemma 8.3.9.

8.3.10 A real analysis proof of the functional equation

We consider again the function Φ defined by

$$\Phi(a) = \sum_{m=1}^{\infty} \frac{1}{m} \cdot \frac{1}{e^{2\pi m a} - 1} \quad \text{for } a > 0.$$

As we saw above (in (8.20)) – and will establish again using real analysis – Φ satisfies the functional equation

$$\Phi(a) - \Phi\left(\frac{1}{a}\right) + \frac{\pi}{12}(a - a^{-1}) = \frac{1}{2}\log a. \tag{8.28}$$

The function Φ thus *almost* satisfies the functional equation

$$\Phi(a) = \Phi\left(\frac{1}{a}\right),$$

which suggests use of the Poisson summation formula[21]

$$\sum_{n\in\mathbb{Z}} f(n\alpha) = \frac{1}{\alpha} \sum_{n\in\mathbb{Z}} \widehat{f}\left(\frac{n}{\alpha}\right), \tag{8.29}$$

with its typical aspect: when we dilate the function f by a factor $\alpha > 0$, the Fourier transform is dilated by $\frac{1}{\alpha}$. However, the function here implicated $\left(x \mapsto \dfrac{1}{x(e^{2\pi x} - 1)}\right)$ is not integrable – it should be corrected – and some parasite terms in a and $\log a$ appear in the difference $\Phi(a) - \Phi\left(\frac{1}{a}\right)$. We are thus going to give a somewhat circuitous proof, where nonetheless the Poisson formula appears in the following form:

$$\frac{1}{e^{2\pi\alpha} - 1} = -\frac{1}{2} + \frac{1}{2\pi\alpha} + \frac{1}{\pi} \sum_{n=1}^{\infty} \frac{\alpha}{n^2 + \alpha^2} \quad \text{for } \alpha > 0. \tag{8.30}$$

---

[21] As we did with the Jacobi θ_0 function in the study of the "other" function of Riemann (see Chapter 7 of this book).

This identity is none other than the Poisson formula (8.29) for the function $f(x) = e^{-2\pi |x|}$ and its Fourier transform

$$\widehat{f}(\xi) = \int_{\mathbb{R}} e^{-2i\pi x\xi} f(x)dx = \frac{1}{\pi(1+\xi^2)}.$$

We will establish, for the partial sum

$$\Phi_N(a) = \sum_{m=1}^{N} \frac{1}{m} \cdot \frac{1}{e^{2\pi ma} - 1},$$

an *approximate* functional equation that, by passage to the limit, will give equation (8.28) above. The essential point is thus the following lemma.

8.3.11 Lemma *For $a > 0$, we have*

$$\Phi_N(a) = f_N(a) + g_N(a) + h_N(a) \qquad (8.31)$$

with

$$\begin{cases} f_N(a) = f_N(a^{-1}), \\ g_N(a) \to \dfrac{\pi}{12a}, \\ h_N(a) - h_N(a^{-1}) \to \dfrac{1}{2} \log a. \end{cases}$$

This lemma clearly gives (8.28). Indeed, it implies

$$\Phi_N(a) - \Phi_N(a^{-1}) = g_N(a) - g_N(a^{-1}) + h_N(a) - h_N(a^{-1}),$$

so that by passage to the limit,

$$\Phi(a) - \Phi(a^{-1}) = \frac{\pi}{12}(a^{-1} - a) + \frac{1}{2} \log a.$$

We still need to prove the lemma. Using (8.30) with $\alpha = ma$, and denoting by H_N the partial sums of the harmonic series, we obtain

$$\Phi_N(a) = -\frac{1}{2} H_N + \frac{1}{2\pi a} \sum_{m=1}^{N} m^{-2} + \frac{1}{\pi} \sum_{m=1}^{N} \sum_{n=1}^{\infty} \frac{1}{m^2 a + n^2 a^{-1}},$$

which we rewrite as

$$\Phi_N(a) = \left(-\frac{1}{2} H_N + \frac{1}{\pi} \sum_{m=1}^{N} \sum_{n=1}^{N} \frac{1}{m^2 a + n^2 a^{-1}} \right)$$

$$+ \left(\frac{1}{2\pi a} \sum_{m=1}^{N} m^{-2} \right) + \left(\frac{1}{\pi} \sum_{m=1}^{N} \sum_{n>N} \frac{1}{m^2 a + n^2 a^{-1}} \right)$$

$$=: f_N(a) + g_N(a) + h_N(a),$$

where f_N and g_N clearly satisfy the relations of the lemma. To estimate the double sum $h_N(a)$, we compare it with the double integral

$$H_N(a) = \frac{1}{\pi} \iint_{0<x<N, y>N} \frac{dx\,dy}{x^2 a + y^2 a^{-1}} = \frac{1}{\pi} \iint_{0<u<1, v>\frac{1}{a}} \frac{du\,dv}{u^2 + v^2},$$

the last equality coming from the change of variables $x = Nu$, $y = Nav$, for which $dx\,dy = N^2 a\,du\,dv$.

It remains to show that

$$H_N(a) - H_N(a^{-1}) = \frac{1}{2} \log a \qquad (8.32)$$

and

$$h_N(a) - H_N(a) \to 0. \qquad (8.33)$$

The relation (8.32) is easy to obtain. Indeed,

$$H_N(a) - H_N(a^{-1}) = \frac{1}{\pi} \iint_{0<u<1, \frac{1}{a}<v<a} \frac{du\,dv}{u^2 + v^2}$$

$$= \frac{1}{\pi} \int_{\frac{1}{a}}^{a} \left(\int_0^1 \frac{du}{u^2 + v^2} \right) dv$$

$$= \frac{1}{\pi} \int_{\frac{1}{a}}^{a} \arctan(v^{-1}) \frac{dv}{v} =: \frac{1}{\pi} I.$$

Using the invariance under inversion of the Haar measure of $\mathbb{R}^*_+$, $\frac{du}{u}$, we also have

$$I = \int_{\frac{1}{a}}^{a} \arctan v \frac{dv}{v},$$

so that

$$2I = \int_{\frac{1}{a}}^{a} \left(\arctan v + \arctan(v^{-1}) \right) \frac{dv}{v} = \int_{\frac{1}{a}}^{a} \frac{\pi}{2} \frac{dv}{v} = \pi \log a.$$

As for (8.33), it ensues from the following inequalities of sum–integral comparisons, valid for a non-negative and non-increasing function g on $\mathbb{R}^+$:

$$\int_{N+1}^{+\infty} g(x)dx \leqslant \sum_{n>N} g(n) \leqslant \int_N^{+\infty} g(x)dx$$

and

$$\int_1^N g(x)dx \leqslant \sum_{n=1}^N g(n) \leqslant \int_0^N g(x)dx.$$

By successively applying the two right-hand inequalities, we obtain

$$h_N(a) = \frac{1}{\pi} \sum_{m=1}^{N} \sum_{n>N} \frac{1}{m^2 a + n^2 a^{-1}}$$

$$\leqslant \frac{1}{\pi} \sum_{m=1}^{N} \int_{N}^{+\infty} \frac{dy}{m^2 a + y^2 a^{-1}}$$

$$\leqslant \frac{1}{\pi} \int_{0}^{N} \left(\int_{N}^{+\infty} \frac{dy}{x^2 a + y^2 a^{-1}} \right) dx = H_N(a).$$

Similarly,

$$h_N(a) \geqslant \frac{1}{\pi} \iint_{1 \leqslant x \leqslant N, y \geqslant N+1} \frac{dx \, dy}{x^2 a + y^2 a^{-1}} =: K_N(a).$$

Now, by omitting the integrands, we can write

$$\pi \left(H_N(a) - K_N(a) \right) = \iint_{0 \leqslant x \leqslant N, y \geqslant N} - \iint_{1 \leqslant x \leqslant N, y \geqslant N+1}$$

$$= \iint_{0 \leqslant x \leqslant 1, y \geqslant N} + \iint_{1 \leqslant x \leqslant N, N \leqslant y \leqslant N+1}$$

$$=: \varepsilon_N(a) + \delta_N(a).$$

We have on the one hand

$$0 \leqslant \varepsilon_N(a) \leqslant \int_{0}^{1} \left(\int_{N}^{+\infty} \frac{dy}{y^2 a^{-1}} \right) dx = \frac{a}{N},$$

and on the other hand, by using the arithmetic–geometric mean inequality,

$$x^2 a + y^2 a^{-1} \geqslant 2xy \geqslant xy,$$

$$0 \leqslant \delta_N(a) \leqslant \int_{1}^{N} \left(\int_{N}^{N+1} \frac{dy}{xy} \right) dx = \log N \cdot \log \left(1 + \frac{1}{N} \right) = o(1).$$

All this shows that

$$H_N(a) + o(1) \leqslant h_N(a) \leqslant H_N(a)$$

and proves the relation (8.33). The real analysis proof is thus complete.

8.4 An equivalent of $p(n)$

In this section we explain – as a warm-up for what is to follow – how to obtain an equivalent of $p(n)$ using only the singularity 1 of the generating function f

and the functional equation (8.14) in the special case where $T = \sigma$, which can be written[22]

$$f(e^{-2\pi z}) = z^{\frac{1}{2}} e^{\frac{\pi}{12z} - \frac{\pi z}{12}} f\left(e^{-\frac{2\pi}{z}}\right) \text{ for Re } z > 0. \tag{8.34}$$

We start with the integral representation of $p(n)$ given by Cauchy's formula:

$$p(n) = \frac{1}{2i\pi} \int_{|x|=r} \frac{f(x)}{x^{n+1}} dx,$$

where r is an element of $]0, 1[$, figuratively close to 1 and *called on to depend on n*. In this integral, what is important occurs in a neighbourhood of the real number r, where the function f "explodes" the most. We thus use the approximation of f provided by (8.16), which can be written, in our particular case, as

$$f(e^{-2\pi z}) \simeq z^{\frac{1}{2}} e^{\frac{\pi}{12z} - \frac{\pi z}{12}} \text{ for } z \text{ close to } 0. \tag{8.35}$$

Let us write $r = e^{-2\pi\rho}$, where ρ is itself an element of $]0, 1[$. A point of the circle $|x| = r$ can be written as

$$x = re^{2i\pi t} = e^{-2\pi z}, \text{ where } z = \rho - it, \text{ with } t \in \left[-\frac{1}{2}, \frac{1}{2}\right].$$

We thus have

$$\frac{dx}{x} = -2\pi dz,$$

so that

$$p(n) = i \int_S f(e^{-2\pi z}) e^{2\pi n z} dz,$$

the letter S denoting the oriented segment of the complex plane joining the point $\rho + \frac{i}{2}$ to the point $\rho - \frac{i}{2}$. The functional equation (8.34) gives

$$p(n) = i \int_S z^{\frac{1}{2}} e^{\frac{\pi}{12z} - \frac{\pi z}{12}} f(z') e^{2\pi n z} dz = i \int_S z^{\frac{1}{2}} e^{2\pi\left(n - \frac{1}{24}\right)z} e^{\frac{\pi}{12z}} f(z') dz, \tag{8.36}$$

where

$$z' = e^{-\frac{2\pi}{z}}.$$

With (8.35), we can approximate $p(n)$ by the integral $P(n)$ derived from (8.36) by replacing $f(z')$ by 1:

$$P(n) := i \int_S z^{\frac{1}{2}} e^{2\pi\left(n - \frac{1}{24}\right)z} e^{\frac{\pi}{12z}} dz.$$

[22] Here, $p = 0$, $q = 1$, $u = 0$ and $\omega_{0,1} = 1$ (see (8.15)).

Let us start by controlling the error thus introduced. It is equal to

$$\Delta(n) := \left| \int_S z^{\frac{1}{2}} e^{2\pi\left(n-\frac{1}{24}\right)z} e^{\frac{\pi}{12z}} \left(f(z') - 1\right) dz \right|.$$

To find an upper bound for $\Delta(n)$, we need to estimate the modulus of z as well as the real part of $\frac{1}{z}$ on the segment S. In all that follows, the letter C will denote an absolute constant that can be different from one line to the next.

On the one hand, if $z = \rho - it \in S$, we have[23]

$$|z| = (\rho^2 + t^2)^{\frac{1}{2}} \leqslant C. \tag{8.37}$$

On the other hand,

$$\operatorname{Re} \frac{1}{z} = \operatorname{Re} \frac{1}{\rho - it} = \frac{\rho}{\rho^2 + t^2} \geqslant \frac{\rho}{\rho^2 + \frac{1}{4}}$$

so that, for $k \geqslant 1$,

$$\left| e^{\frac{\pi}{12z}} z'^k \right| = e^{-2\pi\left(k-\frac{1}{24}\right)\operatorname{Re}\frac{1}{z}} \leqslant e^{-\left(k-\frac{1}{24}\right)\alpha_\rho}, \text{ with } \alpha_\rho = \frac{2\pi\rho}{\rho^2 + \frac{1}{4}}.$$

From this, we obtain

$$\left| e^{\frac{\pi}{12z}} \left(f(z') - 1\right) \right| \leqslant \sum_{k=1}^{\infty} p(k) e^{-\left(k-\frac{1}{24}\right)\alpha_\rho} \leqslant C f(e^{-\alpha_\rho}). \tag{8.38}$$

Finally,

$$\left| e^{2\pi\left(n-\frac{1}{24}\right)z} \right| = e^{2\pi\left(n-\frac{1}{24}\right)\rho} \leqslant e^{2\pi\rho n}. \tag{8.39}$$

We can thus find an upper bound of the error $\Delta(n)$, by using the usual inequality

$$\left| \int_\gamma g(z) dz \right| \leqslant M\ell(\gamma),$$

where g is a holomorphic function, M an upper bound of $|g(z)|$ on the path γ and $\ell(\gamma)$ the length of γ. Here, $\ell(S) = 1$. From (8.37), (8.38) and (8.39), we thus obtain

$$|\Delta(n)| \leqslant C e^{2\pi\rho n} f(e^{-\alpha_\rho}). \tag{8.40}$$

We can then use the elementary upper bound (8.19):

$$f(t) \leqslant e^{\frac{\pi^2}{6(t^{-1}-1)}} \text{ for } t \in \,]0, 1[,$$

---

[23] Remember that $\rho \in \,]0, 1[$ and $|t| \leqslant \frac{1}{2}$.

so that[24]

$$|\Delta(n)| \leqslant C e^{2\pi\rho n + \frac{\pi^2}{6(e^{\alpha\rho}-1)}} \leqslant C e^{2\pi\rho n + \frac{\pi^2}{6\alpha\rho}} = C e^{2\pi\rho n + \frac{\pi}{12}\left(\rho + \frac{1}{4\rho}\right)}.$$

We can optimise this upper bound by choosing

$$\rho = \frac{1}{4\sqrt{6n}}. \tag{8.41}$$

But then,

$$2\pi\rho n + \frac{\pi}{12}\left(\rho + \frac{1}{4\rho}\right) \sim \left(\frac{2\pi}{4\sqrt{6}} + \frac{4\pi\sqrt{6}}{48}\right) n^{\frac{1}{2}} = \frac{\pi}{\sqrt{6}} n^{\frac{1}{2}} \text{ as } n \to +\infty.$$

As $\frac{\pi}{\sqrt{6}} = \frac{K}{2} < K = \pi\sqrt{\frac{2}{3}}$, there exists a constant $H < K$ such that

$$\Delta(n) = O(e^{H\sqrt{n}}). \tag{8.42}$$

The error seems enormous, but we have $H < K$!

The next step consists of estimating the integral

$$P(n) = i \int_S z^{\frac{1}{2}} e^{2\pi\left(n - \frac{1}{24}\right)z} e^{\frac{\pi}{12z}} dz.$$

This could seem quite unattractive to a reader not familiar with the study of special functions, a domain sometimes considered nowadays as preserved in mothballs,[25] but in which Hardy and Rademacher had become masters. In fact, it is a piece of a famous path integral, intimately linked to an integral representation of the function $\frac{1}{\Gamma}$ due to Hankel. More precisely, let us fix a real number $\varepsilon \in\ ]0, 1[$, and consider the path $\mathcal{C}$ represented in Figure 8.2.

We can write

$$I := \int_{\mathcal{C}} z^{\frac{1}{2}} e^{2\pi\left(n - \frac{1}{24}\right)z} e^{\frac{\pi}{12z}} dz = \sum_{j=1}^{7} \int_{\mathcal{C}_j} z^{\frac{1}{2}} e^{2\pi\left(n - \frac{1}{24}\right)z} e^{\frac{\pi}{12z}} dz =: \sum_{j=1}^{7} I_j.$$

In fact, the integral I depends neither on ρ nor on ε, since the integrand is holomorphic in the slit plane $\mathbb{C} \setminus \mathbb{R}_-$ and the integral of a holomorphic function

---

[24] Via the convexity inequality $e^\theta - 1 \geqslant \theta$, valid on $\mathbb{R}$ but evident for $\theta \geqslant 0$ from the power series expansion of the exponential.

[25] A bit like Fourier series at the dawn of the twentieth century, or the holomorphic iteration somewhat later.

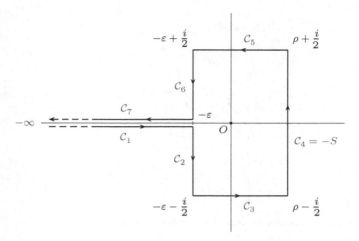

Figure 8.2

over a path is invariant under homotopy. We happen to know its value (we will detail the calculation in Section 8.7):[26]

$$I = \frac{i}{\pi\sqrt{2}} \frac{d}{dn} \left(\frac{\cosh(K\lambda_n)}{\lambda_n} \right), \tag{8.43}$$

where

$$\lambda_n = \left(n - \frac{1}{24}\right)^{\frac{1}{2}} \text{ and } K = \pi\sqrt{\frac{2}{3}}.$$

As for the integrals $I_1, \ldots, I_7$, *a priori* they depend on ρ and/or ε. In what follows, we allow ourselves to occasionally write, for example, $I_j(\varepsilon)$ instead of I_j, in order to visualise the dependence on ε of the integral I_j. So we have[27]

$$P(n) = -iI_4 = -i \left(I - \sum_{j \neq 4} I_j \right). \tag{8.44}$$

The idea is then to let ε tend to 0 in this equality. First of all,

$$I_1 = \int_{\varepsilon}^{+\infty} (-t)^{\frac{1}{2}} e^{-\frac{\pi}{12t}} e^{-2\pi\left(n - \frac{1}{24}\right)t} dt,$$

---

[26] See footnote 10 for the meaning of the notation $\dfrac{d}{dn}$.

[27] The negative sign is due to the opposite orientations of segments S and C_4.

where $(-t)^{\frac{1}{2}} = -i\sqrt{t}$. Similarly,

$$I_7 = -\int_\varepsilon^{+\infty} (-t)^{\frac{1}{2}} e^{-\frac{\pi}{12t}} e^{-2\pi\left(n-\frac{1}{24}\right)t} dt,$$

where this time $(-t)^{\frac{1}{2}} = i\sqrt{t}$. From this,

$$I_1 + I_7 \to -2iV \text{ as } \varepsilon \to 0, \tag{8.45}$$

where

$$V = \int_0^{+\infty} \sqrt{t}\, e^{-\frac{\pi}{12t}} e^{-2\pi\left(n-\frac{1}{24}\right)t} dt.$$

The integral V can also be calculated (see Exercise 8.2):

$$V = -\frac{1}{2\pi\sqrt{2}} \frac{d}{dn}\left(\frac{e^{-K\lambda_n}}{\lambda_n}\right). \tag{8.46}$$

We now show that the contribution of the integrals I_2, I_3, I_5 and I_6 to $P(n)$ is negligible. First, let us seek an upper bound for I_2. On the segment C_2, we can write

$$x = -\varepsilon - iy, \text{ where } y \in \left[0, \frac{1}{2}\right]$$

so that, on this segment:

$$\left| x^{\frac{1}{2}} e^{\frac{\pi}{12x}} e^{2\pi\left(n-\frac{1}{24}\right)x} \right| = (\varepsilon^2 + y^2)^{\frac{1}{4}} e^{\frac{-\pi\varepsilon}{12(\varepsilon^2+y^2)}} e^{-2\pi\left(n-\frac{1}{24}\right)\varepsilon} \leqslant C.$$

Knowing that the length of C_2 is equal to $\frac{1}{2}$, we obtain the inequality

$$|I_2| \leqslant C. \tag{8.47}$$

Of course we have the same bound for the integral I_6.

Let us consider now the integrals I_3 and I_5. On the segment C_3, we can write

$$x = t - \frac{i}{2}, \text{ where } t \in [-\varepsilon, \rho],$$

so that[28]

$$\left| x^{\frac{1}{2}} e^{\frac{\pi}{12x}} e^{2\pi\left(n-\frac{1}{24}\right)x} \right| = \left(t^2 + \frac{1}{4}\right)^{\frac{1}{4}} e^{\frac{\pi t}{12\left(t^2+\frac{1}{4}\right)}} e^{2\pi\left(n-\frac{1}{24}\right)t} \leqslant Ce^{2\pi n\rho}.$$

---

[28] Remember that ρ and ε are elements of $]0, 1[$.

Knowing that the length of C_3 is equal to $\rho + \varepsilon \leqslant 2$, and that we chose $\rho = \dfrac{1}{4\sqrt{6n}}$ (see (8.41)), we obtain

$$|I_3| \leqslant Ce^{2\pi n\rho} = Ce^{\frac{2\pi n}{4\sqrt{6n}}} = Ce^{\frac{\pi\sqrt{n}}{2\sqrt{6}}}, \tag{8.48}$$

and the same bound for $|I_5|$.

Ultimately, we have

$$P(n) = -i\left(I - I_1(\varepsilon) - I_7(\varepsilon)\right) + R(n, \varepsilon), \text{ where } |R(n, \varepsilon)| \leqslant Ce^{\frac{\pi\sqrt{n}}{2\sqrt{6}}}.$$

It ensues that

$$|P(n) + i\,(I + 2iV)| = \overline{\lim_{\varepsilon \to 0}}\,|P(n) + i\,(I - I_1(\varepsilon) - I_7(\varepsilon))| \leqslant Ce^{\frac{\pi\sqrt{n}}{2\sqrt{6}}}.$$

As

$$
\begin{aligned}
I + 2iV &= \frac{i}{\pi\sqrt{2}}\frac{d}{dn}\left(\frac{\cosh(K\lambda_n)}{\lambda_n}\right) - \frac{i}{\pi\sqrt{2}}\frac{d}{dn}\left(\frac{e^{-K\lambda_n}}{\lambda_n}\right) \\
&= \frac{i}{\pi\sqrt{2}}\frac{d}{dn}\left(\frac{\sinh(K\lambda_n)}{\lambda_n}\right),
\end{aligned}
$$

we have thus obtained the estimate

$$P(n) = \frac{1}{\pi\sqrt{2}}\frac{d}{dn}\left(\frac{\sinh(K\lambda_n)}{\lambda_n}\right) + O\left(e^{\frac{\pi\sqrt{n}}{2\sqrt{6}}}\right). \tag{8.49}$$

Finally, (8.42) and (8.49) give

$$p(n) \sim \frac{1}{\pi\sqrt{2}}\frac{d}{dn}\left(\frac{\sinh(K\lambda_n)}{\lambda_n}\right) \sim \frac{1}{2\pi\sqrt{2}}\frac{d}{dn}\left(\frac{e^{K\lambda_n}}{\lambda_n}\right) \sim \frac{1}{2\pi\sqrt{2}}\frac{Ke^{K\lambda_n}}{2\lambda_n^2}$$

and in conclusion,

$$p(n) \sim \frac{1}{4n\sqrt{3}}e^{K\sqrt{n}}. \tag{8.50}$$

8.4.1 Remark If we aim only for the equivalent (8.50) – and not the expansion as a Rademacher series – it is possible to skip the path integral $\int_C z^{\frac{1}{2}}e^{2\pi\left(n-\frac{1}{24}\right)z}e^{\frac{\pi}{12z}}\,dz$ and make a direct asymptotic study of the integral $P(n)$, see for example [168]. The method that we have detailed is more in tune with Hardy and Ramanujan, and consistent with the point of view generally adopted in this book.

8.5 The circle method

8.5.1 Farey sequences

Let N be a positive integer. Denote by $\mathfrak{F}_N$ the finite sequence of rational numbers, between 0 and 1, arranged in increasing order, each of whose denominator in its irreducible representation is $\leqslant N$. Thus,

$$\mathfrak{F}_1: \frac{0}{1}, \frac{1}{1}$$

$$\mathfrak{F}_2: \frac{0}{1}, \frac{1}{2}, \frac{1}{1}$$

$$\mathfrak{F}_3: \frac{0}{1}, \frac{1}{3}, \frac{1}{2}, \frac{2}{3}, \frac{1}{1}$$

$$\mathfrak{F}_4: \frac{0}{1}, \frac{1}{4}, \frac{1}{3}, \frac{1}{2}, \frac{2}{3}, \frac{3}{4}, \frac{1}{1}.$$

In what follows, the rational numbers considered will always be elements of $[0, 1]$, and always be written in their irreducible forms. Given two rational numbers $r = \dfrac{p}{q}$ and $s = \dfrac{p'}{q'}$ such that $0 \leqslant r < s \leqslant 1$, we set

$$t = \frac{p + p'}{q + q'}.$$

One can easily verify that

$$r < t < s.$$

In particular, if r and s are two consecutive elements of $\mathfrak{F}_N$, t cannot belong to $\mathfrak{F}_N$. We thus deduce the following inequalities:

$$\max(q, q') \leqslant N \leqslant q + q' - 1. \tag{8.51}$$

The next proposition will be useful later on. We give a very simple proof, due to Hurwitz.

8.5.2 Proposition *If* $\dfrac{p}{q} < \dfrac{p'}{q'}$ *are two consecutive elements of* $\mathfrak{F}_N$, *then the determinant* $p'q - pq'$ *is equal to 1.*

Proof We reason by induction on N. The result is clear if $N = 1$. Suppose that it is true for rank $N \geqslant 1$. Let $\dfrac{p}{q}$ and $\dfrac{p'}{q'}$ be two consecutive elements of $\mathfrak{F}_N$, and let $\dfrac{p''}{q''}$ be an element of $\mathfrak{F}_{N+1}$ such that

$$\frac{p}{q} < \frac{p''}{q''} < \frac{p'}{q'} \,.$$

There certainly exist *positive* integers λ and μ satisfying

$$\begin{cases} p'' = \lambda p' + \mu p, \\ q'' = \lambda q' + \mu q. \end{cases} \tag{8.52}$$

Indeed, (8.52) is a linear system whose solution (λ, μ) is given by

$$\begin{cases} \lambda = p''q - pq'' > 0, \\ \mu = p'q'' - q'p'' > 0. \end{cases}$$

As $\dfrac{p''}{q''}$ is an element of $\mathfrak{F}_{N+1}$ but not of $\mathfrak{F}_N$, we have

$$\lambda q' + \mu q = N + 1. \tag{8.53}$$

Moreover, the inequalities (8.51) provide

$$q + q' \geqslant N + 1. \tag{8.54}$$

Expressions (8.53) and (8.54) together imply $\lambda = \mu = 1$. In other words

$$\begin{cases} p'' = p + p', \\ q'' = q + q'. \end{cases}$$

This proves two statements.

- First, there exists *at most* one element of $\mathfrak{F}_{N+1} \backslash \mathfrak{F}_N$ between two consecutive elements $\dfrac{p}{q}$ and $\dfrac{p'}{q'}$ of $\mathfrak{F}_N$.

- Next, if there exists such an element $\dfrac{p''}{q''}$, we have

$$p''q - pq'' = p'q'' - p''q' = 1.$$

The proof is hence complete. □

This proof provides a very simple algorithm to calculate the sequence $\mathfrak{F}_N$ from $\mathfrak{F}_1$: given two consecutive elements $\dfrac{p}{q}$ and $\dfrac{p'}{q'}$ of $\mathfrak{F}_N$, we construct their "addition by dummies" $\dfrac{p+p'}{q+q'}$, that we place between $\dfrac{p}{q}$ and $\dfrac{p'}{q'}$ if its denominator is $N+1$.

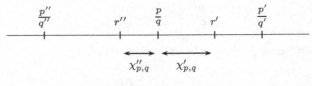

Figure 8.3

We now use the Farey sequences to construct a subdivision of the unit circle. More precisely, let us consider three consecutive elements[29] of $\mathfrak{F}_N$:

$$\frac{p''}{q''} < \frac{p}{q} < \frac{p'}{q'}$$

and set

$$r'' = \frac{p + p''}{q + q''} \text{ and } r' = \frac{p + p'}{q + q'}$$

$$\chi_{p,q}'' = \frac{p}{q} - r'' \text{ and } \chi_{p,q}' = r' - \frac{p}{q}$$

(see Figure 8.3).

Using Proposition 8.5.2, we immediately obtain

$$\chi_{p,q}'' = \frac{1}{q(q + q'')} \text{ and } \chi_{p,q}' = \frac{1}{q(q + q')}.$$

From this, thanks to the inequalities (8.51), we find the bounds

$$\frac{1}{2qN} \leqslant \chi_{p,q}', \chi_{p,q}'' \leqslant \frac{1}{qN}. \tag{8.55}$$

Then, to each element of $\mathfrak{F}_N$ other than 0 and 1, we associate the compact interval

$$I_{p,q} = \left[\frac{p}{q} - \chi_{p,q}'', \frac{p}{q} + \chi_{p,q}' \right].$$

To the rationals $0 = \frac{0}{1}$ and $1 = \frac{1}{1}$, we only associate a half-segment:

$$I_{0,1} = [0, \chi_{0,1}'] \text{ and } I_{1,1} = [1 - \chi_{1,1}'', 1].$$

If r is a fixed positive real number, the images of the different intervals $I_{p,q}$ by the map $t \mapsto re^{2i\pi t}$ provide us with a collection of circular arcs, denoted by $\gamma_{p,q}$: the $\gamma_{p,q}$ cover the circle $|z| = r$, and, taken pairwise, they have at most one point in common. In what follows this collection is called the *Farey subdivision of order N*. The two half-arcs $\gamma_{0,1}$ and $\gamma_{1,1}$, which have 1 as a

[29] With new notations.

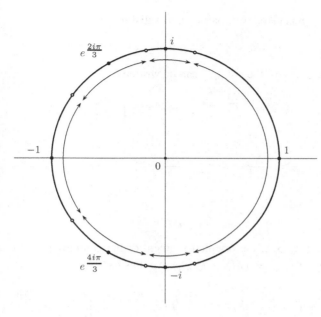

Figure 8.4

common extremity, will be joined into a single arc, again called $\gamma_{0,1}$. With this convention, every point $r e^{\frac{2ip\pi}{q}}$ of the circle $|z| = r$, where $\frac{p}{q} \in \mathfrak{F}_N \cap [0, 1[$, is an interior point of the arc $\gamma_{p,q}$, whose length is controlled by the inequalities (8.55). Figure 8.4 illustrates the Farey subdivision of order 4.

8.5.3 The idea of the method

In what follows, n denotes a fixed positive integer. Let N be a positive integer, $\rho = \rho_N$ a positive real number dependent on N, meant to tend to 0 as $N \to +\infty$, to be adjusted later on. We set $r = e^{-2\pi\rho} \in \,]0, 1[$. Cauchy's formula gives

$$p(n) = \frac{1}{2i\pi} \int_{|x|=r} \frac{f(x)}{x^{n+1}} dx,$$

where f is the generating function of the sequence $(p(n))_{n \geqslant 0}$. We decompose this integral by using the Farey subdivision of order N:

$$p(n) = \sum_{p,q} \frac{1}{2i\pi} \int_{\gamma_{p,q}} \frac{f(x)}{x^{n+1}} dx =: \sum_{p,q} j_{p,q},$$

where the sum is taken over pairs (p, q) satisfying

$$0 \leqslant p < q \leqslant N, \ p \text{ and } q \text{ relatively prime}. \tag{8.56}$$

A typical element of the arc $\gamma_{p,q}$ can be written

$$x = re^{\frac{2i\pi p}{q} + 2i\pi\theta}, \text{ where } \theta \in \left[-\chi''_{p,q}, \chi'_{p,q} \right],$$

or again

$$x = e^{-2\pi\rho + \frac{2i\pi p}{q} + 2i\pi\theta} = e^{\frac{2i\pi p}{q} - \frac{2\pi}{q} q(\rho - i\theta)} = e^{\frac{2i\pi p}{q} - \frac{2\pi z}{q}}$$

with

$$z = q(\rho - i\theta).$$

If we denote by $S_{p,q}$ the oriented segment of the complex plane joining the points $q(\rho + i\chi''_{p,q})$ and $q(\rho - i\chi'_{p,q})$, we thus have

$$j_{p,q} = \frac{1}{2i\pi} \int_{S_{p,q}} e^{-(n+1)\left(\frac{2i\pi p}{q} - \frac{2\pi z}{q}\right)} f\left(e^{\frac{2i\pi p}{q} - \frac{2\pi z}{q}}\right) \left(-\frac{2\pi}{q} e^{\frac{2i\pi p}{q} - \frac{2\pi z}{q}}\right) dz$$

$$= \frac{i}{q} e^{-\frac{2i\pi np}{q}} \int_{S_{p,q}} e^{\frac{2\pi nz}{q}} f\left(e^{\frac{2i\pi p}{q} - \frac{2\pi z}{q}}\right) dz.$$

The functional equation (8.14) gives

$$j_{p,q} = \frac{i}{q} \omega_{p,q} e^{-\frac{2i\pi np}{q}} \int_{S_{p,q}} z^{\frac{1}{2}} e^{\frac{\pi}{12qz} - \frac{\pi z}{12q} + \frac{2\pi nz}{q}} f(z') dz,$$

where

$$z' = e^{-\frac{2i\pi u}{q} - \frac{2\pi}{qz}}.$$

In this integral, z is close to 0, hence so is z', and therefore $f(z')$ is close to 1. The same idea as in Section 8.4 will enable us to approximate $j_{p,q}$ by the integral

$$J_{p,q} = \frac{i}{q} \omega_{p,q} e^{-\frac{2i\pi np}{q}} \int_{S_{p,q}} z^{\frac{1}{2}} e^{\frac{\pi}{12qz} - \frac{\pi z}{12q} + \frac{2\pi nz}{q}} dz.$$

8.5.4 Bounding the error

In what follows, the letter C denotes an absolute constant, possibly different from one line to another.

When approximating $j_{p,q}$ by $J_{p,q}$, we introduce an error bounded by the modulus of the integral

$$J'_{p,q} = \frac{i}{q}\omega_{p,q} e^{-\frac{2i\pi np}{q}} \int_{S_{p,q}} z^{\frac{1}{2}} e^{\frac{\pi}{12qz} - \frac{\pi z}{12q} + \frac{2\pi nz}{q}} \left(f(z') - 1\right) dz$$

where, we recall,

$$z' = e^{-\frac{2i\pi u}{q} - \frac{2\pi}{qz}}.$$

To estimate this integral, we will need orders of magnitude for the modulus of z and for the real part of $\frac{1}{z}$ when $z \in S_{p,q}$. First of all, thanks to (8.55), we have

$$|\theta| \leqslant \max(\chi'_{p,q}, \chi''_{p,q}) \leqslant \frac{1}{qN}, \tag{8.57}$$

so that

$$|z| = q(\rho^2 + \theta^2)^{\frac{1}{2}} \leqslant q\left(\rho^2 + \frac{1}{q^2 N^2}\right)^{\frac{1}{2}} = \left(q^2\rho^2 + \frac{1}{N^2}\right)^{\frac{1}{2}}.$$

If we choose

$$\rho = \frac{1}{N^2},$$

we obtain, recalling that $q \leqslant N$,

$$N^{-1} \leqslant |z| \leqslant \left(q^2\rho^2 + \frac{1}{N^2}\right)^{\frac{1}{2}} \leqslant \sqrt{2}\,N^{-1}. \tag{8.58}$$

Next, thanks to (8.57), we have

$$\frac{1}{q}\operatorname{Re}\frac{1}{z} = \frac{1}{q}\frac{\operatorname{Re}z}{|z|^2} = \frac{\rho}{q^2(\rho^2 + \theta^2)}$$

$$\geqslant \frac{N^{-2}}{q^2(N^{-4} + q^{-2}N^{-2})} = \frac{N^{-2}}{q^2N^{-4} + N^{-2}}$$

$$\geqslant \frac{N^{-2}}{N^{-2} + N^{-2}},$$

so that

$$\frac{1}{q}\operatorname{Re}\frac{1}{z} \geqslant \frac{1}{2}. \tag{8.59}$$

But

$$|z'| = e^{-\frac{2\pi}{q}\operatorname{Re}\frac{1}{z}},$$

hence

$$\left| e^{\frac{\pi}{12qz}} z'^k \right| = e^{\left(\frac{\pi}{12} - 2k\pi \right) \frac{1}{q} \operatorname{Re} \frac{1}{z}} \leqslant e^{\left(\frac{\pi}{24} - k\pi \right)} \quad \text{for } k \geqslant 1$$

and

$$\left| e^{\frac{\pi}{12qz}} \left(f(z') - 1 \right) \right| \leqslant \sum_{k=1}^{\infty} p(k) \left| e^{\frac{\pi}{12qz}} z'^k \right| \leqslant \sum_{k=1}^{\infty} p(k) e^{\left(\frac{\pi}{24} - k\pi \right)} = C.$$

(8.60)

Moreover,[30]

$$\left| e^{-\frac{\pi z}{12q} + \frac{2\pi n z}{q}} \right| = e^{\left(-\frac{\pi}{12q} + \frac{2\pi n}{q} \right) \operatorname{Re} z} = e^{\left(-\frac{\pi}{12} + 2\pi n \right) \rho}$$

$$= e^{\left(-\frac{\pi}{12} + 2\pi n \right) N^{-2}} \leqslant C. \quad (8.61)$$

As the length of the segment $S_{p,q}$ is at most equal to

$$2q \max(\chi'_{p,q}, \chi''_{p,q}) \leqslant \frac{2q}{qN} = 2N^{-1},$$

by (8.58), (8.60) and (8.61), we thus obtain the bound

$$|J'_{p,q}| \leqslant 2N^{-1} \cdot \frac{C}{q} \cdot N^{-\frac{1}{2}} \leqslant \frac{CN^{-\frac{3}{2}}}{q}. \quad (8.62)$$

To control the error resulting from the approximation of $p(n)$ by $\sum_{p,q} J_{p,q}$, it remains to sum (8.62) over all pairs (p, q) satisfying the conditions (8.56):

$$\left| \sum_{p,q} J'_{p,q} \right| \leqslant CN^{-\frac{3}{2}} \sum_{q=1}^{N} \frac{1}{q} \sum_{p} 1 \leqslant CN^{-\frac{3}{2}} \sum_{q=1}^{N} 1 = CN^{-\frac{1}{2}}.$$

Hence, the error term tends to 0 as $N \to +\infty$. Remembering that $p(n)$ is independent of N, we can already conclude that

$$p(n) = \lim_{N \to +\infty} \sum_{p,q} J_{p,q}. \quad (8.63)$$

In the next section, we try to make this limit explicit.

8.5.5 Study of the integral $J_{p,q}$ and completion of the proof of Rademacher's theorem

Remember that we have chosen

$$\rho = N^{-2} \text{ and } r = e^{-2\pi\rho}.$$

---

[30] Remember that n is fixed, while N will tend to $+\infty$.

The change of variable $z = qx$, and a change of orientation, give

$$J_{p,q} = \frac{q^{\frac{1}{2}}}{i} \omega_{p,q} e^{-\frac{2i\pi np}{q}} \int_S x^{\frac{1}{2}} e^{\frac{\pi}{12q^2 x}} e^{2\pi \left(n - \frac{1}{24}\right)x} dx$$

$$=: \frac{q^{\frac{1}{2}}}{i} \omega_{p,q} e^{-\frac{2i\pi np}{q}} K_{p,q},$$

where S denotes the segment – dependent on p and q – joining the points

$$\frac{1}{N^2} - i\chi'_{p,q} \text{ and } \frac{1}{N^2} + i\chi''_{p,q}.$$

We thus obtain the following expression for $p(n)$:

$$p(n) = \lim_{N \to +\infty} \sum_{q=1}^{N} \frac{q^{\frac{1}{2}}}{i} \left(\sum_p \omega_{p,q} e^{-\frac{2i\pi np}{q}} K_{p,q} \right), \qquad (8.64)$$

where the sum $\sum_p$ is taken over the integers p verifying $0 \leqslant p < q$ and relatively prime with q. As in Section 8.4, we consider the integral $K_{p,q}$[31] as the preponderant portion of a path integral that we can calculate. More precisely, let us fix a real number ε such that $0 < \varepsilon < N^{-2}$, and consider the path C illustrated in Figure 8.5.

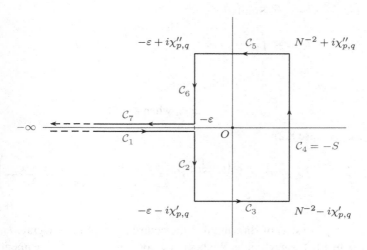

Figure 8.5

---

[31] Which also depends on N.

We write

$$I := \int_{\mathcal{C}} x^{\frac{1}{2}} e^{\frac{\pi}{12q^2 x}} e^{2\pi \left(n - \frac{1}{24}\right) x} dx$$

$$= \sum_{j=1}^{7} \int_{\mathcal{C}_j} x^{\frac{1}{2}} e^{\frac{\pi}{12q^2 x}} e^{2\pi \left(n - \frac{1}{24}\right) x} dx =: \sum_{j=1}^{7} I_j,$$

with (see Section 8.7)

$$I = \frac{i}{\pi \sqrt{2}} \cdot \frac{d}{dn} \left(\frac{1}{\lambda_n} \cosh \frac{K \lambda_n}{q} \right), \tag{8.65}$$

where

$$\lambda_n = \left(n - \frac{1}{24} \right)^{\frac{1}{2}} \text{ and } K = \pi \sqrt{\frac{2}{3}}.$$

We thus have

$$K_{p,q} = I_4 = I - \sum_{j \neq 4} I_j. \tag{8.66}$$

Up to now, ε has been a fixed real number such that $0 < \varepsilon < N^{-2}$. We are now going to let it tend to 0 *before* letting N tend to $+\infty$. First of all,

$$I_1 = \int_{\varepsilon}^{+\infty} (-t)^{\frac{1}{2}} e^{-\frac{\pi}{12q^2 t}} e^{-2\pi \left(n - \frac{1}{24}\right) t} dt,$$

where $(-t)^{\frac{1}{2}} = -i \sqrt{t}$. Similarly,

$$I_7 = - \int_{\varepsilon}^{+\infty} (-t)^{\frac{1}{2}} e^{-\frac{\pi}{12q^2 t}} e^{-2\pi \left(n - \frac{1}{24}\right) t} dt,$$

where this time $(-t)^{\frac{1}{2}} = i \sqrt{t}$. This gives

$$I_1 + I_7 \to -2i V_q \text{ when } \varepsilon \to 0 \tag{8.67}$$

with (see Exercise 8.2)

$$V_q = \int_0^{+\infty} \sqrt{t} \, e^{-\frac{\pi}{12q^2 t}} e^{-2\pi \left(n - \frac{1}{24}\right) t} dt = -\frac{1}{2\pi \sqrt{2}} \cdot \frac{d}{dn} \left(\frac{e^{-\frac{K \lambda_n}{q}}}{\lambda_n} \right). \tag{8.68}$$

The next step consists of showing that the contribution of the integrals I_2, I_3, I_5 and I_6 to $p(n)$ is negligible. We begin by bounding I_2. On the segment $\mathcal{C}_2$, we can write

$$x = -\varepsilon - iy, \text{ where } y \in [0, \chi'_{p,q}].$$

Hence, on this segment,

$$\left| x^{\frac{1}{2}} e^{\frac{\pi}{12q^2 x}} e^{2\pi\left(n-\frac{1}{24}\right)x} \right| = (\varepsilon^2 + y^2)^{\frac{1}{4}} e^{\frac{-\pi\varepsilon}{12q^2(\varepsilon^2+y^2)}} e^{-2\pi\left(n-\frac{1}{24}\right)\varepsilon}$$

$$\leqslant (\varepsilon^2 + y^2)^{\frac{1}{4}}$$

$$\leqslant \left(\frac{1}{N^4} + \frac{1}{q^2 N^2} \right)^{\frac{1}{4}} \quad \text{by (8.55)}$$

$$\leqslant \left(\frac{2}{q^2 N^2} \right)^{\frac{1}{4}} = \frac{2^{\frac{1}{4}}}{(qN)^{\frac{1}{2}}}.$$

Knowing that the length of C_2 is equal to $\chi'_{p,q} \leqslant \frac{1}{qN}$, we obtain the bound

$$|I_2| \leqslant C(qN)^{-\frac{3}{2}}. \tag{8.69}$$

Of course we have the same bound for the integral I_6.

Let us now consider the integrals I_3 and I_5. On the segment C_3, we can write

$$x = t - i\chi'_{p,q}, \text{ where } t \in [-\varepsilon, N^{-2}].$$

Hence

$$\left| x^{\frac{1}{2}} e^{\frac{\pi}{12q^2 x}} e^{2\pi\left(n-\frac{1}{24}\right)x} \right| = (t^2 + \chi'^2_{p,q})^{\frac{1}{4}} e^{\frac{\pi t}{12q^2(t^2+\chi'^2_{p,q})}} e^{2\pi\left(n-\frac{1}{24}\right)t}$$

$$\leqslant \left(\frac{1}{N^4} + \frac{1}{q^2 N^2} \right)^{\frac{1}{4}} e^{\frac{\pi N^{-2}}{12q^2(2qN)^{-2}}} e^{2\pi\left(n-\frac{1}{24}\right)N^{-2}}$$

$$\leqslant C \left(\frac{1}{N^4} + \frac{1}{q^2 N^2} \right)^{\frac{1}{4}}$$

$$\leqslant C(qN)^{-\frac{1}{2}}.$$

Knowing that the length of C_3 is bounded by $2N^{-2}$, we can conclude that

$$|I_3| \leqslant Cq^{-\frac{1}{2}} N^{-\frac{5}{2}}, \tag{8.70}$$

and the same bound holds for $|I_5|$.

From the estimates (8.69) and (8.70), we obtain

$$
\left| \sum_{q=1}^{N} \frac{q^{\frac{1}{2}}}{i} \sum_{p} \omega_{p,q} e^{-\frac{2i\pi np}{q}} \sum_{j \in \{2,3,5,6\}} I_j \right|
$$

$$
\leqslant \sum_{q=1}^{N} q^{\frac{1}{2}} \sum_{p=1}^{q} (|I_2| + |I_3| + |I_5| + |I_6|)
$$

$$
\leqslant C \left(\sum_{q=1}^{N} q^{\frac{1}{2}} \cdot q \cdot \left((qN)^{-\frac{3}{2}} + q^{-\frac{1}{2}} N^{-\frac{5}{2}} \right) \right)
$$

$$
= C \left(N^{-\frac{3}{2}} \sum_{q=1}^{N} 1 + N^{-\frac{5}{2}} \sum_{q=1}^{N} q \right)
$$

$$
\leqslant C N^{-\frac{1}{2}},
$$

where the constant C does not depend on ε. Thanks to (8.66), we thus have

$$
\sum_{q=1}^{N} \frac{q^{\frac{1}{2}}}{i} \left(\sum_{p} \omega_{p,q} e^{-\frac{2i\pi np}{q}} K_{p,q} \right)
$$

$$
= \sum_{q=1}^{N} \frac{q^{\frac{1}{2}}}{i} \left(\sum_{p} \omega_{p,q} e^{-\frac{2i\pi np}{q}} (I - I_1(\varepsilon) - I_7(\varepsilon)) \right) + R(\varepsilon, N),
$$

$$
\tag{8.71}
$$

with

$$
\varlimsup_{\varepsilon \to 0} R(\varepsilon, N) \leqslant C N^{-\frac{1}{2}}.
$$

By letting ε tend to 0, we bound

$$
\left| \sum_{q=1}^{N} \frac{q^{\frac{1}{2}}}{i} \left(\sum_{p} \omega_{p,q} e^{-\frac{2i\pi np}{q}} K_{p,q} \right) \right.
$$

$$
\left. - \sum_{q=1}^{N} \frac{q^{\frac{1}{2}}}{i} \left(\sum_{p} \omega_{p,q} e^{-\frac{2i\pi np}{q}} (I + 2i V_q) \right) \right|
$$

by $CN^{-\frac{1}{2}}$. By finally letting N tend to $+\infty$, we obtain, thanks to (8.64) and (8.65) and (8.68),

$$
\begin{aligned}
p(n) &= \lim_{N \to +\infty} \sum_{q=1}^{N} \frac{q^{\frac{1}{2}}}{i} \left(\sum_{p} \omega_{p,q} e^{-\frac{2i\pi np}{q}} K_{p,q} \right) \\
&= \sum_{q=1}^{\infty} \frac{q^{\frac{1}{2}}}{i} \left(\sum_{p} \omega_{p,q} e^{-\frac{2i\pi np}{q}} \left(I + 2i V_q \right) \right) \\
&= \frac{1}{\pi \sqrt{2}} \sum_{q=1}^{\infty} q^{\frac{1}{2}} \frac{d}{dn} \left(\frac{1}{\lambda_n} \sinh \frac{K \lambda_n}{q} \right) \left(\sum_{p} \omega_{p,q} e^{-\frac{2i\pi np}{q}} \right).
\end{aligned}
$$

This completes the proof of Rademacher's theorem.

8.5.6 Theorem [Rademacher, 1936] *For $n \geqslant 1$, we have the convergent series expansion*

$$
p(n) = \sum_{q=1}^{\infty} L_q(n) \psi_q(n),
$$

where

$$
L_q(n) = \sum_{p} \omega_{p,q} e^{-\frac{2i\pi np}{q}},
$$

$$
K = \pi \sqrt{\frac{2}{3}}, \; \lambda_n = \left(n - \frac{1}{24} \right)^{\frac{1}{2}}
$$

and

$$
\psi_q(n) = \frac{q^{\frac{1}{2}}}{\pi \sqrt{2}} \frac{d}{dn} \left(\frac{1}{\lambda_n} \sinh \frac{K \lambda_n}{q} \right),
$$

the sum $\sum_p$ being taken over the integers p satisfying

$$
0 \leqslant p < q, \text{ with } p \text{ and } q \text{ relatively prime,}
$$

while the roots of unity $\omega_{p,q}$ were defined in (8.14).

8.6 Asymptotic developments and numerical calculations

Rademacher's theorem is obviously a remarkable result on its own. We will see that this is true in more than one way, since it can be used for numerical calculations as well as for asymptotic estimations. We begin by bounding the tail of the Rademacher series. We use the brute-force estimate $|L_q(n)| \leqslant q$, as well as the sum–integral comparison inequality

$$\sum_{q=N+1}^{\infty} \frac{1}{q^{\alpha}} \leqslant \int_{N}^{+\infty} \frac{dt}{t^{\alpha}} = \frac{N^{1-\alpha}}{\alpha - 1},$$

valid for $\alpha > 1$. This gives

$$\left| \sum_{q=N+1}^{\infty} L_q(n) \psi_q(n) \right| \leqslant \frac{1}{\pi\sqrt{2}} \sum_{q=N+1}^{\infty} q^{\frac{3}{2}} \frac{d}{dn} \sum_{k=0}^{\infty} \frac{1}{(2k+1)!} \left(\frac{K}{q} \right)^{2k+1} \lambda_n^{2k}$$

$$= \frac{1}{\pi\sqrt{2}} \sum_{q=N+1}^{\infty} q^{\frac{3}{2}} \sum_{k=1}^{\infty} \frac{k}{(2k+1)!} \left(\frac{K}{q} \right)^{2k+1} \lambda_n^{2k-2}$$

$$= \frac{1}{\pi\sqrt{2}} \sum_{k=1}^{\infty} \frac{k}{(2k+1)!} K^{2k+1} \lambda_n^{2k-2} \sum_{q=N+1}^{\infty} \frac{1}{q^{2k-\frac{1}{2}}}$$

$$\leqslant \frac{1}{\pi\sqrt{2}} \sum_{k=1}^{\infty} \frac{k}{(2k+1)!} K^{2k+1} \lambda_n^{2k-2} \frac{N^{\frac{3}{2}-2k}}{2k - \frac{3}{2}}.$$

Now, if $k \geqslant 1$, we have

$$\frac{k}{2k - \frac{3}{2}} \leqslant 2,$$

so that

$$\left| \sum_{q=N+1}^{\infty} L_q(n) \psi_q(n) \right| \leqslant \frac{\sqrt{2}}{\pi} \sum_{k=1}^{\infty} \frac{K^{2k+1} \lambda_n^{2k-2} N^{\frac{3}{2}-2k}}{(2k+1)!}$$

$$\leqslant \frac{\sqrt{2}\, K^2 N^{\frac{1}{2}}}{\pi \lambda_n} \sum_{k=1}^{\infty} \frac{\left(\dfrac{K\lambda_n}{N} \right)^{2k-1}}{(2k-1)!}$$

and hence finally

$$\left| \sum_{q=N+1}^{\infty} L_q(n) \psi_q(n) \right| \leqslant \frac{\sqrt{2}\, K^2 N^{\frac{1}{2}}}{\pi \lambda_n} \sinh\left(\frac{K\lambda_n}{N} \right). \tag{8.72}$$

From this, first we can fix N and then let n tend to $+\infty$. We thus obtain the following Hardy–Ramanujan asymptotic expansion with a finite number of singularities:

$$p(n) = \frac{1}{2\pi\sqrt{2}} \sum_{q=1}^{N-1} q^{\frac{1}{2}} L_q(n) \frac{d}{dn} \left(\frac{e^{\frac{K\lambda_n}{q}}}{\lambda_n} \right) + O\left(e^{\frac{K\lambda_n}{N}} \right).$$

For $N = 2$, of course we find the equivalent obtained in Section 8.4.

We can also choose N dependent on n, for example N of the order $\sqrt{n}$. This gives

$$p(n) = \sum_{q \leqslant n^{\frac{1}{2}}} L_q(n) \psi_q(n) + O\left(n^{-\frac{1}{4}} \right),$$

and we recover the asymptotic estimation (8.4) of Hardy–Ramanujan.

We can finally fix n and use (8.72) to calculate the exact value of $p(n)$. It suffices to wait until the tail of the Rademacher series has modulus $< \frac{1}{2}$.

8.7 Appendix: Calculation of an integral

To finish, let us return to the integral

$$I = \int_C x^{\frac{1}{2}} e^{\frac{\pi}{12q^2x}} e^{2\pi\left(n - \frac{1}{24}\right)x} dx,$$

where the path C has been defined in Section 8.5.5. As in the Rademacher Theorem 8.5.6, we set

$$\lambda_n = \left(n - \frac{1}{24} \right)^{\frac{1}{2}} \text{ and } K = \pi\sqrt{\frac{2}{3}}.$$

We have

$$I = \int_C \sum_{k=0}^{\infty} \frac{1}{k!} x^{\frac{1}{2}} \left(\frac{\pi}{12q^2x} \right)^k e^{2\pi\lambda_n^2 x} dx$$

$$= \sum_{k=0}^{\infty} \frac{1}{k!} \left(\frac{\pi}{12q^2} \right)^k \int_C x^{\frac{1}{2}-k} e^{2\pi\lambda_n^2 x} dx$$

$$= \sum_{k=0}^{\infty} \frac{1}{k!} \left(\frac{\pi}{12q^2} \right)^k \left(2\pi\lambda_n^2 \right)^{k-\frac{3}{2}} \int_{C'} x^{\frac{1}{2}-k} e^x dx,$$

where the path C' is the image of C under the homothety $x \mapsto \sqrt{2\pi}\,\lambda_n x$.

We recall that the Γ function, initially defined by

$$\Gamma(s) = \int_0^{+\infty} t^{s-1} e^{-t} dt \text{ for Re } s > 0,$$

admits a meromorphic extension to the entire complex plane, that has no zeros. The function $\frac{1}{\Gamma}$ is hence entire, with zeros at 0 and at the negative integers. Hankel's formula (see Exercise 8.3) provides the integral representation[32]

$$\frac{1}{\Gamma(s)} = \frac{1}{2i\pi} \int_{C'} x^{-s} e^x dx \text{ for } s \in \mathbb{C}.$$

From this,

$$I = 2i\pi \sum_{k=0}^{\infty} \frac{\left(\dfrac{\pi}{12q^2}\right)^k \left(2\pi\lambda_n^2\right)^{k-\frac{3}{2}}}{k! \, \Gamma\left(k - \dfrac{1}{2}\right)}.$$

Next, for $k \geqslant 1$, we have

$$\begin{aligned}
\Gamma\left(k - \frac{1}{2}\right) &= \left(k - \frac{3}{2}\right)\left(k - \frac{5}{2}\right) \cdots \frac{1}{2} \cdot \Gamma\left(\frac{1}{2}\right) \\
&= \left(k - \frac{3}{2}\right)\left(k - \frac{5}{2}\right) \cdots \frac{1}{2} \cdot \sqrt{\pi} \\
&= \frac{(2k-3)(2k-5)\cdots 1}{2^{k-1}} \sqrt{\pi} \\
&= \frac{(2k-2)!}{2^{2k-2}(k-1)!} \sqrt{\pi} \\
&= \frac{(2k)!}{(2k-1)2^{2k-1}k!} \sqrt{\pi},
\end{aligned}$$

the last expression remaining valid if $k = 0$. Hence,

$$\begin{aligned}
I &= 2i\sqrt{\pi} \sum_{k=0}^{\infty} (2k-1) \frac{2^{2k-1}\left(\dfrac{\pi}{12q^2}\right)^k \left(2\pi\lambda_n^2\right)^{k-\frac{3}{2}}}{(2k)!} \\
&= 2i\sqrt{\pi} \cdot \frac{\sqrt{\pi}}{2q\sqrt{3}} \cdot \left(\lambda_n\sqrt{2\pi}\right)^{-2} \sum_{k=0}^{\infty} (2k-1) \frac{1}{(2k)!} \left(2 \cdot \frac{\sqrt{\pi}}{2q\sqrt{3}} \cdot \lambda_n\sqrt{2\pi}\right)^{2k-1} \\
&= \frac{i\lambda_n^{-2}}{2q\sqrt{3}} \sum_{k=0}^{\infty} (2k-1) \frac{1}{(2k)!} \left(\frac{K\lambda_n}{q}\right)^{2k-1}.
\end{aligned}$$

---

[32] The path of integration is not the same as in Exercise 8.3, but this is of little importance by homotopy.

Now,

$$\frac{d}{dx}\frac{\cosh x}{x} = \frac{d}{dx}\sum_{k=0}^{\infty}\frac{x^{2k-1}}{(2k)!} = \sum_{k=0}^{\infty}(2k-1)\frac{x^{2k-2}}{(2k)!} = \frac{1}{x}\sum_{k=0}^{\infty}(2k-1)\frac{x^{2k-1}}{(2k)!}.$$

Hence, by setting $x = \dfrac{K\lambda_n}{q}$,

$$\frac{d}{dn}\frac{\cosh x}{x} = \frac{dx}{dn}\cdot\frac{d}{dx}\frac{\cosh x}{x} = \frac{K}{2q\lambda_n}\cdot\frac{d}{dx}\frac{\cosh x}{x}.$$

We thus obtain

$$
\begin{aligned}
I &= \frac{i\lambda_n^{-2}}{2q\sqrt{3}}\cdot x\cdot\frac{d}{dx}\frac{\cosh x}{x}\\[2mm]
&= \frac{i\lambda_n^{-2}}{2q\sqrt{3}}\cdot\frac{K\lambda_n}{q}\cdot\frac{2q\lambda_n}{K}\cdot\frac{d}{dn}\frac{\cosh x}{x}\\[2mm]
&= \frac{i}{q\sqrt{3}}\frac{d}{dn}\frac{\cosh x}{x}\\[2mm]
&= \frac{i}{q\sqrt{3}}\frac{d}{dn}\frac{\cosh\dfrac{K\lambda_n}{q}}{\dfrac{K\lambda_n}{q}}\\[2mm]
&= \frac{i}{K\sqrt{3}}\frac{d}{dn}\left(\frac{1}{\lambda_n}\cosh\left(\frac{K\lambda_n}{q}\right)\right),
\end{aligned}
$$

so that finally

$$I = \frac{i}{\pi\sqrt{2}}\frac{d}{dn}\frac{\cosh\dfrac{K\lambda_n}{q}}{\lambda_n}.$$

Exercises

8.1. Our purpose is to give an elementary proof of the existence of two constants $A, B > 0$ such that

$$e^{A\sqrt{n}} \leqslant p(n) \leqslant e^{B\sqrt{n}}\text{ for }n \geqslant 1.$$

Recall that the notation f denotes the generating function of the sequence $(p(n))_{n\geqslant 0}$.

(a) Show that there exist constants $C, D > 0$ such that[33]

$$e^{\frac{C}{t}} \leqslant f(e^{-t}) \leqslant e^{\frac{D}{t}} \quad \text{for } t \in \,]0, 1].$$

(One can even choose $D = \frac{\pi^2}{6}$.)

(b) Show that, for $n \geqslant 1$ and $t > 0$,

$$p(n)e^{-nt} \leqslant e^{\frac{D}{t}},$$

and deduce the existence of the constant B.

(c) Show that, for $n \geqslant 1$, we have

$$p(n) \geqslant \frac{1}{n+1} \left(e^{\frac{C\sqrt{n}}{2B}} - \sum_{k=n+1}^{\infty} e^{-B\sqrt{k}} \right),$$

and deduce the existence of A.

8.2. For a, b positive real numbers, set

$$f(a, b) = \int_0^{+\infty} e^{-a^2 t^2 - \frac{b^2}{t^2}} \, dt, \quad g(a, b) = \int_0^{+\infty} t^{-\frac{1}{2}} e^{-at - \frac{b}{t}} \, dt$$

$$\text{and } h(a, b) = \int_0^{+\infty} t^{\frac{1}{2}} e^{-at - \frac{b}{t}} \, dt.$$

(a) Let $u \in L^1(\mathbb{R})$. Prove that

$$\int_{-\infty}^{+\infty} u\left(t - \frac{1}{t}\right) dt = \int_{-\infty}^{+\infty} u(x) dx.$$

(b) Deduce that

$$f(a, b) = \frac{\sqrt{\pi}}{2a} e^{-2ab}.$$

(c) Show that

$$g(a, b) = \sqrt{\frac{\pi}{a}} e^{-2\sqrt{ab}} \quad \text{and} \quad h(a, b) = -\sqrt{\pi} \frac{d}{da} \left(\frac{e^{-2\sqrt{ab}}}{\sqrt{a}} \right).$$

(d) Finally, prove formulas (8.46) and (8.68).

8.3. Hankel's integral. Let δ and ε be positive real numbers. Denote by $\gamma_{\delta, \varepsilon}$ the path in the complex plane formed by:

---

[33] The inequality on the right holds for all $t > 0$.

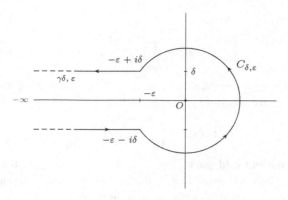

Figure 8.6

- the half-line parametrised by $z = t - i\delta$, t varying from $-\infty$ to $-\varepsilon$.
- the arc of circle $C_{\delta,\varepsilon}$ centred at 0, oriented counter-clockwise, and joining the points $-\varepsilon - i\delta$ and $-\varepsilon + i\delta$;
- the half-line parametrised by $z = t + i\delta$, t varying from $-\varepsilon$ to $-\infty$

(see Figure 8.6).

For $s \in \mathbb{C}$, we set

$$f(s) = \frac{1}{2i\pi} \int_{\gamma_{\delta,\varepsilon}} z^{-s} e^z dz,$$

where $z^{-s} = e^{-s \log z}$, log denoting the principal branch of the logarithm, that is,

$$\log z = \log r + i\theta \text{ if } z = re^{i\theta}, r > 0, -\pi < \theta < \pi.$$

(a) Show that f is an entire function, independent of δ and ε.
(b) Show that, if $\operatorname{Re} s < 1$,

$$\int_{C_{\delta,\varepsilon}} z^{-s} e^z dz \to 0 \text{ as } \delta \to 0 \text{ and } \varepsilon \to 0.$$

(c) Show that[34]

$$f(s) = \frac{1}{\Gamma(s)} \text{ for } s \in \mathbb{C}.$$

8.4. Restricted partitions. Let S be a finite subset of $\mathbb{N}^*$. For $n \geqslant 1$, we let $p_S(n)$ be the number of partitions of n using only the elements of S as "atoms".

---

[34] Recall that the function $\frac{1}{\Gamma}$ admits a holomorphic extension to the whole of $\mathbb{C}$. See, for example, [168].

(a) Show that

$$\sum_{n=0}^{\infty} p_S(n)z^n = \prod_{n \in S} \frac{1}{1 - z^n} \text{ for } |z| < 1.$$

(b) Furthermore, suppose that $S = \{a_1, a_2, \ldots, a_p\}$, the integers $a_1, \ldots, a_p$ being mutually relatively prime. Show that

$$p_S(n) \sim \frac{n^{p-1}}{(p-1)!a_1a_2 \cdots a_p}.$$

8.5. Partitions into odd parts. For $n \geqslant 1$, denote by $q(n)$ the number of partitions of n into odd parts, that is, the number of representations of n as a sum of *odd* positive integers, with the convention $q(0) = 1$.

(a) Show that

$$\sum_{n=0}^{\infty} q(n)z^n = \prod_{n=0}^{\infty} \frac{1}{1 - z^{2n+1}} = \frac{f(z)}{f(z^2)} =: g(z) \text{ for } |z| < 1,$$

where f denotes the generating function of the sequence $(p(n))_{n \geqslant 0}$.

(b) Give an approximation of $g(e^{-2\pi z})$ for z close to 0.

(c) Deduce an equivalent of $q(n)$ as $n \to +\infty$.

9

The approximate functional equation of the function θ_0

In a very rich article published in 1914 [81], Hardy and Littlewood investigated the order of magnitude of sums of the type

$$S_n(x) = \sum_{k=0}^{n} e^{i\pi k^2 x}$$

as $n \to +\infty$, where x denotes a fixed irrational real number. The sums S_n appear as partial sums of the series defining the Jacobi function θ_0, defined as

$$\theta_0(\tau) = \sum_{n \in \mathbb{Z}} e^{i\pi n^2 \tau} \text{ for } \tau \in \mathbb{C} \text{ such that } \operatorname{Im} \tau > 0.$$

This function satisfies the following functional equation:

$$\theta_0(\tau) = \frac{e^{i\pi/4}}{\sqrt{\tau}} \theta_0\left(-\frac{1}{\tau}\right), \tag{9.1}$$

which is a direct consequence of the Poisson formula (see Theorem 7.3.6 on p. 188). In the study of the behaviour of S_n, we cannot profit from (9.1), since we are only dealing here with partial sums. To remedy this, Hardy and Littlewood started by establishing an *approximate functional equation*, to be used as a substitute for (9.1): this is the purpose of the first section of this chapter. In the second section, we will explain how to use it to derive asymptotic estimations of

$$\left| \sum_{k=0}^{n} e^{i\pi k^2 x} \right|$$

at a fixed irrational x, and then of

$$\int_0^1 \left| \sum_{k=0}^{n} e^{i\pi k^2 x} \right| dx,$$

thanks to the continued fraction expansion of irrational numbers.

In what follows, in fact, we will be interested in somewhat more general sums than $S_n(x)$ and henceforth we set

$$s_n(x, t) = \sum_{k=0}^{n} e^{i\pi k^2 x + 2i\pi kt},$$

where n is a non-negative integer, *or even a non-negative real number*,[1] x and t being real numbers with

$$0 < x < 2 \text{ and } 0 \leqslant t \leqslant 1.$$

9.1 The approximate functional equation

The aim of this section is to establish the following result.

9.1.1 Theorem [approximate functional equation]   *For $0 < x < 2$ and $0 \leqslant t \leqslant 1$, we have*

$$s_n(x, t) = \frac{e^{i\pi/4}}{\sqrt{x}} e^{-i\pi t^2/x} s_{nx}\left(-\frac{1}{x}, \frac{t}{x}\right) + O\left(\frac{1}{\sqrt{x}}\right),$$

where the notation $O\left(\dfrac{1}{\sqrt{x}}\right)$ *denotes here a function of n, x and t bounded in modulus by* $\dfrac{C}{\sqrt{x}}$, *where C is an* absolute *constant*.[2]

The proof presented here, simpler than that of Hardy and Littlewood, is due to the Anglo-American number theorist Louis Mordell [130].

Proof   Throughout this proof, *x and t are fixed as in the theorem*. The starting point consists of the following remark: if $g : \mathbb{C} \to \mathbb{C}$ is an entire function and if $k \in \mathbb{N}$, then $g(k)$ is none other than the residue at $z = 0$ of the meromorphic function

$$z \mapsto 2i\pi \frac{g(z+k)}{e^{2i\pi z} - 1}.$$

Hence, if $n \geqslant 1$, $\sum_{k=0}^{n-1} g(k)$ is equal to $2i\pi$ times the residue at 0 of the function

$$f : z \mapsto \frac{1}{e^{2i\pi z} - 1} \sum_{k=0}^{n-1} g(z+k).$$

---

[1] In this case, the sum is to be understood as $\sum_{k=0}^{[n]}$.
[2] That is, dependent on nothing.

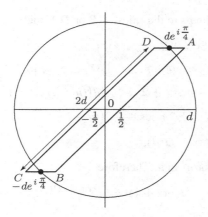

Figure 9.1

Of course, in the case that interests us, we choose

$$g : \mathbb{C} \to \mathbb{C}, \ z \mapsto e^{i\pi z^2 x + 2i\pi z t}.$$

By the residue theorem, if γ is a closed contour such that the index of 0 with respect to γ is 1, the other poles of f (i.e., the non-zero integers) being of index zero, we have

$$s_{n-1}(x, t) = \sum_{k=0}^{n-1} g(k) = \int_{\gamma} f(z) \, dz. \tag{9.2}$$

The choice of γ is primordial; we choose a parallelogram $DCBA$ with sides DC and BA of length $2d > 0$ inclined at $45°$ with respect to the coordinate axes, and with horizontal sides AD and CB of length 1 (Figure 9.1).

More precisely:

- CB is parametrised by $z = -e^{i\pi/4}d + u, \ -\frac{1}{2} \leqslant u \leqslant \frac{1}{2}$;
- DA is parametrised by $z = e^{i\pi/4}d + u, \ -\frac{1}{2} \leqslant u \leqslant \frac{1}{2}$;
- CD is parametrised by $-\frac{1}{2} + e^{i\pi/4}u, \ -d \leqslant u \leqslant d$;
- BA is parametrised by $\frac{1}{2} + e^{i\pi/4}u, \ -d \leqslant u \leqslant d.$

Now, if $k \in \mathbb{N}$,

$$\frac{g(z + k)}{e^{2i\pi z} - 1} = \frac{e^{i\pi(z+k)^2 x + 2i\pi(z+k)t}}{e^{2i\pi z} - 1}.$$

In particular, if z belongs to the sides[3] CB or DA, then

$$\begin{aligned}
\operatorname{Re}&\left(i\pi(z+k)^2x + 2i\pi(z+k)t\right)\\
&= \operatorname{Re}\left(i\pi(\pm e^{i\pi/4}d + u + k)^2x + 2i\pi(\pm e^{i\pi/4}d + u + k)t\right)\\
&= \operatorname{Re}\left(-\pi d^2x \pm 2\pi e^{3i\pi/4}d((u+k)x + t)\right)\\
&= -\pi d^2x \pm 2\pi \cos\frac{3\pi}{4}d((u+k)x + t)\\
&= -\pi d^2x + O(d),
\end{aligned}$$

the $O(d)$ being uniform[4] in u. Therefore,

$$|g(z+k)| = e^{-x\pi d^2 + O(d)}.$$

Moreover,

$$|e^{2i\pi z}| = e^{-2\pi \operatorname{Im} z} = e^{\sigma\sqrt{2}\pi d},$$

where

$$\sigma = \begin{cases} 1 & \text{if } z \in CB,\\ -1 & \text{if } z \in DA. \end{cases}$$

Hence,

$$|e^{2i\pi z} - 1| \underset{d\to+\infty}{\longrightarrow} \begin{cases} +\infty & \text{if } z \in CB,\\ 1 & \text{if } z \in DA, \end{cases}$$

the limit being uniform in u. From these estimates, we deduce that

$$\int_{CB\cup AD} f(z)\,dz \underset{d\to+\infty}{\longrightarrow} 0. \tag{9.3}$$

Assertions (9.2) and (9.3) thus give

$$s_{n-1}(x,t) \underset{d\to+\infty}{=} \int_{BA} f(z)\,dz - \int_{CD} f(z)\,dz + o(1).$$

Since $BA = CD + 1$, we also have

$$\begin{aligned}
s_{n-1}(x,t) &\underset{d\to+\infty}{=} \int_{CD} f(z+1)\,dz - \int_{CD} f(z)\,dz + o(1)\\
&\underset{d\to+\infty}{=} \int_{CD} (f(z+1) - f(z))\,dz + o(1).
\end{aligned}$$

Moreover, as $e^{2i\pi(z+1)} = e^{2i\pi z}$,

$$f(z+1) - f(z) = \frac{g(z+n) - g(z)}{e^{2i\pi z} - 1},$$

---

[3] It is here that we profit from the 45° inclination of $DCBA$!

[4] Remember that the range of u is the compact interval $\left[-\frac{1}{2}, \frac{1}{2}\right]$.

which finally gives, on letting $d \to +\infty$:

$$s_{n-1}(x, t) = \underbrace{\int_{-\frac{1}{2} - e^{i\pi/4}\infty}^{-\frac{1}{2} + e^{i\pi/4}\infty} \frac{g(z+n)}{e^{2i\pi z} - 1} dz}_{:= I_1} - \underbrace{\int_{-\frac{1}{2} - e^{i\pi/4}\infty}^{-\frac{1}{2} + e^{i\pi/4}\infty} \frac{g(z)}{e^{2i\pi z} - 1} dz}_{:= I_2}.$$

$$(9.4)$$

We estimate I_1 and I_2 separately, and in the process show the existence of these two integrals.

We begin by bounding I_1. In this integral,

$$z = z(u) = -\frac{1}{2} + u e^{i\pi/4}, \quad u \text{ increasing from } -\infty \text{ to } +\infty.$$

However,

$$\begin{aligned}
&\mathrm{Re}\left(i\pi(z+n)^2 x + 2i\pi(z+n)t\right) \\
&= \mathrm{Re}\left(i\pi\left(-\frac{1}{2} + u e^{i\pi/4} + n\right)^2 x\right) \\
&\quad + \mathrm{Re}\left(2i\pi\left(-\frac{1}{2} + u e^{i\pi/4} + n\right)t\right) \\
&= -\pi x u^2 + 2\pi\left(n - \frac{1}{2}\right)ux \cos\frac{3\pi}{4} + 2\pi \cos\frac{3\pi}{4} ut \\
&= -\pi x u^2 - \sqrt{2}\pi u\theta
\end{aligned}$$

by setting $\theta = \left(n - \frac{1}{2}\right)x + t$. Therefore,

$$\left|\frac{g(z+n)}{e^{2i\pi z} - 1}\right| = \left|\frac{e^{i\pi(z+n)^2 x + 2i\pi(z+n)t}}{e^{2i\pi z} - 1}\right| = e^{-\pi x u^2} \frac{e^{-\sqrt{2}\pi u\theta}}{\left|e^{2i\pi z} - 1\right|}.$$

For simplicity, we suppose until further notice that

$$\boxed{0 \leqslant \theta \leqslant 1} \qquad (9.5)$$

and we will see at the end of the proof how to overcome this condition.

As $|e^{2i\pi z}| = e^{-2\pi \,\mathrm{Im}\, z} = e^{-\sqrt{2}\pi u}$, $\dfrac{1}{\left|e^{2i\pi z} - 1\right|}$ tends to 0 (resp. 1) when u tends to $-\infty$ (resp. $+\infty$). Since the function $u \mapsto \dfrac{1}{\left|e^{2i\pi z} - 1\right|}$ is continuous, there exists an absolute constant C such that

$$\frac{1}{\left|e^{2i\pi z} - 1\right|} \leqslant C \text{ for all } u \in \mathbb{R}.$$

But then, if $u \geqslant 0$,

$$\left|\frac{g(z+n)}{e^{2i\pi z} - 1}\right| \leqslant C e^{-\pi x u^2}.$$

If, on the contrary, $u \leqslant 0$, then $|e^{2i\pi z}| \geqslant 1$, so that

$$\left| \frac{g(z+n)}{e^{2i\pi z} - 1} \right| = e^{-\pi x u^2} \frac{|e^{2i\pi z}|^{\theta}}{|e^{2i\pi z} - 1|} \leqslant e^{-\pi x u^2} \frac{|e^{2i\pi z}|}{|e^{2i\pi z} - 1|} \leqslant C' e^{-\pi x u^2},$$

since $\theta \in [0, 1]$ and the function $u \mapsto \dfrac{|e^{2i\pi z}|}{|e^{2i\pi z} - 1|}$ is continuous and has finite limits at $\pm\infty$. Finally,

$$|I_1| \leqslant C'' \int_{-\infty}^{+\infty} e^{-\pi x u^2} du = \frac{C''}{\sqrt{x}}, \tag{9.6}$$

with C'' an absolute constant.

It remains to estimate I_2, which will be done using similar methods, even though the reasoning is nonetheless a bit more delicate. We fix a non-negative integer λ, to be adjusted later on. We have

$$\frac{g(z)}{1 - e^{2i\pi z}} = \sum_{k=0}^{\lambda-1} g(z) e^{2ik\pi z} + \frac{e^{2i\lambda\pi z} g(z)}{1 - e^{2i\pi z}},$$

so that

$$-I_2 = \sum_{k=0}^{\lambda-1} \int_{-1/2 - e^{i\pi/4}\infty}^{-1/2 + e^{i\pi/4}\infty} g(z) e^{2ik\pi z} dz + \underbrace{\int_{-1/2 - e^{i\pi/4}\infty}^{-1/2 + e^{i\pi/4}\infty} \frac{e^{2i\lambda\pi z} g(z)}{1 - e^{2i\pi z}} dz}_{:= I_3}.$$

First of all,

$$\int_{-1/2 - e^{i\pi/4}\infty}^{-1/2 + e^{i\pi/4}\infty} g(z) e^{2ik\pi z} dz = \int_{-1/2 - e^{i\pi/4}\infty}^{-1/2 + e^{i\pi/4}\infty} e^{i\pi z^2 x + 2i\pi(t+k)z} dz$$

$$= e^{-i\pi(t+k)^2/x} \int_{-1/2 - e^{i\pi/4}\infty}^{-1/2 + e^{i\pi/4}\infty} e^{i\pi x \left(z + \frac{t+k}{x}\right)^2} dz$$

$$= e^{-i\pi(t+k)^2/x} \int_{-\infty}^{+\infty} e^{i\pi x \left(u e^{i\frac{\pi}{4}} - \frac{1}{2} + \frac{t+k}{x}\right)^2} e^{i\frac{\pi}{4}} du$$

$$= e^{i\frac{\pi}{4}} e^{-i\pi(t+k)^2/x} \int_{-\infty}^{+\infty} e^{-\pi x(u+v)^2} du,$$

where v is a suitable complex number. However, the function

$$\Phi : v \mapsto \int_{-\infty}^{+\infty} e^{-\pi x(u+v)^2} du$$

is entire, and if v is real, its value is

$$\int_{-\infty}^{+\infty} e^{-\pi x u^2} du = \frac{1}{\sqrt{x}}.$$

By analytic extension, Φ is everywhere equal to $\dfrac{1}{\sqrt{x}}$. Finally,

$$\int_{-1/2-e^{i\pi/4}\infty}^{-1/2+e^{i\pi/4}\infty} g(z)e^{2ik\pi z}dz = \frac{e^{i\pi/4}}{\sqrt{x}}e^{-i\pi(t+k)^2/x}.$$

We have thus shown that

$$-I_2 = \frac{e^{i\pi/4}}{\sqrt{x}}\sum_{k=0}^{\lambda-1}e^{-i\pi(t+k)^2/x} + I_3. \tag{9.7}$$

We now focus on I_3. While estimating I_1, we had obtained

$$|g(z+n)| = e^{-\pi xu^2}e^{-\sqrt{2}\pi u\theta}, \text{ with } \theta = t+nx-\frac{x}{2}. \tag{9.8}$$

Now, we must estimate

$$|e^{2i\pi\lambda z}g(z)| = e^{-\pi xu^2}e^{-\sqrt{2}\pi u\theta'}, \text{ with this time } \theta' = t+\lambda-\frac{x}{2}. \tag{9.9}$$

Earlier, (9.8) had allowed us to obtain $I_1 = O\left(\dfrac{1}{\sqrt{x}}\right)$, valid under the supposition $0 \leqslant \theta \leqslant 1$. The comparison of (9.8) and (9.9) suggests choosing λ such that

$$0 \leqslant t+\lambda-\frac{x}{2} \leqslant 1. \tag{9.10}$$

However, by (9.5),

$$0 \leqslant t+nx-\frac{x}{2} \leqslant 1.$$

This prompts us to choose λ on the order of nx, or rather $[nx]$. However, *one of the two integers $[nx]$ and $[nx]+1$ is certainly an element of the interval* $\left[-t+\dfrac{x}{2}, 1-t+\dfrac{x}{2}\right]$ (of length 1). Indeed, if this was not the case, we would have

- either $1-t+\dfrac{x}{2} < [nx]$, so that $1-t+\dfrac{x}{2} < nx$, which (9.5) excludes,
- or $[nx]+1 < -t+\dfrac{x}{2}$, hence $nx < -t+\dfrac{x}{2}$: another contradiction.

Then we select an integer λ, between $[nx]$ and $[nx]+1$, satisfying (9.10). The study of I_1, henceforth applicable, thus leads to

$$I_3 = O\left(\frac{1}{\sqrt{x}}\right), \tag{9.11}$$

the O being absolute. Then, by (9.4), (9.6), (9.7) and (9.11),

$$s_{n-1}(x,t) = \frac{e^{i\pi/4}}{\sqrt{x}}\sum_{k=0}^{\lambda-1}e^{-i\pi(t+k)^2/x} + O\left(\frac{1}{\sqrt{x}}\right).$$

After letting $O\left(\dfrac{1}{\sqrt{x}}\right)$ swallow, if necessary, a finite (and absolute) number

of exponentials of modulus 1, we can replace[5] $\displaystyle\sum_{k=0}^{\lambda-1}$ by $\displaystyle\sum_{k=0}^{nx}$ and $s_{n-1}(x,t)$ by

$s_n(x,t)$. This gives

$$s_n(x,t) = \frac{e^{i\pi/4}}{\sqrt{x}} \sum_{k=0}^{nx} e^{-i\pi(t+k)^2/x} + O\left(\frac{1}{\sqrt{x}}\right). \tag{9.12}$$

To achieve a complete proof of the theorem, it remains to overcome the draconian condition (9.5):

$$0 \leqslant \left(n - \frac{1}{2}\right)x + t \leqslant 1.$$

There certainly exists an integer p such that

$$0 \leqslant \left(n - \frac{1}{2}\right)x + t + p \leqslant 1. \tag{9.13}$$

We then set $y = t + p$. By (9.12),

$$\sum_{k=0}^{n} e^{i\pi k^2 x + 2i\pi ky} = \frac{e^{i\pi/4}}{\sqrt{x}} \sum_{k=0}^{nx} e^{-i\pi(y+k)^2/x} + O\left(\frac{1}{\sqrt{x}}\right),$$

which can also be written

$$\sum_{k=0}^{n} e^{i\pi k^2 x + 2i\pi kt} = \frac{e^{i\pi/4}}{\sqrt{x}} \sum_{k=p}^{p+nx} e^{-i\pi(t+k)^2/x} + O\left(\frac{1}{\sqrt{x}}\right).$$

However, p is close to $-nx$, and $p + nx$ is close to -1; more precisely, by (9.13),

$$\begin{cases} p + nx \leqslant 1 - t + \dfrac{x}{2} \leqslant 2, \\ p + nx \geqslant -t + \dfrac{x}{2} \geqslant -1, \end{cases}$$

since $0 \leqslant t \leqslant 1$ and $0 < x < 2$. Consequently, after changing $O\left(\dfrac{1}{\sqrt{x}}\right)$ again

if necessary, we can replace $\displaystyle\sum_{k=p}^{p+nx}$ by $\displaystyle\sum_{k=-nx}^{0}$. All this gives

---

[5] We have already explained how to interpret this sum if nx is not an integer.

$$s_n(x, t) = \frac{e^{i\pi/4}}{\sqrt{x}} \sum_{k=-nx}^{0} e^{-i\pi(t+k)^2/x} + O\left(\frac{1}{\sqrt{x}}\right)$$

$$= \frac{e^{i\pi/4}}{\sqrt{x}} \sum_{k=0}^{nx} e^{-i\pi(t-k)^2/x} + O\left(\frac{1}{\sqrt{x}}\right)$$

$$= \frac{e^{i\pi/4}}{\sqrt{x}} \sum_{k=0}^{nx} e^{-i\pi k^2/x + 2i\pi kt/x - i\pi t^2/x} + O\left(\frac{1}{\sqrt{x}}\right)$$

$$= \frac{e^{i\pi/4}}{\sqrt{x}} e^{-i\pi t^2/x} s_{nx}\left(-\frac{1}{x}, \frac{t}{x}\right) + O\left(\frac{1}{\sqrt{x}}\right),$$

which completes the proof. $\qquad\square$

9.2 Other forms of the approximate functional equation and applications

In the rest of this chapter, we suppose that x is an irrational real number in the interval $]0, 1[$, and t an element of $[0, 1]$. For $n \geqslant 1$, set

$$s_n(x, t) = \sum_{k=0}^{n} e^{i\pi k^2 x + 2i\pi kt}.$$

We would like to study the behaviour of $s_n(x, t)$ when $n \to +\infty$, in various ways. In particular, we attempt to obtain estimates that are either *uniform in t*, that is for

$$\sup_{0 \leqslant t \leqslant 1} |s_n(x, t)|$$

or *averaged over x*, that is for

$$\int_0^1 |s_n(x, t)| dx.$$

The methods we use combine two essential ingredients: the approximate functional equation of Theorem 9.1.1 that we will give in a more practical form, well suited to an iteration and the continued fraction expansion of irrational numbers that we described briefly in Chapter 7 (see in particular Section 7.4.11). We extend these rudiments with a few useful complements in the following section.

9.2.1 Complements on continued fractions

Recall the notations: given an irrational $x \in]0, 1[$, we write

$$
\begin{cases}
x & = \omega_0 \\[4pt]
\dfrac{1}{\omega_0} & = a_1' = a_1 + \omega_1 & \text{with } a_1 \in \mathbb{N}^*,\ 0 < \omega_1 < 1, \\[4pt]
\dfrac{1}{\omega_1} & = a_2' = a_2 + \omega_2 & \text{with } a_2 \in \mathbb{N}^*,\ 0 < \omega_2 < 1, \\[4pt]
\ \vdots & \\[4pt]
\dfrac{1}{\omega_n} & = a_{n+1}' = a_{n+1} + \omega_{n+1} & \text{with } a_{n+1} \in \mathbb{N}^*,\ 0 < \omega_{n+1} < 1, \\[4pt]
\ \vdots &
\end{cases}
$$

The a_n are called the *partial quotients* of the continued fraction expansion of x. The *convergents* of this expansion are the rationals $\dfrac{p_n}{q_n}$ $(n \geqslant 0)$ defined by the following two sequences of integers $(p_n)_{n \geqslant -1}$ and $(q_n)_{n \geqslant -1}$:

$$
\begin{cases}
p_{-1} = 1, & p_0 = 0 & \text{and} & p_{n+1} = a_{n+1} p_n + p_{n-1} & \text{for } n \geqslant 0, \\
q_{-1} = 0, & q_0 = 1 & \text{and} & q_{n+1} = a_{n+1} q_n + q_{n-1} & \text{for } n \geqslant 0.
\end{cases}
$$

For $n \geqslant 1$, we have

$$
\frac{p_n}{q_n} = \cfrac{1}{a_1 + \cfrac{1}{a_2 + \cfrac{1}{a_3 + \cfrac{\ddots}{a_{n-1} + \cfrac{1}{a_n}}}}}.
$$

The following result will be very useful; it relates the ω_n to the convergents.

9.2.2 Lemma   *For all $n \geqslant 0$, we have*

$$
(-1)^n \omega_0 \omega_1 \times \cdots \times \omega_n = q_n x - p_n.
$$

Proof   Set $\beta_n = (-1)^n \omega_0 \omega_1 \times \cdots \times \omega_n$ for $n \geqslant 0$. Then, for $n \geqslant 1$, we have

$$
\begin{aligned}
\beta_{n+1} &= -\beta_n \omega_{n+1} \\
&= -\beta_n \left(\frac{1}{\omega_n} - a_{n+1} \right) \\
&= a_{n+1} \beta_n - \frac{\beta_n}{\omega_n} \\
&= a_{n+1} \beta_n + \beta_{n-1}.
\end{aligned}
$$

So, the sequence $(\beta_n)_{n \geqslant 0}$ satisfies the same recurrence relation as the sequences $(p_n)_{n \geqslant 0}$ and $(q_n)_{n \geqslant 0}$, hence also as the sequence $(q_n x - p_n)_{n \geqslant 0}$. Moreover, as $\beta_0 = \omega_0 = x$, we have

$$\beta_0 = q_0 x - p_0 \text{ and } \beta_1 = -x\omega_1 = -x\left(\frac{1}{x} - a_1\right) = a_1 x - 1 = q_1 x - p_1,$$

and the result follows. $\square$

9.2.3 Corollary *For all $n \geqslant 0$, we have*

$$\frac{1}{2q_{n+1}} \leqslant \omega_0 \omega_1 \times \cdots \times \omega_n \leqslant \frac{1}{q_{n+1}}.$$

Proof By Lemma 9.2.2 and equation (7.20), we have

$$\omega_0 \omega_1 \times \cdots \times \omega_n = |q_n x - p_n| = \frac{1}{a'_{n+1} q_n + q_{n-1}}.$$

However,

$$a'_{n+1} q_n + q_{n-1} \geqslant a_{n+1} q_n + q_{n-1} = q_{n+1}$$

and

$$a'_{n+1} q_n + q_{n-1} \leqslant (1 + a_{n+1}) q_n + q_{n-1} = q_n + q_{n+1} \leqslant 2q_{n+1},$$

since the sequence $(q_n)_{n \geqslant 0}$ increases. This implies the result. $\square$

9.2.4 A practical form of the approximate functional equation

In this section, we intend to give an estimation, as precise as possible, of

$$\|s_n\|_\infty = \sup_{0 \leqslant t \leqslant 1} |s_n(x, t)|,$$

where x is a fixed irrational in $]0, 1[$. It is easy to find a lower bound for this norm using Parseval's identity.[6] This gives

$$\|s_n\|_\infty \geqslant \|s_n\|_2 = \left(\int_0^1 |s_n(x, t)|^2 dt\right)^{1/2} = \sqrt{n+1},$$

since $|\widehat{s_n}(k)| = |e^{i\pi k^2 x}| = 1$. As we will see further on, for numbers such as $\sqrt{2}$ or the golden ratio $\dfrac{1 + \sqrt{5}}{2}$, or more generally for all quadratic irrational numbers, *this estimate in $\sqrt{n}$ is "the right one"*.

Using the beginning of the continued fraction expansion of x, we start by obtaining a more practical form of the approximate functional equation, expressed in the following lemma.

---

[6] We then consider s_n as a 1-periodic function of the variable t.

9.2.5 Lemma *For all* $t \in [0, 1]$, *there exists a* $u \in \mathbb{C}$ *of modulus* 1 *and* $x_1 \in [0, 1]$ *such that*

$$s_n(x, t) = \frac{u}{\sqrt{x}} \overline{s_{nx}(\omega_1, x_1)} + O\left(\frac{1}{\sqrt{x}}\right),$$

the O being absolute.

Proof By Theorem 9.1.1, we have

$$s_n(x, t) = \frac{e^{i\pi/4}}{\sqrt{x}} e^{-i\pi t^2/x} s_{nx}\left(-\frac{1}{x}, \frac{t}{x}\right) + O\left(\frac{1}{\sqrt{x}}\right)$$

$$= \frac{u}{\sqrt{x}} s_{nx}(-a_1 - \omega_1, -\theta) + O\left(\frac{1}{\sqrt{x}}\right),$$

with

$$u = e^{i\pi/4 - i\pi t^2/x} \in \mathbb{T} \text{ and } \theta = -\frac{t}{x} \in \mathbb{R}.$$

Let us observe that s_n satisfies another functional equation (this one non-approximate!),

$$s_n(x - 1, t) = \sum_{k=0}^{n} e^{i\pi k^2(x-1) + 2i\pi kt} = \sum_{k=0}^{n} (-1)^k e^{i\pi k^2 x + 2i\pi kt}$$

$$= \sum_{k=0}^{n} e^{i\pi k^2 x + 2i\pi k(t+1/2)} = s_n\left(x, t + \frac{1}{2}\right),$$

based on the following amazing fact: $(-1)^{k^2} = (-1)^k$ if k is an integer. We thus deduce that

$$s_n(x, t) = \frac{u}{\sqrt{x}} s_{nx}\left(-\omega_1, -\theta + \frac{a_1}{2}\right) + O\left(\frac{1}{\sqrt{x}}\right)$$

$$= \frac{u}{\sqrt{x}} s_{nx}(-\omega_1, -x_1) + O\left(\frac{1}{\sqrt{x}}\right),$$

where

$$x_1 = \theta - \frac{a_1}{2} - \left[\theta - \frac{a_1}{2}\right],$$

since s_n is 1-periodic with respect to its second variable. Finally,

$$s_n(x, t) = \frac{u}{\sqrt{x}} \overline{s_{nx}(\omega_1, x_1)} + O\left(\frac{1}{\sqrt{x}}\right). \qquad \square$$

9.2.6 Remark Before continuing, let us make a remark that will be useful later: Lemma 9.2.5 holds *even if n is not an integer*. Indeed, if n is a non-negative real number and p its integer part, we have

$$s_n(x, t) = s_p(x, t)$$

$$= \frac{u}{\sqrt{x}} \overline{s_{px}(\omega_1, x_1)} + O\left(\frac{1}{\sqrt{x}}\right)$$

$$= \frac{u}{\sqrt{x}} \left(\overline{s_{nx}(\omega_1, x_1)} + O(1)\right) + O\left(\frac{1}{\sqrt{x}}\right)$$

$$= \frac{u}{\sqrt{x}} \overline{s_{nx}(\omega_1, x_1)} + O\left(\frac{1}{\sqrt{x}}\right),$$

the O being absolute.

We are now in a position to iterate the approximate functional equation, and to obtain the following fundamental theorem.

9.2.7 Theorem *Let x be an irrational in the interval $]0, 1[$ and $t \in [0, 1]$. Then, for $n, s \geqslant 1$, we have*

$$|s_n(x, t)| \leqslant \frac{n}{\sqrt{q_s}} + C\sqrt{q_s},$$

where $C > 0$ is an absolute constant, q_s denoting the denominator of the sth convergent of the continued fraction expansion of x.

Proof The idea of Hardy and Littlewood, simplified by Zalcwasser [193], is *to iterate* Lemma 9.2.5, which is allowed by Remark 9.2.6. In the following calculations, u denotes a complex number of modulus 1 that can vary from one line to another, and s^* denotes either s or $\bar{s}$ depending on the parity of the number of iterations. By Lemma 9.2.5, we have, since $x = \omega_0$,

$$s_n(x, t) = \frac{u}{\sqrt{x}} s_{nx}^*(\omega_1, x_1) + O\left(\frac{1}{\sqrt{x}}\right)$$

$$= \frac{u}{\sqrt{\omega_0}} \left(\frac{u'}{\sqrt{\omega_1}} s_{n\omega_0\omega_1}^*(\omega_2, x_2) + O\left(\frac{1}{\sqrt{\omega_1}}\right)\right) + O\left(\frac{1}{\sqrt{x}}\right)$$

$$= \frac{u}{\sqrt{\omega_0\omega_1}} s_{n\omega_0\omega_1}^*(\omega_2, x_2) + \left(\frac{1}{\sqrt{\omega_0}} + \frac{1}{\sqrt{\omega_0\omega_1}}\right) O(1),$$

where $O(1)$ simply denotes an absolute constant that can vary from one line to another. Step by step, we obtain, for $p \geqslant 0$,

$$s_n(x, t) = \frac{u}{\sqrt{\omega_0\omega_1 \times \cdots \times \omega_p}} s_{n\omega_0\omega_1 \times \cdots \times \omega_p}^*(\omega_{p+1}, x_{p+1})$$

$$+ \left(\frac{1}{\sqrt{\omega_0}} + \frac{1}{\sqrt{\omega_0\omega_1}} + \cdots + \frac{1}{\sqrt{\omega_0\omega_1 \times \cdots \times \omega_p}}\right) O(1).$$

We can then apply the triangle inequality:

$$|s_n(x,t)| \leqslant \frac{n\omega_0\omega_1 \times \cdots \times \omega_p}{\sqrt{\omega_0\omega_1 \times \cdots \times \omega_p}} + \left(\frac{1}{\sqrt{\omega_0}} + \frac{1}{\sqrt{\omega_0\omega_1}} + \cdots + \frac{1}{\sqrt{\omega_0\omega_1 \times \cdots \times \omega_p}} \right) O(1)$$

$$\leqslant n\sqrt{\omega_0\omega_1 \times \cdots \times \omega_p} + \frac{1}{\sqrt{\omega_0\omega_1 \times \cdots \times \omega_p}}$$

$$\times \left(1 + \sqrt{\omega_p} + \sqrt{\omega_p\omega_{p-1}} + \cdots + \sqrt{\omega_p\omega_{p-1} \times \cdots \times \omega_1} \right) O(1).$$

We can find a simple bound for the ω_j by noting that

$$\omega_j = \frac{1}{a_{j+1} + \omega_{j+1}} \leqslant \frac{1}{1 + \omega_{j+1}},$$

so that

$$\omega_j\omega_{j+1} \leqslant \frac{\omega_{j+1}}{1 + \omega_{j+1}} \leqslant \frac{1}{2}, \qquad (9.14)$$

since the function $u \mapsto \frac{u}{1+u}$ is increasing on $\mathbb{R}_+$ and $0 \leqslant \omega_{j+1} < 1$. From this and (9.14):

$$|s_n(x,t)| \leqslant n\sqrt{\omega_0\omega_1 \times \cdots \times \omega_p} + \frac{O(1)}{\sqrt{\omega_0\omega_1 \times \cdots \times \omega_p}}$$

$$\times \left(1 + 1 + \frac{1}{\sqrt{2}} + \frac{1}{\sqrt{2}} + \frac{1}{2} + \frac{1}{2} + \frac{1}{2\sqrt{2}} + \frac{1}{2\sqrt{2}} + \cdots \right)$$

$$\leqslant n\sqrt{\omega_0\omega_1 \times \cdots \times \omega_p} + \frac{O(1)}{\sqrt{\omega_0\omega_1 \times \cdots \times \omega_p}}$$

as the series $\sum \left(\frac{1}{\sqrt{2}} \right)^n$ is convergent. Finally,

$$|s_n(x,t)| \leqslant n\sqrt{\omega_0\omega_1 \times \cdots \times \omega_p} + \frac{c}{\sqrt{\omega_0\omega_1 \times \cdots \times \omega_p}}, \qquad (9.15)$$

where c is an absolute constant. We now select $p = s - 1 \geqslant 0$ and use Corollary 9.2.3 to obtain the bound

$$|s_n(x,t)| \leqslant \frac{n}{\sqrt{q_s}} + C\sqrt{q_s},$$

with $C = c\sqrt{2}$. This completes the proof of Theorem 9.2.7. $\qquad\square$

9.2.8 Estimations for the uniform norm

We are now in a position to prove the following two theorems, they provide non-trivial upper bounds for

$$\sup_{0 \leqslant t \leqslant 1} |s_n(x,t)|$$

when $x \in\ ]0, 1[$ is a fixed irrational number, depending on the "diophantine" properties of x, and are two typical applications of Theorem 9.2.7.

9.2.9 Theorem   *If x has bounded partial quotients, that is, if the sequence $(a_p)_{p \geqslant 1}$ is bounded,[7] then*

$$\sup_{0 \leqslant t \leqslant 1} |s_n(x, t)| = O(\sqrt{n}) \text{ as } n \to +\infty.$$

9.2.10 Theorem   *If x is irrational, then*

$$\sup_{0 \leqslant t \leqslant 1} |s_n(x, t)| = o(n) \text{ as } n \to +\infty.$$

Proof   Let n be an integer $\geqslant 1$. As $n \geqslant q_0 = 1$, there exists an integer $s \geqslant 1$ such that $q_{s-1} \leqslant n < q_s$. Next, we have

$$q_s = a_s q_{s-1} + q_{s-2} \leqslant 2a_s q_{s-1} \leqslant 2a_s n.$$

The upper bound of the fundamental Theorem 9.2.7 now gives us

$$|s_n(x, t)| \leqslant \sqrt{n} + C\sqrt{q_s} \leqslant \sqrt{n}\left(1 + C\sqrt{2a_s}\right). \tag{9.16}$$

If we suppose furthermore that the a_j are bounded by M, then

$$|s_n(x, t)| \leqslant \sqrt{n}(1 + C\sqrt{2M}),$$

and Theorem 9.2.9 follows, in the following form:

$$\sup_{0 \leqslant t \leqslant 1} |s_n(x, t)| = O(\sqrt{nM}),$$

where M denotes an upper bound of the partial quotients of the continued fraction expansion of x, the O being absolute.

In the general case, set

$$M_n = \sup_{0 \leqslant t \leqslant 1} |s_n(x, t)| \text{ and } L = \varlimsup_{n \to \infty} \frac{M_n}{n}.$$

Theorem 9.2.7 gives, for $n, t \geqslant 1$:

$$\frac{M_n}{n} \leqslant \frac{1}{\sqrt{q_t}} + \frac{C\sqrt{q_t}}{n},$$

---

[7]   This is certainly true if x is a quadratic irrational, according to a famous theorem of Lagrange [87, 138].

so that, with t fixed:

$$L \leqslant \frac{1}{\sqrt{q_t}}.$$

It remains to let t tend to infinity to obtain $L = 0$.   □

9.2.11 Remarks   (1) We cannot expect to say much more, as a_s can make the bound of (9.16) substantially greater than $\sqrt{n}$ if the partial quotients of x are gigantic. In the exercises we will see other variations on this theme, according to the size of the partial quotients.

(2) We cannot emphasise enough how the estimate of Theorem 9.2.7 is remarkable. For example, in addition to the preceding Theorem 9.2.9, whereby

$$\sup_{0 \leqslant t \leqslant 1} |s_n(x, t)| = O(\sqrt{n})$$

when x is an irrational with bounded partial quotients, our estimate will also give us, among other results, the almost everywhere convergence of the series

$$\sum \frac{e^{i\pi n^2 x}}{(\ln n)^\alpha \sqrt{n}}$$

with $\alpha > \dfrac{3}{2}$. For all this, see Exercise 9.7.

(3) One can show [87] that the set of irrational numbers whose sequence of partial quotients is bounded is negligible in the sense of Lebesgue.

(4) It is possible to give a very elementary proof of

$$s_n(x, 0) = o(n) \text{ as } n \to +\infty$$

when x is irrational (see [80], p. 117), which means that the sequence $(n^2 x)_{n \geqslant 0}$ is equidistributed modulo 1. More generally, H. Weyl showed that if $P \in \mathbb{R}[X]$ is a polynomial with at least one coefficient of index $\geqslant 1$ irrational, then the sequence $(P(n))_{n \geqslant 0}$ is equidistributed modulo 1.

9.2.12 Estimations of the L^1-norm

We now seek to obtain an estimate of the L^1-norm of the 2-periodic function defined by

$$S_n(x) = s_n(x, 0) = \sum_{k=0}^{n} e^{i\pi k^2 x}.$$

As $S_n(2 - x) = \overline{S_n(x)}$, it suffices to study the order of magnitude of

$$\int_0^1 |S_n(x)|dx.$$

To achieve this, we once again use the form of the Hardy–Littlewood approximate functional equation due to Zalcwasser [193], as given in Theorem 9.2.7. In the case of uniform estimations, we "froze" the variable x and let t vary. This time, we "freeze" the variable t at the value 0, let x vary and then estimate the L^1-norm of $S_n(x) = s_n(x, 0)$. This L^1-estimate is also due to Zalcwasser [193].

First, from our first-rate Theorem 9.2.7 we are going to derive an upper bound for the tail distribution of $\dfrac{|S_n|}{\sqrt{n}}$. For this, we denote by $\mathbf{P}$ the Lebesgue measure on $[0, 1]$.

9.2.13 Proposition *There exists an absolute constant C such that*

$$\mathbf{P}(|S_n| > t\sqrt{n}) \leqslant Ct^{-4} \text{ for } t > 0.$$

Proof Let c be the constant appearing in Theorem 9.2.7 on p. 279. We fix $t > 0$ and $n \in \mathbb{N}^*$, and denote by E the set of irrational numbers x in $]0, 1[$ such that $|S_n(x)| > t\sqrt{n}$. Let $x \in E$. We distinguish two cases (as previously, the q_j are the denominators of the convergents in the continued fraction expansion of x).

- If $q_1 > \dfrac{t^2 n}{4c^2}$, then

$$x < \frac{1}{a_1} = \frac{1}{q_1} \leqslant \frac{4c^2}{t^2 n},$$

in other words, $x \in \left]0, \dfrac{4c^2}{t^2 n}\right]$.

- If $q_1 \leqslant \dfrac{t^2 n}{4c^2}$, by the hypothesis and Theorem 9.2.7, we have, for $s \geqslant 1$:

$$t\sqrt{n} < |S_n(x)| \leqslant \frac{n}{\sqrt{q_s}} + c\sqrt{q_s}.$$

Therefore, either $\dfrac{n}{\sqrt{q_s}} > \dfrac{t\sqrt{n}}{2}$ or $c\sqrt{q_s} > \dfrac{t\sqrt{n}}{2}$. Equivalently, for all $s \geqslant 1$, we have

$$\text{either } q_s < \frac{4n}{t^2} \text{ or } q_s > \frac{t^2 n}{4c^2}. \tag{9.17}$$

Moreover, as $q_1 \leqslant \dfrac{t^2 n}{4c^2}$ and $q_s \to +\infty$ as $s \to +\infty$, there exists an integer $s_0 \geqslant 1$ such that

$$q_{s_0} \leqslant \frac{t^2 n}{4c^2} < q_{s_0+1}.$$

In this case, by (9.17),

$$q_{s_0} < \frac{4n}{t^2}.$$

But then,[8]

$$\left| x - \frac{p_{s_0}}{q_{s_0}} \right| \leqslant \frac{1}{q_{s_0} q_{s_0+1}} \leqslant \frac{1}{q_{s_0}} \frac{4c^2}{t^2 n},$$

so that, since $q_{s_0} < \dfrac{4n}{t^2}$:

$$x \in \bigcup_{1 \leqslant p \leqslant q \leqslant \frac{4n}{t^2}} \left[\frac{p}{q} - \frac{1}{q} \frac{4c^2}{t^2 n}, \frac{p}{q} + \frac{1}{q} \frac{4c^2}{t^2 n} \right].$$

Finally, joining the two preceding cases end-to-end, and noting that the set of rationals has measure zero, we obtain

$$
\begin{aligned}
\mathbf{P}(|S_n| > t\sqrt{n}) &\leqslant \frac{4c^2}{t^2 n} + \sum_{1 \leqslant p \leqslant q \leqslant \frac{4n}{t^2}} \frac{2}{q} \frac{4c^2}{t^2 n} \\
&\leqslant \frac{4c^2}{t^2 n} + \sum_{1 \leqslant q \leqslant \frac{4n}{t^2}} \frac{8c^2}{t^2 n} \\
&\leqslant \frac{4c^2}{t^2 n} + \frac{4n}{t^2} \frac{8c^2}{t^2 n} = \frac{4c^2}{t^2 n} + \frac{32c^2}{t^4}.
\end{aligned}
$$

Again, we distinguish two cases.

- If $t \leqslant 2\sqrt{n}$, then $\mathbf{P}(|S_n| > t\sqrt{n}) \leqslant \dfrac{48c^2}{t^4}$.
- If $t > 2\sqrt{n}$, then $\mathbf{P}(|S_n| > t\sqrt{n}) \leqslant \mathbf{P}(|S_n| > 2n) = 0$ since $|S_n| \leqslant n + 1$.

This completes the proof of Proposition 9.2.13, with the constant $C = 48c^2$. $\qquad\square$

We can easily deduce the following two corollaries.

---

[8] If necessary, see equality (7.20) on p. 205.

9.2.14 Corollary *There exists a constant $C > 0$ such that*

$$\int_0^1 |S_n(x)|dx \geqslant C\sqrt{n} \text{ for } n \geqslant 1.$$

Proof The L^p-norms that we will use are relative to the probability space $[0, 1]$ equipped with its Borel algebra and the Lebesgue measure $\mathbf{P}$. We set $X = \dfrac{|S_n|}{\sqrt{n+1}}$ and fix a real number $p \in]2, 4[$. By Proposition 9.2.13 and the usual formula of integration by parts, we have

$$\begin{aligned}
\|X\|_p^p &= \int_{[0,1]} X^p d\mathbf{P} \\
&= \int_0^{+\infty} pt^{p-1}\mathbf{P}(X > t)\,dt \\
&\leqslant \int_0^1 pt^{p-1}dt + \int_1^{+\infty} Cpt^{p-5}dt \\
&\leqslant 1 + \frac{Cp}{4-p} =: C_p^p.
\end{aligned}$$

Next, we interpolate 2 between 1 and p to find a lower bound for $\|X\|_1$; we thus write

$$\frac{1}{2} = \frac{1-\theta}{1} + \frac{\theta}{p}$$

with $0 < \theta < 1$, and the interpolation inequality of Exercise 9.5 gives

$$\|X\|_2 \leqslant \|X\|_1^{1-\theta}\|X\|_p^\theta \leqslant \|X\|_1^{1-\theta}C_p^\theta.$$

However, $|S_n|$ is an even function, hence

$$\int_0^1 |S_n(x)|^2 dx = \frac{1}{2}\int_{-1}^1 |S_n(x)|^2 dx = n + 1$$

by Parseval's identity. Thus, $\|X\|_2 = 1$ and the above inequality becomes

$$1 \leqslant C_p^\theta \|X\|_1^{1-\theta},$$

so that

$$\|X\|_1 \geqslant C_p^{-\frac{\theta}{1-\theta}} =: C > 0.$$

In other words, $\|S_n\|_1 \geqslant C\sqrt{n+1}$, which proves Corollary 9.2.14. $\qquad\square$

9.2.15 Corollary *There exists a numerical constant $C > 0$ such that*

$$\frac{1}{2\pi}\int_{-\pi}^\pi \left|\sum_{k=0}^n e^{ik^2 x}\right|dx \geqslant C\sqrt{n} \text{ for } n \geqslant 1.$$

Proof Set

$$\phi(x) = \sum_{k=0}^{n} e^{ik^2 x}.$$

The function $|\phi|$ is even, so that

$$\frac{1}{2\pi} \int_{-\pi}^{\pi} |\phi(x)|dx = \frac{1}{\pi} \int_{0}^{\pi} |\phi(x)|dx$$

or, by making the change of variable $x = \pi u$:

$$\frac{1}{2\pi} \int_{-\pi}^{\pi} |\phi(x)|dx = \int_{0}^{1} \left| \sum_{k=0}^{n} e^{i\pi k^2 u} \right| du = \int_{0}^{1} |S_n(u)|du \geq C\sqrt{n},$$

by Corollary 9.2.14. $\qquad\square$

9.2.16 Remark Corollary 9.2.15 is interesting in relation to the Littlewood L^1 conjecture (see Chapter 10).

Exercises

9.1. Let $\alpha \in \mathbb{R}$ be fixed.
(a) Suppose that $\alpha > \dfrac{1}{2}$. Show that the series $\sum \dfrac{e^{i\pi n^2 \sqrt{2}}}{n^\alpha}$ is convergent. On the contrary, it diverges if $\alpha \leq \dfrac{1}{2}$ (see Exercise 7.10, on p. 216).

(b) Suppose that $\alpha > 1$. Show that the series $\sum \dfrac{e^{i\pi n^2 \sqrt{2}}}{\sqrt{n}(\ln n)^\alpha}$ is convergent.

9.2. Let $\sum a_n z^n$ be a power series with radius of convergence 1, such that the sequence $(|a_n|)_{n \geq 0}$ is non-decreasing. For $|z| < 1$, set $f(z) = \sum_{n=0}^{\infty} a_n z^n$. Finally, let δ be a fixed positive real number.
(a) Suppose that there exists a positive constant C such that, for $|z| < 1$,

$$|f(z)| \leq C(1 - |z|)^{-1-\delta}.$$

Show that

$$a_n = O(n^{\delta+1/2}) \text{ as } n \to +\infty. \tag{9.18}$$

(b) Show that the estimate (9.18) is, in general, optimal for a given δ.

9.3. Let $\sum a_n z^n$ be a power series with radius of convergence 1. For $|z| < 1$, set $f(z) = \sum_{n=0}^{\infty} a_n z^n$, and suppose that f is bounded in the open unit disk. Finally, fix $r \in \,]0, 1[$.

(a) Show that if $\rho \in]r, 1[$, we have

$$\sum_{n=0}^{\infty} |a_n| r^n \leqslant \|f\|_\infty \left(1 - \left(\tfrac{r}{\rho}\right)^2\right)^{-1/2},$$

and then that

$$\sum_{n=0}^{\infty} |a_n| r^n \leqslant \|f\|_\infty (1 - r^2)^{-1/2}$$

(Landau's inequality).

(b) Show that if C_r is a real constant C_r such that

$$\sum_{n=0}^{\infty} |a_n| r^n \leqslant C_r \|f\|_\infty$$

for all functions f satisfying the hypothesis of the exercise, then

$$C_r \geqslant C(1 - r^2)^{-1/2},$$

where C is an absolute constant. Bombieri and Bourgain [27] showed that the best constant C_r is equivalent to $(1-r^2)^{-1/2}$ as $r \to 1$.

9.4. Let $(c_k)_{k \in \mathbb{Z}}$ be a sequence of complex numbers of modulus 1, such that

$$\sup_{t \in \mathbb{R}} \left| \sum_{k=-n}^{n} c_k e^{ikt} \right| = o(n) \text{ as } n \to +\infty$$

(for example: $c_k = e^{i\pi k^2 x}$ with x irrational, by Theorem 9.2.10). We intend to show that there does not exist any Borel measure on the circle $\mathbb{T}$ such that

$$\widehat{\mu}(k) := \int_{\mathbb{T}} e^{-ikt} d\mu(t) = c_k \text{ for all } k \in \mathbb{Z}.$$

(a) Show that if μ is a Borel measure on $\mathbb{T}$ and $P(t) = \sum_{k=-n}^{n} a_k e^{ikt}$ a trigonometric polynomial, then

$$\int_{\mathbb{T}} P(-t) d\mu(t) = \sum_{k \in \mathbb{Z}} a_k \widehat{\mu}(k).$$

(b) Conclude.

9.5. Let $p < r < q$ be three positive real numbers, and $\theta \in]0, 1[$ such that

$$\frac{1}{r} = \frac{1-\theta}{p} + \frac{\theta}{q}.$$

We intend to prove the following inequality, valid on an arbitrary measure space $(\Omega, \mathcal{A}, \mu)$:

$$\|X\|_r \leqslant \|X\|_p^{1-\theta} \|X\|_q^{\theta}$$

where, of course, $X : \Omega \to \mathbb{C}$ is measurable and

$$\|X\|_p = \left(\int_\Omega |X|^p d\mu \right)^{1/p} \in \mathbb{R}_+ \cup \{+\infty\}.$$

(a) Suppose first that $\|X\|_p = \|X\|_q = 1$. Show that $\|X\|_r \leqslant 1.$[9]

(b) We move to the general case. Show that the change of X to $\dfrac{X}{a}$ and of μ to $\dfrac{\mu}{b}$, where a and b are two positive constants to be adjusted, brings us back to the previous case, and conclude.

9.6. Given an element x of $[0, 1[$, denote by $\big(a_n(x)\big)_{n \geqslant 1}$ the sequence of partial quotients of its continued fraction expansion.[10] Also, define the Gauss transformation as

$$T : [0, 1[\to [0, 1[, x \mapsto \begin{cases} \dfrac{1}{x} - \left[\dfrac{1}{x} \right] & \text{if } x \neq 0 \\ 0 & \text{if } x = 0 \end{cases}$$

and the Gauss measure as

$$d\mu = \frac{1}{\ln 2} \frac{dx}{1+x}.$$

(a) Show that, if $x \in [0, 1[$,

$$a_1(x) = \left[\frac{1}{x} \right] \text{ if } x \neq 0$$

and

$$a_n(x) = a_1(T^{n-1}(x)) \text{ for all } n \geqslant 1.$$

(b) Show that T preserves the measure μ, that is to say, $\mu\big(T^{-1}(A)\big) = \mu(A)$ for all Borel sets $A \subset [0, 1[$.

(c) Let $(M_n)_{n \geqslant 1}$ be a sequence of positive real numbers such that $\sum_{n=1}^{\infty} \dfrac{1}{M_n} < \infty$. Show that, if $n \geqslant 1$,

$$\mu(a_n > M_n) \leqslant \frac{C}{M_n},$$

---

[9] A method by "tensorisation" is indicated on the web page of Terence Tao at
   http://terrytao.wordpress.com/2008/08/25/tricks-wiki-article-the-tensor-product-trick/.

[10] If x is rational, the $a_n(x)$ are zero beyond a certain rank.

where C is a positive constant, and thus deduce that for μ-almost all $x \in [0, 1[$, there exists an integer $n_0(x) \geqslant 1$ such that

$$a_n(x) \leqslant M_n \text{ for } n \geqslant n_0(x).$$

(d) Show that the result of part (c) holds for dx-almost all $x \in [0, 1[$. One can show ([21], pp. 324–325) that if $\sum_{n=1}^{\infty} \dfrac{1}{M_n} = +\infty$ then, for dx-almost all $x \in [0, 1[$, $a_n(x) > M_n$ for an infinity of values of n.

9.7. We keep the notations of the preceding exercise.

(a) Let x be an irrational number in $[0, 1]$, and $(q_n)_{n \geqslant 0}$ the sequence of denominators of its convergents. Show that

$$q_n \geqslant \theta^{n-1} \text{ for all } n \geqslant 0,$$

where $\theta = \dfrac{1 + \sqrt{5}}{2}$ is the golden ratio.

(b) Fix $\varepsilon > 0$. Show that there exists a Borel set $E \subset [0, 1]$, with Lebesgue measure 1, such that

$$a_n(x) = O\big(n(\ln n)^{1+\varepsilon}\big) \text{ for each } x \in E.$$

(c) We want to use Theorem 9.2.7 of the present chapter: if $n, s \geqslant 1$, we have

$$|S_n(x)| \leqslant \frac{n}{\sqrt{q_s}} + C\sqrt{q_s},$$

where C is a positive constant, x an irrational in $]0, 1[$, q_s the denominator of the sth convergent of the continued fraction expansion of x and

$$S_n(x) := \sum_{k=0}^{n} e^{i\pi k^2 x}.$$

For this, we fix $x \in E$, $n \geqslant 1$ and denote by s the integer $\geqslant 1$ such that $q_{s-1} \leqslant n < q_s$. Show that

$$q_s \leqslant 2a_s n,$$

and then that

$$S_n(x) = O(\sqrt{n \ln n (\ln \ln n)^{1+\varepsilon}}).$$

(d) Show that the series $\sum \dfrac{e^{i\pi n^2 x}}{\sqrt{n}(\ln n)^{\alpha}}$ converges for almost all $x \in [0, 1]$ if $\alpha > \dfrac{3}{2}$. By using a deep result of Carleson, one can show that the result holds for $\alpha > \dfrac{1}{2}$.

Our purpose in the following two exercises is to develop a method of Little-wood, perfected by Kuzmin and Landau, which, for a non-trivial special case, leads to an inequality of van der Corput, without resorting to integrals and a series–integral comparison.

9.8. Littlewood–Kuzmin–Landau. Let $(\Phi_n)_{1 \leqslant n \leqslant N}$ be a non-decreasing sequence of real numbers and let $a < b$ be integers $\geqslant 1$. Set

$$h_n = \Phi_n - \Phi_{n-1}, \; \psi_n = e^{2i\pi\Phi_n}, \; \Delta_n = \psi_n - \psi_{n-1} \text{ and } S(a,b) = \sum_{a < n \leqslant b} \psi_n.$$

Suppose also that

$$p + \theta \leqslant h_{a+1} \leqslant h_{a+2} \leqslant \cdots \leqslant h_b \leqslant p + 1 - \theta, \text{ with } p \in \mathbb{Z} \text{ and } 0 < \theta < \frac{1}{2}.$$

(a) Observe that $\psi_n = \dfrac{\Delta_n}{2i} \cot(\pi h_n) + \dfrac{1}{2}\Delta_n$ (see Chapter 1).

(b) By performing an Abel transformation, show that

$$|S(a,b)| \leqslant 1 + (|\cot(\pi h_{a+1})| + |\cot(\pi h_b)|) .$$

(c) Prove the inequality $|S(a,b)| \leqslant \dfrac{2}{\theta}$.

9.9. Van der Corput. With the notations of the preceding exercise, suppose that $b \leqslant 2a$ and take $\Phi_n = f(n)$, where $f(x) = x \ln x + tx$, with t fixed in $[0, 1]$. Set $I =]a, b]$, $\theta = \dfrac{1}{\sqrt{a}}$ and let $p \in \mathbb{Z}$ such that $p \leqslant h_{a+1} < p + 1$.

(a) Let $n \in I$. Show that

$$\frac{1}{3a} \leqslant \frac{1}{n+1} \leqslant h_{n+1} - h_n \leqslant \frac{1}{n-1} \leqslant \frac{1}{a}.$$

Deduce that

$$h_b \leqslant h_{a+1} + \frac{b-a}{a} < p + 2.$$

(b) The sets of integers $I_1, \ldots, I_5$ are defined by[11]

$$I_1 = \{n \in I / p \leqslant h_n < p + \theta\},$$
$$I_2 = \{n \in I / p + \theta \leqslant h_n \leqslant p + 1 - \theta\},$$
$$I_3 = \{n \in I / p + 1 - \theta < h_n < p + 1 + \theta\},$$
$$I_4 = \{n \in I / p + 1 + \theta \leqslant h_n \leqslant p + 2 - \theta\},$$
$$I_5 = \{n \in I / p + 2 - \theta < h_n < p + 2\}.$$

Correspondingly, set

$$S_j = \sum_{n \in I_j} \psi_n,$$

---

[11] Draw a picture!

so that[12] $S(a, b) = S_1 + \cdots + S_5$. Also, write $I_1 =]a, c]$. Show that

$$h_c - h_{a+1} \geqslant \frac{c - a - 1}{3a}$$

and thus deduce that $|I_1| \leqslant C\theta a$, where C is a numerical constant, possibly varying in what follows. Similarly, bound I_3 and I_5.

(c) Show that $\max(|S_1|, |S_3|, |S_5|) \leqslant C\theta a$.

(d) Show that $\max(|S_2|, |S_4|) \leqslant \dfrac{C}{\theta}$.

(e) Conclude that $|S(a, b)| \leqslant C\sqrt{a}$.

(f) Set $a_n = e^{2i\pi n \ln n}$ (so that $|a_n| = 1$) and let

$$S_N(z) = \sum_{n=1}^{N} a_n z^n$$

be the partial sum of order N of the power series $\sum a_n z^n$. Show that

$$\sqrt{N} \leqslant \|S_N\|_\infty := \sup_{z \in \overline{D}} |S_N(z)| \leqslant C\sqrt{N} \text{ for all } N \geqslant 1.$$

9.2.17 Remark The method of Littlewood–Kuzmin–Landau thus provides an example of a power series with coefficients a_n of modulus 1, all of whose partial sums have an L^∞-norm with an order of magnitude as small as possible. Another example was seen in this chapter:

$$a_n = e^{in^2\pi\sqrt{2}},$$

but it is much less elementary than that of the exercise.

---

[12] Note that certain sums could be empty.

10

The Littlewood conjecture

10.1 Introduction

In 1948, a year after Hardy's death, an article [83] authored by Hardy and Littlewood was published in the *Journal of the London Mathematical Society*. In this article, a problem of rearrangement of Fourier coefficients[1] led to the following question.

Let $\lambda_1 < \cdots < \lambda_N$ be a sequence of N distinct integers. Set

$$\phi(t) = \sum_{k=1}^{N} e^{i\lambda_k t} \text{ and } \|\phi\|_1 = \frac{1}{2\pi} \int_{-\pi}^{\pi} |\phi(t)| dt.$$

Is it true that we always have $\|\phi\|_1 \geqslant c \ln N$, for a positive numerical constant c?

The question is partly motivated by the affirmative answer when the λ_k are consecutive integers (provided by elementary properties of the Dirichlet kernel). This apparently innocent question is a rather special problem in harmonic analysis, as was the question of the differentiability of the Riemann function. However, it appears to us to be one of the great problems in analysis of the twentieth century, and has become famous under the name of the "Littlewood conjecture", not least for the following reasons.

(1) It was posed (firstly around 1930, then, as mentioned above, in 1948) by great mathematicians – Littlewood and also Hardy – and it was studied by great mathematicians. We only lend to the rich. . .

(2) Despite its very elementary statement, it put up unexpected resistance for several decades.

(3) Its solution, highly non-trivial, was finally found almost simultaneously (1981) by Konyagin [111] then McGehee *et al.* [127], using different methods.

---

[1] See Section 10.2 for more details.

We cite a few partial results obtained before the definitive solution.

- In 1955, Salem [164] showed that if $\ln \lambda_k = O(\ln k)$, then

$$\overline{\lim_{N \to \infty}} \frac{1}{\sqrt{\ln N}} \left\| \sum_{k=1}^{N} e^{i\lambda_k t} \right\|_1 > 0.$$

- Olevskii in 1967 [139] and Bočhkarëv in 1975 [25] showed, among other things, that for any increasing sequence $(\lambda_k)_{k \geq 1}$ of integers, we have

$$\overline{\lim_{N \to \infty}} \frac{1}{\ln N} \left\| \sum_{k=1}^{N} e^{i\lambda_k t} \right\|_1 > 0.$$

This is better in two respects than Salem's result, but it remains an average result, which says nothing about the individual behaviour of $\|\phi\|_1$.

- The first individual result was obtained by P. Cohen in 1960 [42], namely:

$$\left\| \sum_{k=1}^{N} e^{i\lambda_k t} \right\|_1 \geq c \left(\frac{\ln N}{\ln \ln N} \right)^{1/8}.$$

- The same year, Davenport [44] was able to replace the exponent $1/8$ by $1/4$. In this regard, the following anecdote is instructive:[2] the Greek mathematician S. K. Pichorides [144], who won the Salem prize for his remarkable progress on the Littlewood conjecture, recounted how he became interested in this problem. One day, in a colleague's office at the Paris 11 University (Orsay), he happened to come across p. 73 of W. Rudin's *Fourier Analysis on Groups* [159], containing the following passage, where by the way the exponent $1/4$ is incorrectly attributed to Cohen:

If $n_1, \ldots, n_k$ are distinct integers and if

$$d\mu(x) = \sum_{j=1}^{k} e^{in_j x} \frac{dx}{2\pi} \qquad (10.1)$$

then μ is an idempotent measure on the circle group $\mathbb{T}$, and

$$\|\mu\| = \frac{1}{2\pi} \int_{-\pi}^{\pi} \left| e^{in_1 x} + \cdots + e^{in_k x} \right| dx. \qquad (10.2)$$

It is an interesting problem to determine the order of magnitude of $m(k)$, the greatest lower bound of the numbers (2) for all possible choices of $n_1, \ldots, n_k$. The best result in this direction so far is that

---

[2] In the mathematical community, it is fitting to suggest that the problems we tackle are natural: "*It is a fundamental question in the theory of . . .* ". Reality is sometimes different, even if, as we will see later on, this problem is effectively connected to results on the Hardy spaces and on the rearrangement of Fourier coefficients.

$$m(k) > A \left(\frac{\ln k}{\ln \ln k} \right)^{1/4},$$

where A is an absolute constant (Cohen [2]). If the integers
$n_1, \ldots, n_k$ *are in arithmetic progression, then (2) is asymptotic to*
$A \ln k$ *and it is conceivable that this is the right order of magnitude*
of $m(k)$.

The problem intrigued him, and little by little came to fascinate him, to
the point of leading him to an almost complete solution, as we mentioned
above. Indeed, in 1980 Pichorides [145] succeeded in obtaining the following
individual lower bound:

$$\|\phi\|_1 \geqslant c \frac{\ln N}{(\ln \ln N)^2},$$

which almost completely answered the initial question.

It is the complete solution of McGehee *et al.* that we will present here. This
solution is a prime example of what the obstinacy of a community can achieve
in the face of a problem that appears "impregnable".[3]

We first need to recall a few facts about Fourier series, and introduce a few
notations. As usual, $\mathbb{T}$ denotes the compact multiplicative group of complex
numbers with modulus 1, identified via $t \mapsto e^{it}$ with the quotient $\mathbb{R}/2\pi\mathbb{Z}$. Its
normalised Haar measure m is defined by

$$\int_{\mathbb{T}} f \, dm = \int_0^{2\pi} f(e^{it}) \frac{dt}{2\pi},$$

if $f : \mathbb{T} \to \mathbb{C}$ is continuous.

Now, the Lebesgue spaces L^p, $1 \leqslant p \leqslant \infty$, are defined with respect to the
measure m, and the Hardy space H^p is the closed subspace of L^p made up of
L^p functions whose negative Fourier coefficients are zero, that is

$$\widehat{f}(n) = \int_{\mathbb{T}} f e_{-n} \, dm = 0 \quad \text{for any } n \in \mathbb{Z} \text{ such that } n < 0,$$

the function e_r ($r \in \mathbb{Z}$) being the imaginary exponential defined by $e_r(t) = e^{irt}$
for $t \in \mathbb{R}$ (e_r is a character of $\mathbb{T}$). Another way to say that a function belongs to
H^p uses the notion of the *spectrum* of a function, which will turn out to
be useful later: if $f \in L^1$, the spectrum of f (denoted by Sp f) is defined
by (see [159])

$$\text{Sp} f = \{n \in \mathbb{Z}/\widehat{f}(n) \neq 0\}.$$

More generally, if μ is a bounded Borel measure on $\mathbb{T}$ and if

$$\widehat{\mu}(n) = \int_{\mathbb{T}} e_{-n} d\mu,$$

---

[3] *"À la septième fois, les murailles tombèrent..."* ([93], *Les Châtiments*, "Sonnez, sonnez
toujours...").

the spectrum of μ, denoted by Sp μ, is defined by

$$\text{Sp}\,\mu = \{n \in \mathbb{Z}/\widehat{\mu}(n) \neq 0\}.$$

With this terminology, H^p is none other than the set of L^p functions whose spectrum is contained in $\mathbb{N}$. These spaces H^p also play an important role in Chapter 12, and the summary on the spaces H^p given there (p. 360) will be useful in what follows, especially the following result:

$$H^\infty \text{ is a closed subalgebra of } L^\infty. \tag{10.3}$$

The closed nature of H^∞ is an obvious consequence of the inequality

$$\|\widehat{f}\|_\infty \leqslant \|f\|_1 \leqslant \|f\|_\infty.$$

As for its subalgebra nature, it is evident in terms of analytic functions, but can also be seen without reference to these functions, thanks to the following useful property:

$$\text{if } f, g \in L^2, \quad \text{then } \text{Sp}(fg) \subset \text{Sp}\,f + \text{Sp}\,g, \tag{10.4}$$

where the notation $A + B$ denotes the set of sums $a + b$, with $a \in A, b \in B$. Indeed, let $n \in \text{Sp}\,(fg)$. Since[4]

$$\widehat{fg}(n) = \sum_{u+v=n} \widehat{f}(u)\widehat{g}(v) \neq 0,$$

there exist $u, v \in \mathbb{Z}$ such that $n = u + v$ and $\widehat{f}(u) \neq 0$, $\widehat{g}(v) \neq 0$; in other words,

$$u \in \text{Sp}\,f, \quad v \in \text{Sp}\,g \text{ and } n \in \text{Sp}\,f + \text{Sp}\,g,$$

hence (10.4), and (10.3), since $\mathbb{N} + \mathbb{N} \subset \mathbb{N}$.

The following proposition may seem totally out of place here, but it brings together two properties that will be extremely useful in the rest of this chapter.

10.1.1 Proposition [majorant property in H^1]

(1) *If $f \in H^1$, there exists an $F \in H^1$ such that $|\widehat{f}(n)| \leqslant \widehat{F}(n)$ for all $n \in \mathbb{Z}$, and*

$$\|F\|_1 \leqslant \|f\|_1.$$

(2) *If $H \in H^\infty$ verifies $\text{Re}\,H \geqslant 0$, then $e^{-H} \in H^\infty$, and*

$$\|e^{-H} - 1\|_2 \leqslant \|H\|_2.$$

---

[4] Note that the series (indexed by $\mathbb{Z}$) with general term $\widehat{f}(u)\widehat{g}(n-u)$ is absolutely convergent, with n fixed, by the Cauchy–Schwarz inequality. The equality follows from Parseval's identity applied to the two L^2 functions f and $g_n(x) = \overline{g(x)}e^{inx}$, since $\widehat{g_n}(u) = \overline{\widehat{g}(n-u)}$.

Proof For point (1), we write f (see Chapter 12) as the product of two H^2 functions with controlled norms:

$$f = gh, \quad \text{with } \|g\|_2 = \|h\|_2 = \|f\|_1^{1/2}.$$

Let G and H be H^2 functions such that $\widehat{G} = |\widehat{g}|$ and $\widehat{H} = |\widehat{h}|$. Such functions exist by the Riesz–Fischer theorem, and they satisfy

$$\|G\|_2 = \|g\|_2 \text{ and } \|H\|_2 = \|h\|_2.$$

Hence, $F := GH$ is an element of H^1, with

$$\|F\|_1 \leqslant \|G\|_2 \|H\|_2 = \|g\|_2 \|h\|_2 = \|f\|_1.$$

Moreover, for $n \in \mathbb{N}$, we have

$$|\widehat{f}(n)| = \left| \sum_{u+v=n} \widehat{g}(u)\widehat{h}(v) \right| \leqslant \sum_{u+v=n} |\widehat{g}(u)| \, |\widehat{h}(v)| = \sum_{u+v=n} \widehat{G}(u)\widehat{H}(v) = \widehat{F}(n).$$

Let us now consider point (2). The partial sums $\sum_{k=0}^{n}(-1)^k \dfrac{H^k}{k!}$ of e^{-H} are elements of H^∞, since H^∞ is an algebra, and they converge uniformly to e^{-H}, as H is bounded, so that $e^{-H} \in H^\infty$ by (10.3). Finally, if $z \in \mathbb{C}$ and $\operatorname{Re} z \geqslant 0$, we have

$$|e^{-z} - 1| = \left| \int_0^1 ze^{-tz}dt \right| \leqslant \int_0^1 |z|e^{-t\operatorname{Re}z}dt \leqslant |z|.$$

In the current situation, this gives

$$|e^{-H(t)} - 1| \leqslant |H(t)|$$

for (almost) all $t \in \mathbb{T}$, and by integration we obtain the desired inequality. $\qquad \square$

10.1.2 Remarks (1) The second point of Proposition 10.1.1 remains true if $H \in H^2$ and $\operatorname{Re} H \geqslant 0$, by a simple argument of holomorphic functions; we will not need this generalisation in Section 10.3, but will need it in Section 10.4, for the version of Nazarov in the case of real frequencies.

(2) Let f, g be two trigonometric polynomials, or more generally, let f be a function with absolutely convergent Fourier series and g an integrable function. Parseval's identity is usually written

$$\sum_{n \in \mathbb{Z}} \widehat{f}(n)\overline{\widehat{g}(n)} = \int_{\mathbb{T}} f\overline{g}\, dm,$$

but we can write it in a more convenient way:

$$\sum_{n \in \mathbb{Z}} \widehat{f}(n)\widehat{g}(-n) = \int_{\mathbb{T}} fg\, dm. \tag{10.5}$$

The verification is immediate, the formal calculations being justified by the hypotheses on f and g:

$$\sum_{n\in\mathbb{Z}} \widehat{f}(n)\widehat{g}(-n) = \sum_{n\in\mathbb{Z}} \widehat{f}(n)\int_{\mathbb{T}} e_n g\, dm$$

$$= \int_{\mathbb{T}}\Big(\sum_{n\in\mathbb{Z}} \widehat{f}(n)e_n\Big)g\, dm = \int_{\mathbb{T}} fg\, dm.$$

Note that the equality (10.5) also holds if f and g are square integrable; in this case, (10.5) can be deduced immediately from the usual Parseval identity, applied to the functions f and $\overline{g}$.

Finally, let us recall the properties of the Poisson kernel P_r and the Fejér kernel K_N, and those of convolution. Set

$$P_r(t) = \sum_{n\in\mathbb{Z}} r^{|n|}e^{int} = \frac{1-r^2}{|1-re^{it}|^2} \quad \text{for } 0 \leqslant r < 1 \text{ and } t \in \mathbb{R}.$$

Immediately:

$$\|P_r\|_1 = \widehat{P_r}(0) = 1, \text{ since } P_r \text{ is non-negative.}$$

On the contrary, for $N \in \mathbb{N}^*$ and $t \in \mathbb{R}$,

$$K_N(t) := \sum_{n=-N}^{N}\Big(1 - \frac{|n|}{N+1}\Big)e^{int} = \frac{1}{N+1}\left(\frac{\sin\big((N+1)\frac{t}{2}\big)}{\sin\big(\frac{t}{2}\big)}\right)^2.$$

We also have

$$\|K_N\|_1 = \widehat{K_N}(0) = 1.$$

Now, if $f \in L^1$, μ is a bounded measure on $\mathbb{T}$ and

$$(f * \mu)(x) = \int_{\mathbb{T}} f(x-t)d\mu(t),$$

then Fubini's theorem implies

$$f * \mu \in L^1, \ \widehat{f * \mu} = \widehat{f}\,\widehat{\mu} \text{ and } \|f * \mu\|_1 \leqslant \|f\|_1\|\mu\|. \tag{10.6}$$

This inequality will be applied later on to progress from the trigonometric polynomials (for which we will have done most of the work, using the principle of *a priori* inequalities) to arbitrary measures.

10.2 Properties of the L^1-norm and the Littlewood conjecture

Let us consider the *analytic* Dirichlet kernel, defined by

$$D_N(t) = \sum_{k=1}^{N} e^{ikt}.$$

A well-known fact in harmonic analysis [104], at the origin of many counter-examples about Fourier series, is the tendency of the quantity $\|D_N\|_1$ to be large when $N \to \infty$. More precisely,

$$\|D_N\|_1 \sim \frac{4}{\pi^2} \ln N \text{ as } N \to +\infty. \tag{10.7}$$

This large magnitude of $\|D_N\|_1$ is related to (at least) two general notions: Hardy's inequality and idempotent measures, briefly described below.

10.2.1 Theorem [Hardy's inequality] *Let $f \in H^1$. Then, we have*

$$\sum_{n=1}^{\infty} \frac{|\widehat{f}(n)|}{n} \leqslant \pi \|f\|_1. \tag{10.8}$$

Proof Proposition 10.1.1 allows us to linearise the problem in the following manner: let F be as described in this proposition and $G \in L^\infty$ defined by the formula

$$G(t) = i(\pi - t) \text{ if } 0 \leqslant t < 2\pi.$$

Then G clearly satisfies

$$\widehat{G}(n) = \frac{1}{n} \text{ if } n \geqslant 1 \text{ and } \|G\|_\infty = \pi.$$

For $0 \leqslant r < 1$, by setting

$$F_r(t) = (F * P_r)(t) = \sum_{n=0}^{\infty} r^n \widehat{F}(n) e^{int},$$

we thus have

$$\sum_{n=1}^{\infty} \frac{r^n |\widehat{f}(n)|}{n} \leqslant \sum_{n=1}^{\infty} \frac{r^n \widehat{F}(n)}{n} = \sum_{n=1}^{\infty} \widehat{F_r}(n) \widehat{G}(n) = \sum_{n \in \mathbb{Z}} \widehat{F_r}(n) \widehat{G}(n),$$

since $\widehat{G}(0) = 0$ and $\widehat{F}(n) = 0$ if $n < 0$, or again, by using Parseval's identity (10.5) with an evident change of sign:

$$\sum_{n=1}^{\infty} \frac{r^n |\widehat{f}(n)|}{n} \leqslant \int_{\mathbb{T}} F_r(-t) G(t) \, dt.$$

However,

$$\|F_r\|_1 \leqslant \|F\|_1 \|P_r\|_1 = \|F\|_1 \leqslant \|f\|_1,$$

hence

$$\sum_{n=1}^{\infty} \frac{r^n |\widehat{f}(n)|}{n} \leqslant \|F_r\|_1 \|G\|_\infty \leqslant \pi \|f\|_1.$$

We now let r tend to 1, and use Fatou's lemma to obtain (10.8). $\qquad\square$

10.2.2 Remarks (1) Here is another form of Hardy's inequality, valid for $f \in H^1$:

$$\sum_{n=0}^{\infty} \frac{|\widehat{f}(n)|}{n+1} \leqslant \pi \|f\|_1.$$

This is an immediate consequence of Theorem 10.2.1 applied to the function $g \in H^1$ defined by $g(t) = e^{it} f(t)$, for which we have

$$\widehat{g}(n) = \widehat{f}(n-1) \quad \text{and} \quad \|g\|_1 = \|f\|_1.$$

(2) With this new form, we can legitimately wonder: "Why $\dfrac{1}{n+1}$?". The following generalisation helps understand that Hardy's inequality, via the majorant property of Proposition 10.1.1, is nothing but the duality between H^1 and its dual $X = L^\infty / \overline{H_0^\infty}$, where

$$\overline{H_0^\infty} = \{h \in H^\infty / \widehat{h}(n) = 0 \text{ if } n \geqslant 0\},$$

and where X is equipped with the quotient norm

$$\|G\|_X = \inf_{h \in \overline{H_0^\infty}} \|G - h\|_\infty.$$

Indeed, let $G \in L^\infty$ be such that $\widehat{G}(n) \geqslant 0$ for $n \geqslant 0$. Then we will see that

$$\sum_{n=0}^{\infty} |\widehat{f}(n)| \widehat{G}(n) \leqslant \|G\|_X \|f\|_1. \tag{10.9}$$

We prove (10.9) by imitating the proof of Theorem 10.2.1: let $r < 1$, F and F_r be as in that proof, and let $h \in \overline{H_0^\infty}$. We have the chain of inequalities

$$\sum_{n=0}^{\infty} r^n |\widehat{f}(n)||\widehat{G}(n)| \leqslant \sum_{n=0}^{\infty} r^n \widehat{F}(n)\widehat{G}(n)$$

$$= \sum_{n=0}^{\infty} r^n \widehat{F}(n)\big(\widehat{G}(n) - \widehat{h}(n)\big)$$

$$= \sum_{n \in \mathbb{Z}} r^{|n|} \widehat{F}(n)\big(\widehat{G}(n) - \widehat{h}(n)\big)$$

$$= \int_{\mathbb{T}} F_r(-t)(G - h)(t)\, dt$$

$$\leqslant \|F_r\|_1 \|G - h\|_\infty$$

$$\leqslant \|F\|_1 \|G - h\|_\infty$$

$$\leqslant \|f\|_1 \|G - h\|_\infty,$$

and we obtain (10.9) by passing to the lower bound on h and letting r tend to 1. The fact that the dual of H^1 is isometric to X is explained in Section 12.2 of Chapter 12: indeed, the dual of H^1 is none other than the quotient of L^∞ by the orthogonal of H^1 for the duality between L^1 and L^∞ defined by

$$(f, g) = \int_{\mathbb{T}} f(-t)g(t)\, dm(t),$$

and this orthogonal is exactly the space $\overline{H_0^\infty}$ described above. As we will see in Chapter 12, the dual of H^1 is also isomorphic to the space *BMO* of functions with bounded mean oscillation, and we could replace the quotient norm in X by the *BMO* norm in (10.9), up to a constant. In Hardy's inequality, the quotient norm is exactly π, and we cannot obtain more.

Let us return to the Littlewood conjecture. Observe that an immediate consequence of (10.8) is a slightly weakened form of (10.7). Indeed:

$$\|D_N\|_1 \geqslant \frac{1}{\pi} \sum_{n=1}^{\infty} \frac{1}{n} |\widehat{D}_N(n)| = \frac{1}{\pi} \sum_{n=1}^{N} \frac{1}{n} \geqslant c \ln N,$$

where c is a positive constant.

Moreover, $\widehat{D}_N$ takes values in $\{0, 1\}$, hence $\widehat{D}_N$ is idempotent, that is, $\big(\widehat{D}_N\big)^2 = \widehat{D}_N$. Therefore, D_N is idempotent for the operation of convolution:

$$D_N * D_N = D_N.$$

The measures that are idempotent for the operation of convolution play an important role, as they can be associated with a projection operator P in L^1 defined as follows: if μ is a measure on the circle $\mathbb{T}$ such that $\mu * \mu = \mu$, we set

$$Pf = f * \mu \text{ for } f \in L^1,$$

so that $P^2 = P$. Then the measure μ tends to have a large L^1-norm; for example, if $\mu \neq 0$, we have

$$\|\mu\| = \|\mu * \mu\| \leqslant \|\mu\|^2,$$

hence

$$\|\mu\| \geqslant 1. \tag{10.10}$$

Note that (10.10) can even be an equality (for example, if $\mu = \delta_0$ or $\mu = m$), but it can be shown [159] that if $\|\mu\| > 1$, then automatically $\|\mu\| \geqslant \dfrac{\sqrt{5}}{2}$, and the estimate (10.7) leads in that direction. *Another idempotent measure*, similar to D_N, would be the measure (or the function) $\phi\,dm$, where

$$\phi(t) = \sum_{k=1}^{N} e^{i\lambda_k t}, \text{ with } \lambda_1 < \lambda_2 < \cdots < \lambda_N, \ \lambda_j \in \mathbb{Z} \tag{10.11}$$

(the dependence of ϕ on N being implicit). The norm $\|\phi\|_1$ will tend to be large, but how large? Hardy's inequality is not very precise: only

$$\|\phi\|_1 \geqslant \frac{1}{\pi} \sum_{k=1}^{N} \frac{1}{\lambda_k},$$

a quantity that could be bounded as N grows, for example if $\lambda_k = k^2$. Moreover,

$$\|\phi\|_1 \leqslant \|\phi\|_2 = N^{1/2},$$

and $N^{1/2}$ is in general the correct order of magnitude of $\|\phi\|_1$. For example, it can be shown that [120]:

- If $(\lambda_k)_{1 \leqslant k \leqslant N}$ satisfies $\lambda_1 \geqslant 1$ and $\dfrac{\lambda_{k+1}}{\lambda_k} \geqslant q > 1$ (lacunarity in the sense of Hadamard), then

$$\|\phi\|_1 \geqslant c\|\phi\|_2 = cN^{1/2},$$

the constant c depending only on q, and not on N. More generally, such an inequality holds if $(\lambda_k)_{1 \leqslant k \leqslant N}$ is a Sidon set, or a $\Lambda(2)$ set [120, 159].

- If $(\lambda_k)_{1 \leqslant k \leqslant N}$ is the sequence of the first N perfect squares ($\lambda_k = k^2$), we again have $\|\phi\|_1 \geqslant cN^{1/2}$. This more difficult result is due to Zalcwasser [30, 193]; its proof was detailed in Chapter 9.
- If the λ_k, $1 \leqslant k \leqslant N$, are chosen at random, equitably, among the $2N$ first integers, then we have again, most probably, $\|\phi\|_1 \geqslant cN^{1/2}$ (see Exercise 5.2 in Chapter 5, devoted to Uchiyama's theorem).

Thus, the case $\lambda_k = k$ appears to be relatively exceptional,[5] but Littlewood (and don't forget Hardy!) had conjectured [83]:

Littlewood conjecture *Nothing can be "worse" than this exception, i.e. if ϕ is as in (10.11), we always have*

$$\|\phi\|_1 \geqslant c \ln N, \tag{10.12}$$

where c is a positive constant.

The precise motivation of the authors stemmed from a problem of the rearrangement of coefficients, where the Heisenberg uncertainty principle starts to appear (we will return to this later). Let

$$f(t) = \sum_{n=0}^{\infty} c_n e^{int}$$

be a trigonometric polynomial, and let $c_0^*, \ldots, c_N^*, \ldots$ be the sequence of moduli of the c_n, rearranged in non-increasing order, and then

$$f^*(t) = \sum_{n=0}^{\infty} c_n^* e^{int}.$$

For example, if $f(t) = -1 + 3ie^{4it} + 2e^{7it}$, $f^*(t) = 3 + 2e^{it} + e^{2it}$. In the article cited above, Hardy and Littlewood proved, for $1 < p \leqslant 2$, the existence of a constant A_p such that

$$\|f^*\|_p \leqslant A_p \|f\|_p,$$

and logically asked if this still holds for $p = 1$:

$$\|f^*\|_1 \leqslant A_1 \|f\|_1.$$

In particular, if

$$\phi(t) = \sum_{k=1}^{N} e^{i\lambda_k t},$$

---

[5] If the λ_k were consecutive elements in a fixed arithmetical progression $a\mathbb{Z} + b$, we would still have $\|\phi\|_1 = \|D_N\|_1$.

then ϕ^* is none other than the Dirichlet kernel D_N, and the question becomes: do we have

$$\|\phi\|_1 \geqslant A_1^{-1} \|D_N\|_1 ?$$

This brings us back to (10.12).

We could even wonder if we always have $\|\phi\|_1 \geqslant \|D_N\|_1$. To this day, the question remains open, but after many efforts of the mathematical community and numerous partial results as indicated in Section 10.1, the weakened form (10.12) was proved in 1981, independently, by Konyagin [111] on the one hand and McGehee *et al.* [127] on the other hand. It is the proof of the latter authors that we will present in Section 10.3. In fact, following the good mathematical principle stating *Who can not do more can not do less*, the authors proved much more than (10.12), namely a generalisation of Hardy's inequality, stated as follows:[6]

$$\phi(t) = \sum_{k=1}^{N} a_k e^{i\lambda_k t} \implies \|\phi\|_1 \geqslant c \sum_{k=1}^{N} \frac{|a_k|}{k}. \qquad (10.13)$$

It is simultaneously more difficult to show (in principle) than (10.12), since we allow arbitrary coefficients a_k, and much more natural, since the inequality (10.13) appears to be linear with respect to the a_k, once we have fixed the frequencies λ_k.

10.3 Solution of the Littlewood conjecture

McGehee *et al.* [127] proved the following result, which immediately implies (10.12).

10.3.1 Theorem [McGehee–Pigno–Smith (MPS) solution of the Littlewood conjecture] *There exists a constant $A \geqslant 1$ ($A = 128$ will do) such that, for all finite sequences of integers $\lambda_1 < \lambda_2 < \cdots < \lambda_N$ and all functions*

$$\phi(t) = \sum_{k=1}^{N} a_k e^{i\lambda_k t},$$

we have

$$\sum_{k=1}^{N} \frac{|a_k|}{k} \leqslant A \|\phi\|_1. \qquad (10.14)$$

---

[6] Later, Stegeman showed that we can take $c = \dfrac{4}{\pi^3}$ in (10.13).

Before starting the proof, we remark that Hardy's inequality (10.8) is established in *two steps*. The first step consists of linearising the problem using Proposition 10.1.1: without loss of generality, we can suppose (when $\lambda_k = k$) that $\widehat{\phi} \geqslant 0$. The second step consists of producing a function $G \in L^\infty$ such that $\widehat{G}(k) = \dfrac{1}{k}$ for $k \geqslant 1$. We can then conclude with Parseval's identity.

What can we do here? The first step works in the same way, and we could suppose that $a_k \geqslant 0$. It is the second step that causes a problem: this time, we would need to produce $G \in L^\infty$ such that

$$\widehat{G}(\lambda_k) = \frac{1}{k} \text{ if } 1 \leqslant k \leqslant N \text{ and } \|G\|_\infty \leqslant A,$$

for a constant A independent of N. The existence of such a function is not at all evident, and in fact will be deduced[7] from Theorem 10.3.1, which we prove differently.

Let us return to an independent proof of (10.14). First, we linearise the problem by writing $|a_k| = a_k u_k$ with $|u_k| = 1$, and setting

$$T_0(t) = \sum_{k=1}^{N} \frac{u_k}{k} e^{-i\lambda_k t}.$$

Then

$$S := \sum_{k=1}^{N} \frac{|a_k|}{k} = \sum_{k=1}^{N} \widehat{\phi}(\lambda_k)\widehat{T_0}(-\lambda_k) = \int_{\mathbb{T}} \phi T_0 \, dm$$

thanks to Parseval's identity (10.5), so that

$$S \leqslant \|\phi\|_1 \|T_0\|_\infty.$$

The problem is that, normally, we have no control over the L^∞-norm of T_0, other than the explosive and trivial control with $\sum_{k=1}^{N} k^{-1}$. The idea will be to replace the test-function T_0 by another test-function T_1, thus *killing two birds with one stone*:

(1) this time, the L^∞-norm of the corrected function T_1 is controlled;
(2) $\widehat{T_1}(-\lambda_k)$ is little different from $\widehat{T_0}(-\lambda_k)$ for $1 \leqslant k \leqslant N$ (for $n \neq -\lambda_k$, $\widehat{T_1}(n)$ will have the right to do as it pleases).

However, satisfying these two conditions simultaneously goes precisely against the grain of a good old principle, known to harmonic analysts under the name of the *Heisenberg uncertainty principle* (even if a physicist might

---

[7] See Corollary 10.3.7.

not recognise this here): basically, it says that it is difficult to "touch" a function without exploding its Fourier transform. A classic example is that of the cosine function, which is its own Fourier series:

$$\cos x = \cos x \dots$$

Imagine that we would like to make this function non-negative, and consider instead the function $|\cos|$; by doing this, we have hardly touched the function (one arch out of two of the cosine is replaced by its symmetric reflection with respect to the real axis), and yet the Fourier series of $|\cos|$ no longer has anything to do with the original Fourier series: $\cos x = \cos x$ becomes

$$|\cos x| = \frac{2}{\pi} + \frac{4}{\pi} \sum_{k=1}^{\infty} (-1)^{k-1} \frac{\cos(2kx)}{4k^2 - 1}.$$

Therefore, claiming to achieve (1) and (2) simultaneously is *a priori* quite exorbitant. In the MPS proof, the tour de force for our three mathematicians is precisely to have succeeded in this, in the form of a key lemma, whose proof gives the impression that we are passing a camel through the eye of a needle. No wonder that the following key lemma had long eluded researchers!

10.3.2 Lemma *There exists a function $T_1 \in L^\infty$ such that:*

(1) $\|T_1\|_\infty \leqslant C$, *where C is a numerical constant ($C = 64$ will do);*
(2) $|\widehat{T_1}(-\lambda_k) - \widehat{T_0}(-\lambda_k)| \leqslant \frac{1}{2} |\widehat{T_0}(-\lambda_k)| = \frac{1}{2k}$, *for $1 \leqslant k \leqslant N$.*

Proof We first show that this key lemma immediately implies Theorem 10.3.1. Indeed, let

$$S = \sum_{k=1}^{N} \frac{|a_k|}{k} = \int_{\mathbb{T}} \phi T_0 \, dm.$$

Via Parseval's identity (10.5), we have

$$\left| S - \int_{\mathbb{T}} \phi T_1 \, dm \right| = \left| \int_{\mathbb{T}} \phi (T_0 - T_1) \, dm \right| = \left| \sum_{k=1}^{N} \widehat{\phi}(\lambda_k) \left(\widehat{T_0}(-\lambda_k) - \widehat{T_1}(-\lambda_k) \right) \right|$$

$$\leqslant \sum_{k=1}^{N} |\widehat{\phi}(\lambda_k)| \, |\widehat{T_0}(-\lambda_k) - \widehat{T_1}(-\lambda_k)| \leqslant \sum_{k=1}^{N} \frac{1}{2k} |a_k|$$

$$= \frac{1}{2} S,$$

so that

$$S \leqslant \frac{1}{2}S + \left| \int_{\mathbb{T}} \phi T_1 \, dm \right|,$$

thus

$$S \leqslant 2 \left| \int_{\mathbb{T}} \phi T_1 \, dm \right| \leqslant 2 \|\phi\|_1 \|T_1\|_\infty \leqslant 2C \|\phi\|_1,$$

and (10.14) follows, with $A = 2C$. $\qquad\qquad\qquad\qquad\qquad\qquad\square$

To prove the key lemma, we start by decomposing T_0 into a sum of dyadic blocks on which the amplitude $|\widehat{T_0}(-\lambda_k)| = \frac{1}{k}$ is more or less constant (it varies at most by a factor of two), that is, we set

$$I_j = [\![2^j, 2^{j+1}[\![, \quad f_j = \sum_{\substack{k \in I_j \\ k \leqslant N}} \frac{u_k}{k} e^{-i\lambda_k t} \quad \text{and} \quad T_0 = f_0 + \cdots + f_m,$$

where m is the integer such that $2^m \leqslant N < 2^{m+1}$. Observe that we have

$$\|f_j\|_\infty \leqslant 1 \quad \text{and} \quad \|f_j\|_2 \leqslant 2^{-j/2}. \qquad\qquad (10.15)$$

Indeed, on the one hand,

$$\|f_j\|_\infty \leqslant \sum_{k \in I_j} \frac{1}{k} \leqslant \frac{|I_j|}{2^j} = 1,$$

where $|I_j|$ denotes the cardinality of I_j; on the other hand, by Parseval's identity,

$$\|f_j\|_2^2 \leqslant \sum_{k \in I_j} \frac{1}{k^2} \leqslant \frac{|I_j|}{4^j} = 2^{-j}.$$

The function T_0 now appears as the partial sum of order m of the series $\sum f_j$. We correct the partial sums

$$f_0, \ f_0 + f_1, \ \ldots, \ f_0 + f_1 + \cdots + f_j, \ \ldots, \ f_0 + f_1 + \cdots + f_m$$

of this series to

$$F_0, \ F_1, \ \ldots, \ F_j, \ \ldots, \ F_m$$

in a way that keeps the L^∞-norm of these new "partial sums" under control. We start with $F_0 = f_0$. Suppose that we have reached F_j, how do we define F_{j+1}? As we want to mimic the partial sums, we feel like taking $F_{j+1} = F_j + f_{j+1}$, but that wouldn't change a thing, and $\|F_j\|_\infty$ is at risk of exploding. The idea of MPS is hence the following: first, we *damp F_j slightly* by multiplying it by the exponential of a function with a negative real part; and only then do we add on f_{j+1}. In other words, we adopt the following recurrence relation:

$$F_0 = f_0, \quad F_{j+1} = F_j e^{-\varepsilon h_{j+1}} + f_{j+1}, \text{ with } \operatorname{Re} h_{j+1} = |f_{j+1}| \text{ and } \varepsilon > 0.$$
$$(10.16)$$

Here, ε is a parameter of "smallness"[8] to be adjusted later on, and h_{j+1} is a function whose real part is $|f_{j+1}|$, to be specified later. We thus take, for the "corrected" test-function,

$$T_1 = F_m.$$

Suppose that for some $j \in [\![0, m-1]\!]$, $\|F_j\|_\infty \leqslant C$. Then, taking into account (10.15) and (10.16), for each $t \in \mathbb{T}$ we have

$$|F_{j+1}(t)| \leqslant |F_j(t)| \exp\left(-\varepsilon \operatorname{Re} h_{j+1}(t)\right) + |f_{j+1}(t)|.$$

Thus, by setting $x = |f_{j+1}(t)| \in [0, 1]$,

$$|F_{j+1}(t)| \leqslant C e^{-\varepsilon x} + x.$$

This quantity will remain less than or equal to C (and hence will not explode) if $C e^{-\varepsilon x} + x \leqslant C$, that is, if

$$C \geqslant \sup_{0 < x \leqslant 1} \frac{x}{1 - e^{-\varepsilon x}} =: C_\varepsilon.$$

Note that $C_\varepsilon < +\infty$ since $\dfrac{x}{1 - e^{-\varepsilon x}} \to \dfrac{1}{\varepsilon}$ as $x \to 0$, and in passing, we see that

$$C_\varepsilon \geqslant \frac{1}{\varepsilon}$$

and

$$C_\varepsilon \leqslant \frac{2}{\varepsilon} \text{ for } \varepsilon \leqslant \frac{1}{2},$$

this second estimate being a consequence of the elementary inequality

$$e^{-y} \leqslant 1 - \frac{y}{2} \text{ for } 0 \leqslant y \leqslant \frac{1}{2}.$$

It thus suffices to choose, for $\varepsilon \in]0, 1[$ fixed, $C = C_\varepsilon \geqslant 1$ and the preceding calculation guarantees that

$$\|T_1\|_\infty \leqslant C_\varepsilon.$$

---

[8] Don't forget that we need to satisfy both conditions (1) *and* (2) of p. 304.

It remains to adjust ε to satisfy, if possible, condition (2) of the key Lemma 10.3.2. Let us examine more closely what T_1 looks like. From (10.16), simple induction gives

$$T_1 = F_m = \sum_{j=0}^{m} f_j g_j, \qquad (10.17)$$

where

$$g_j = e^{-\varepsilon H_j} \text{ and } H_j = \begin{cases} h_{j+1} + \cdots + h_m & \text{if } j < m, \\ 0 & \text{if } j = m. \end{cases} \qquad (10.18)$$

At this stage, T_1 appears as a T_0 seriously perturbed (by the g_j) and remains somewhat arbitrary (Re $h_j = |f_j|$, but h_j is not completely determined), and it is difficult to see *a priori* why condition (2) of the key lemma would stand a chance of being satisfied. But a *new* element enters the picture, in the form of the following lemma.

10.3.3 Lemma *There exists a function h_j satisfying the following conditions:*

(i) Re $h_j = |f_j|$;
(ii) h_j *is analytic[9] and bounded, in other words,* $h_j \in H^\infty$;
(iii) h_j *satisfies* $\|h_j\|_2 \leqslant \sqrt{2}\|f_j\|_2$ *for* $0 \leqslant j \leqslant m$.

Proof The function f_j is a trigonometric polynomial, hence $|f_j|$ has an absolutely convergent Fourier series, for example because $|f_j|$ is Lipschitz:[10]

$$|f_j(t)| = \sum_{n \in \mathbb{Z}} c_n e^{int} \text{ and } \sum_{n \in \mathbb{Z}} |c_n| < \infty.$$

Moreover,

$$c_{-n} = \overline{c_n},$$

as $|f_j|$ is real. We thus set

$$h_j(t) = c_0 + 2 \sum_{n=1}^{\infty} c_n e^{int}.$$

Of course, $h_j \in H^\infty$ and

$$\overline{h_j(t)} = c_0 + 2 \sum_{n=1}^{\infty} c_{-n} e^{-int},$$

---

[9] That is, its spectrum is contained in $\mathbb{N}$.
[10] Which allows us to invoke Bernstein's theorem (see [153], Theorem 5.1.1, p. 121 or Exercise 10.1 of this chapter); for a variant, see Exercise 10.2.

hence

$$\operatorname{Re} h_j = c_0 + \sum_{n \neq 0} c_n e^{int} = |f_j(t)|.$$

Moreover,

$$\|h_j\|_2^2 = |c_0|^2 + 4 \sum_{n=1}^{\infty} |c_n|^2$$

$$\leqslant 2\left(|c_0|^2 + 2 \sum_{n=1}^{\infty} |c_n|^2\right)$$

$$= 2 \sum_{n \in \mathbb{Z}} |c_n|^2 = 2\|f_j\|_2^2. \qquad \square$$

If H_j and g_j are defined as in (10.18), we also have $H_j, g_j \in H^{\infty}$ by Proposition 10.1.1, which here reveals its full force, and the analyticity of the g_j leads to the following *decisive result*:

$$\text{if } k \in I_\ell = [\![2^\ell, 2^{\ell+1}[\![, \text{ then } \widehat{f_j g_j}(-\lambda_k) = 0 \text{ if } j < \ell. \qquad (10.19)$$

Indeed, we must show that $-\lambda_k \notin \operatorname{Sp}(f_j g_j)$. Let us introduce the following notation: if A is a subset of the interval of integers $[\![1, N]\!]$, we denote by Λ_A the set of λ_j, $j \in A$. Thus, $-\lambda_k \in -\Lambda_{I_\ell}$, whereas, by (10.4), we have

$$\operatorname{Sp}(f_j g_j) \subset \operatorname{Sp} f_j + \operatorname{Sp} g_j \subset -\Lambda_{I_j} + \mathbb{N}.$$

However, since $j < \ell$, $-\Lambda_{I_\ell}$ is completely to the left of $-\Lambda_{I_j}$, and *a fortiori* to the left of $-\Lambda_{I_j} + \mathbb{N}$ (Figure 10.1).

Once we have (10.19), it is easy to find an upper bound for $\widehat{T}_1(-\lambda_k) - \widehat{T}_0(-\lambda_k)$: let $k \in [\![1, N]\!]$, and $\ell \leqslant m$ such that $k \in I_\ell$. Then, by (10.19):

$$\widehat{T}_1(-\lambda_k) = \sum_{j=\ell}^{m} \widehat{f_j g_j}(-\lambda_k),$$

whereas

$$\widehat{T}_0(-\lambda_k) = \widehat{f_\ell}(-\lambda_k) = \sum_{j=\ell}^{m} \widehat{f_j}(-\lambda_k),$$

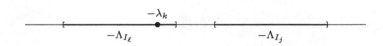

Figure 10.1

hence

$$\widehat{T_1}(-\lambda_k) - \widehat{T_0}(-\lambda_k) = \sum_{j=\ell}^{m} f_j \widehat{(g_j - 1)}(-\lambda_k).$$

At this stage, we can indulge in a brute-force bound:

$$
\begin{aligned}
|f_j \widehat{(g_j - 1)}(-\lambda_k)| &\leqslant \|f_j(g_j - 1)\|_1 \\
&\leqslant \|f_j\|_2 \|g_j - 1\|_2 = \|f_j\|_2 \|e^{-\varepsilon H_j} - 1\|_2 \\
&\leqslant \varepsilon \|f_j\|_2 \|H_j\|_2 \text{ by Proposition 10.1.1} \\
&\leqslant \varepsilon \|f_j\|_2 \sum_{r=j+1}^{m} \|h_r\|_2 \\
&\leqslant \varepsilon 2^{-j/2} \sqrt{2} \sum_{r=j+1}^{\infty} 2^{-r/2} \text{ by (10.15) and Lemma 10.3.3} \\
&= \varepsilon 2^{-j}(\sqrt{2} + 1)\sqrt{2} \leqslant 4\varepsilon 2^{-j}.
\end{aligned}
$$

It ensues that

$$|\widehat{T_1}(-\lambda_k) - T_0(-\lambda_k)| \leqslant 4\varepsilon \sum_{j=\ell}^{m} 2^{-j} \leqslant 4\varepsilon \sum_{j=\ell}^{\infty} 2^{-j} = 8\varepsilon 2^{-\ell} \leqslant 16\varepsilon |\widehat{T_0}(-\lambda_k)|$$

as $k \in I_\ell$, hence

$$|\widehat{T_0}(-\lambda_k)| = \frac{1}{k} > 2^{-\ell-1}.$$

Then, if we take $\varepsilon = \dfrac{1}{32}$ and $C = \dfrac{2}{\varepsilon} = 64 \geqslant C_\varepsilon$, we complete the proof of the key lemma, and in turn that of the MPS theorem.

10.3.4 Remarks (1) As we have more or less indicated in Section 10.1, the MPS theorem is not restricted to trigonometric polynomials, but holds for bounded measures with unidirectional spectrum. The following statement is an easy corollary of Theorem 10.3.1.

10.3.5 Theorem *Let μ be a bounded measure with unidirectional spectrum:*

$$\mathrm{Sp}\,\mu = \{\lambda_1 < \cdots < \lambda_N < \ldots\}.$$

Then, we have the inequality

$$\sum_{k=1}^{\infty} \frac{|\widehat{\mu}(\lambda_k)|}{k} \leqslant A\|\mu\|,$$

where A is the same as in Theorem 10.3.1.

Proof We denote by K_p the Fejér kernel of index p. Theorem 10.3.1, applied to the trigonometric polynomial

$$\mu * K_p = \sum_{k=1}^{\infty} \left(1 - \frac{|\lambda_k|}{p+1}\right)^+ \widehat{\mu}(\lambda_k) e_{\lambda_k},$$

gives

$$\sum_{k=1}^{\infty} \left(1 - \frac{|\lambda_k|}{p+1}\right)^+ \frac{|\widehat{\mu}(\lambda_k)|}{k} \leqslant A\|\mu * K_p\|_1 \leqslant A\|K_p\|_1\|\mu\| = A\|\mu\|$$

(see (10.6)). The passage to the limit when $p \to \infty$ thus gives Theorem 10.3.5, via Fatou's lemma. □

(2) We will see in Exercise 10.6 that the preceding theorem does not hold for measures with bidirectional spectrum.[11] Moreover, according to a theorem of F. and M. Riesz, the measure μ in this theorem is always absolutely continuous with respect to m.

(3) Theorem 10.3.5 admits the following two interesting corollaries.

10.3.6 Corollary *If* $\phi(t) = \sum\limits_{k=1}^{N} \varepsilon_k e^{i\lambda_k t}$ *with* $|\varepsilon_k| \geqslant 1$, *we have*

$$\|\phi\|_1 \geqslant c \ln N.$$

Proof Theorem 10.3.1 or 10.3.5 gives

$$\|\phi\|_1 \geqslant A^{-1} \sum_{k=1}^{N} \frac{|\varepsilon_k|}{k} \geqslant A^{-1} \sum_{k=1}^{N} \frac{1}{k} \geqslant c \ln N,$$

where c is a positive constant. □

As we have already mentioned, the following corollary is *equivalent* to the solution of the Littlewood conjecture.

10.3.7 Corollary *Let* $S = \{\lambda_1 < \cdots < \lambda_N < \ldots\}$ *be a unidirectional sequence of integers, and* $(w_k)_{k \geqslant 1}$ *a sequence of complex numbers such that* $|w_k| \leqslant \frac{1}{k}$ *for all* $k \geqslant 1$. *Then, there exists some* $g \in L^\infty$ *realising the interpolation*

$$\widehat{g}(\lambda_k) = w_k \text{ for all } k \geqslant 1, \text{ with } \|g\|_\infty \leqslant A.$$

---

[11] See nonetheless the final remarks of this chapter, notably those concerning the result of Klemes.

Proof Let L_S^1 be the subspace of L^1 constituted of the functions with spectrum in S, and let φ be the linear functional on L_S^1 defined by

$$\varphi(f) = \sum_{k=1}^{\infty} \widehat{f}(\lambda_k) w_k.$$

This expression is well-defined because, by Theorem 10.3.5, we have for $f \in L_S^1$:

$$\sum_{k=1}^{\infty} |\widehat{f}(\lambda_k) w_k| \leqslant \sum_{k=1}^{\infty} \frac{|\widehat{f}(\lambda_k)|}{k} \leqslant A \|f\|_1,$$

hence *a fortiori* $|\varphi(f)| \leqslant A \|f\|_1$. By the Hahn–Banach theorem, φ admits a continuous linear extension to L^1 with norm $\leqslant A$. The dual of L^1 is L^{∞}, which means that there exists a function $g \in L^{\infty}$ such that

$$\|g\|_{\infty} \leqslant A \quad \text{and} \quad \varphi(f) = \int_{\mathbb{T}} f(-x) g(x) \, dm(x) \text{ for all } f \in L^1.$$

By specialising to $f(x) = e^{i\lambda_k x} \in L_S^1$, a function for which $\varphi(f) = w_k$, we obtain

$$w_k = \int_{\mathbb{T}} g(x) e^{-i\lambda_k x} \, dm(x) = \widehat{g}(\lambda_k) \text{ for } k \geqslant 1,$$

which completes the proof of the corollary. $\qquad\qquad\qquad\qquad\qquad\qquad\square$

10.4 Extension to the case of real frequencies

Theorem 10.3.1 can be reformulated as follows: let $\lambda_1 < \cdots < \lambda_N$ be integers, and let $a_1, \ldots, a_N$ be complex numbers. We then have the inequality

$$\sum_{k=1}^{N} \frac{|a_k|}{k} \leqslant A \int_{-1/2}^{1/2} \Big| \sum_{k=1}^{N} a_k e^{2i\pi\lambda_k t} \Big| dt.$$

With this form, it is tempting to extend the result to the case of *real* frequencies $\lambda_1 < \cdots < \lambda_N$, still well separated (i.e., such that $\lambda_{k+1} - \lambda_k \geqslant 1$). An additional serious difficulty seems to appear: the loss of orthogonality of the $e^{2i\pi\lambda_k t}$. However, this difficulty can be overcome (see Exercise 10.10), and Nazarov [131] succeeded in obtaining the following generalisation of the McGehee–Pigno–Smith–Konyagin theorem.

10.4.1 Theorem [Nazarov]   *Let $\delta > 0$ and $\Delta = \delta + 1$. Then there exists a positive constant A_δ such that, for all sequences $\lambda_1 < \cdots < \lambda_N$ of real numbers* *with* $\lambda_{k+1} - \lambda_k \geqslant 1$, *and all sequences* $a_1, \ldots, a_N$ *of complex numbers, we have*

$$\sum_{k=1}^{N} \frac{|a_k|}{k} \leqslant A_\delta \int_{-\Delta/2}^{\Delta/2} \left| \sum_{k=1}^{N} a_k e^{2i\pi\lambda_k t} \right| dt. \tag{10.20}$$

The proof of this theorem is technically difficult, even if the ideas are quite close to those of the preceding section, and the reader can skip it for the moment. Nonetheless, we thought it important to include this proof for several reasons. First, it makes us reflect on the case of integer frequencies. Next, the result is very interesting. Finally, Nazarov's original article is difficult to follow and does not give many details, but is fairly easy to understand for a reader perfectly familiar with the case of integer frequencies.

Let us begin by *carefully re-examining* the MPS proof of the Littlewood conjecture, in terms of 1-periodic functions. First, suppose that $\lambda_1 < \cdots < \lambda_N$ are integers. Write

$$S = \sum_{k=1}^{N} \frac{|a_k|}{k}, \quad |a_k| = a_k u_k \text{ where } |u_k| = 1,$$

and also

$$\phi(t) = \sum_{k=1}^{N} a_k e^{2i\pi\lambda_k t} \text{ and } T_0(t) = \sum_{k=1}^{N} \frac{u_k}{k} e^{-2i\pi\lambda_k t}.$$

Then, we proceed in two steps.

Step 1. We have

$$S = \int_0^1 T_0 \phi \, dt.$$

This corresponds simply to the orthonormality of the characters $t \mapsto e^{2i\pi\lambda_k t}$, and more precisely to the Fourier formula

$$\frac{u_k}{k} = \int_0^1 T_0(t) e^{2i\pi\lambda_k t} dt.$$

It then remains only to multiply by a_k and take the sum.

Step 2. We correct T_0 to T_1 to obtain $\|T_1\|_\infty \leqslant A$ (where A is a constant), and also

$$\left| \int_0^1 (T_1 - T_0) e^{2i\pi\lambda_k t} dt \right| = \left| \widehat{T_1}(-\lambda_k) - \widehat{T_0}(-\lambda_k) \right| \leqslant \frac{\alpha}{k} \text{ for } 1 \leqslant k \leqslant N,$$

where $\alpha < 1$ (we had taken $\alpha = 1/2$ to fix the ideas). Next, multiplying by a_k and adding, we obtain

$$\left| \int_0^1 T_1 \phi - \int_0^1 T_0 \phi \right| \leqslant \alpha S.$$

Combining the two steps gives the result

$$S = \int_0^1 T_1 \phi + \int_0^1 (T_0 - T_1) \phi,$$

hence

$$S \leqslant \alpha S + \left| \int_0^1 T_1 \phi \right| \leqslant \alpha S + \|T_1\|_\infty \|\phi\|_1,$$

and finally

$$S \leqslant \frac{A}{1 - \alpha} \|\phi\|_1.$$

If we follow the same approach here, we encounter a problem *right from the start*: we no longer have $S = \int_0^1 T_0 \phi$, as the functions $t \mapsto e^{2i\pi \lambda_k t}$ are no longer orthogonal. Hence, the Fourier formulas

$$\frac{u_k}{k} = \int_0^1 T_0(t) e^{2i\pi \lambda_k t} \, dt$$

are no longer valid. We should replace them by

$$\frac{u_k}{k} = \lim_{T \to \infty} \frac{1}{2T} \int_{-T}^T T_0(t) e^{2i\pi \lambda_k t} \, dt$$

and then

$$S = \lim_{T \to \infty} \frac{1}{2T} \int_{-T}^T T_0 \phi,$$

formulas that involve the values of ϕ over the whole real line. Fortunately, by using *localised* Fourier formulas, inspired by the theory of wavelets,[12] Nazarov [131] found the following method to circumvent this difficulty. Let us consider the interval

$$I = \left[-\frac{\Delta}{2}, \frac{\Delta}{2} \right],$$

of length $\Delta = 1 + \delta$, and the function θ defined by

$$\theta(t) = \begin{cases} \cos(\pi t) & \text{if } |t| \leqslant \dfrac{1}{2}, \\ 0 & \text{if } |t| > \dfrac{1}{2}. \end{cases}$$

---

[12] See the proof by Kahane–Izumi–Izumi of Freud's Tauberian theorem [153].

Let us calculate the Fourier transform of θ. We have

$$
\begin{aligned}
\widehat{\theta}(\lambda) &= \int_{-\infty}^{+\infty} \theta(t) e^{-2i\pi\lambda t} \, dt \\
&= 2 \int_{0}^{+\infty} \theta(t) \cos(2\pi\lambda t) \, dt \\
&= \int_{0}^{1/2} 2\cos(\pi t) \cos(2\pi\lambda t) \, dt \\
&= \int_{0}^{1/2} \big(\cos((1+2\lambda)\pi t) + \cos((1-2\lambda)\pi t) \big) \, dt \\
&= \frac{\cos(\lambda\pi)}{\pi} \Big(\frac{1}{1+2\lambda} + \frac{1}{1-2\lambda} \Big) \\
&= \frac{2}{\pi} \frac{\cos(\lambda\pi)}{1-4\lambda^2}.
\end{aligned}
$$

We then fix ψ, a C^∞ function that is non-negative, even, with support in $\left[-\frac{\delta}{2}, \frac{\delta}{2} \right]$, such that $\widehat{\psi}(0) = 1$, and finally we denote by φ the *localising function*

$$
\frac{\pi}{2} \theta * \psi,
$$

that satisfies the following conditions:

$$
\varphi = \frac{\pi}{2} \theta * \psi, \quad \widehat{\varphi}(\lambda) = \frac{\cos(\lambda\pi)}{1-4\lambda^2} \widehat{\psi}(\lambda), \quad \widehat{\varphi}(0) = 1 \text{ and } \operatorname{supp}\varphi \subset I. \quad (10.21)
$$

Indeed,

$$
\widehat{\varphi} = \frac{\pi}{2} \widehat{\theta}\widehat{\psi} \quad \text{and} \quad \operatorname{supp}\varphi \subset \operatorname{supp}\theta + \operatorname{supp}\psi.
$$

For technical reasons that will become apparent as we proceed, we will not try to find an upper bound right away for the sum

$$
S = \sum_{k=1}^{N} \frac{|a_k|}{k},
$$

but instead for

$$
S_\delta = \sum_{k=1}^{N} \frac{|a_k|}{k + N_\delta},
$$

where N_δ is a large integer dependent only on δ, to be adjusted later. With all the L^p-norms referring now to the Lebesgue measure *on I*, we show that

$$
S_\delta \leqslant B_\delta \|\phi\|_1,
$$

where B_δ depends only on δ. Next, the evident inequality

$$
k + N_\delta \leqslant (1 + N_\delta)k \text{ for all } k \geqslant 1
$$

will show that

$$S \leqslant (1 + N_\delta)S_\delta \leqslant (1 + N_\delta)B_\delta \|\phi\|_1,$$

and (10.20) will ensue, with $A_\delta = (1 + N_\delta)B_\delta$.

This being said, to lighten the notation in what follows, we simply set

$$S = \sum_{k=1}^{N} \frac{|a_k|}{k + N_\delta}, \quad |a_k| = a_k u_k$$

where $|u_k| = 1$ and $T_0(t) = \sum_{k=1}^{N} \frac{u_k}{k + N_\delta} e^{-2i\pi\lambda_k t}$.

We fix $\alpha \in \,]0, 1[$, to be adjusted. Steps 1 and 2 will be replaced respectively by the following steps, which establish the approximate Fourier formulas.

Step 3. We have the inequality

$$\left| \int_I T_0(t)e^{2i\pi\lambda_k t}\varphi(t)\,dt - \frac{u_k}{k + N_\delta} \right| \leqslant \frac{1 - \alpha}{k + N_\delta} \quad \text{for } 1 \leqslant k \leqslant N. \quad (10.22)$$

After multiplying by a_k and adding, we thus obtain

$$\left| S - \int_I T_0\phi\varphi \right| \leqslant (1 - \alpha)S. \quad (10.23)$$

In other words, S is almost equal[13] to $\int_I T_0\phi\varphi$.

Step 4. We correct T_0 to T_1 with controlled L^∞-norm, such that

$$\left| \int_I (T_0(t) - T_1(t))e^{2i\pi\lambda_k t}\varphi(t)\,dt \right| \leqslant \frac{\frac{2\alpha}{3}}{k + N_\delta} \quad \text{for } 1 \leqslant k \leqslant N. \quad (10.24)$$

After multiplying by a_k and adding, we obtain

$$\left| \int_I (T_1 - T_0)\phi\varphi \right| \leqslant \frac{2\alpha}{3} S. \quad (10.25)$$

Clearly, combining steps 3 and 4 gives the result. Indeed, by adding (10.23) and (10.25), we get

$$\left| S - \int_I T_1\phi\varphi \right| \leqslant \left(1 - \frac{\alpha}{3}\right)S,$$

so that

$$S \leqslant \left(1 - \frac{\alpha}{3}\right)S + \left| \int_I T_1\phi\varphi \right|,$$

---

[13] Throughout the proof, we have the following underlying principle: an unknown quantity S is "almost equal" to a known quantity R if we have

$$|S - R| \leqslant \beta S,$$

where β is a constant < 1.

hence

$$\frac{\alpha}{3} S \leqslant \left| \int_I T_1 \phi \varphi \right| \leqslant \|T_1\|_\infty \|\varphi\|_\infty \|\phi\|_1,$$

and finally

$$S \leqslant B_\delta \|\phi\|_1,$$

with $B_\delta = \dfrac{3}{\alpha} \|T_1\|_\infty \|\varphi\|_\infty$.

We now tackle the proof of inequalities (10.22) and (10.24).

Proof of inequality (10.22). First, note that this case presents *an additional difficulty* compared with the case of integer λ_k, for which we have exactly

$$\int_0^1 T_0(t) e^{2i\pi \lambda_k t} dt = \frac{u_k}{k + N_\delta}.$$

This inequality will follow from the following lemma.[14]

10.4.2 Lemma *For any $\delta > 0$, there exists an integer $N_\delta \geqslant 1$ and a real number $\alpha \in \,]0, 1[$ such that*

$$\sum_{1 \leqslant j \leqslant N, j \neq k} \frac{|\widehat{\varphi}(\lambda_j - \lambda_k)|}{j + N_\delta} \leqslant \frac{1 - \alpha}{k + N_\delta} \text{ for } 1 \leqslant k \leqslant N. \tag{10.26}$$

Proof First we fix $\alpha = \alpha_\delta \in \,\left]0, \dfrac{1}{3}\right[$ such that

$$|\widehat{\psi}(\lambda)| \leqslant 1 - 3\alpha \text{ for } |\lambda| \geqslant 1.$$

Such an α exists because on the one hand $\widehat{\psi}(\lambda) \to 0$ as $|\lambda| \to +\infty$ and on the other hand, by the triangle inequality and the non-negativeness of ψ:

$$|\widehat{\psi}(\lambda)| = \left| \int_{-\infty}^{+\infty} \psi(t) e^{-2i\pi \lambda t} dt \right| < \int_{-\infty}^{+\infty} \psi(t)\, dt = \widehat{\psi}(0) = 1, \text{ if } \lambda \neq 0.$$

We thus have

$$|\widehat{\varphi}(\lambda)| \leqslant \frac{1 - 3\alpha}{4\lambda^2 - 1} \text{ for } |\lambda| \geqslant 1, \tag{10.27}$$

according to (10.21). If E denotes the left-hand side of (10.26), we can write $E = E_1 + E_2$, with

$$E_1 = \sum_{j + N_\delta < (1-\alpha)(k+N_\delta)} \cdots\cdots \quad \text{and} \quad E_2 = \sum_{j + N_\delta \geqslant (1-\alpha)(k+N_\delta), j \neq k} \cdots\cdots,$$

---

[14] Which would not be needed if we had $\widehat{\varphi}(x) = 0$ for $|x| \geqslant 1$; but φ and $\widehat{\varphi}$ cannot simultaneously be compactly supported!

N_δ being a positive integer, to be specified in a moment. Within E_1, the denominators $j + N_\delta$ may be small, while $\geqslant 1$, but the numerators $|\widehat{\varphi}(\lambda_j - \lambda_k)|$ are also small. Indeed, let $A = A_\delta > 0$ such that

$$|\widehat{\varphi}(\lambda)| \leqslant |\lambda|^{-3} \text{ if } |\lambda| \geqslant A.$$

Such an A exists because $\widehat{\varphi}$ is rapidly decreasing. We then choose a non-negative integer m_δ such that $N_\delta := 2^{m_\delta}$ satisfies

$$\alpha N_\delta \geqslant A.$$

If j is an index appearing in the sum E_1, we thus have

$$|\lambda_k - \lambda_j| \geqslant |k - j| = (k + N_\delta) - (j + N_\delta) \geqslant \alpha(k + N_\delta) \geqslant \alpha N_\delta \geqslant A,$$

hence

$$
\begin{aligned}
E_1 &\leqslant \sum_{j+N_\delta<(1-\alpha)(k+N_\delta)} |\widehat{\varphi}(\lambda_j - \lambda_k)| \\
&\leqslant \sum_{j+N_\delta<(1-\alpha)(k+N_\delta)} \frac{1}{(\lambda_k - \lambda_j)^3} \\
&\leqslant \sum_{j+N_\delta<(1-\alpha)(k+N_\delta)} \frac{1}{(\alpha(k + N_\delta))^3} \\
&\leqslant \frac{\alpha^{-3}}{(k + N_\delta)^2},
\end{aligned}
$$

since the sum E_1 is of length $\leqslant k + N_\delta$, as the j over which it runs satisfy $j \leqslant j + N_\delta \leqslant k + N_\delta$. From this:

$$E_1 \leqslant \frac{\alpha^{-3}}{N_\delta} \frac{1}{(k + N_\delta)} \leqslant \frac{\alpha}{k + N_\delta}$$

if we choose N_δ even larger:

$$N_\delta \geqslant \alpha^{-4}.$$

Next, in the sum E_2, since $j \neq k$, we have

$$j + N_\delta \geqslant (1 - \alpha)(k + N_\delta) \text{ and } |\lambda_j - \lambda_k| \geqslant 1.$$

Therefore, thanks to (10.27) and the fact that

$$1 - 3\alpha \leqslant (1 - \alpha)(1 - 2\alpha),$$

we have

$$E_2 \leqslant \frac{1}{(1-\alpha)(k+N_\delta)} \sum_{1 \leqslant j \leqslant N, j \neq k} \frac{1-3\alpha}{4(\lambda_j - \lambda_k)^2 - 1}$$

$$\leqslant \frac{1-3\alpha}{(1-\alpha)(k+N_\delta)} \sum_{1 \leqslant j \leqslant N, j \neq k} \frac{1}{4(j-k)^2 - 1}$$

$$\leqslant 2 \frac{1-2\alpha}{k+N_\delta} \sum_{\ell=1}^{\infty} \frac{1}{4\ell^2 - 1}.$$

Finally,

$$E_2 \leqslant \frac{1-2\alpha}{k+N_\delta} \sum_{\ell=1}^{\infty} \left(\frac{1}{2\ell-1} - \frac{1}{2\ell+1} \right) = \frac{1-2\alpha}{k+N_\delta},$$

which completes the proof of Lemma 10.4.2. $\qquad\square$

Now, we deduce (10.22) as follows:

$$\int_I T_0(t) e^{2i\pi\lambda_k t} \varphi(t)\, dt = \sum_{j=1}^{N} \frac{u_j}{j+N_\delta} \int_I e^{-2i\pi\lambda_j t} e^{2i\pi\lambda_k t} \varphi(t)\, dt$$

$$= \sum_{j=1}^{N} \frac{u_j}{j+N_\delta} \widehat{\varphi}(\lambda_j - \lambda_k)$$

$$= \frac{u_k}{k+N_\delta} + \sum_{1 \leqslant j \leqslant N, j \neq k} \frac{u_j}{j+N_\delta} \widehat{\varphi}(\lambda_j - \lambda_k)$$

since supp $\varphi \subset I$ and $\widehat{\varphi}(0) = 1$, hence (10.22) thanks to Lemma 10.4.2.

Proof of inequality (10.24). We must first describe how to define T_1 starting from T_0. We decompose T_0 into dyadic blocks f_j on which the amplitudes $\frac{1}{r+N_\delta}$ have values close to 2^{-j}; in other words, we set

$$I_j = [\![2^j, 2^{j+1} [\![\quad \text{and} \quad f_j(t) = \sum_{r+N_\delta \in I_j} \frac{u_r}{r+N_\delta} e^{-2i\pi\lambda_r t}. \tag{10.28}$$

We always have

$$\|f_j\|_{L^2(I)} \leqslant C_\delta 2^{-j/2},$$

thanks to a lemma due to Salem.[15] In terms of f_j, we have

$$T_0 = \sum_{m_\delta \leqslant j \leqslant m} f_j,$$

---

[15] See Exercises 10.9, 10.10 and 10.11.

where m is the integer such that

$$2^m \leqslant N + N_\delta < 2^{m+1}.$$

The sum begins at $j = m_\delta$, since, as $N_\delta = 2^{m_\delta}$, we have, for $r \geqslant 1$:

$$r + N_\delta \geqslant 1 + 2^{m_\delta} > 2^{m_\delta}.$$

We then denote by $\tilde{f}_j$ the Δ-periodic function coinciding with $|f_j|$ on $\left[-\frac{\Delta}{2}, \frac{\Delta}{2} \right[$, and

$$\sum_{s \in \mathbb{Z}} a_{sj} e^{\frac{2is\pi}{\Delta} t}$$

its Fourier expansion in $L^2(I)$, and h_j the function of $L^2(I)$ whose Fourier expansion is

$$a_{0j} + 2 \sum_{s=1}^{\infty} a_{sj} e^{\frac{2is\pi}{\Delta} t}.$$

Then, everything works as in the periodic case, that is

$$h_j \in H^2(I), \quad \mathrm{Re}\, h_j = |f_j| \text{ a.e. on } I \quad \text{and} \quad \|h_j\|_2 \leqslant \sqrt{2} \|f_j\|_2,$$

the L^p-norms referring again to the Lebesgue measure on I. We thus define T_1 *exactly as in the case of integer frequencies*, that is $T_1 = F_m$, where the sequence (F_j) is defined by

$$F_{m_\delta} = f_{m_\delta} \quad \text{and} \quad F_{j+1} = F_j e^{-\varepsilon h_{j+1}} + f_{j+1}.$$

We still have

$$\|T_1\|_\infty \leqslant \sup_{0 < x \leqslant 1} \frac{x}{1 - e^{-\varepsilon x}} =: K_\varepsilon,$$

and also

$$T_1 = \sum_{j=m_\delta}^{m} f_j g_j \text{ on } I,$$

where

$$g_j = e^{-\varepsilon H_j}, \quad \text{with } H_j = h_{j+1} + \cdots + h_m,$$

which we can write

$$T_1 = \sum_{j \geqslant m_\delta} f_j g_j,$$

with $f_j = 0$ and $g_j = 1$ for $j > m$. The function g_j is analytic, that is an element of $H^\infty(I)$, by the remark following Proposition 10.1.1. Moreover, H_j satisfies

$$\|H_j\|_2 \leqslant \sum_{r > j} \|h_r\|_2 \leqslant \sqrt{2} \sum_{r > j} \|f_r\|_2 \leqslant C_\delta \sqrt{2} \sum_{r > j} 2^{-r/2} \leqslant D_\delta 2^{-j/2},$$

which, thanks to Proposition 10.1.1, leads to

$$\|g_j - 1\|_2 \leqslant \varepsilon \|H_j\|_2 \leqslant \varepsilon D_\delta 2^{-j/2}.$$

However, on I, in the sense of L^2 convergence, we have

$$g_j(t) - 1 = \sum_{s \geqslant 0} c_{sj} e^{\frac{2is\pi}{\Delta}t}, \qquad (10.29)$$

so that with $\varepsilon > 0$ small enough ($\varepsilon \leqslant 2^{m_\delta/2} D_\delta^{-1}$), we have

$$\left(\sum_{s \geqslant 0} |c_{sj}|^2 \right)^{1/2} = \frac{1}{\sqrt{\Delta}} \|g_j - 1\|_2 \leqslant \varepsilon D_\delta 2^{-j/2} \leqslant 1,$$

and in particular

$$|c_{sj}| \leqslant 1.$$

We then fix $k \in [\![1, N]\!]$, and denote by ℓ the index such that $k + N_\delta \in I_\ell$, that is

$$2^\ell \leqslant k + N_\delta < 2^{\ell+1}.$$

Let R be the left-hand side of (10.24). Then

$$R = \int_I (T_1(t) - T_0(t)) e^{2i\pi\lambda_k t} \varphi(t)\, dt$$

$$= \int_I \sum_{m_\delta \leqslant j \leqslant \ell-2} f_j(t)(g_j(t) - 1) e^{2i\pi\lambda_k t} \varphi(t)\, dt + \int_I \sum_{\ell-1 \leqslant j \leqslant m} \cdots$$

$$=: R_1 + R_2.$$

The portion R_2 *can be bounded as in the case of integer frequencies*, thanks to the properties of the tail of a geometric series:

$$|R_2| \leqslant \sum_{\ell-1 \leqslant j \leqslant m} \int_I |f_j(t)||g_j(t) - 1||\varphi(t)| dt$$

$$\leqslant \|\varphi\|_\infty \sum_{\ell-1 \leqslant j \leqslant m} \|f_j\|_2 \|g_j - 1\|_2$$

$$\leqslant \|\varphi\|_\infty \sum_{j \geqslant \ell-1} C_\delta 2^{-j/2} \varepsilon D_\delta 2^{-j/2}$$

$$\leqslant \varepsilon E_\delta 2^{-\ell} \leqslant \frac{2\varepsilon E_\delta}{k + N_\delta},$$

since $2^{\ell+1} \geqslant k + N_\delta$. By adjusting ε small enough, we have

$$|R_2| \leqslant \frac{\frac{\alpha}{3}}{k + N_\delta}. \qquad (10.30)$$

It remains to bound the portion R_1, *which presents an additional difficulty compared with the case of integer frequencies*, where it was zero! But hang on – we are almost there.

First of all, the sum R_1 can be calculated by applying, for j fixed in the interval $[\![m_\delta, \ell - 2]\!]$, the Parseval identity (10.5) to the functions $g_j - 1$ on the one hand, and $t \mapsto f_j(t)\varphi(t)e^{2i\pi\lambda_k t}$ on the other hand. The Fourier coefficients of the latter, considered as an element of $L^2(I)$, are easily calculated, and for $s \in \mathbb{Z}$, we have

$$\int_I f_j(t)\varphi(t)e^{2i\pi\lambda_k t}e^{-\frac{2i\pi s}{\Delta}t}\,dt$$
$$= \sum_{r+N_\delta \in I_j} \frac{u_r}{r + N_\delta} \int_I \varphi(t)e^{-2i\pi\lambda_r t + 2i\pi\lambda_k t - \frac{2i\pi s}{\Delta}t}\,dt$$
$$= \sum_{r+N_\delta \in I_j} \frac{u_r}{r + N_\delta}\widehat{\varphi}\left(\lambda_r - \lambda_k + \frac{s}{\Delta}\right).$$

We thus obtain successively

$$\int_I f_j(t)(g_j(t) - 1)e^{2i\pi\lambda_k t}\varphi(t)\,dt = \sum_{s=0}^{\infty} c_{sj} \sum_{r+N_\delta \in I_j} \frac{u_r}{r + N_\delta}\widehat{\varphi}\left(\lambda_r - \lambda_k - \frac{s}{\Delta}\right)$$
$$= \sum_{r+N_\delta \in I_j} \frac{u_r}{r + N_\delta}\sum_{s=0}^{\infty} c_{sj}\widehat{\varphi}\left(\lambda_r - \lambda_k - \frac{s}{\Delta}\right),$$

then

$$R_1 = \sum_{m_\delta \leqslant j \leqslant \ell - 2} \sum_{r+N_\delta \in I_j} \frac{u_r}{r + N_\delta}\sum_{s=0}^{\infty} c_{sj}\widehat{\varphi}\left(\lambda_r - \lambda_k - \frac{s}{\Delta}\right)$$
$$= \sum_{2^{m_\delta} \leqslant r+N_\delta < 2^{\ell-1}} \frac{u_r}{r + N_\delta}\sum_{s=0}^{\infty} c_s(r)\widehat{\varphi}\left(\lambda_r - \lambda_k - \frac{s}{\Delta}\right),$$

by setting

$$c_s(r) = c_{sj} \text{ if } r + N_\delta \in I_j.$$

Finally, taking into account the parity of $\widehat{\varphi}$, we obtain

$$R_1 = \sum_{2^{m_\delta} \leqslant r+N_\delta < 2^{\ell-1}} \frac{u_r}{r + N_\delta}\underbrace{\sum_{s=0}^{\infty} c_s(r)\widehat{\varphi}\left(\lambda_k - \lambda_r + \frac{s}{\Delta}\right)}_{=:E_r}.$$

However, we can find a $B = B_\delta > 0$ such that

$$|\widehat{\varphi}(\lambda)| \leqslant \frac{\alpha}{24\Delta^3|\lambda|^3} \text{ if } |\lambda| \geqslant B,$$

since $\widehat{\varphi}$ is rapidly decreasing. Moreover, in the preceding sum, because of the non-negativeness of s (which follows from the *analytic* character of the g_j), *the presence of the "spacing" term* $N_\delta = 2^{m_\delta}$ and the inequalities

$$r + N_\delta < 2^{\ell-1} \text{ and } 2^\ell \leqslant k + N_\delta < 2^{\ell+1},$$

we have

$$\left| \lambda_k - \lambda_r + \frac{s}{\Delta} \right| \geqslant \lambda_k - \lambda_r \geqslant k - r = (k + N_\delta) - (r + N_\delta)$$
$$> 2^\ell - 2^{\ell-1} = 2^{\ell-1} \geqslant 2^{m_\delta-1},$$

which will be $\geqslant B$ if m_δ is large enough. Since $\Delta > 1$, we then have

$$\left| \widehat{\varphi}\left(\lambda_k - \lambda_r + \frac{s}{\Delta} \right) \right| \leqslant \frac{\alpha}{24\Delta^3 \left| \lambda_k - \lambda_r + \frac{s}{\Delta} \right|^3} = \frac{\alpha}{24 |\Delta(\lambda_k - \lambda_r) + s|^3}$$
$$\leqslant \frac{\alpha}{24 |\lambda_k - \lambda_r + s|^3} \leqslant \frac{\alpha}{24 (k - r + s)^3}.$$

As the moduli of the $c_s(r)$ are bounded by 1 if ε is chosen small enough (see p. 321), we can thus bound the sum E_r appearing in the expression of R_1 as follows:

$$|E_r| \leqslant \frac{\alpha}{24} \sum_{s \geqslant 0} \frac{1}{(k - r + s)^3} = \frac{\alpha}{24} \sum_{n \geqslant k-r} \frac{1}{n^3}$$
$$\leqslant \frac{\alpha}{24} \int_{k-r-1}^{+\infty} \frac{dt}{t^3} \leqslant \frac{\alpha}{48} \frac{1}{(k - r - 1)^2}.$$

However, we have seen above that

$$k - r - 1 \geqslant 2^{\ell-1} \geqslant \frac{1}{4}(k + N_\delta),$$

hence

$$|E_r| \leqslant \frac{\alpha}{48} \frac{16}{(k + N_\delta)^2} = \frac{\alpha}{3} \frac{1}{(k + N_\delta)^2}.$$

Finally, we have the brute-force estimate

$$|R_1| \leqslant \sum_{2^{m_\delta} \leqslant r+N_\delta < 2^{\ell-1}} |E_r|,$$

where this sum contains at most $2^{\ell-1} \leqslant k + N_\delta$ terms, each smaller than $\frac{\alpha}{3} \frac{1}{(k + N_\delta)^2}$, so that

$$|R_1| \leqslant \frac{\alpha}{3} \frac{1}{k + N_\delta}. \tag{10.31}$$

By adding (10.30) and (10.31), we finally obtain

$$|R| \leqslant \frac{\frac{2\alpha}{3}}{k + N_\delta},$$

which is exactly the estimate (10.24), and completes the proof of the Nazarov Theorem 10.4.1. Phew!

10.4.3 Remarks (1) Thus, the veracity of the Littlewood conjecture does not rely on rigid properties of orthogonality; it holds for much more general quasi-orthogonal systems. Moreover, the construction of $T_1 = \sum f_j g_j$ starting from $T_0 = \sum f_j$ vaguely suggests a *martingale transformation* (in the sense of Burkholder). In this vein of dyadic martingales and their transforms, Nazarov gives a simplified proof of a lemma of Bočhkarëv, quite in tune with the MPS solution of the Littlewood conjecture: we will not detail it here, but simply refer the reader to [131].

(2) A nice presentation, with many comments, of the proof of Theorem 10.4.1 and of Salem's lemma can be found in D. Bellay's Master's thesis [16].

(3) We do not know if we can let $\delta = 0$ in (10.20), in other words, if the optimal constant A_δ in (10.20) does not explode as $\delta \to 0$.

(4) Another interesting generalisation, this time of Hardy's inequality for H^1, was proposed by Klemes [109]. As we have already said, Hardy's inequality no longer holds for an arbitrary L^1 function. Nevertheless, if we "complete" the right-hand side of this inequality, we obtain the following inequality, valid this time for *any* L^1 function, in which c denotes a numerical constant:

$$\sum_{j=1}^{\infty} \left(4^{-j} \sum_{4^{j-1} \leqslant n < 4^j} |\widehat{f}(n)|^2 \right)^{1/2} \leqslant c \|f\|_1 + c \sum_{j=1}^{\infty} \left(4^{-j} \sum_{4^{j-1} \leqslant n < 4^j} |\widehat{f}(-n)|^2 \right)^{1/2}.$$

The Cauchy–Schwarz inequality shows that this is a substantial improvement of Hardy's inequality. Klemes' proof uses a variant of the proof MPS presented in this chapter. However, it seems that the following problem is still open: does there exist a constant $c > 0$ such that, for all $f \in L^1$, we have

$$\sum_{n=1}^{\infty} \frac{|\widehat{f}(n)|}{n} \leqslant c \|f\|_1 + c \sum_{n=1}^{\infty} \frac{|\widehat{f}(-n)|}{n} \; ?$$

We refer the interested reader to [109].

Exercises

10.1. Let $f : \mathbb{R} \to \mathbb{C}$ be a 2π-periodic function and α-Hölder with $\frac{1}{2} < \alpha \leqslant 1$, in the following sense:

$$|f(x) - f(y)| \leqslant C|x - y|^\alpha \text{ for } x, y \in \mathbb{R}.$$

Let us set $c_n = \widehat{f}(n)$ for $n \in \mathbb{Z}$ and let h be a positive real number. In what follows, A denotes a constant, possibly different from one line to another.
(a) Show that

$$\sum_{n \in \mathbb{Z}} |c_n|^2 \sin^2(nh) \leqslant Ah^{2\alpha}.$$

(b) Suppose that $\alpha = 1$. Show that

$$\sum_{n \in \mathbb{Z}} n^2 |c_n|^2 < \infty.$$

(c) Suppose that $\frac{1}{2} < \alpha < 1$. Show that

$$\sum_{N < |n| \leqslant 2N} |c_n|^2 \leqslant AN^{-2\alpha},$$

and then that

$$\sum_{n \in \mathbb{Z}} |n|^{2\beta} |c_n|^2 < \infty \text{ for all } \beta < \alpha.$$

(d) We return to the general case $\frac{1}{2} < \alpha \leqslant 1$. Use the preceding results to conclude that

$$\sum_{n \in \mathbb{Z}} |n|^\delta |c_n| < \infty \text{ for } 0 \leqslant \delta < \alpha - \frac{1}{2}.$$

In particular, $\sum |c_n| < \infty$. The result of this exercise was used in the proof of Lemma 10.3.3.
(e) Take $\alpha = 1$. Do we always have $\sum |n|^{1/2} |c_n| < \infty$? (See also Exercise 5.10 on p. 146 of Chapter 5.)

10.2. The aim of this exercise is to show that if f is a trigonometric polynomial, then $|f|$ is an element of W, the Wiener algebra of absolutely convergent Fourier series (see Chapter 2 for the definition). This gives an alternative proof to part of Exercise 10.1. Note that J. P. Kahane [100] showed that there exist functions $f \in W$ such that $|f| \notin W$.
(a) Let $a \in \mathbb{C}$ be such that $|a| \leqslant 1$. Show that $|1 - ae^{it}| \in W$.
(b) Show that $|e^{it} - a| \in W$ for all $a \in \mathbb{C}$.
(c) Conclude.

10.3. Detail the proof by induction of relation (10.17).

10.4. The Davenport–Erdős–Levêque lemma. Let $(a_n)_{n \geqslant 1}$ be a sequence of non-negative real numbers such that $\sum_{n=1}^{\infty} \frac{a_n}{n} < \infty$, and let $(\lambda_n)_{n \geqslant 1}$ be a sequence of real numbers greater than 1, increasing to infinity, such that $\sum_{n=1}^{\infty} \frac{\lambda_n a_n}{n} < \infty$. Such a sequence always exists, by a classic lemma due to du Bois–Reymond.

(a) Let $(p_k)_{k \geqslant 1}$ be an increasing sequence of integers $\geqslant 1$ such that

$$\lambda_{p_k}(p_{k+1} - p_k) \geqslant p_{k+1},$$

for example (and we will stick to this)

$$\lambda_{p_1} > 1 \quad \text{and} \quad p_{k+1} = \left[\frac{p_k \lambda_{p_k}}{\lambda_{p_k} - 1} \right] + 1,$$

where $[\cdot]$ denotes the integer part. For each $k \geqslant 1$, fix an integer n_k satisfying

$$p_k < n_k \leqslant p_{k+1} \quad \text{and} \quad a_{n_k} = \min_{p_k < u \leqslant p_{k+1}} a_u.$$

Show that

$$a_{n_k} \leqslant \sum_{p_k < n \leqslant p_{k+1}} \frac{\lambda_n a_n}{n} \quad \text{and} \quad \frac{p_{k+1}}{p_k} \to 1 \text{ as } k \to +\infty.$$

(b) Conclude that

$$\sum_{k=1}^{\infty} a_{n_k} < \infty \quad \text{and} \quad \frac{n_{k+1}}{n_k} \to 1 \text{ as } k \to +\infty.$$

10.5. Let $f \in H^1$. Show that there exists an increasing sequence $(n_k)_{k \geqslant 1}$ of integers $\geqslant 1$ so that we have simultaneously

$$\sum_{k=1}^{\infty} |\widehat{f}(n_k)| < \infty \quad \text{and} \quad \frac{n_{k+1}}{n_k} \to 1 \text{ as } k \to +\infty.$$

Can such a sequence be independent of f?

10.6. Show that Hardy's inequality is false in general for measures with bidirectional spectrum. For example, consider the Fejér kernel, or the function

$$f(t) = \sum_{n=2}^{\infty} \frac{\cos(nt)}{\ln n} \in L^1,$$

whose Fourier coefficients are precisely the $\frac{1}{\ln n}$.

10.7. Let S be a subset of $\mathbb{Z}$, and let L^1_S be the set of $f \in L^1$ such that $\operatorname{Sp} f \subset S$. We say that L^1_S is *complemented in* L^1 if there exists a continuous linear projection Q from L^1 onto L^1_S (see Chapter 13).

(a) Show that if Q exists, then the formula

$$Pf = \int_{\mathbb{T}} (T_{-a} Q T_a)(f) \, dm(a)$$

defines a continuous linear projection from L^1 onto L^1_S, that commutes with the translations T_a, defined by $T_a f(t) = f(t + a)$.

(b) Show that, under the hypothesis of part (a), there exists a measure μ on $\mathbb{T}$ such that $\widehat{\mu} = \mathbf{1}_S$, the indicator function of S.

(c) Show that L^1_S is complemented in L^1 *if and only if* there exists a measure μ such that $\widehat{\mu} = \mathbf{1}_S$.

(d) Thus prove again the fact that H^1 is not complemented in L^1; do the same for L^1_S if S is unidirectionally infinite.

10.8. Let $a_1, \ldots, a_N$ be complex numbers. Set

$$\phi(t) = \sum_{j=1}^{N} a_j e^{i 2^j t} \quad \text{for } t \in \mathbb{R}.$$

(a) Show that if $(j, k) \neq (j', k')$, $j \neq k$ and $j' \neq k'$, then

$$2^j - 2^k \neq 2^{j'} - 2^{k'}.$$

By using Parseval's identity, deduce that

$$\|\phi\|_4 \leqslant 2^{1/4} \|\phi\|_2.$$

(b) Let $\theta \in \,]0, 1[$ be such that $\dfrac{1}{2} = \dfrac{1 - \theta}{1} + \dfrac{\theta}{4}$. Show that

$$\|\phi\|_2 \leqslant \|\phi\|_1^{1-\theta} \|\phi\|_4^{\theta}.$$

(c) Show that if $|a_j| \geqslant 1$ for all j, then $\|\phi\|_1 \geqslant \delta \sqrt{N}$, where δ is a numerical constant.

10.9. Let $\lambda_1, \ldots, \lambda_N$ be real numbers such that $|\lambda_j - \lambda_k| \geqslant \delta$ if $j \neq k$, and let $z_1, \ldots, z_N$ be arbitrary complex numbers. Show (or admit!) the generalised Hilbert inequality (see [129]):

$$\left| \sum_{1 \leqslant j \neq k \leqslant N} \frac{z_j \overline{z_k}}{\lambda_j - \lambda_k} \right| \leqslant \frac{\pi}{\delta} \sum_{j=1}^{N} |z_j|^2.$$

10.10. Let $\lambda_1 < \cdots < \lambda_N$ be real numbers, with $\lambda_{j+1} - \lambda_j \geqslant 1$. Show that, if $a < b$ are real numbers and $c_1, \ldots, c_N$ complex numbers, then

$$\int_a^b \left| \sum_{j=1}^N c_j e^{2i\pi\lambda_j t} \right|^2 dt \leqslant (b - a + 1) \sum_{j=1}^N |c_j|^2.$$

This inequality constitutes *Salem's lemma*: deduce it here from Exercise 10.9 (see also [16]). We used this lemma in the proof of the Nazarov Theorem 10.4.1.

10.11. Alternate proof of Salem's lemma. We keep the notations of Exercise 10.10. Remember that the Fourier transform of a function φ of $L^1(\mathbb{R})$ is defined by

$$\widehat{\varphi}(x) = \int_{\mathbb{R}} \varphi(t) e^{-2i\pi x t} dt \text{ for } x \in \mathbb{R},$$

and set $A = [a, b]$.

(a) Show that there exists an even function $\varphi \in L^1(\mathbb{R})$ such that

$$\varphi \geqslant \mathbf{1}_A \text{ and } \widehat{\varphi}(x) = 0 \text{ if } |x| \geqslant 1.$$

(b) Show that

$$I := \int_a^b \left| \sum_{j=1}^N c_j e^{2i\pi\lambda_j t} \right|^2 dt \leqslant \int_{\mathbb{R}} \left| \sum_{j=1}^N c_j e^{2i\pi\lambda_j t} \right|^2 \varphi(t) \, dt$$

$$= \sum_{1 \leqslant j,k \leqslant N} c_j \overline{c_k} \widehat{\varphi}(\lambda_j - \lambda_k)$$

$$= \widehat{\varphi}(0) \sum_{j=1}^N |c_j|^2.$$

(c) Conclude that there exists a constant $C(A) > 0$, depending only on A, such that for all choices of $c_1, \ldots, c_N$, we have

$$I \leqslant C(A) \sum_{j=1}^N |c_j|^2.$$

11

Banach algebras

The purpose of this chapter is essentially to facilitate the comprehension of other chapters of this book, notably those devoted to Wiener's Tauberian theorem and especially to Carleson's corona theorem. For this reason, we limit ourselves here to a few basic notions, while being keen to illustrate them with numerous examples.

The origins of the theory of Banach algebras date to the 1940s, with the Russian mathematician I. Gelfand [67] and his collaborators (notably D. Raikov and G. Shilov). Situated at the junction of algebra and topology, this theory in fact originates in harmonic analysis: its persistent ideal was to situate the study of the convolution algebra of integrable functions on a locally compact group in a framework of abstract algebra. Its immediate success can be explained by the elegance of its methods, which allowed non-trivial theorems to be obtained "automatically", the first of these certainly being Wiener's lemma (Theorem 11.3.9).

We call a *Banach algebra* any complex algebra $(A, +, \times, \cdot)$ – always supposed not reduced to $\{0\}$ – equipped with a norm $\| \cdot \|$ for which A is a Banach space, and which satisfies

$$\|xy\| \leqslant \|x\| \, \|y\| \text{ for } (x, y) \in A^2. \tag{11.1}$$

If, furthermore, A admits a unit element, often denoted by e, A is said to be *unital*. In this case, (11.1) implies automatically that $\|e\| \geqslant 1$. In fact (see Exercise 11.1), replacing $\| \cdot \|$ by an equivalent norm if necessary, we can always suppose that $\|e\| = 1$: *we will do so in all that follows*. In this chapter, most of the algebras encountered will also be *commutative*.

Here are a few common examples.

- The space $C(X, \mathbb{C})$ of continuous functions from a compact topological space X to $\mathbb{C}$, equipped with $\| \cdot \|_\infty$.

329

- *The disk algebra* of functions from D (the open unit disk of $\mathbb{C}$) to $\mathbb{C}$ that are continuous on $\overline{D}$ and holomorphic in D, equipped with $\| \cdot \|_\infty$.
- *The Wiener algebra* W of continuous functions $f : \mathbb{T} \to \mathbb{C}$ ($\mathbb{T}$ being the unit circle of $\mathbb{C}$) such that $\sum_{n \in \mathbb{Z}} |\widehat{f}(n)| < \infty$, where

$$\widehat{f}(n) = \frac{1}{2\pi} \int_0^{2\pi} f(e^{it}) e^{-int} \, dt,$$

equipped with the norm defined by $\|f\| = \sum_{n \in \mathbb{Z}} |\widehat{f}(n)|$.

- The algebra $H^\infty(D)$ of bounded holomorphic functions in D, equipped with $\| \cdot \|_\infty$. This algebra, more complicated than those mentioned previously, will be studied thoroughly in Chapter 12.
- The convolution algebra $L^1(\mathbb{R}, \mathbb{C})$, equipped with $\| \cdot \|_1$, defined by

$$\|f\|_1 = \int_{-\infty}^{+\infty} |f(t)| \, dt.$$

This algebra is not unital.

Any Banach algebra can be embedded isometrically into a unital Banach algebra in the following way: let A be a non-unital Banach algebra. We set $B = \mathbb{C} \times A$, and equip it with the operations and norm defined below (with evident notations):

$$\begin{cases} (\lambda, x) + (\mu, y) &= (\lambda + \mu, x + y), \\ (\lambda, x)(\mu, y) &= (\lambda\mu, xy + \lambda y + \mu x), \\ \alpha(\lambda, x) &= (\alpha\lambda, \alpha x), \\ \|(\lambda, x)\| &= |\lambda| + \|x\|. \end{cases}$$

It is easy to verify that B is then a unital Banach algebra $\big($the unit being $(1, 0)\big)$. We can identify A with the set of pairs $(0, x)$, $x \in A$, and A appears then as a maximal ideal of B (see the definitions below).

11.1 Spectrum of an element in a Banach algebra

In what follows, A denotes a unital Banach algebra.

11.1.1 Definition and first properties

Given $x \in A$, we call the *spectrum* of x, denoted by $\sigma(x)$, the set of complex numbers λ such that $x - \lambda e$ is not invertible. A fundamental tool in the study of the invertible elements of A, and hence of the spectrum of its elements, is the *Neumann series* $\sum x^n$, absolutely convergent for all elements x of A such that $\|x\| < 1$. In this case, $e - x$ is invertible, and

$$(e - x)^{-1} = \sum_{n=0}^{\infty} x^n.$$

From this, we can easily deduce that the set U of invertible elements of A is an open subset of A, and that the map $U \to U$, $x \mapsto x^{-1}$ is differentiable, its differential at x being the map

$$A \to A, \ h \mapsto -x^{-1}hx^{-1}. \tag{11.2}$$

We can now establish two essential properties of the spectrum.

11.1.2 Proposition *If $x \in A$, the spectrum $\sigma(x)$ is a non-empty compact subset of $\mathbb{C}$.*

Proof Let $\lambda \in \mathbb{C}$ be such that $|\lambda| > \|x\|$. Then,

$$x - \lambda e = -\lambda(e - \lambda^{-1}x), \ \text{with} \ \|\lambda^{-1}x\| < 1,$$

so that $x - \lambda e$ is invertible. Hence, $\sigma(x)$ is contained in the closed disk centred at 0 and with radius $\|x\|$. Moreover, if $x - \lambda e$ is invertible, so is $x - \mu e$ for μ close to λ, by the continuity of $\mu \mapsto x - \mu e$ and the openness of the set of invertible elements of A. Thus, $\sigma(x)$ is closed in $\mathbb{C}$. Finally, supposing for an instant that $\sigma(x)$ is empty, we fix a continuous linear functional $f : A \to \mathbb{C}$, and for any $\lambda \in \mathbb{C}$, set $u(\lambda) = f\left((x - \lambda e)^{-1}\right)$. Because of (11.2), u is an entire function. Moreover, as $u(\lambda) \to 0$ when $|\lambda| \to +\infty$, u is null by Liouville's theorem, hence $f\left((x - \lambda e)^{-1}\right) = 0$ for all continuous linear functionals on A and all $\lambda \in \mathbb{C}$. By the Hahn–Banach theorem, $(x - \lambda e)^{-1} = 0$ for all λ, which of course is absurd. $\qquad\square$

11.1.3 Theorem [Gelfand–Mazur] *A unital Banach algebra that is a field is isometric to $\mathbb{C}$.*

Proof Indeed, in this case, if $x \in A$, then there exists a $\lambda(x) \in \mathbb{C}$ such that $x - \lambda(x)e$ is not invertible, *that is to say, it is null*. The map

$$\lambda : A \to \mathbb{C}, \ x \mapsto \lambda(x)$$

is thus an isometry from A onto $\mathbb{C}$ (remember that we supposed $\|e\| = 1$). $\quad\square$

If x is an element of A, we define the *spectral radius* of x as the real number

$$r(x) = \sup_{\lambda \in \sigma(x)} |\lambda|.$$

11.1.4 Theorem [Beurling–Gelfand formula for spectral radius]   *For all* $x \in A$,

$$\|x^n\|^{1/n} \to r(x) \text{ as } n \to +\infty.$$

Proof   First of all, let $\lambda \in \sigma(x)$. Then, for $n \geqslant 1$, $\lambda^n \in \sigma(x^n)$ since

$$x^n - \lambda^n e = (x - \lambda e) \sum_{k=0}^{n-1} \lambda^{n-1-k} x^k.$$

From this, $|\lambda| \leqslant \|x^n\|^{1/n}$, and hence

$$r(x) \leqslant \varliminf_{n \to +\infty} \|x^n\|^{1/n}. \tag{11.3}$$

Let us fix a continuous linear functional f on A, and again set

$$u(\lambda) = f\left((x - \lambda e)^{-1}\right) \text{ for } \lambda \in \mathbb{C} \setminus \sigma(x).$$

The function u is holomorphic in the open set where it is defined. Moreover, if $\lambda \in \mathbb{C}$ satisfies $|\lambda| > \|x\|$, we have

$$u(\lambda) = -\lambda^{-1} f\left((e - \lambda^{-1} x)^{-1}\right) = -\frac{1}{\lambda} \sum_{n=0}^{\infty} \frac{f(x^n)}{\lambda^n}.$$

According to the Cauchy–Laurent theory, the expansion above in fact holds for $|\lambda| > r(x)$. We then fix λ satisfying this last condition. In particular,

$$\left\{ f\left(\frac{x^n}{\lambda^n}\right), n \geqslant 0 \right\}$$

is a bounded subset of $\mathbb{C}$. As this is true for all continuous linear functionals, by the Banach–Steinhaus theorem, there exists a positive constant $K(\lambda)$ such that

$$\left\| \frac{x^n}{\lambda^n} \right\| \leqslant K(\lambda) \text{ for } n \geqslant 0.$$

But then, $\|x^n\|^{1/n} \leqslant |\lambda| \, K(\lambda)^{1/n}$, hence $\varlimsup_{n \to +\infty} \|x^n\|^{1/n} \leqslant |\lambda|$ and finally

$$\varlimsup_{n \to +\infty} \|x^n\|^{1/n} \leqslant r(x). \tag{11.4}$$

The inequalities (11.3) and (11.4) provide the desired result. $\square$

11.1.5 Passage to a closed subalgebra

Let us now examine what happens when an element a of A belongs to a closed subalgebra B of A (always supposed to contain e). In this case, $\sigma_A(a)$ (resp.

$\sigma_B(a)$) can be defined as the spectrum of a considered as an element of A (resp. B). Clearly, we have

$$\sigma_A(a) \subset \sigma_B(a).$$

For an inclusion in the opposite sense, we have the following result.

11.1.6 Proposition *The boundary*

$$\partial\sigma_B(a) = \sigma_B(a) \setminus \overset{\circ}{\sigma_B}(a)$$

of $\sigma_B(a)$ is contained in $\sigma_A(a)$.

Proof Replacing if necessary a by $a - \lambda e$ (with a suitable $\lambda \in \mathbb{C}$), it suffices to show that if $0 \in \partial\sigma_B(a)$, then a is not invertible in A. Suppose, on the contrary, that $0 \in \partial\sigma_B(a)$ and that a is invertible in A. Then there exists a sequence $(\lambda_n)_{n\geqslant 0}$ converging to 0, such that, for every $n \geqslant 0$, $a - \lambda_n e$ is invertible in B. However, in A,

$$(a - \lambda_n e)^{-1} \to a^{-1} \text{ as } n \to +\infty.$$

As the subalgebra B is closed in A, $a^{-1} \in B$, which contradicts the hypothesis that $0 \in \sigma_B(a)$. $\qquad\square$

We have this important consequence: let Ω be a connected component of $\mathbb{C} \setminus \sigma_A(a)$; it is an open subset of $\mathbb{C}$. Set $F = \Omega \cap \sigma_B(a)$. Then F is a closed subset of Ω. Let us show that F is also an open subset of Ω and hence of $\mathbb{C}$. If ever $x \in F$ is not in the interior of $\sigma_B(a)$, then x is in the boundary of $\sigma_B(a)$, hence in $\sigma_A(a)$ according to Proposition 11.1.6, which is absurd. Now, as Ω is connected, either $F = \varnothing$ or $F = \Omega$. In the first case, $\Omega \cap \sigma_B(a) = \varnothing$, while in the second case, $\Omega \subset \sigma_B(a)$.

In other words, $\sigma_B(a)$ is obtained from $\sigma_A(a)$ by *plugging the holes*, that is, adding some (bounded) connected components of $\mathbb{C} \setminus \sigma_A(a)$. For example, if $\sigma_A(a) = \mathbb{T}$, the only possibilities for $\sigma_B(a)$ are $\mathbb{T}$ or $\overline{D}$.

11.2 Characters of a Banach algebra

A *character* of a unital Banach algebra A is defined as a homomorphism of $\mathbb{C}$-algebras from A to $\mathbb{C}$, in other words, as a linear functional $\varphi : A \to \mathbb{C}$ such that

$$\varphi(xy) = \varphi(x)\varphi(y) \text{ for all } (x, y) \in A^2 \text{ and } \varphi(e) = 1.$$

The mere existence of a character is not evident, *and could even be lacking if A is not commutative* (see Exercise 11.10). In what follows, $\mathfrak{M}$ denotes the set of characters of A. The set $\mathfrak{M}$ is called the *spectrum* of the algebra A. The aim of this section is to show that, *in the commutative case*, the spectrum is non-empty, and that the invertibility of an element of A *can be derived by studying its images under the characters of A*. This fact is based on a fundamental correspondence, discovered by Gelfand, between the maximal ideals of A and the characters of A.

11.2.1 Maximal ideals

An *ideal* of A is a linear subspace M of A satisfying in addition $ax \in M$ and $xa \in M$ for all $(a, x) \in A \times M$. An ideal M of A is said to be *maximal* if it satisfies the following two conditions:

(i) $M \neq A$,
(ii) if M' is an ideal of A containing M, then either $M' = M$ or $M' = A$.

If A is commutative, it is equivalent to saying that the quotient ring A/M is a field.[1] We can easily deduce from Zorn's lemma that any ideal of A distinct from A is contained in a maximal ideal. Another important fact is the following.

11.2.2 Proposition *Any maximal ideal of A is closed.*

Proof Let M be a maximal ideal of A. We can easily verify that $\overline{M}$, the closure of M in A, is in turn an ideal of A, which of course contains M. Now, if we call U the group of invertible elements of A, we have $M \cap U = \varnothing$, thus $\overline{M} \cap U = \varnothing$ since U is an open subset of A. In particular, $\overline{M}$ is distinct from A, hence $\overline{M} = M$ by the maximality of M. $\qquad\square$

11.2.3 Correspondence between maximal ideals and characters

In this section, A is a unital and *commutative* Banach algebra. We denote by A^* the set of continuous linear functionals on A. It is remarkable that, in the framework of Banach algebras, the purely algebraic definition of characters forces their continuity.

---

[1] A field is by definition a *non-zero* ring in which all elements distinct from 0 are invertible. It can be seen that if $n \geqslant 2$, the two-sided ideal $\{0\}$ is maximal in the algebra $\mathcal{M}_n(\mathbb{C})$, but the corresponding quotient is not a field.

11.2.4 Proposition   *The set $\mathfrak{M}$ of characters of A is contained in the unit sphere of A^*.*

Proof Let $x \in A$, and $\lambda \in \mathbb{C}$ such that $|\lambda| > \|x\|$. Then $e - \lambda^{-1}x$ is invertible, so that $\varphi(e - \lambda^{-1}x) \neq 0$ (since $\varphi(e) = 1$), that is, $\varphi(x) \neq \lambda$. We thus deduce that $|\varphi(x)| \leqslant \|x\|$. The linear functional φ is hence continuous, and $\|\varphi\| \leqslant 1$. As $\|e\| = 1$ and $\varphi(e) = 1$, we have in fact $\|\varphi\| = 1$. $\square$

11.2.5 Theorem   *The map $\varphi \mapsto \ker \varphi$ is a bijection from the set of characters of A onto the set of maximal ideals of A. In particular, the maximal ideals of A are (closed) hyperplanes.*

Proof We first note that, for any $\varphi \in \mathfrak{M}$, the kernel $\ker \varphi$ is a maximal ideal of A, since $A/\ker \varphi$ is isomorphic, as a ring, to the range of φ, which is $\mathbb{C}$. Suppose then that two characters φ and ψ have the same kernel. Given that these are linear functionals on A, φ and ψ are proportional. Since $\varphi(e) = \psi(e) = 1$, this requires $\varphi = \psi$. Finally, let M be a maximal ideal of A. Set $\mathbb{K} = A/M$. For $x \in A$, we denote by $\overline{x}$ the class of x modulo M, and set

$$\|\overline{x}\| = \inf_{u \in \overline{x}} \|u\|.$$

We can verify that this defines a norm on $\mathbb{K}$ (as M is closed), which makes it a Banach algebra (for details, see [160]). Moreover, $\mathbb{K}$ is a field since M is maximal (here, we use commutativity!) By the Gelfand–Mazur theorem, there exists an isometric isomorphism of $\mathbb{C}$-algebras $i : \mathbb{K} \to \mathbb{C}$. We then denote by $\pi : A \to \mathbb{K}$, $x \mapsto \overline{x}$ the canonical surjection, and set $\varphi = i \circ \pi$. It is clear that $\varphi \in \mathfrak{M}$, and that $\ker \varphi = M$. $\square$

As A certainly admits at least one maximal ideal, we can state the following result.

11.2.6 Theorem   *The spectrum of a unital commutative Banach algebra is non-empty.*

If $x \in A$ is a non-invertible element of A, x generates a proper ideal of A, this ideal being itself contained in a maximal ideal. Conversely, if x is invertible, it does not belong to any maximal ideal. Taking Theorem 11.2.5 into account, this gives the following theorem.

11.2.7 Theorem   *Let A be a commutative Banach algebra, and x an element of A. Then, x is invertible if and only if, for all characters φ of A, $\varphi(x) \neq 0$.*

11.2.8 Topology on the spectrum

For more detailed developments on weak topologies, we refer the reader to [31, 152].

Let A be a unital commutative Banach algebra. We have already seen that the spectrum $\mathfrak{M}$ of A is a subset of the unit sphere S of A^*. We then equip A^* with the weak-* topology, that is, the topology of pointwise convergence, or again the coarsest topology on A^* that makes continuous the evaluations

$$\delta_x : A^* \to \mathbb{C}, f \mapsto f(x), \; x \text{ varying over } A.$$

A base of this topology is made up of subsets of $\mathfrak{M}$ of the form

$$\bigcap_{i=1}^{n} \delta_{x_i}^{-1}(U_i), \tag{11.5}$$

where $n \geqslant 1$, $x_i \in A$, the U_i being open subsets of $\mathbb{C}$. If we fix $\varphi \in \mathfrak{M}$, a base of neighbourhoods of φ is formed by the subsets of $\mathfrak{M}$ of the form

$$\{\psi \in \mathfrak{M}/|\psi(x_i) - \varphi(x_i)| < \varepsilon \text{ for } 1 \leqslant i \leqslant n\},$$

where $\varepsilon > 0$, $n \geqslant 1$ and $x_1, \ldots, x_n \in A$.

By the Banach–Alaoglu theorem, the closed unit ball $\overline{B}$ of A^*, equipped with the topology induced by the weak-* topology on A^*, is compact. Moreover, $\mathfrak{M}$ is none other than the set of $f \in \overline{B}$ satisfying $\delta_e(f) = 1$ and

$$\begin{cases} (\delta_{\lambda x + \mu y} - \lambda \delta_x - \mu \delta_y)(f) = 0, \\ (\delta_{xy} - \delta_x \delta_y)(f) = 0, \end{cases}$$

for $(x, y) \in A^2$ and $(\lambda, \mu) \in \mathbb{C}^2$. As the δ_x are continuous, this proves that $\mathfrak{M}$ is a closed subset of $\overline{B}$. Thus we have proved the following important result.

11.2.9 Proposition *Equipped with the topology induced by the weak topology of A^*, the spectrum of A is compact.*

The ability to test the continuity of an $\mathfrak{M}$-valued function f defined on a topological space $\mathcal{T}$ will turn out to be useful.

11.2.10 Proposition *A map $f : \mathcal{T} \to \mathfrak{M}$ is continuous if and only if, for each $x \in A$, the function $\delta_x \circ f$ is continuous on $\mathcal{T}$.*

Proof The direct implication is easy, the δ_x being continuous on $\mathfrak{M}$. Conversely, suppose that $\delta_x \circ f$ is continuous for all $x \in X$. Let V be an open subset of $\mathfrak{M}$ of the form (11.5): $V = \bigcap_{i=1}^{n} \delta_{x_i}^{-1}(U_i)$. Then,

$$f^{-1}(V) = \bigcap_{i=1}^{n} f^{-1}(\delta_{x_i}^{-1}(U_i)) = \bigcap_{i=1}^{n} (\delta_{x_i} \circ f)^{-1}(U_i),$$

which is indeed an open subset of $\mathcal{T}$ by the continuity of $\delta_x \circ f$. □

11.2.11 The Gelfand transform

We keep the hypotheses of the preceding section. For any $a \in A$, we define the map

$$\widehat{a} : \mathfrak{M} \to \mathbb{C}, \ \varphi \mapsto \varphi(a),$$

which is none other than the evaluation at a of the characters of A. The map $\widehat{a}$ is called the *Gelfand transform of* a. The basic properties of the Gelfand transform are the following.

11.2.12 Proposition *For any $a \in A$, the range of $\widehat{a}$ is the spectrum $\sigma(a)$ of a.*

Proof For all $\lambda \in \mathbb{C}$,

$a - \lambda e$ is non-invertible $\Leftrightarrow$ there exists $\varphi \in \mathfrak{M}$ such that $\varphi(a - \lambda e) = 0$

$\Leftrightarrow$ there exists $\varphi \in \mathfrak{M}$ such that $\lambda = \widehat{a}(\varphi)$

$\Leftrightarrow$ λ belongs to the range of $\widehat{a}$. □

11.2.13 Proposition *For any $a \in A$, the map $\widehat{a}$ is continuous on $\mathfrak{M}$.*

Proof The map $\widehat{a}$ is the restriction to $\mathfrak{M}$ of the evaluation at a, which is continuous on A^* by definition of the weak-* topology. □

We thus have at hand the map

$$\Gamma : A \to C(\mathfrak{M}, \mathbb{C}), \ a \mapsto \widehat{a},$$

called the *Gelfand transformation*. The map Γ is a homomorphism of algebras. Moreover, for any $a \in A$,

$$\|\widehat{a}\|_\infty = \sup_{\varphi \in \mathfrak{M}} |\widehat{a}(\varphi)| \stackrel{\text{Prop. 11.2.12}}{=} \sup_{\lambda \in \sigma(a)} |\lambda| = r(a),$$

where $r(a)$ denotes the spectral radius of a. As $r(a) \leqslant \|a\|$, the map Γ is continuous.

11.2.14 Proposition *For any $a \in A$, $\|\widehat{a}\|_\infty \leqslant \|a\|$.*

The kernel of the Gelfand transform consists of the elements a of A with spectral radius null, in other words, such that

$$\|a^n\|^{1/n} \to 0 \text{ as } n \to +\infty.$$

In conclusion, we point out a particularly interesting case: that of *uniform algebras*. These are Banach algebras that are unital and commutative, satisfying in addition $\|a^2\| = \|a\|^2$ for all $a \in A$. In this case, for all $n \geqslant 0$,

$$\|a\| = \|a^{2^n}\|^{1/2^n} \to r(a) = \|\widehat{a}\|_\infty.$$

The Gelfand transform is then *an isometry from A to a closed subalgebra of* $C(\mathfrak{M}, \mathbb{C})$: the algebra A can be seen as an algebra of continuous functions on a compact space.

11.3 Examples

11.3.1 Continuous functions on a compact space

Let X be a compact topological space, and $A = C(X, \mathbb{C})$.

11.3.2 Theorem *The characters of A are exactly the evaluations*

$$e_x : A \to \mathbb{C}, \ f \mapsto f(x), \ x \in X.$$

Proof Let M be a maximal ideal of A. Let us suppose that, for any $x \in X$, there exists an $f \in M$ such that $f(x) \neq 0$. Then, for each $x \in X$, we can select $f_x \in M$ and V_x a neighbourhood of x such that f_x vanishes at no point of V_x. As X is compact, a finite number of V_x, that we denote by $V_{x_1}, \ldots, V_{x_n}$, cover X. Let $f_{x_1}, \ldots, f_{x_n}$ be the associated f_x, and define

$$f = \sum_{i=1}^n |f_{x_i}|^2 = \sum_{i=1}^n \overline{f_{x_i}} f_{x_i}.$$

The function f is then an invertible element of M, which is absurd. Hence, there exists an $x \in X$ such that $M \subset \ker e_x$, an inclusion which is in fact an equality by the maximality of M. We conclude thanks to Theorem 11.2.5. $\square$

11.3.3 Proposition *The map $e : X \to \mathfrak{M}, x \mapsto e_x$ is a homeomorphism.*

Proof To show that *e* is continuous we use Proposition 11.2.10: for any $f \in A$, we have

$$\delta_f \circ e : X \to \mathbb{C}, \; x \mapsto e_x(f) = f(x).$$

The composition $\delta_f \circ e$ coincides with *f* and is thus continuous on *X*. Moreover, as *A* separates the points of *X* according to Urysohn's lemma (see [152]), the map *e* is injective. Finally, surjectivity of *e* had already been proved, and the continuity of e^{-1} is automatic because of the compactness of *X*. □

Thanks to this homeomorphism, we can identify the spectrum of *A* with *X*. Then, *each element of A is equal to its Gelfand transform.*

11.3.4 The disk algebra

In this section, we denote by *A* the disk algebra (defined in Section 11.1).

11.3.5 Lemma *The algebra A is the closure in $C(\overline{D}, \mathbb{C})$ of the set of polynomial functions on $\overline{D}$.*

Proof Obviously, *A* contains the set of polynomial functions, hence also its closure. Conversely, let $f \in A$. For $0 \leqslant r < 1$, we define the function

$$f_r : \overline{D} \to \mathbb{C}, \; z \mapsto f(rz).$$

If $f(z) = \sum_{n=0}^{\infty} a_n z^n$ for $|z| < 1$, we have

$$f_r(z) = \sum_{n=0}^{\infty} a_n r^n z^n \text{ for } |z| \leqslant 1.$$

Set $P_{r,N}(z) = \sum_{n=0}^{N} a_n r^n z^n$. For $N \geqslant 0$ and $|z| \leqslant 1$, we have

$$|f_r(z) - P_{r,N}(z)| \leqslant \sum_{n=N+1}^{\infty} |a_n| r^n \to 0 \text{ as } N \to +\infty,$$

which proves that *for r fixed, f_r is a uniform limit of polynomials on $\overline{D}$.*

Now, as *f* is uniformly continuous on $\overline{D}$, *f_r converges uniformly to f when $r \nearrow 1$.* The lemma follows immediately from the two statements in italics. □

11.3.6 Theorem *The characters of A are exactly the* $e_z : A \to \mathbb{C}$ *with* $f \mapsto f(z)$, *z varying over* $\overline{D}$.

Proof Let φ be a character of A, and $e : \overline{D} \to \mathbb{C}, z \mapsto z$. Set $u = \varphi(e)$. We have $|u| = |\varphi(e)| \leqslant \|e\| = 1$, hence $u \in \overline{D}$. For all polynomials P on $\overline{D}$, we thus have $\varphi(P) = P(u)$. As the polynomials on $\overline{D}$ are dense in A, for all $f \in A, \varphi(f) = f(u) = e_u(f)$. $\qquad\square$

11.3.7 Wiener algebra

We denote by W the Wiener algebra of continuous functions on $\mathbb{T}$ with absolutely convergent Fourier series, equipped with the norm defined by

$$\|f\| = \sum_{n \in \mathbb{Z}} |\widehat{f}(n)|.$$

Note that if $f \in W$, then f is the sum of its Fourier series, and

$$\|f\|_\infty \leqslant \|f\|.$$

11.3.8 Theorem *The characters of W are exactly the* $e_z : W \to \mathbb{C}$ *with* $f \mapsto f(z)$, *z varying over* $\mathbb{T}$.

Proof Let φ be a character of W. Consider the function $e : \mathbb{T} \to \mathbb{T}, z \mapsto z$ and set $u = \varphi(e)$. As e is invertible[2] in $W, u \neq 0$. Moreover, for all $n \in \mathbb{Z}$,

$$|u|^n = |\varphi(e^n)| \leqslant \|e^n\| = 1,$$

which imposes $u \in \mathbb{T}$. Let then $f \in W$. If we denote by $S_N = \sum_{|n| \leqslant N} \widehat{f}(n)e^n$ the partial sums of the Fourier series of f, we have

$$\|f - S_N\| = \sum_{n \in \mathbb{Z}} |\widehat{f}(n) - \widehat{S_N}(n)| = \sum_{|n| > N} |\widehat{f}(n)| \to 0 \text{ as } N \to +\infty.$$

As φ is continuous, it ensues that

$$\varphi(f) = \sum_{n \in \mathbb{Z}} \widehat{f}(n)u^n = f(u),$$

hence the result. $\qquad\square$

We can immediately deduce the following result, which was the first great historical success of Gelfand theory.

---

[2] While the function $z \mapsto z$ of the disk algebra is not.

11.3.9 Theorem [Wiener's lemma]   *If $f \in W$ does not vanish at any point of $\mathbb{T}$, then $\dfrac{1}{f} \in W$.*

It is instructive to compare this "automatic" proof with Wiener's original argument presented in Chapter 2 of this book, in which the seeds of the theory of Banach algebras appear.

Wiener's lemma is a special case of a more general result, which states that *the holomorphic functions operate on the set (denoted by $\widehat{A}$) of the Gelfand transforms of elements of an arbitrary commutative and unital Banach algebra A.* The following theorem states this more precisely.

11.3.10 Theorem [Wiener–Lévy]   *If $x \in A$ and if f is holomorphic in an open neighbourhood Ω of $\sigma(x)$ in $\mathbb{C}$, then there exists an element y of A such that $f \circ \widehat{x} = \widehat{y}$.*

Proof   The proof is based on the use of the Riemann integral of functions with values in a Banach space [104], as well as on an "improved" Cauchy formula: there exists[3] a closed contour γ contained in $\Omega \setminus \sigma(x)$, such that all elements of $\sigma(x)$ have index 1 with respect to γ. Then, for $z \in \sigma(x)$, we have

$$ f(z) = \frac{1}{2i\pi} \int_\gamma \frac{f(\zeta)}{\zeta - z} d\zeta. $$

We then set

$$ y = \frac{1}{2i\pi} \int_\gamma f(\zeta)(\zeta e - x)^{-1} d\zeta, $$

which is an element of A. Then, for $\varphi \in \mathfrak{M}$, we have

$$ \widehat{y}(\varphi) = \varphi(y) = \frac{1}{2i\pi} \int_\gamma \frac{f(\zeta)}{\zeta - \varphi(x)} d\zeta = f(\varphi(x)) = f \circ \widehat{x}(\varphi). \qquad \square $$

In the case where A is the Wiener algebra W, each element of W can be identified with its Gelfand transform. Consequently, if $f(= \widehat{f}) \in W$ here and F is holomorphic in a neighbourhood of $f(\mathbb{T})$ (spectrum of f) in $\mathbb{C}$, then $F \circ f \in W$. For a converse of this theorem, see [101, 103, 104].

We end by pointing out that in 1975, D. J. Newman gave a proof of Wiener's lemma based uniquely on the Neumann series (see [133] or Chapter 2).

These first three examples could lead one to believe that when A is an algebra of functions on a set X, the evaluations at the points of X exhaust the spectrum of A. Nothing could be more false, as the counter-example of

---

[3] The existence of γ, intuitively evident, is a bit difficult to show properly; see [160].

H^∞ shows (see Exercises 11.6 and 11.8, and also Chapter 12). This being said, when this occurs (for example, continuous functions on X, disk algebra, Wiener algebra), then the spectrum of an element f of A is exactly its image $f(X)$.

11.3.11 The convolution algebra $L^1(\mathbb{R}, \mathbb{C})$

The convolution algebra $A = L^1(\mathbb{R}, \mathbb{C})$ is not unital, but a full description of the spectrum of the algebra B obtained by adjoining a unit to A is available.

11.3.12 Theorem   *The spectrum of B consists of the map*

$$(\lambda, f) \mapsto \lambda$$

and of the maps

$$(\lambda, f) \mapsto \lambda + \mathfrak{F}f(x),$$

with x varying over $\mathbb{R}$, and $\mathfrak{F}f(x) = \displaystyle\int_{-\infty}^{+\infty} f(t)e^{-ixt}\,dt$.

Proof   See Chapter 2.   □

Thus, none of the elements of A are invertible in B (which seems reasonable, as A is a maximal ideal of B) and, if $\lambda \neq 0$, (λ, f) is invertible in B if and only if the Fourier transform of f does not take the value $-\lambda$. It is remarkable that these simple facts alone provide a quasi-automatic proof of Wiener's Tauberian theorem (see Chapter 2).

It is also interesting to remark that the Gelfand transform is a generalisation of the Fourier transform. Indeed, let us fix $g \in A$. If φ is a character of B not identically zero on A, there exists a unique real number x such that $\varphi(\lambda, f) = \lambda + \mathfrak{F}f(x)$ for $(\lambda, f) \in B$. Then,

$$\widehat{g}(\varphi) = \varphi(g) = \mathfrak{F}g(x).$$

If we identify φ with the real number x, $\widehat{g}$ is none other than the Fourier transform of g.

11.4 C^*-algebras

A C^*-algebra is a unital Banach algebra A, not necessarily commutative, equipped with an anti-linear involution $x \mapsto x^*$:

$$(\lambda x + y)^* = \overline{\lambda}x^* + y^* \text{ and } (x^*)^* = x \text{ for } x, y \in A \text{ and } \lambda \in \mathbb{C},$$

which also satisfies

$$(xy)^* = y^*x^* \text{ and } \|x^*x\| = \|x\|^2 \text{ for } x, y \in A.$$

11.4.1 Remark We thus automatically have $e^* = e$. Indeed, if $x \in A$,

$$x^*e^* = (ex)^* = x^* \text{ and similarly } e^*x^* = x^*,$$

so e^* is a neutral element for multiplication, hence the result.

11.4.2 Example The algebra $C(X, \mathbb{C})$, where X is a compact space, is a C^*-algebra, the semi-linear involution being complex conjugation.

11.4.3 Example Let H be a complex Hilbert space, and $\mathcal{L}(H)$ the set of continuous linear operators of H. Equipped with the operator norm induced by that of H, defined by

$$\|h\| = \sup_{\|x\| \leqslant 1} \|h(x)\|,$$

the algebra $\mathcal{L}(H)$ is a Banach algebra. One can easily verify that

$$\|h^*h\| = \|h\|^2 \text{ for } h \in \mathcal{L}(H),$$

where h^* is the adjoint operator of h.

11.4.4 Example The disk algebra A, equipped with the involution

$$A \to A, f \mapsto f^* \text{ where } f^* : z \mapsto \overline{f(\bar{z})},$$

is not a C^*-algebra. For example, if $f : z \mapsto z + i$, then $f^* : z \mapsto z - i$, hence $\|f^*f\|_\infty = 2$ while $\|f\|_\infty^2 = 4$.

In what follows, A denotes a C^*-algebra.

11.4.5 Basic properties

An element a of A is said to be *Hermitian* (resp. *normal*) if

$$a^* = a \text{ (resp. } a^*a = aa^*\text{).}$$

11.4.6 Proposition *For all $a \in A$, $\|a^*\| = \|a\|$.*

Proof We have $\|a\|^2 = \|a^*a\| \leqslant \|a^*\| \|a\|$, hence $\|a\| \leqslant \|a^*\|$ and finally $\|a^*\| = \|a\|$. $\qquad\square$

11.4.7 Proposition   *If $a \in A$ is Hermitian, then $r(a) = \|a\|$.*

Proof   We have $\|a\| = \|a^*a\|^{1/2} = \|a^2\|^{1/2}$. Then, by induction,

$$\|a\| = \left\|a^{2^n}\right\|^{1/2^n} \text{ for all } n \geqslant 0.$$

Letting $n \to +\infty$, we obtain the result thanks to the spectral radius formula.
$\square$

11.4.8 Proposition   *If $a \in A$ is Hermitian, and $\varphi \in \mathfrak{M}$, then $\varphi(a) \in \mathbb{R}$.*

Proof   Write $\varphi(a) = u + iv$, with u and v real numbers. As $\|\varphi\| = 1$, we have, for all $\lambda \in \mathbb{R}$,

$$\|a + i\lambda e\|^2 \geqslant |\varphi(a + i\lambda e)|^2 = u^2 + (v + \lambda)^2.$$

Moreover,

$$\begin{aligned}
\|a + i\lambda e\|^2 &= \|(a + i\lambda e)^*(a + i\lambda e)\| \\
&= \|(a - i\lambda e)(a + i\lambda e)\| \text{ since } e^* = e \\
&= \|a^2 + \lambda^2 e\| \\
&\leqslant \|a^2\| + \lambda^2.
\end{aligned}$$

Thus, for all $\lambda \in \mathbb{R}$, $u^2 + v^2 + 2\lambda v \leqslant \|a^2\|$. This requires $v = 0$. $\square$

11.4.9 Corollary   *If $a \in A$ is Hermitian, $\sigma(a) \subset \mathbb{R}$.*

11.4.10 Corollary   *If $a \in A$ and $\varphi \in \mathfrak{M}$, $\varphi(a^*) = \overline{\varphi(a)}$ (i.e., $\widehat{a^*} = \overline{\widehat{a}}$).*

Proof   Write $a = b + ic$, with $b = \dfrac{1}{2}(a + a^*)$ and $c = \dfrac{1}{2i}(a - a^*)$. Defined as such, b and c are Hermitian. We thus have $a^* = b - ic$, hence

$$\varphi(a^*) = \varphi(b) - i\varphi(c) = \overline{\varphi(a)}$$

as $\varphi(b)$ and $\varphi(c)$ are real numbers. $\square$

11.4.11 Invariance of the spectrum

Let B be a closed subalgebra of A invariant under the involution $x \mapsto x^*$ (i.e., a sub-C^*-algebra of A), and a an element of B. As we have seen above, we certainly have $\sigma_A(a) \subset \sigma_B(a)$. In fact, in the case of C^*-algebras, we obtain even more, as stated in the following result.

11.4.12 Proposition $\sigma_A(a) = \sigma_B(a)$.

Proof   To show the non-trivial inclusion, it suffices to show that if $a \in B$ is invertible in A, then $a^{-1} \in B$. For this, we note that a^*a is a Hermitian element of the C^*-algebra B. As we have seen above, $\sigma_B(a^*a) \subset \mathbb{R}$; $\sigma_B(a^*a)$ thus has empty interior in $\mathbb{C}$, so that

$$\sigma_B(a^*a) = \partial\sigma_B(a^*a) \subset \sigma_A(a^*a).$$

As a is invertible in A (thus so is a^*a), $0 \notin \sigma_A(a^*a)$. *A fortiori*, $0 \notin \sigma_B(a^*a)$; therefore, a^*a, and hence also a, are invertible in B. $\square$

11.4.13 The Gelfand–Naimark representation theorem

11.4.14 Theorem [Gelfand–Naimark]   *Let A be a commutative C^*-algebra, with spectrum $\mathfrak{M}$. The Gelfand transformation Γ provides an isometric isomorphism from A onto $C(\mathfrak{M}, \mathbb{C})$.*

Proof   We already know that Γ is a homomorphism of algebras from A to $C(\mathfrak{M}, \mathbb{C})$. Let us fix $a \in A$. We have

$$\|\widehat{a}\|_\infty^2 = \sup_{\varphi \in \mathfrak{M}} |\varphi(a)|^2 = \sup_{\varphi \in \mathfrak{M}} \overline{\varphi(a)}\varphi(a) = \sup_{\varphi \in \mathfrak{M}} \varphi(a^*a) = r(a^*a)$$

and, as a^*a is Hermitian, $r(a^*a) = \|a^*a\| = \|a\|^2$. Consequently now, $\|\widehat{a}\|_\infty = \|a\|$ and Γ is an isometry. In particular, the image $\Gamma(A)$ of Γ is a closed subalgebra of $C(\mathfrak{M}, \mathbb{C})$. Moreover, $\Gamma(A)$ contains the constant function 1, is stable by conjugation, and separates the points of $\mathfrak{M}$. Let us justify this last point: let φ_1 and φ_2 be two distinct elements of $\mathfrak{M}$. We can thus fix $a \in A$ such that $\varphi_1(a) \neq \varphi_2(a)$. But this means exactly $\widehat{a}(\varphi_1) \neq \widehat{a}(\varphi_2)$. By the Stone–Weierstrass theorem, $\Gamma(A)$ is dense in $C(\mathfrak{M}, \mathbb{C})$. Ultimately, $\Gamma(A) = C(\mathfrak{M}, \mathbb{C})$. $\square$

11.4.15 Application to normal elements of A

Let a be a normal element of A. Denote by B the closed subalgebra of A generated by a and a^*. The algebra B is thus a commutative sub-C^*-algebra of A. Let $\mathfrak{M}$ be the spectrum of B. By the Gelfand–Naimark theorem, the algebras B and $C(\mathfrak{M}, \mathbb{C})$ are isometric, where the involution $b \mapsto b^*$ corresponds to complex conjugation. Consequently,

$$a^* = a \Leftrightarrow \widehat{a^*} = \widehat{a}$$
$$\Leftrightarrow \overline{\widehat{a}} = \widehat{a}$$
$$\Leftrightarrow \widehat{a} \text{ is real-valued}$$
$$\Leftrightarrow \sigma(a) \subset \mathbb{R}.$$

The element a is thus Hermitian if and only if its spectrum is real. We now try to identify $\mathfrak{M}$. For this, we define

$$u : \mathfrak{M} \to \sigma(a), \ \varphi \mapsto \varphi(a).$$

The map u is well-defined, surjective and continuous. Let us show that it is injective. If φ_1 and φ_2 are two elements of $\mathfrak{M}$ satisfying $\varphi_1(a) = \varphi_2(a)$, we have $\overline{\varphi_1(a)} = \overline{\varphi_2(a)}$, hence $\varphi_1(a^*) = \varphi_2(a^*)$. But then, by definition of B and continuity of φ_1 and φ_2, we have $\varphi_1 = \varphi_2$.

Finally, as $\mathfrak{M}$ is compact, u is a homeomorphism from $\mathfrak{M}$ to $\sigma(a)$. We can thus construct an *isometric isomorphism*

$$v : B \to C\big(\sigma(a), \mathbb{C}\big), \ b \mapsto \widehat{b} \circ u^{-1}.$$

We finish with an application of the Gelfand–Naimark theorem: it shows the existence and uniqueness of the square root of a positive Hermitian element of A (i.e., with spectrum contained in $\mathbb{R}_+$).

11.4.16 Proposition   *Let a be a Hermitian element of A with $\sigma(a) \subset \mathbb{R}_+$. Then, there exists a unique positive Hermitian $b \in A$ such that $b^2 = a$.*

Proof   Let us start with the existence, and denote by B the closed subalgebra of A generated by a, and $\mathfrak{M}$ its spectrum. By the Gelfand–Naimark theorem, B is isomorphic to the $\mathbb{C}$-algebra $C(\mathfrak{M}, \mathbb{C})$ via the Gelfand transformation Γ, and the image of a in $C(\mathfrak{M}, \mathbb{C})$ is a positive continuous function; it thus admits a positive square root g. Then, $b = \Gamma^{-1}(g)$ is an answer to the question.

Now we consider uniqueness: let $c \in A$ be Hermitian and positive such that $c^2 = a$. Then, c commutes with a, thus with all element of B, hence with b. Then we conclude by considering the closed subalgebra of A generated by a, b and c: b and c have the same Gelfand transform (namely $\sqrt{\widehat{a}}$), and hence are equal. $\square$

Exercises

11.1. Let A be a unital Banach algebra. For each $x \in A$, define the map

$$L_x : A \to A, \ y \mapsto xy$$

(left multiplication by x). The map L_x is a continuous linear endomorphism of A, which allows us to set $\|x\|' = \|L_x\|$ for all $x \in A$. Show that $\| \cdot \|'$ is a norm on A, equivalent to $\| \cdot \|$, satisfying $\|xy\|' \leqslant \|x\|'\|y\|'$ for $x, y \in A$, and that $\|e\|' = 1$.

11.2. Let A be the disk algebra, and $f_1, \ldots, f_n$ be elements of A without common zeros. Show that there exist elements $g_1, \ldots, g_n$ of A such that $\sum_{k=1}^{n} f_k g_k = 1$ (this is Bézout's theorem in the disk algebra).

11.3. Closed ideals of $L^1(\mathbb{T})$. For any subset A of $\mathbb{Z}$, define

$$I_A = \{f \in L^1(\mathbb{T}) / \widehat{f}(n) = 0 \text{ for all } n \in A\}.$$

(a) Show that I_A is a closed ideal of the convolution algebra $L^1(\mathbb{T})$.

(b) Conversely, let I be a closed ideal of $L^1(\mathbb{T})$. Denote by A the set of all $n \in \mathbb{Z}$ such that $\widehat{f}(n) = 0$ for all $f \in I$.

   (i) Show that I contains all trigonometric polynomials P with spectrum contained in $\mathbb{Z} \setminus A$, that is, such that $\widehat{P}(n) = 0$ for all $n \in A$.

   (ii) Show that $I = I_A$.

11.4. Closed ideals of continuous functions. Let K be a compact topological space. Set $A = C(K, \mathbb{C})$. If F is a closed subset of K, denote by I_F the set of elements of A that are identically zero on F.

(a) Show that I_F is a closed ideal of A.

(b) Conversely, let I be a closed ideal of A. Denote by Z the set of zeros common to all elements of I.

   (i) Let U be an open subset of K containing Z. Show that there exists a function $\varphi \in I$ that does not vanish at any point in a neighbourhood of $K \setminus U$.

   (ii) Deduce that I contains all the elements of A that are identically zero on U.

   (iii) Let $f \in I_Z$. For any $\varepsilon > 0$, set $U_\varepsilon = \{x \in X / |f(x)| < \varepsilon\}$. Show that there exists a $g \in A$ identically zero on U_ε such that $\|f - g\|_\infty \leqslant 2\varepsilon$.

   (iv) Finally, show that $I = I_Z$.

11.5. Equip $A = C^1([0, 1], \mathbb{C})$ with the norm $\|f\| = \|f\|_\infty + \|f'\|_\infty$.

(a) Verify that A, equipped with $\| \cdot \|$, is a unital Banach algebra.

(b) Determine the spectrum of A.

(c) Fix $a \in [0, 1]$ and define $I = \{f \in A / f(a) = f'(a) = 0\}$. Show that I is an ideal of A, and that I cannot be written as an intersection of maximal ideals of A.

11.6. Let K be a non-empty compact subset of $\mathbb{C}$. Denote by $P(K)$ the set of functions from K to $\mathbb{C}$ that are a uniform limit on K of a sequence of polynomials. Recall the statement of Runge's theorem: if $\mathbb{C} \setminus K$ is connected, and if $a \in \mathbb{C} \setminus K$, the function $z \mapsto \dfrac{1}{z - a}$ is an element of $P(K)$ (see [152]).

(a) Show that $P(K)$, equipped with $\| \cdot \|_\infty$, is a unital Banach algebra.

(b) Verify that $\mathbb{C} \setminus K$ possesses a unique non-bounded connected component. In what follows, we denote by $\widehat{K}$ the union of K and of the bounded connected components of $\mathbb{C} \setminus K$. Verify that $\widehat{K}$ is compact, and that its boundary is contained in K.

(c) Show that any element f of $P(K)$ admits a unique extension $\widetilde{f}$ to $\widehat{K}$ that is continuous on $\widehat{K}$ and holomorphic in the interior of $\widehat{K}$.

(d) Show that if $a \notin \widehat{K}$, there exists a sequence $(Q_n)_{n \geqslant 0}$ of polynomials such that $Q_n(a) = 1$ for all $n \geqslant 0$ and $\sup_K |Q_n| \to 0$ as $n \to +\infty$. Thus, deduce that $\widehat{K}$ is the *polynomially convex hull* of K, that is, the set of all $z \in \mathbb{C}$ such that, for every polynomial P,

$$|P(z)| \leqslant \sup_K |P|.$$

(e) Show that the characters of $P(K)$ are the functions

$$P(K) \to \mathbb{C}, \ f \mapsto \widetilde{f}(a), \ a \text{ varying over } \widehat{K}.$$

The spectrum of $P(K)$ can thus be identified with $\widehat{K}$.

11.7. Let $A = L^1([0, 1], \mathbb{C})$. For $f, g \in A$ and $t \in [0, 1]$, define

$$f * g(t) = \int_0^t f(t - u)g(u)\, du.$$

(a) Show that A, equipped with this "truncated" convolution product and with the usual norm $\| \cdot \|_1$, is a commutative (non-unital) Banach algebra.

(b) Let e be the function identically equal to 1. Calculate

$$\lim_{n \to +\infty} \| \underbrace{e * \cdots * e}_{n \text{ times}} \|_1^{1/n}.$$

(c) Let B be the algebra obtained by adjoining a unit to A, and Γ the associated Gelfand transformation. Show that the kernel of Γ contains all polynomials, and thus deduce that, for any $f \in A$,

$$\| f^n \|^{1/n} \to 0 \text{ as } n \to +\infty,$$

and then that any multiplicative linear functional on A is identically zero. Is this in contradiction with Theorem 11.2.6?

11.8. Consider the Banach algebra H^∞ of bounded holomorphic functions in the unit disk D. Denote by $\mathfrak{M}$ its spectrum, and define

$$e : D \to D, z \mapsto z.$$

(a) Fix $u \in \mathbb{T}$, and denote by I the set of elements f of H^∞ such that

$$f\big((1 - 1/n)u\big) \to 0 \text{ as } n \to +\infty.$$

Show that there exists a $\varphi \in \mathfrak{M}$ such that $\varphi(f) = 0$ for all $f \in I$, and that $\varphi(e) = u$.

(b) Show that φ is not an evaluation at a single point of D. This proves that the spectrum of H^∞ contains much more than the points of D. Carleson's corona theorem asserts that D is dense in $\mathfrak{M}$ (see Chapter 12).

11.9. Let X and Y be two compact topological spaces. Show that the following statements are equivalent:[4]

(a) X and Y are homeomorphic;

(b) the Banach algebras $C(X, \mathbb{C})$ and $C(Y, \mathbb{C})$ are isometric;

(c) these algebras are isomorphic, that is, if there exists a continuous isomorphism of algebras from one to the other (with its inverse automatically continuous according to Banach's theorem).

11.10. Show that if $n \geqslant 2$, there does not exist a non-zero linear functional $\varphi : \mathcal{M}_n(\mathbb{C}) \to \mathbb{C}$ such that $\varphi(MN) = \varphi(M)\varphi(N)$ for $M, N \in \mathcal{M}_n(\mathbb{C})$.

11.11. Let A be the disk algebra, and let B be the set of all continuous functions $f : \mathbb{T} \to \mathbb{C}$ such that $\widehat{f}(n) = 0$ for $n < 0$, both equipped with the norm $\| \cdot \|_\infty$.

(a) Show that if $f, g \in L^2(\mathbb{T})$, then $fg \in L^1(\mathbb{T})$ and

$$\widehat{fg}(n) = \sum_{k \in \mathbb{Z}} \widehat{f}(k)\widehat{g}(n - k) \text{ for } n \in \mathbb{Z}.$$

Thus deduce that B is a Banach algebra.

(b) Show that the restriction map

$$A \to B, f \mapsto f_{|\mathbb{T}}$$

is an isomorphism of algebras and an isometry, and describe the spectrum of B. Compare with Exercise 11.6.

11.12. Let H be a Hilbert space, and A the C^*-algebra of continuous linear operators of H. Show that an element a of A is Hermitian (i.e., verifies $a = a^*$) if and only if $\|e^{ita}\| = 1$ for all $t \in \mathbb{R}$.

11.13. The aim of this exercise is to determine the points of $\mathcal{M}_n(\mathbb{C})$ at which the exponential matrix is a local diffeomorphism. We fix $A \in \mathcal{M}_n(\mathbb{C})$.

---

[4] The situation is different if we consider the *Banach spaces* $C(X, \mathbb{C})$ and $C(Y, \mathbb{C})$. One can show that if they are isometric, then X and Y are homeomorphic; this holds if they are almost linearly isometric (in the sense that their Banach–Mazur distance is < 2). However, this last result becomes false if the distance in question is $\geqslant 2$.

(a) Let $H \in \mathcal{M}_n(\mathbb{C})$. By solving the Cauchy problem

$$\begin{cases} X' = (A + H)X \\ X(0) = I_n \end{cases}$$

in two different ways (where the unknown function X is matrix-valued), show that

$$\exp(A + H) - \exp(A) = \int_0^1 \exp((1 - t)A)H \exp(t(A + H))dt.$$

(b) Deduce that exp is differentiable at A, and give an integral expression of $d(\exp)(A)(H)$ for $H \in \mathcal{M}_n(\mathbb{C})$.

(c) Denote by $L = L_A$ and $R = R_A$ the operators of left and right multiplication by A:

$$L(M) = AM \text{ and } R(M) = MA \text{ for } M \in \mathcal{M}_n(\mathbb{C}).$$

Show that, in the Banach algebra of operators of $\mathcal{M}_n(\mathbb{C})$, we have

$$d(\exp)(A) = \exp(L)\Phi(R - L),$$

where Φ is the entire function defined by

$$\Phi(z) = \frac{e^z - 1}{z} \text{ for } z \in \mathbb{C}^*.$$

(d) Show that $d(\exp)(A)$ is invertible if and only if A does not have two distinct eigenvalues whose difference is in $2i\pi\mathbb{Z}$.

The next three exercises require a few notions on entire functions of exponential type, that is, the functions f holomorphic in the whole of $\mathbb{C}$ and satisfying a moderate growth inequality of the form

$$|f(z)| \leqslant ae^{b|z|} \text{ for all } z \in \mathbb{C}, \tag{11.6}$$

where a and b are positive constants. An entire function satisfying (11.6) and bounded on the real axis satisfies Bernstein's inequality [196], namely

$$\|f'\|_\infty \leqslant b\|f\|_\infty,$$

where the L^∞-norm of course refers to the least upper bound on the real axis. We also need a generalisation of the maximum principle, known as the Phragmén–Lindelöf principle [11], stated as follows. Let S be an open sector of $\mathbb{C}$, with vertex 0 and angle $\alpha < 2\pi$, and f a holomorphic function in S and continuous on $\overline{S}$. Suppose that

- *$|f(z)| \leqslant M$ for all $z \in \partial S$,*
- *$|f(z)| \leqslant ae^{b|z|^\beta}$ for all $z \in S$,*
- *$\alpha\beta < \pi$,*

where a, b and β are positive constants. Then,

$$|f(z)| \leqslant M \text{ for all } z \in S.$$

11.14. Let f be an entire function.

(a) Suppose that f satisfies (11.6) and is bounded in modulus by M on the real axis. By applying the Phragmén–Lindelöf principle to the entire function $g(z) = f(z)e^{\pm ibz}$ in each of the four quadrants $S_1, \ldots, S_4$, which are open sectors of angle $\frac{\pi}{2}$, conclude that

$$|f(z)| \leqslant M e^{b|\operatorname{Im} z|} \text{ for all } z \in \mathbb{C}.$$

(b) Suppose that f is bounded in modulus by M on the real axis, and verifies

$$\forall \varepsilon > 0, \exists C_\varepsilon > 0 \text{ such that } |f(z)| \leqslant C_\varepsilon e^{\varepsilon |z|} \text{ for all } z \in \mathbb{C}.$$

Show that $|f(z)| \leqslant M$ for all $z \in \mathbb{C}$, and then that f is constant.

11.15. V. E. Katsnelson. We saw in Exercise 11.12 that the Hermitian elements a of the C^*-algebra $A = \mathcal{L}(H)$, where H is a Hilbert space, are also the elements a that satisfy

$$\|e^{ita}\| = 1 \text{ for all } t \in \mathbb{R}. \tag{11.7}$$

By extension, let A be an arbitrary unital Banach algebra, with unit e. We say that $a \in A$ is Hermitian if (11.7) holds. We intend to show that the Hermitian elements a with this generalised definition continue to satisfy

$$\|a\| = r(a),$$

where $r(a)$ denotes the spectral radius of a. Thus let $a \in A$ be Hermitian and let $L \in A^*$ such that $\|L\| = 1$. Set $f(z) = L(e^{iza})$ for $z \in \mathbb{C}$.

(a) Show that f is an entire function satisfying

$$\forall \varepsilon > 0, \exists C_\varepsilon > 0 \text{ such that } |f(z)| \leqslant C_\varepsilon e^{(r(a)+\varepsilon)|z|} \text{ for all } z \in \mathbb{C},$$

$$|f(t)| \leqslant 1 \text{ for all } t \in \mathbb{R},$$

and

$$f'(0) = iL(a).$$

(b) By using Bernstein's inequality, show that $|f'(0)| \leqslant r(a)$.

(c) By using the Hahn–Banach theorem, conclude that $\|a\| = r(a)$.

11.16. I. M. Gelfand. Let A be a unital Banach algebra with unit e and let B be its closed unit ball. We aim to show that e is an extremal point of B, that is, if $w \in A$ and $e \pm w \in B$, then $w = 0$. Thus, let w be an element verifying this property.

(a) Show that $\widehat{w} = 0$ and that $r(w) = 0$.

(b) Show that if $-1 \leqslant x \leqslant 1$, then $\|e + xw\| \leqslant 1$. Thus, deduce that

$$\|e^{tw}\| = \left\| \lim_{n \to \infty} \left(e + \frac{tw}{n} \right)^n \right\| \leqslant 1 \text{ for all } t \in \mathbb{R}.$$

(c) Let $L \in A^*$ such that $\|L\| = 1$, and $f(z) = L(e^{zw})$ for $z \in \mathbb{C}$. Show that f is an entire function satisfying

$$\forall \varepsilon > 0, \exists C_\varepsilon > 0 \text{ such that } |f(z)| \leqslant C_\varepsilon e^{\varepsilon |z|} \text{ for all } z \in \mathbb{C}$$

and

$$|f(t)| \leqslant 1 \text{ for all } t \in \mathbb{R}.$$

By using Exercise 11.14, show that f is constant.

(d) Show that e is an extremal point of B.

12

The Carleson corona theorem

12.1 Introduction

The aim of this chapter is to present a complete proof of Carleson's corona theorem, as well as two related fundamental results:

- Beurling's[1] theorem on the subspaces invariant under the shift operator, a little jewel of twentieth-century functional analysis.
- The characterisation by Carleson (before his solution of the corona problem) of the interpolating sequences of H^∞, again a marvel, but also a very natural question, as it is no more than Lagrangian interpolation ... with additional constraints. Carleson's solution allowed him to highlight the importance of what is today known as "Carleson measures", a capital notion for the solution of the corona problem and many other problems concerning operators.

The solution of the corona problem, a relatively recent (1962) and extremely difficult result, is at the crossroads of multiple domains (Hankel operators, Toeplitz operators, etc.) that are still today the object of intense research. It is thus not possible to have the same perspective on the subject as, for example, the 1911 Littlewood theorem, and our only ambition is to give a proof as clear and self-contained as possible, thus engaging the reader to delve into more extensive works [53, 64, 66, 90, 112, 136].

In this chapter, the basic notions of Banach algebras, as presented in Chapter 11 of this book, are assumed to be known.

In Section 12.2, we present the prerequisites for the Hörmander and Wolff proof of Section 12.6. Section 12.7 touches on Carleson's initial proof and its context (interpolating sequences, Carleson measures), a context to be found

---

[1] Beurling was in fact Carleson's thesis advisor.

later in the works of Fefferman and Stein, and an inspiration for T. Wolff. Finally, Section 12.8 presents a few recent developments.

12.2 Prerequisites

12.2.1 The differential operators

The differential operators

$$\partial = \frac{1}{2}\left(\frac{\partial}{\partial x} - i\frac{\partial}{\partial y}\right) \text{ and } \overline{\partial} = \frac{1}{2}\left(\frac{\partial}{\partial x} + i\frac{\partial}{\partial y}\right)$$

are assumed to be known, as well as their elementary properties

- $\partial(fg) = f\,\partial g + g\,\partial f$ and $\overline{\partial}(fg) = f\,\overline{\partial}g + g\,\overline{\partial}f$,
- $\overline{\partial f} = \partial\,\overline{f}$,
- $\partial f = f'$ and $\overline{\partial}f = 0$ if f is holomorphic,
- $\partial(g \circ f) = (g' \circ f)\partial f$ if g is a differentiable function of a real variable and f a real function of two real variables (the chain rule).

Also supposed to be known is the Laplacian

$$\Delta = \frac{\partial^2}{\partial x^2} + \frac{\partial^2}{\partial y^2} = 4\partial\overline{\partial} = 4\overline{\partial}\partial.$$

We pass from $\dfrac{\partial}{\partial x}$, $\dfrac{\partial}{\partial y}$ to $\dfrac{\partial}{\partial r}$, $\dfrac{\partial}{\partial \theta}$ (where r and θ denote the usual polar coordinates) using the classical formulas

$$\begin{cases} \dfrac{\partial}{\partial x} = \cos\theta\,\dfrac{\partial}{\partial r} - \dfrac{\sin\theta}{r}\dfrac{\partial}{\partial\theta}, \\[2mm] \dfrac{\partial}{\partial y} = \sin\theta\,\dfrac{\partial}{\partial r} + \dfrac{\cos\theta}{r}\dfrac{\partial}{\partial\theta}, \\[2mm] \dfrac{\partial}{\partial r} = \cos\theta\,\dfrac{\partial}{\partial x} + \sin\theta\,\dfrac{\partial}{\partial y}, \\[2mm] \dfrac{\partial}{\partial\theta} = -r\sin\theta\,\dfrac{\partial}{\partial x} + r\cos\theta\,\dfrac{\partial}{\partial y}. \end{cases}$$

We also have

$$\begin{cases} \dfrac{\partial}{\partial r} = e^{-i\theta}\overline{\partial} + e^{i\theta}\partial, \\[2mm] \overline{\partial} = \dfrac{e^{i\theta}}{2}\dfrac{\partial}{\partial r} + \dfrac{ie^{i\theta}}{2r}\dfrac{\partial}{\partial\theta}, \\[2mm] \partial = \dfrac{e^{-i\theta}}{2}\dfrac{\partial}{\partial r} - \dfrac{ie^{-i\theta}}{2r}\dfrac{\partial}{\partial\theta}. \end{cases} \tag{12.1}$$

We denote by D the open unit disk of $\mathbb{C}$. A holomorphic function in D is none other than a function $f \in C^1(D)$ that satisfies the partial differential equation (PDE) $\overline{\partial}f = 0$. This PDE, as well as the associated non-homogeneous PDE

$\overline{\partial} f = \varphi$, will play an essential role in Hörmander's algebraic reduction, which led to the triumph of the Cauchy–Riemann point of view, as opposed to that of Weierstrass based on power series expansions.

12.2.2 The Stokes formula

The Stokes formula, describing the boundary of D–interior of D interaction, will also play an essential role, in the form

$$\int_{\partial\Omega} P\,dz + Q d\overline{z} = 2i \iint_{\Omega} (\overline{\partial} P - \partial Q) d\lambda(z). \tag{12.2}$$

In this formula,

- Ω is a bounded open subset of $\mathbb{C}$, with a positively oriented C^1 boundary,
- P and Q are C^1 functions in a neighbourhood of $\overline{\Omega}$,
- $d\lambda(z) = dx\,dy$ is the Lebesgue measure on $\mathbb{R}^2$.

We will only need to use this formula when Ω is an annulus $a < |z| < b$, with $a > 0$. Then, with

$$z = re^{i\theta},\, dz = ire^{i\theta} d\theta,\, \overline{z} = re^{-i\theta} \text{ and } d\overline{z} = -ire^{-i\theta} d\theta,$$

the first term of (12.2) appears as

$$\left(\int_0^{2\pi} P(be^{i\theta}) i b e^{i\theta} d\theta + \int_0^{2\pi} Q(be^{i\theta})(-ibe^{-i\theta}) d\theta \right)$$
$$-\left(\int_0^{2\pi} P(ae^{i\theta}) i a e^{i\theta} d\theta + \int_0^{2\pi} Q(ae^{i\theta})(-iae^{-i\theta}) d\theta \right).$$

In this particular case, (12.2) is easy to check. For example, via (12.1) and integration by parts:

$$2 \iint_{\Omega} \overline{\partial} P d\lambda(z) = \iint_{\substack{a<r<b\\0<\theta<2\pi}} \left(e^{i\theta} \frac{\partial P}{\partial r} + \frac{ie^{i\theta}}{r} \frac{\partial P}{\partial \theta} \right) r\,dr\,d\theta$$

$$= \iint e^{i\theta} r \frac{\partial P}{\partial r} dr\,d\theta + \iint i e^{i\theta} \frac{\partial P}{\partial \theta} dr\,d\theta$$

$$= \int_0^{2\pi} e^{i\theta} \left[r P(re^{i\theta}) \right]_a^b d\theta - \iint e^{i\theta} P(re^{i\theta}) dr\,d\theta$$

$$+ i \int_a^b \left[e^{i\theta} P(re^{i\theta}) \right]_0^{2\pi} dr - i \iint i e^{i\theta} P(re^{i\theta}) dr\,d\theta.$$

The second and fourth terms cancel, the third is zero by periodicity, and we are left with

$$\int_0^{2\pi} \left(bP(be^{i\theta}) - aP(ae^{i\theta}) \right) e^{i\theta} d\theta,$$

hence (12.2) in the case $Q = 0$, after multiplication by i. Similarly,

$$2i \iint_\Omega -\partial Q d\lambda(z) = \int_{\partial\Omega} Q d\bar{z},$$

which proves (12.2) in the general case.

Here are some important consequences of Stokes' formula.

12.2.3 Proposition [Green's formula]

$$\iint_\Omega (u\Delta v - v\Delta u) d\lambda(z) = \int_{\partial\Omega} \left(u \frac{\partial v}{\partial n} - v \frac{\partial u}{\partial n} \right) ds, \qquad (12.3)$$

where u and v are C^2 in a neighbourhood of $\overline{\Omega}$, $\dfrac{\partial}{\partial n}$ denotes the derivative in the direction of the outer unit normal and s a curvilinear abscissa on $\partial\Omega$.

Proof Consider the case where Ω is the preceding annulus, then $\dfrac{\partial}{\partial n} = \pm \dfrac{\partial}{\partial r}$, $ds = r \, d\theta$ and Green's formula is an easy consequence of that of Stokes. Indeed, provisionally setting $\delta = \partial\bar{\partial}$ and using (12.1), we have, with $\dfrac{\partial}{\partial n} = \dfrac{\partial}{\partial r}$:

$$\int_{|z|=b} \left(u \frac{\partial v}{\partial n} - v \frac{\partial u}{\partial n} \right) ds = \int_0^{2\pi} \left(u(e^{-i\theta}\bar{\partial}v + e^{i\theta}\partial v) - v(e^{-i\theta}\bar{\partial}u + e^{i\theta}\partial u) \right) r \, d\theta$$

$$= \int_0^{2\pi} (u\partial v - v\partial u) r e^{i\theta} d\theta + \int_0^{2\pi} (u\bar{\partial}v - v\bar{\partial}u) r e^{-i\theta} d\theta$$

$$= -i \int_{|z|=b} P \, dz + Q d\bar{z},$$

where $P = u\partial v - v\partial u$ and $Q = v\bar{\partial}u - u\bar{\partial}v$.

Similarly, with this time $\dfrac{\partial}{\partial n} = -\dfrac{\partial}{\partial r}$, we have

$$\int_{|z|=a} \left(u \frac{\partial v}{\partial n} - v \frac{\partial u}{\partial n} \right) ds = i \int_{|z|=a} P \, dz + Q d\bar{z}.$$

By summing the two formulas, and with $\partial\Omega$ the *oriented* boundary of Ω, we obtain

$$\int_{\partial\Omega} \left(u \frac{\partial v}{\partial n} - v \frac{\partial u}{\partial n} \right) ds = -i \int_{\partial\Omega} P \, dz + Q d\bar{z}.$$

However, an easy calculation gives

$$\overline{\partial}P - \partial Q = \overline{\partial}u\partial v - \overline{\partial}v\partial u + u\delta v - v\delta u - (\partial v\overline{\partial}u - \partial u\overline{\partial}v + v\delta u - u\delta v)$$
$$= 2(u\,\delta v - v\,\delta u).$$

Therefore, by (12.2), the right-hand term of (12.3) is

$$-i\int_{\partial\Omega} P\,dz + Q\,d\overline{z} = 2\iint_{\Omega}(\overline{\partial}P - \partial Q)d\lambda(z)$$
$$= 4\iint_{\Omega}(u\,\delta v - v\,\delta u)d\lambda(z)$$
$$= \iint_{\Omega}(u\,\Delta v - v\,\Delta u)d\lambda(z). \qquad \square$$

For what follows, it will be useful to denote by

- m the normalised Haar measure of the compact Abelian group $\mathbb{T}$ of complex numbers with modulus 1, defined by

$$\int_{\mathbb{T}} f(z)\,dm(z) = \int_0^{2\pi} f(e^{i\theta})\frac{d\theta}{2\pi}$$

  for all continuous functions $f : \mathbb{T} \to \mathbb{C}$, the first integral often being – abusively – written

$$\int_{\mathbb{T}} f(e^{i\theta})\,dm(\theta),$$

- $dA(z) = \pi^{-1}d\lambda(z)$ the normalised Lebesgue measure on the open unit disk D of $\mathbb{C}$,
- $d\lambda_1(z) = 2\ln\frac{1}{|z|}dA(z)$.

The measure λ_1 is a probability measure on D, that will play a very important role in this chapter (see (12.5) below).

With these notations, Green's formula in turn leads to the important consequence below.

12.2.4 Proposition [Riesz formula] *If u is a C^2 function in a neighbourhood of $\overline{D}$, then*

$$u(0) = \int_{\mathbb{T}} u(e^{i\theta})\,dm(\theta) - \frac{1}{4}\int_D \Delta u(z)d\lambda_1(z). \qquad (12.4)$$

Proof To obtain (12.4), we restrict Green's formula to the case where $b = 1$, $0 < a < 1$ and $v(z) = \ln\frac{1}{|z|}$, a function that is harmonic ($\Delta v = 0$), zero on the external boundary of Ω and satisfies $\frac{\partial v}{\partial r} = -\frac{1}{r}$. We thus obtain

$$\iint_{a<|z|<1} -v\Delta u\, d\lambda(z) = \int_0^{2\pi} u(e^{i\theta})(-1)d\theta$$

$$-\int_0^{2\pi}\left(u(ae^{i\theta})\left(-\frac{1}{a}\right) - \ln\frac{1}{a}\,\frac{\partial u}{\partial r}\left(ae^{i\theta}\right)\right)a\, d\theta$$

$$= -\int_0^{2\pi} u(e^{i\theta})d\theta + \int_0^{2\pi} u(ae^{i\theta})d\theta$$

$$+a\ln\frac{1}{a}\int_0^{2\pi}\frac{\partial u}{\partial r}(ae^{i\theta})d\theta.$$

As $a \to 0$, the second integral tends to $2\pi u(0)$, and the third to 0. Therefore,

$$\iint_D -v\Delta u\, d\lambda(z) = -\int_0^{2\pi} u(e^{i\theta})d\theta + 2\pi u(0),$$

which gives the formula (12.4). $\square$

12.2.5 Proposition [Littlewood identity]   *If f is a holomorphic function in a neighbourhood of $\overline{D}$, then*

$$\|f\|_2^2 := \int_{\mathbb{T}}|f|^2\, dm = |f(0)|^2 + \iint_D |f'(z)|^2 d\lambda_1(z). \tag{12.5}$$

Proof   We apply the Riesz formula to $u = |f|^2 = f\overline{f}$. It suffices to calculate Δu. However, $\partial u = \overline{f}\partial f + f\partial\overline{f} = \overline{f}\partial f$ since $\partial\overline{f} = \overline{\partial}\overline{f} = 0$. Then, $\overline{\partial}\partial u = \partial f\overline{\partial}\,\overline{f} = \partial f\overline{\partial f} = |\partial f|^2 = |f'|^2$, hence $\Delta u = 4|f'|^2$, which gives the identity (12.5) as a consequence of (12.4). $\square$

12.2.6 Resolution of d-bar

Let $\varphi : \mathbb{C} \to \mathbb{C}$ be a C^∞ function with compact support. We would like to solve the non-homogeneous PDE $\overline{\partial}w = \varphi$, where the unknown function w is C^∞. As the function φ is bounded, we can set, for all $a \in \mathbb{C}$:

$$w(a) = \int \frac{\varphi(z)}{a - z}dA(z), \tag{12.6}$$

the integral being extended to $\mathbb{C}$, which makes sense as the function $z \mapsto \dfrac{1}{z-a}$ is integrable over all disks centred at a. We also have

$$w(a) = \int \frac{\varphi(a - z)}{z}dA(z),$$

which allows us to differentiate under the integral sign. We obtain

$$\overline{\partial}w(a) = \int \frac{\overline{\partial}\varphi(a - z)}{z}dA(z).$$

First suppose that $a = 0$. Then

$$\overline{\partial} w(0) = - \int \frac{\overline{\partial} \varphi(z)}{z} \, dA(z)$$

$$= - \lim_{\substack{\varepsilon \to 0 \\ R \to +\infty}} \int_{\varepsilon \leqslant |z| \leqslant R} \frac{\overline{\partial} \varphi(z)}{z} \, dA(z)$$

$$= - \lim_{\substack{\varepsilon \to 0 \\ R \to +\infty}} \int_{\varepsilon \leqslant |z| \leqslant R} \overline{\partial} \left(\frac{\varphi(z)}{z} \right) dA(z),$$

and by Stokes' formula:

$$\overline{\partial} w(0) = - \frac{1}{2i\pi} \lim_{\substack{\varepsilon \to 0 \\ R \to +\infty}} \left(\int_0^{2\pi} \frac{\varphi(Re^{i\theta})}{Re^{i\theta}} i Re^{i\theta} \, d\theta - \int_0^{2\pi} \frac{\varphi(\varepsilon e^{i\theta})}{\varepsilon e^{i\theta}} i \varepsilon e^{i\theta} \, d\theta \right)$$

$$= \frac{1}{2i\pi} \lim_{\varepsilon \to 0} \int_0^{2\pi} \varphi(\varepsilon e^{i\theta}) i \, d\theta = \varphi(0),$$

since $\varphi(Re^{i\theta}) = 0$ for large R. The general case follows because differentiation and convolution commute with the translations: if $\phi(z) = \varphi(z + a)$ and

$$W(b) = w(b + a) = \int \frac{\phi(z)}{b - z} \, dA(z),$$

we have

$$\overline{\partial} w(a) = \overline{\partial} W(0) = \phi(0) = \varphi(a).$$

In fact, the explicit form (12.6) of the solution of $\overline{\partial} w = \varphi$ will be of no use to us: we simply need to know that a solution exists.

12.2.7 Blaschke products

Let $(z_n)_{n \geqslant 1}$ be a sequence of points of D (as always, D is the open unit disk of $\mathbb{C}$) such that

$$\sum_{n=1}^{\infty} (1 - |z_n|) < \infty.$$

For $z \in D$, we set (with the convention $\dfrac{|z_n|}{z_n} = -1$ if $z_n = 0$)

$$B(z) = \prod_{n=1}^{\infty} \frac{|z_n|}{z_n} \frac{z_n - z}{1 - \overline{z}_n z}.$$

The function B is then holomorphic in D and bounded (by 1), and its zeros are precisely the z_n. These products will only be used in Sections 12.7 and 12.8, but are nonetheless[2] the first non-trivial elements[3] of H^∞, which in fact they generate, according to Marshall's theorem [66]. In particular, they provide a proof of the non-separability of this algebra (see [120], pp. 379–380).

12.2.8 The Hardy spaces

A brief overview will be sufficient here, and we refer to [53, 66, 90, 112] for the proofs. These spaces are denoted by H^p, and for us, $p \in \{1, 2, \infty\}$. They can be considered either as function spaces on the boundary $\mathbb{T}$ of D, the open unit disk of $\mathbb{C}$, or as spaces of holomorphic functions in D.

- We denote by $H^\infty(D)$ the space of all bounded holomorphic functions on D. Let us now suppose that $p \in \{1, 2\}$, and let $F : D \to \mathbb{C}$ be a holomorphic function. For $0 < r < 1$ and $t \in \mathbb{R}$, we set $F_r(t) = F(re^{it})$ and

$$\|F_r\|_p = \left(\int_\mathbb{T} |F_r|^p \, dm \right)^{1/p}.$$

The space $H^p(D)$ is made up of the F for which

$$\sup_{0<r<1} \|F_r\|_p = \lim_{r \nearrow 1} \|F_r\|_p := \|F\|_{H^p} < +\infty.$$

- Let us now explain why $H^p(D)$ is isometric to a space of functions defined on the circle. If $F \in H^p(D)$, one can show that for almost all t, $F(re^{it})$ admits a finite limit $f^*(e^{it})$ as $r \nearrow 1$, the function f^* being an element of the subspace $H^p(\mathbb{T})$ of $L^p(\mathbb{T})$ consisting of functions $f \in L^p(\mathbb{T})$ such that $\widehat{f}(n) = 0$ for $n < 0$.[4] The space $H^p(\mathbb{T})$ is equipped with the norm $\|f\|_{H^p(\mathbb{T})} := \|f\|_{L^p(\mathbb{T})}$. Hence, we have at hand a linear map

$$H^p(D) \to H^p(\mathbb{T}), \quad F \mapsto f^*.$$

It is bijective, essentially because F can be recovered from f^* by the Poisson extension formula

$$F(z) = \sum_{n=0}^{\infty} \widehat{f^*}(n) z^n.$$

Thus, the Fourier coefficients of f^* are the coefficients of the power series defining F. Moreover, one can verify that

$$\|F\|_{H^p(D)} = \|f^*\|_{H^p(\mathbb{T})}.$$

---

[2] Along with the singular inner function $S(z) = \exp\left(-\dfrac{1+z}{1-z} \right)$.

[3] That is to say, not in the subalgebra $A(D)$.

[4] As usual, the notation $\widehat{f}(n)$ denotes the nth Fourier coefficient of f.

A few specific results about H^1 and H^2 will be necessary. The inner product of $L^2(\mathbb{T})$ – linear on the left! – will be denoted by

$$\langle f, g \rangle = \int_{\mathbb{T}} f(e^{i\theta}) \overline{g(e^{i\theta})} dm(\theta).$$

12.2.9 Theorem   *The space $H^1(\mathbb{T})$ (resp. $H^2(\mathbb{T})$) is the closure in $L^1(\mathbb{T})$ (resp. $L^2(\mathbb{T})$) of the subspace $\mathcal{P}_+$ generated by the $e_n : t \mapsto e^{int}$, $n \geqslant 0$. In particular, $H^2(\mathbb{T})$, equipped with the inner product of $L^2(\mathbb{T})$, is a Hilbert space.*

Proof   First of all, if $f \in H^1(\mathbb{T})$, f is the limit in the sense of Cesàro, in $L^1(\mathbb{T})$, of the sequence of partial sums of its Fourier series, which are in $\mathcal{P}_+$. Conversely, if f is the limit in $L^1(\mathbb{T})$ of a sequence $(p_k)_{k \geqslant 0}$ of elements of $\mathcal{P}_+$, we have, for $n < 0$, $\int_0^{2\pi} p_k(e^{it}) e^{-int} dt = 0$, hence $\widehat{f}(n) = 0$ by passage to the limit. For $H^2(\mathbb{T})$, it is even simpler: the family $(e_n)_{n \in \mathbb{Z}}$, where $e_n(t) = e^{int}$, is a Hilbertian basis of $L^2(\mathbb{T})$, and $H^2(\mathbb{T})$ appears as the orthogonal complement of $\mathrm{Vect}(e_n)_{n<0}$, the latter subspace being none other than the orthogonal complement of $\overline{\mathcal{P}_+}$.   □

12.2.10 Theorem [factorisation theorems]   *(1) Let $f \in H^p$ ($p \in \{1, 2, \infty\}$), f not identically zero, and let $(z_n)_{n \geqslant 1}$ be the sequence of its zeros (counted with multiplicity). Then the sequence $(z_n)_{n \geqslant 1}$ satisfies the Blaschke condition*

$$\sum_{n=1}^{\infty} (1 - |z_n|) < +\infty.$$

Moreover, denoting by B the Blaschke product associated with the sequence $(z_n)_{n \geqslant 1}$, there exists a function $g \in H^p$ satisfying

$$f = Bg \text{ and } \|f\|_p = \|g\|_p.$$

(2) Let $f \in H^1$. Then, there exist $g, h \in H^2$ such that

$$f = gh \text{ and } \|g\|_2 = \|h\|_2 = \|f\|_1^{1/2}.$$

Proof   See [66], p. 51 for the first point (or Exercise 12.5 for the case $p = \infty$). The second point follows easily: we write $f = Bu$, with $u \in H^1$ without zeros, and then $u = g^2$, where $g \in H^2$. It remains to set $h = Bg$.   □

12.2.11 Theorem [Littlewood–Paley identity]   *Let $f \in H^2$. Then, we have*

$$\boxed{\|f\|_2^2 = |f(0)|^2 + \int_D |f'(z)|^2 d\lambda_1(z).} \tag{12.7}$$

Proof The identity (12.7) was shown when f is holomorphic in a neighbourhood of $\overline{D}$ (see (12.5)). Let us detail the generalisation to H^2: write $f(z) = \sum_{n=0}^{\infty} a_n z^n$ and set, for $0 < r < 1$, $f_r(z) = f(rz)$. By (12.5) and Parseval, we have

$$|f(0)|^2 + \int_D r^2 |f'(rz)|^2 d\lambda_1(z) = \int_{\mathbb{T}} |f_r|^2 \, dm = \sum_{n=0}^{\infty} |a_n|^2 r^{2n}.$$

On the contrary,

$$\int_D r^2 |f'(rz)|^2 d\lambda_1(z) = \int_D r^2 |f'(rz)|^2 2 \ln \frac{1}{|z|} dA(z)$$

$$= \int_{rD} |f'(u)|^2 2 \ln \frac{r}{|u|} dA(u)$$

(after the change of variable $u = rz$). By letting $r \to 1$, and using the Beppo Levi theorem, we obtain

$$|f(0)|^2 + \int_D |f'(u)|^2 2 \ln \frac{1}{|u|} dA(u) = \sum_{n=0}^{\infty} |a_n|^2 = \|f\|_2^2,$$

in other words (12.7). $\square$

12.2.12 Duality in Banach spaces

Let X be a Banach space with dual space X^*, E a closed subspace of X with orthogonal space

$$E^{\perp} = \{\psi \in X^* / \psi(e) = 0 \text{ for all } e \in E\},$$

and $X^*/E^{\perp}$ the Banach quotient space of X^* by $E^{\perp}$, the norm on $X^*/E^{\perp}$ being defined by the formula

$$\|\overline{\varphi}\|_{X^*/E^{\perp}} = \inf_{\psi \in E^{\perp}} \|\varphi + \psi\|.$$

The subspace $E^{\perp}$ is none other than the kernel of the restriction map

$$\rho : X^* \to E^*, \varphi \mapsto \varphi_{|E}.$$

The operator ρ thus induces a linear injection

$$\overline{\rho} : X^*/E^{\perp} \to E^*, \overline{\varphi} \mapsto \varphi_{|E},$$

which is bijective by the Hahn–Banach theorem.

The following identity shows that $\bar{\rho}$ is in fact an isometry:

$$\text{if } \varphi \in X^*, \quad \|\bar{\varphi}\|_{X^*/E^\perp} = \sup_{\substack{x \in E \\ \|x\| \leqslant 1}} |\varphi(x)|. \tag{12.8}$$

Proof Denote by σ (resp. τ) the left-hand (resp. right-hand) term of (12.8). With $\psi \in E^\perp$ and $x \in E$ satisfying $\|x\| \leqslant 1$, we have

$$|\varphi(x)| = |\varphi(x) + \psi(x)| \leqslant \|\varphi + \psi\|,$$

hence successively $\tau \leqslant \|\varphi + \psi\|$ and $\tau \leqslant \sigma$. On the contrary, the Hahn–Banach theorem guarantees the existence of $\chi \in X^*$ such that $\|\chi\| = \tau$ and $\chi = \varphi$ on E (χ is a norm-preserving extension of $\varphi_{|E}$). Thus, $\psi = \chi - \varphi \in E^\perp$, and

$$\sigma \leqslant \|\varphi + \psi\| = \|\chi\| = \tau. \qquad \square$$

12.2.13 Remark This identity can be expressed by saying that "the dual space of a subspace is a quotient space". We also have the expected dual expression: "the dual space of a quotient space is a subspace", which translates the fact that the Banach spaces $(X/E)^*$ and $E^\perp$ are isometric. Hence, the formula

$$\|\bar{x}\|_{X/E} = \sup_{\substack{\varphi \in E^\perp \\ \|\varphi\| \leqslant 1}} |\varphi(x)| \text{ for all } x \in X.$$

This identity will be used, notably in Exercise 12.10, and its proof is quite similar to that of (12.8).

12.3 Beurling's theorem

Today the proof of Beurling's theorem is simple and well understood, but over time it turned out to be of fundamental importance and has influenced the whole of operator theory in the second half of the twentieth century, not to mention the recent developments.

First we need a definition. A function $u \in H^2$ is said to be an *inner function* if

$$|u(\zeta)| = 1 \text{ almost everywhere on } \mathbb{T} = \partial D$$

(where the function and its value on the boundary are denoted in the same way). Two typical examples are the Blaschke products and the singular function $u(z) = e^{-\frac{1+z}{1-z}}$. There is a variety of inner functions, and we can describe them all (see [160], p. 342).

12.3.1 Theorem   *The inner functions u are exactly the functions of the form*

$$u(z) = cB(z)\exp\left(-\int_{\mathbb{T}} \frac{\zeta + z}{\zeta - z}d\mu(\zeta)\right),$$

where c is a constant with modulus 1, B a Blaschke product, and μ a positive Borel measure on the circle $\mathbb{T}$, singular with respect to the Lebesgue measure[5] m on that same circle.

The preceding example corresponds to $c = 1$, $B = 1$, $\mu = \delta_1$. We can now state and prove Beurling's theorem ([160], p. 348).

12.3.2 Theorem [Beurling, 1949] *Let S be the "forward shift" on H^2, defined by $Sf(z) = zf(z)$. Then, the closed subspaces of H^2 invariant under S are exactly the subspaces uH^2, where $u \in H^2$ is an inner function.*

Proof   First, remember that we can write the inner product of $f, g \in H^2$ using the boundary values of f and g:

$$\langle f, g \rangle = \int_{\mathbb{T}} f\left(e^{i\theta}\right)\overline{g\left(e^{i\theta}\right)}dm(\theta).$$

This inner product is thus the restriction of that of $L^2(\mathbb{T})$. In particular, if $f(z) = \sum_{n=0}^{\infty} a_n z^n$, we have

$$a_n = \langle f, z^n \rangle = \int_{\mathbb{T}} f\left(e^{i\theta}\right)e^{-in\theta}dm(\theta) = \widehat{f}(n),$$

the nth Fourier coefficient of the boundary value function of f.

If $u \in H^2$ is inner, we have the following important identity, valid for $F, G \in L^2(\mathbb{T})$:

$$\langle uF, uG \rangle = \int_{\mathbb{T}} F\overline{G}|u|^2 dm = \int_{\mathbb{T}} F\overline{G}dm = \langle F, G \rangle. \qquad (12.9)$$

Now let M be a closed and invariant subspace of H^2. We can suppose that $M \neq \{0\}$, which allows us to fix a non-zero element f of M. The proof is broken into *two steps*.

• We first note that

$$zM \subsetneq M.$$

Indeed, if $zM = M$, then $z^n M = M$ for all integers $n \geqslant 0$. In particular, we can write $f = z^n g$ with $g \in H^2$. But the non-zero analytic f cannot

---

[5] This means that μ is concentrated on a Borel set of Lebesgue measure zero.

have a zero of multiplicity arbitrarily large at 0, hence the strict inclusion above. Moreover, zM is closed in M: indeed, as z has modulus one, the map $f \mapsto zf$ is an isometry from M into itself, hence its image is closed in M. The theory of Hilbert spaces thus guarantees the existence of $u \in M$ such that $\|u\|_2 = 1$ and $u \perp zM$ (symbolically, $u \in M \ominus zM$). We still denote by $u : \mathbb{T} \to \mathbb{C}$ the boundary value function of u, and set $v = |u|^2 \in L^1(\mathbb{T})$. If $n \geqslant 1$ is an integer, we have

$$0 = \langle u, z^n u \rangle = \int_{\mathbb{T}} u(\zeta) \, \overline{\zeta^n} \, \overline{u(\zeta)} dm(\zeta) = \int_{\mathbb{T}} v(\zeta) \, \overline{\zeta^n} dm(\zeta) = \widehat{v}(n).$$

Since the function v is real, we also have $\widehat{v}(-n) = \overline{\widehat{v}(n)} = 0$, and the injectivity of the Fourier transformation implies that v is (almost everywhere) equal to $\widehat{v}(0)$. Moreover, $\widehat{v}(0) = \int_{\mathbb{T}} v(\zeta) dm(\zeta) = \|u\|_2^2 = 1$. In other words, $|u| = 1$ almost everywhere on the boundary, and u is an inner function. Furthermore, we obviously have

$$uH^2 \subset M \tag{12.10}$$

since the polynomials are dense in H^2, and since M is invariant and closed.
• We have in fact

$$uH^2 = M. \tag{12.11}$$

To see this, first note that uH^2 is closed in H^2, as it is the image of the isometry $f \mapsto uf$. Next (one never changes a winning tactic!), let $f \in M$ again be such that $f \perp uH^2$. We have

$$\langle f\overline{u}, z^n \rangle = \langle f, z^n u \rangle = 0 \text{ for } n \geqslant 0,$$

and

$$\langle f\overline{u}, z^{-n} \rangle = \langle z^n f, u \rangle = 0 \text{ for } n \geqslant 1.$$

For the first equation, we used the fact that $f \perp uH^2$; for the second, the fact that $u \perp zM$. We also used the unimodularity of u and z^n on the boundary, as well as (12.9). It ensues that $\widehat{f\overline{u}}(n) = 0$ for $n \in \mathbb{Z}$, hence $f\overline{u} = 0$ and $f = 0$ in $L^2(\mathbb{T})$, as $|u| = 1$ almost everywhere on $\mathbb{T}$. Hence, $M \ominus uH^2 = \{0\}$, and this indeed implies (12.11) as announced. The proof (orthogonality twice) is complete. $\qquad\square$

12.3.3 Remark The proof of Beurling's theorem gives the impression of a simple checkmate in two moves, but behind it lies the whole theory of Hardy spaces – notably the inner and outer functions, developed (by Beurling) essentially to find *all* the subspaces of H^2 invariant under the shift operator.

12.3.4 Remark A word about outer functions: we have just seen that the spaces "that do not change" under the action of S are given by the inner functions. Conversely, which are the vectors f that "change a lot" under this action, that is, if they exist, the vectors f whose iterates by S, the $S^n f$ $(n \geqslant 0)$, generate a dense subspace in H^2? The answer is provided by another theorem of Beurling ([160], p. 344).

12.3.5 Theorem [Beurling, 1949]   *Let S be the* forward shift *on H^2, defined by $Sf(z) = zf(z)$. Then, the vectors $f \in H^2$ that are cyclic for S are exactly the outer functions, that is, the zero-free functions of H^2 satisfying the case of equality in Jensen's inequality:*

$$\log |f(0)| = \int_{\mathbb{T}} \log |f(\zeta)| dm(\zeta). \tag{12.12}$$

We point out that, for $f \in H^1$ not identically zero, we always have Jensen's inequality (see [160], p. 344):

$$\log |f(0)| \leqslant \int_{\mathbb{T}} \log |f(\zeta)| dm(\zeta).$$

Here are some examples of outer functions and a few remarks.

- The function $f(z) = (1 - z)^\alpha$, where $\alpha > -1/2$, is cyclic. Indeed, the two quantities of (12.12) are 0. A direct verification is possible when α is an integer (see Exercise 12.17), but appears difficult for the other cases.
- The function $f(z) = (1 - z)e^{-\frac{1+z}{1-z}}$ is not cyclic. Indeed, the left-hand side of (12.12) is then -1 and the right-hand side is 0. Yet f is without zeros in D and continuous on $\overline{D}$.
- A good player is always lucky, but only up to a point. The complete description of subspaces of the Bergman space $\mathcal{B}^2$ or the Dirichlet space $\mathcal{D}$ that are invariant under the shift operator is unknown to this day.
- As we will see further on with the applications of interpolating sequences, the orthogonal complements in H^2 of invariant subspaces uH^2, that is, the *model* spaces $K_u = H^2 \ominus uH^2$, are exactly the invariant subspaces of S^*, the adjoint of the shift. Their study is at the core of the work of N. Nikolskii and his students.
- Sarason [165] observed that, modulo the Paley–Wiener theorem, the knowledge of all the invariant subspaces of H^2 under S implied that of all the invariant subspaces of $L^2([0, 1])$ under the Volterra operator V, defined by $Vf(x) = \int_0^x f(t)dt$. These are the subspaces of the form $L^2([a, 1])$[6],

---

[6] We mean by this the set of functions of $L^2([0, 1])$ that are zero (almost everywhere) on $[0, a]$.

$0 \leqslant a \leqslant 1$. Even if we could prove this theorem differently [50], a nice consequence is the following theorem (see [191], p. 166 or [118], p. 280), all the more remarkable since rings of functions are very rarely integral domains (except in the case of (quasi-) analytic functions).

12.3.6 Theorem [Titchmarsh, 1942]   *Let f, g be two continuous functions on $\mathbb{R}^+$ and $f * g$ their Volterra convolution product, defined by*

$$f * g(x) = \int_0^x f(x - t)g(t)dt \text{ for } x \geqslant 0.$$

If $f * g = 0$, then $f = 0$ or $g = 0$.

12.4 The Lagrange–Carleson problem for an infinite sequence

12.4.1 Lagrange and Schwarz–Pick interpolation

Let us recall the principle of Lagrange interpolation: we consider a finite sequence $z_1, \ldots, z_n$ of complex numbers, pairwise distinct, the *base points*. Let $w_1, \ldots, w_n$ be n other complex numbers known as the *data*. We seek a polynomial f such that $f(z_j) = w_j$ for $1 \leqslant j \leqslant n$. To do this, Lagrange starts with a *mother function*, the polynomial $L(z) = \prod_{p=1}^n (z - z_p)$, and then considers its *children*, the polynomials $L_k(z) = \dfrac{L(z)}{(z - z_k)L'(z_k)}$, that satisfy $L_k(z_j) = \delta_{k,j}$ (Kronecker symbol). Finally he interpolates the sequence $w_1, \ldots, w_n$ by the polynomial $f = \sum_{k=1}^n w_k L_k$, of degree $\leqslant n - 1$, for which we indeed have

$$f(z_j) = \sum_{k=1}^n w_k \delta_{k,j} = w_j \text{ for } 1 \leqslant j \leqslant n.$$

Now let $(z_j)_{j \geqslant 1}$ be an *infinite* sequence of distinct points in the open unit disk D of $\mathbb{C}$, without any limit point in D or, if we prefer, such that $|z_j| \to 1$: the principle of isolated zeros makes this restriction indispensable. Given an *arbitrary* sequence $(w_j)_{j \geqslant 1}$ of complex numbers, it is always possible (the Mittag–Leffler problem) to find a function f, analytic in D, such that $f(z_j) = w_j$ for all $j \geqslant 1$.

For this, we can mimic Lagrange's method by adding a convergence factor, as done in the proof of Borel's theorem on the C^∞ functions (see Exercise 12.9 for details). This problem of interpolation with an infinite sequence of data

$(w_j)_{j \geqslant 1}$ thus has a solution, but this time the unknown function f is sought within the *analytic functions* on D, and not only within the polynomials.

Matters become seriously more difficult as soon as we try to quantify the result, that is, to impose f bounded when the sequence $(w_j)_{j \geqslant 1}$ is itself bounded. It is worth mentioning a first fundamental interpolation result ([66], p. 7); in fact, Carleson began his 1958 article by citing it. Note that, for $n = 2$, the following condition corresponds to the Schwarz–Pick inequality for an analytic function $f : D \to D$, namely $d(f(z_1), f(z_2)) \leqslant d(z_1, z_2)$, where d is the pseudo-hyperbolic distance on D.

12.4.2 Theorem [Schwarz–Pick]   *Let $z_1, \ldots, z_n \in D$ be pairwise distinct and $w_1, \ldots, w_n \in \mathbb{C}$. A necessary and sufficient condition for the interpolation problem*

$$f(z_j) = w_j \text{ for } 1 \leqslant j \leqslant n$$

to have a solution f in the unit ball of H^{∞} (i.e., $|f(z)| \leqslant 1$ for $z \in D$) is for the Hermitian matrix

$$H := \left[\frac{1 - w_i \overline{w_j}}{1 - z_i \overline{z_j}} \right]_{1 \leqslant i, j \leqslant n} =: [h_{ij}]$$

to be positive, in the sense that $\sum_{1 \leqslant i, j \leqslant n} h_{i,j} \lambda_i \overline{\lambda_j} \geqslant 0$ for all $\lambda_1, \ldots, \lambda_n \in \mathbb{C}$.

We thus see that interpolation without increasing the norm imposes severe constraints on the data $w_1, \ldots, w_n$. Let us return to our problem of infinite interpolation, without imposing constraints other than the boundedness of the sequence of data $(w_j)_{j \geqslant 1}$, but with a weakened condition on the norm of the interpolating function: it has the right to grow, but by no more than a constant factor.

The least we can do is to impose what is known as the *Blaschke condition* (as we do in the remainder of this section):

$$\sum_{j=1}^{\infty} (1 - |z_j|) < \infty. \tag{12.13}$$

Indeed, if we wish to interpolate the sequence w defined by $w_1 = 1$ and $w_j = 0$ for $j \geqslant 2$ by a bounded analytic function f, condition (12.13) is necessary – as verified by the zeros of a bounded analytic function on D (or even by a function of H^p, $p > 0$), not identically zero (see Theorem 12.2.10), for instance $(z - z_1)f$.

Thus let us fix, once and for all, a sequence $(z_j)_{j \geq 1}$ satisfying (12.13), and introduce a few classical notations and a property before getting to the heart of the matter.

- We denote by d the pseudo-hyperbolic distance on D, defined by

$$d(a, b) = \left| \frac{a - b}{1 - \bar{a}b} \right|.$$

We always have $d(a, b) < 1$.

- Let B be the infinite Blaschke product associated with the z_j, b_n its nth factor, and $B_n = \prod_{j \neq n} b_j$ the product B deprived of its nth factor b_n. We also set $\beta_n = c_n b_1 \cdots b_n$, the normalised partial product of order n of B where c_n is a constant of modulus 1. Thus:

$$B(z) = \prod_{j=1}^{\infty} \frac{|z_j|}{z_j} \frac{z_j - z}{1 - \bar{z}_j z}, \quad B_n(z) = \prod_{j \neq n} \frac{|z_j|}{z_j} \frac{z_j - z}{1 - \bar{z}_j z}$$

and

$$\beta_n(z) = \prod_{j=1}^{n} \frac{z_j - z}{1 - \bar{z}_j z}.$$

Recall that the *parasite* factor $\dfrac{|z_j|}{z_j}$ is necessary to force convergence of B or B_n, but can disappear when we consider β_n. If $z_j = 0$, the jth factor of the product B is replaced by z.

- The quantity $\delta_n := \prod_{j \neq n} d(z_j, z_n) = |B_n(z_n)|$ denotes the product of the pseudo-hyperbolic distances between z_n and the other points of the sequence.

- We have the property

$$\delta_n = (1 - |z_n|^2)|B'(z_n)|. \tag{12.14}$$

Indeed, since $B(z_n) = 0$, we can write

$$\frac{B(z) - B(z_n)}{z_n - z} = \frac{|z_n|}{z_n(1 - \bar{z}_n z)} \prod_{j \neq n} \frac{|z_j|}{z_j} \frac{z_j - z}{1 - \bar{z}_j z}$$

and the result follows by taking the modulus and letting $z \to z_n$.

- The sequence $(z_j)_{j \geq 1}$ is said to be *separated* (resp. *uniformly separated*) if it satisfies $\inf_{j \neq k} d(z_j, z_k) > 0$ (resp. $\inf_{n \geq 1} \delta_n =: \delta > 0$).

12.4.3 The Carleson interpolation theorem

With the notations of the preceding section, we can state and prove a fundamental result of Carleson ([66], p. 278; see also [35]), which shows how the interpolation problem ceases to be formal (it is the same with Borel's C^∞ theorem) as soon as we impose growth conditions on the interpolating function f.

12.4.4 Theorem [Carleson, 1958] *Let $(z_j)_{j \geqslant 1}$ be an infinite sequence of distinct points of D. The following conditions are equivalent:*

(i) *any* bounded *sequence* $w = (w_j)_{j \geqslant 1}$ *can be interpolated by a* bounded *analytic function f, that is, there exists $f \in H^\infty$ satisfying $f(z_j) = w_j$ for $j \geqslant 1$;*

(ii) *the sequence $(z_j)_{j \geqslant 1}$ is* uniformly separated, *in the sense that*

$$\delta := \inf_{n \geqslant 1} \delta_n = \inf_{n \geqslant 1} \prod_{j \neq n} d(z_j, z_n) > 0.$$

When either of the equivalent conditions of Theorem 12.4.4 is satisfied, we say that $(z_j)_{j \geqslant 1}$ is an *interpolating sequence*.

The implication (i) $\Rightarrow$ (ii) is a gem of an application of the open mapping theorem ([153], p. 200). Indeed, the hypothesis states that the operator $T :$ $H^\infty \to \ell^\infty$ defined by $T(f) = (f(z_j))_{j \geqslant 1}$ is surjective.[7] Thus, by the cited theorem, there exists a constant $C > 0$ such that, for all $w \in \ell^\infty$, the equation $T(f) = w$ has at least one solution f satisfying $\|f\|_\infty \leqslant C \|w\|_\infty$. Let us test this information on a w "à la Lagrange": $w_n = 1$ and $w_j = 0$ if $j \neq n$. We obtain a function $f_n \in H^\infty$ such that

$$f_n(z_n) = 1, \ f_n(z_j) = 0 \text{ if } j \neq n \text{ and } \|f_n\|_\infty \leqslant C.$$

By Theorem 12.2.10 (or Exercise 12.5), we can write $f_n = B_n g_n$, where $g_n \in H^\infty$ and $\|g_n\|_\infty \leqslant C$. We thus have

$$1 = |f_n(z_n)| = |B_n(z_n)| \cdot |g_n(z_n)| \leqslant C |B_n(z_n)|,$$

hence $|B_n(z_n)| \geqslant C^{-1}$. But $|B_n(z_n)|$ is none other than the quantity that we have called δ_n, and hence $\delta = \inf_{n \geqslant 1} \delta_n \geqslant C^{-1}$.

This δ is called the *uniform separation constant* of the sequence $(z_n)_{n \geqslant 1}$ (we always have $\delta < 1$), whereas the best C is called the *interpolation constant* of $(z_n)_{n \geqslant 1}$.

---

[7] Observe that by the definition of the norms involved here, we have $\|T\| \leqslant 1$.

The implication (ii) $\Rightarrow$ (i) is considerably more difficult. The tendency today is to use proofs that are purely Hilbertian. The space H^∞ disappears, and one resolves the following interpolation problem: given a sequence $(w_j)_{j\geqslant 1}$ such that $\sum_{j=1}^\infty (1 - |z_j|^2)|w_j|^2 < \infty$, find $f \in H^2$ such that $f(z_j) = w_j$ for $j \geqslant 1$. We restrict ourselves here to (a variant of) the initial proof, which appears direct and instructive. This proof requires the notion of a Carleson measure, but we can now do without the Carleson embedding Theorem 12.7.2, in the special case that interests us [132]. For this, we need the following crucial lemma.

12.4.5 Lemma [Carleson]   *Let the sequence* $(z_n)_{n\geqslant 1}$ *be uniformly separated, with uniform separation constant* δ, *and set* $d_j = 1 - |z_j|^2$ *for* $j \geqslant 1$. *Then the discrete measure* $\mu = \sum_{j=1}^\infty (1 - |z_j|^2)\delta_{z_j}$ *is a Carleson measure, that is, there exists a constant* $C_\delta > 0$ *(we can take* $C_\delta = 32\,\delta^{-4}$*) such that the following inequalities hold:*

$$\int_D |F|^2 d\mu = \sum_{j=1}^\infty d_j |F(z_j)|^2 \leqslant C_\delta \|F\|_2^2 \text{ for } F \in H^2, \tag{12.15}$$

$$\int_D |F| d\mu = \sum_{j=1}^\infty d_j |F(z_j)| \leqslant C_\delta \|F\|_1 \text{ for } F \in H^1. \tag{12.16}$$

Proof (according to Neville). First, note that the mere δ-*separation* of the sequence $(z_n)_{n\geqslant 1}$[8] is sufficient to imply the following inequality, interesting in itself:

$$\sum_{j=1}^\infty d_j^3 |F'(z_j)|^2 \leqslant 32\,\delta^{-2} \|F\|_2^2 \text{ for } F \in H^2. \tag{12.17}$$

To see this, note that

$$|z_j - z_k| \geqslant \frac{\delta}{2}(d_j + d_k) \text{ for } j \neq k. \tag{12.18}$$

Indeed, the hypothesis of δ-separation on the z_j gives

$$|z_j - z_k| \geqslant \delta|1 - \overline{z_j}z_k| \geqslant \delta(1 - |z_j|\,|z_k|) \geqslant \frac{\delta}{2}(d_j + d_k)$$

since $2|z_j|\,|z_k| \leqslant |z_j|^2 + |z_k|^2$. The Euclidean disks $\Delta_j := D\left(z_j, \frac{\delta}{4}d_j\right)$ are thus pairwise disjoint, and contained in D as $d_j < 2(1 - |z_j|)$ and

$$|z_j| + \frac{\delta}{4}d_j \leqslant |z_j| + \frac{d_j}{2} < |z_j| + 1 - |z_j| = 1.$$

---

[8]   In the sense where $d(z_j, z_k) \geqslant \delta > 0$ for $j \neq k$.

Now let $F \in H^2$. By setting $G = F'^2 \in H(D)$, we have[9]

$$|G(a)| \leqslant \frac{1}{r^2} \int_{D(a,r)} |G(z)| dA(z) \text{ for all disks } D(a,r) \subset D,$$

$dA(z) = \pi^{-1} d\lambda(z)$ denoting the normalised Lebesgue measure on D. In particular,

$$d_j^3 |F'(z_j)|^2 \leqslant 16 \delta^{-2} d_j \int_{\Delta_j} |F'(z)|^2 dA(z)$$

$$\leqslant 32 \delta^{-2} \int_{\Delta_j} (1 - |z|^2) |F'(z)|^2 dA(z). \qquad (12.19)$$

Let us justify this latter inequality. If $z \in \Delta_j$, we have

$$
\begin{aligned}
d_j &= 1 - |z|^2 + |z|^2 - |z_j|^2 \\
&\leqslant 1 - |z|^2 + \left(|z| + |z_j| \right) \left| |z| - |z_j| \right| \\
&\leqslant 1 - |z|^2 + 2|z - z_j| \\
&\leqslant 1 - |z|^2 + \frac{1}{2} d_j,
\end{aligned}
$$

and hence $d_j \leqslant 2(1 - |z|^2)$. By taking into account the pairwise disjointness of the Δ_j, we sum up the inequalities (12.19) to obtain

$$\sum_{j=1}^{\infty} d_j^3 |F'(z_j)|^2 \leqslant 32 \delta^{-2} \int_D (1 - |z|^2) |F'(z)|^2 dA(z).$$

Then, to deduce (12.17), we observe that

$$\int_D (1 - |z|^2) |F'(z)|^2 dA(z) \leqslant \|F\|_2^2.$$

This is, in a way, an application of the Littlewood–Paley identity (12.7), but we can obtain this inequality directly via Parseval's formula. For this, we write $F(z) = \sum_{n=0}^{\infty} c_n z^n$. Then, Parseval and a passage to polar coordinates give

[9] Indeed, Cauchy's formula gives, for $\rho < r$:

$$|G(a)| = \left| \frac{1}{2\pi} \int_0^{2\pi} G(a + \rho e^{it}) dt \right| \leqslant \frac{1}{2\pi} \int_0^{2\pi} |G(a + \rho e^{it})| dt$$

(this is the *subharmonicity of* $|G|$). Multiplying by ρ, and then integrating in ρ over $[0, r[$ and using polar coordinates, we obtain

$$\frac{r^2}{2} |G(a)| \leqslant \frac{1}{2\pi} \int_0^r \int_0^{2\pi} |G(a + re^{it})| \rho \, dt \, d\rho = \frac{1}{2\pi} \int_{D(a,r)} |G(z)| d\lambda(z),$$

hence the result.

$$\int_D (1 - |z|^2)|F'(z)|^2 dA(z) = \int_0^1 2(1 - r^2) \sum_{n=1}^\infty n^2 |c_n|^2 r^{2n-1} dr$$

$$= \sum_{n=1}^\infty \frac{n^2 |c_n|^2}{n(n+1)} \leqslant \sum_{n=1}^\infty |c_n|^2 \leqslant \|F\|_2^2,$$

which proves (12.17).

When the sequence $(z_j)_{j\geqslant 1}$ is in addition *uniformly* separated, *Neville's idea* is to apply (12.17) not to F, but to $G = BF \in H^2$ when $F \in H^2$. By taking into account the relation $G'(z_j) = B'(z_j)F(z_j)$, where the derivative no longer involves F, we obtain

$$\sum_{j=1}^\infty d_j^3 |B'(z_j)|^2 |F(z_j)|^2 \leqslant 32\,\delta^{-2}\|G\|_2^2 = 32\,\delta^{-2}\|F\|_2^2.$$

However, according to (12.14) we have $|B'(z_j)| \geqslant \dfrac{\delta}{d_j}$, hence we obtain (12.15):

$$\sum_{j=1}^\infty d_j |F(z_j)|^2 \leqslant 32\,\delta^{-4}\|F\|_2^2.$$

Next, we prove (12.16) as follows. Write $F \in H^1$ in the form $F = GH$ with $G, H \in H^2$ such that $\|F\|_1 = \|G\|_2\|H\|_2$. With this, Cauchy–Schwarz and (12.15) imply

$$\sum_{j=1}^\infty d_j |F(z_j)| \leqslant \left(\sum_{j=1}^\infty d_j |G(z_j)|^2\right)^{1/2} \left(\sum_{j=1}^\infty d_j |H(z_j)|^2\right)^{1/2}$$

$$\leqslant 32\,\delta^{-4}\|G\|_2\|H\|_2 = 32\,\delta^{-4}\|F\|_1. \qquad \square$$

12.4.6 Remark More generally, a *Carleson measure* for H^p when $p = 1$ or 2 is a positive measure on D such that

$$\|f\|_{L^p(\mu)} \leqslant C_\delta \|f\|_p \text{ for } f \in H^p.$$

Thus the lemma shows, for a uniformly separated sequence $(z_j)_{j\geqslant 1}$ ($\delta > 0$), that the discrete measure $\mu = \sum_{j=1}^\infty (1 - |z_j|^2)\delta_{z_j}$ (or the equivalent measure $\mu = \sum_{j=1}^\infty (1 - |z_j|)\delta_{z_j}$) is a Carleson measure. It is in fact while reflecting on this special case that Carleson derived the general notion of a Carleson measure for a Hilbert space of analytic functions on D, a central notion appearing today in many problems where complex analysis and functional analysis interact.

We are now going to exploit Lemma 12.4.5 using the duality formula (12.8):

$$\|\overline{\varphi}\|_{X^*/E^\perp} = \sup_{\substack{x \in E \\ \|x\| \leqslant 1}} |\varphi(x)| \text{ for } \varphi \in X^*,$$

applied to the spaces $X = L^1$, $X^* = L^\infty$, $E = H^1$ and $E^\perp = H^\infty$. Here we use the following $L^1 - L^\infty$ duality:

$$\langle f, F \rangle = \frac{1}{2i\pi} \int_{\mathbb{T}} f(z)F(z)dz = \frac{1}{2\pi} \int_0^{2\pi} f(e^{it})F(e^{it})e^{it} dt$$

for $f \in L^\infty$ and $F \in L^1$. Now, if $e_n \in L^1$ is defined by $e_n(t) = e^{int}$, the relation $\langle f, e_n \rangle = \widehat{f}(-n-1)$ shows that

$$f \in E^\perp \Leftrightarrow \widehat{f}(-n-1) = 0 \text{ for } n \geqslant 0$$
$$\Leftrightarrow f \in H^\infty.$$

Now let $w = (w_j)_{j \geqslant 1} \in \ell^\infty$. We seek $f \in H^\infty$ such that $f(z_j) = w_j$ for $j \geqslant 1$. The formula "à la Lagrange" (with the mother function B)

$$f(z) = \sum_{k=1}^\infty w_k \frac{B(z)}{(z - z_k)B'(z_k)}$$

does not pose many problems of convergence since[10] $|B'(z_k)|^{-1} \leqslant \frac{d_k}{\delta}$. However, it poses serious problems for the L^∞-norm of f. We thus proceed differently: when the integer n is fixed, we can of course solve within H^∞ (and even within the polynomials) the *finite* interpolation problem

$$f_n(z_j) = w_j \text{ for } 1 \leqslant j \leqslant n. \tag{12.20}$$

But we are going to solve it with uniform control of the norm: $\|f_n\|_\infty \leqslant M$, where the constant M does not depend on n. Next, by Montel's theorem for normal families, we can suppose, modulo extraction, that the f_n converge uniformly on all compact subsets of D to $f \in H^\infty$ such that $\|f\|_\infty \leqslant M$. A passage to the limit in (12.20), when $n \to \infty$ with j fixed, gives $f(z_j) = w_j$ for $j \geqslant 1$.

Thus, for a fixed $n \geqslant 1$, let $g = g_n \in H^\infty$ be a solution of (12.20), for example a well-chosen polynomial. We denote by E_n the set of functions of H^∞ satisfying (12.20). It is clear that $E_n = g + \beta_n H^\infty := \{g + \beta_n h, \ h \in H^\infty\}$. By definition of the quotient norm, since $|\beta_n| = 1$ on $\mathbb{T}$ and since (by the

---

[10] Remember that $\sum_{k=1}^\infty (1 - |z_k|) < +\infty$.

maximum principle) the L^∞-norm can be evaluated equally well on the circle or in the disk, we have

$$M_n := \inf_{f \in E_n} \|f\|_\infty = \inf_{h \in H^\infty} \left\| \frac{g}{\beta_n} + h \right\|_\infty = \left\| \frac{g}{\beta_n} \right\|_{L^\infty / H^\infty}.$$

However, by the duality relation mentioned above, and by the residue theorem,[11] this quotient norm is none other than the quantity

$$M_n = \sup_{\substack{F \in H^1, \\ \|F\|_1 \leqslant 1}} \left| \frac{1}{2i\pi} \int_{\mathbb{T}} \frac{g(z) F(z)}{\beta_n(z)} dz \right|$$

$$= \sup_{\substack{F \in H^1, \\ \|F\|_1 \leqslant 1}} \left| \sum_{j=1}^n \frac{g(z_j) F(z_j)}{\beta'_n(z_j)} \right|$$

$$= \sup_{\substack{F \in H^1, \\ \|F\|_1 \leqslant 1}} \left| \sum_{j=1}^n \frac{w_j F(z_j)}{\beta'_n(z_j)} \right|.$$

However, the same calculation as for (12.14) shows that, with $1 \leqslant j \leqslant n$,

$$(1 - |z_j|^2)|\beta'_n(z_j)| = \prod_{k \leqslant n, k \neq j} d(z_k, z_j) \geqslant \delta_j \geqslant \delta$$

and we thus obtain:[12]

$$M_n \leqslant \delta^{-1} \|w\|_\infty \sup_{\substack{F \in H^1, \\ \|F\|_1 \leqslant 1}} \sum_{j=1}^n d_j |F(z_j)| \leqslant \delta^{-1} \|w\|_\infty \sup_{\substack{F \in H^1, \\ \|F\|_1 \leqslant 1}} \sum_{j=1}^\infty d_j |F(z_j)|.$$

At this stage, an inequality of the type

$$\sum_{j=1}^\infty d_j |F(z_j)| \leqslant C \|F\|_1 \text{ for } F \in H^1 \qquad (12.21)$$

would be most useful. Precisely, this inequality is provided by Lemma 12.4.5 with $C = 32\,\delta^{-4}$, and we finally obtain $M_n \leqslant 32\,\delta^{-5} \|w\|_\infty$, which completes the proof of the Carleson interpolation theorem.

12.4.7 Examples

The interest of the Carleson condition is that it is (sometimes) verifiable. For example, here is a sufficient condition for the z_j to form an interpolating sequence (see [90], p. 203).

---

[11] So the residue theorem is more than just a tool to calculate integrals by the dozen!
[12] Remember that $d_j = 1 - |z_j|^2$.

12.4.8 Theorem [Hayman–Newman]   *Let* $(z_j)_{j \geqslant 1}$ *be a sequence of points of D such that*

$$\frac{1 - |z_{j+1}|}{1 - |z_j|} \leqslant c < 1 \text{ for } j \geqslant 1.$$

Then, the sequence $(z_j)_{j \geqslant 1}$ *is an interpolating sequence and its uniform separation constant* δ *satisfies*

$$\delta \geqslant \prod_{n=1}^{\infty} \left(\frac{1 - c^n}{1 + c^n} \right)^2 =: P_c. \tag{12.22}$$

In particular, the sequence $(1 - c^j)_{j \geqslant 1}$ *is an interpolating sequence.*

Proof   Let j and n be integers such that $j > n \geqslant 1$. First of all, we have

$$1 - |z_j| \leqslant c^{j-n}(1 - |z_n|),$$

hence

$$|z_j| - |z_n| \geqslant (1 - |z_n|)(1 - c^{j-n})$$

and

$$\begin{aligned}
1 - |z_j| \, |z_n| &\leqslant 1 - |z_n| \left(1 - c^{j-n}(1 - |z_n|) \right) \\
&= 1 - |z_n| + c^{j-n}|z_n|(1 - |z_n|) \\
&\leqslant 1 - |z_n| + c^{j-n}(1 - |z_n|) \\
&= (1 - |z_n|)(1 + c^{j-n}).
\end{aligned}$$

We then deduce that

$$d\left(|z_j|, |z_n| \right) \geqslant \frac{1 - c^{j-n}}{1 + c^{j-n}} \text{ for } j > n \geqslant 1.$$

By using the inequality[13]

$$d(a, b) \geqslant d(|a|, |b|) \text{ for } a, b \in D$$

---

[13] Easy to obtain:

$$1 - d(a, b)^2 = \frac{\left(1 - |a|^2 \right) \left(1 - |b|^2 \right)}{|1 - \bar{a}b|^2} \leqslant \frac{\left(1 - |a|^2 \right) \left(1 - |b|^2 \right)}{(1 - |a|\,|b|)^2} = 1 - d(|a|, |b|)^2.$$

and by permuting the roles of n and j when $j < n$, we thus obtain

$$\delta_n = \prod_{j<n} d(z_j, z_n) \cdot \prod_{j>n} d(z_j, z_n)$$

$$\geq \prod_{j<n} d(|z_j|, |z_n|) \cdot \prod_{j>n} d(|z_j|, |z_n|)$$

$$\geq \prod_{j<n} \left(\frac{1 - c^{n-j}}{1 + c^{n-j}} \right) \cdot \prod_{j>n} \left(\frac{1 - c^{j-n}}{1 + c^{j-n}} \right) \geq P_c,$$

which completes the proof of the theorem. □

The preceding condition is only sufficient. It becomes necessary and sufficient when the z_j are on the same radius of D (see [90], p. 204).

12.4.9 Theorem [Hayman–Newman] *Let $(z_j)_{j \geq 1}$ be an increasing sequence in $]0, 1[$. Then it is an interpolating sequence if and only if there exists a constant $c \in]0, 1[$ such that*

$$\frac{1 - z_{j+1}}{1 - z_j} \leq c \text{ for } j \geq 1.$$

Proof If $(z_j)_{j \geq 1}$ is an interpolating sequence, it is uniformly separated, thus *a fortiori* separated, and there exists a $\delta \in]0, 1[$ such that

$$\delta \leq \frac{z_{j+1} - z_j}{1 - z_{j+1}z_j} \text{ for } j \geq 1.$$

A fortiori, we have $\delta \leq \dfrac{z_{j+1} - z_j}{1 - z_j}$, or $z_{j+1} \geq z_j + \delta(1 - z_j)$, and hence

$$\frac{1 - z_{j+1}}{1 - z_j} \leq 1 - \delta \text{ for } j \geq 1.$$

The converse results from Theorem 12.4.8. □

The interest of the preceding result is also to provide examples of sequences of points of D that, while tending quite quickly to the boundary, are not interpolating sequences such as, for example, the sequences $z_j = 1 - j^{-2}$ or $z_j = 1 - e^{-\sqrt{j}}$ (both of which, nonetheless, satisfy the Blaschke condition $\sum_{j=1}^{\infty} (1 - |z_j|) < \infty$). We see that not just any old sequence can be an interpolating sequence! That said, a very fine result of Naftalevitch (see Exercise 12.14) gives us the following theorem.

12.4.10 Theorem [Naftalevitch]   *Let* $(z_j)_{j \geqslant 1}$ *be a Blaschke sequence of D, that is, satisfying* $\sum_{j=1}^{\infty}(1 - |z_j|) < \infty$. *Then there exists an interpolating sequence* $(z'_j)_{j \geqslant 1}$ *such that* $|z'_j| = |z_j|$ *for* $j \geqslant 1$.

12.4.11 General remarks on interpolation sequences

12.4.12 Remark   In [160] (p. 313), Rudin qualifies as *surprisingly clear-cut* the necessary and sufficient condition $\sum_{n=1}^{\infty} \dfrac{1}{\lambda_n} = \infty$ of the density theorem of Müntz–Szasz. We could make the same remark about Carleson's interpolation condition, that we could otherwise be tempted to replace by two even simpler conditions: "$(z_n)_{n \geqslant 1}$ is separated and is a Blaschke sequence".

  (i) $\inf_{j \neq k} d(z_j, z_k) > 0.$
 (ii) $\sum_{j=1}^{\infty}(1 - |z_j|) < \infty.$

Exercise 12.12 shows that this is not possible. However, by reinforcing the Blaschke condition we have the following result.

12.4.13 Theorem   *Let* $(z_n)_{n \geqslant 1}$ *be a sequence of distinct points of D. The following conditions are equivalent:*

  (i) the sequence $(z_n)_{n \geqslant 1}$ *is an interpolating sequence;*
 (ii) the sequence $(z_n)_{n \geqslant 1}$ *is separated and the finite measure* $\mu = \sum_{j=1}^{\infty}$
       $(1 - |z_j|^2)\delta_{z_j}$ *is a Carleson measure for* H^2.

Proof   If $(z_n)_{n \geqslant 1}$ is an interpolating sequence, it is uniformly separated, and so *a fortiori* separated, as we saw in the proof of Theorem 12.4.4. Moreover, the measure $\mu = \sum_{j=1}^{\infty}(1 - |z_j|^2)\delta_{z_j}$ is a Carleson measure, as shown by Neville's method.

  Conversely, suppose that the sequence $(z_n)_{n \geqslant 1}$ is separated and μ is a Carleson measure. Let us test the Carleson character of μ on the normalised reproducing kernel[14] of H^2, that is, the functions $f_n(z) = \dfrac{\sqrt{1 - |z_n|^2}}{1 - \overline{z_n}z}$, that are in the unit ball of H^2. With C_μ a positive constant, we obtain

$$\int_D |f_n|^2 d\mu = \sum_{j=1}^{\infty}(1 - |z_j|^2)|f_n(z_j)|^2 = \sum_{j=1}^{\infty} \frac{(1 - |z_n|^2)(1 - |z_j|^2)}{|1 - \overline{z_n}z_j|^2} \leqslant C_\mu + 1,$$

---

[14]   This is what is known as the reproducing kernel thesis (RKT). Indeed, testing the Carleson nature of μ does not need more, even though the closed convex hull of the f_n is very far from filling the unit ball of H^2.

or again[15]

$$\sum_{j \neq n} \left(1 - d(z_j, z_n)^2\right) \leqslant C_\mu$$

and *a fortiori*

$$\sum_{j \neq n} \left(1 - d(z_j, z_n)\right) \leqslant C_\mu.$$

By hypothesis, there exists a $\delta \in \,]0, 1[$ such that

$$d(z_j, z_n) \geqslant \delta \text{ for } j \neq n,$$

and we know that there exists a constant $C_\delta > 0$ such that

$$\frac{\log(1 - x)}{x} \geqslant -C_\delta \text{ for } 0 < x \leqslant 1 - \delta.$$

Then, with $x = 1 - d(z_j, z_n)$, we deduce that

$$\log \frac{1}{d(z_j, z_n)} \leqslant C_\delta \left(1 - d(z_j, z_n)\right) \text{ for } j \neq n.$$

It ensues that

$$\sum_{j \neq n} \log \frac{1}{d(z_j, z_n)} \leqslant C_\delta C_\mu,$$

or again

$$\prod_{j \neq n} d(z_j, z_n) \geqslant e^{-C_\delta C_\mu}.$$

This completes the proof of Theorem 12.4.13. $\qquad\qquad\qquad\qquad\square$

12.4.14 Remark The inequality (12.21) is at first sight so little evident that Newman (see [90], p. 196) added it as a necessary condition to obtain the interpolation! It was Carleson that showed this condition to be superfluous, and a consequence of the hypothesis $\delta = \inf_{n \geqslant 1} \delta_n > 0$. His proof is nontrivial, and, as we have already said, led him to defining the concept of what is today called a Carleson measure (see Section 12.7).

12.4.15 Remark When the sequence $(z_j)_{j \geqslant 1}$ satisfies the uniform separation condition $\delta = \inf_{n \geqslant 1} \delta_n > 0$, a gem of a theorem of Pehr Beurling (see

---

[15] The term of index n in the last sum above is equal to 1.

[66], pp. 285–288) states that we can find a constant C_δ and a sequence $(f_k)_{k \geqslant 1}$ of H^∞ such that

$$f_k(z_j) = \delta_{k,j} \text{ for } j, k \geqslant 1 \text{ and } \sum_{k=1}^\infty |f_k(z)| \leqslant C_\delta \text{ for } z \in D. \qquad (12.23)$$

Then, the series $\sum f_k$ converges normally on all compact subsets of D[16] and if $w = (w_j)_{j \geqslant 1} \in \ell^\infty$, the Lagrange-type formula

$$g := R(w) := \sum_{k=1}^\infty w_k f_k$$

defines a function $g \in H^\infty$ such that $g(z_j) = w_j$ for $j \geqslant 1$, in other words, such that $T(g) = w$, with the notations of the beginning of the proof of Carleson's theorem. In the language of the next chapter, we have thus constructed a continuous right inverse $R : \ell^\infty \to H^\infty$ of the continuous linear surjection

$$T : H^\infty \to \ell^\infty, f \mapsto \left(f(z_j) \right)_{j \geqslant 1}$$

(we also speak of a *continuous lifting*: indeed, we have $TR = I$, where I is the identity of ℓ^∞). A nice consequence of the existence of these Beurling functions is the following [18].

12.4.16 Theorem　*The Cartesian product of two interpolating sequences $A = (a_j)_{j \geqslant 1}$ and $B = (b_k)_{k \geqslant 1}$ of D is an interpolating sequence of D^2: if $w = (w_{j,k})_{j,k \geqslant 1} \in \ell^\infty(\mathbb{N}^{*2})$, there exists an $f \in H^\infty(D^2)$ such that*

$$f(a_j, b_k) = w_{j,k} \text{ for } j, k \geqslant 1.$$

Proof　Indeed, it suffices to take

$$f(z, w) = \sum_{j,k \geqslant 1} w_{j,k} f_j(z) g_k(w),$$

where $(f_j)_{j \geqslant 1}$ and $(g_k)_{k \geqslant 1}$ are the Beurling sequences associated respectively with A and B. □

12.4.17 Remark　Beurling's proof, detailed in [66] (pp. 285–288),[17] uses non-linear optimisation. An explicit construction of the sequence $(f_k)_{k \geqslant 1}$, which corrects the children $\dfrac{B(z)}{(z - z_k)B'(z_k)}$ of the mother function B, was given by P. Jones, who exhibited a "magic formula" for these f_k. See [188], pp. 182–183 or [141], pp. 76–78.

---

[16]　Why? The Cauchy formula can help.

[17]　See also p. 288 of the same book for an alternative and more general proof, due to Drury.

12.4.18 Remark Let $(z_j)_{j \geqslant 1}$ be a uniformly separated sequence, with uniform separation constant $\delta = \inf_{n \geqslant 1} \delta_n$, and C the best constant (the interpolation constant) such that all $w \in \ell^\infty$ can be written

$$w = T(f) := \big(f(z_j)\big)_{j \geqslant 1} \text{ with } \|f\|_\infty \leqslant C\|w\|_\infty.$$

The proof of the implication (i) $\Rightarrow$ (ii) of Theorem 12.4.4 showed that $C \geqslant \delta^{-1}$ (in the form $\delta \geqslant C^{-1}$), and the proof of the implication (ii) $\Rightarrow$ (i) showed that $C \leqslant 32\,\delta^{-5}$. In fact, one can prove (see [66], p. 278) that

$$\frac{1}{\delta} \leqslant C \leqslant \frac{a}{\delta}\left(1 + \log\frac{1}{\delta}\right),$$

where a is an absolute constant.

12.4.19 Remark The preceding upper bound is *optimal*, as the following example ([66], p. 284) shows. Let $N \geqslant 1$ be an integer, and $r \in \,]0, 1[$ to be adjusted. Set $\omega = e^{\frac{2i\pi}{N}}$ and consider the interpolation problem $f(z_j) = w_j$ ($j \geqslant 1$), for which

$$z_j = r\omega^j, \; w_j = \omega^{-j} \text{ for } 1 \leqslant j \leqslant N. \tag{12.24}$$

The (finite) Blaschke product associated with z_j is, up to the sign, none other than

$$B(z) = \prod_{p=1}^{N} \frac{z - z_p}{1 - \overline{z_p}z} = \frac{z^N - r^N}{1 - r^N z^N}.$$

The uniform separation constant δ of the z_j is (see (12.14))

$$\delta = \inf_{1 \leqslant j \leqslant N}(1 - |z_j|^2)|B'(z_j)| = Nr^{N-1}\frac{1 - r^2}{1 - r^{2N}}. \tag{12.25}$$

Indeed, since $z_j^N = r^N$, we have $B'(z_j) = \dfrac{Nz_j^{N-1}}{1 - r^{2N}}$. To calculate the interpolation constant m associated with z_j and w_j, we admit the following result ([66], p. 132), reminiscent of the fact that the Lagrange interpolation polynomial for a given N is of degree $\leqslant N - 1$.

12.4.20 Theorem *Let* $z_1, \ldots, z_N \in D$ *be distinct points and let* $w_1, \ldots, w_N \in \mathbb{C}$. *Then there exists a* unique *function* $f \in H^\infty$ *of minimal norm realising the interpolation* $f(z_j) = w_j$, $1 \leqslant j \leqslant N$. *Moreover, f is a constant multiple of a Blaschke product of length* $\leqslant N - 1$.

We are going to deduce that

$$m = r^{1-N}. \tag{12.26}$$

Temporarily admitting this relation, we can conclude as follows. Let N tend to infinity, and select $r = e^{-\frac{\log N}{N}}$ satisfying $r^N = \frac{1}{N}$. Thus, by (12.25) and (12.26), we have[18]

$$\delta \sim N r^N \cdot 2(1 - r) \sim 2(1 - r) \sim 2\frac{\log N}{N} \text{ and } m \sim r^{-N} = N.$$

But then $m \sim \frac{2}{\delta} \log \frac{1}{\delta}$ and we cannot avoid the logarithmic factor!

It remains to show (12.26). Let $f \in H^\infty$ be the function of minimal norm m realising the interpolation, and let $g \in H^\infty$ be defined by $g(z) = \omega f(\omega z)$.

We see that

$$g(z_j) = \omega f(\omega z_j) = \omega f(z_{j+1}) = \omega w_{j+1} = w_j$$

(and this explains the choice of the w_j).

Moreover, $\|g\|_\infty = \|f\|_\infty$. According to the assumed theorem, we have $g = f$. The zeros of the Blaschke product f are thus invariant under the rotation $z \mapsto \omega z$. If a is one of these zeros, so are $\omega a, \ldots, \omega^N a$, which are distinct if $a \neq 0$. But the Blaschke product f has at most $p \leqslant N - 1$ zeros. This forces $a = 0$ and hence $f(z) = cmz^p$ with c a constant of modulus 1. The functional equation $g = f$ gives $\omega^{p+1} = 1$, hence $p = N - 1$. Then, the interpolation relation $f(r\omega) = \omega^{-1}$ implies $cmr^{N-1}\omega^{N-1} = \omega^{-1}$, hence $c = 1$ and $m = r^{1-N}$, which proves (12.26). By the way, even if it is no longer worth it, note that $f(z) = \left(\frac{z}{r}\right)^{N-1}$.

12.5 Applications to functional analysis

12.5.1 Riesz systems

In Section 12.7 we will see an application of interpolating sequences to the corona problem. Here are two others.

A sequence $(x_n)_{n \geqslant 1}$ of a Hilbert space H is said to be a *Riesz system* if there exist constants $c_1, c_2 > 0$ such that for all finite sequences $(\lambda_n)_{1 \leqslant n \leqslant N}$ of scalars, we have a quasi-Pythagorean theorem

---

[18] Taking into account that $1 + r \to 2$ and $1 - r^{2N} \to 1$ as $N \to +\infty$.

$$c_1^2 \sum_{n=1}^{N} |\lambda_n|^2 \|x_n\|^2 \leqslant \left\| \sum_{n=1}^{N} \lambda_n x_n \right\|^2 \leqslant c_2^2 \sum_{n=1}^{N} |\lambda_n|^2 \|x_n\|^2.$$

Recall that the inner product of $L^2(\mathbb{T})$ is defined by

$$\langle f, g \rangle = \int_{\mathbb{T}} f(e^{i\theta}) \overline{g(e^{i\theta})} dm(\theta).$$

In what follows, the norm associated with this inner product will be denoted by $\| \cdot \|$. Let $a \in D$ and let $K_a \in H^2$, the *reproducing kernel* of H^2 at the point a, defined by

$$K_a(z) = \frac{1}{1 - \overline{a}z} \text{ for } z \in D.$$

It satisfies[19]

$$f(a) = \langle f, K_a \rangle \text{ for } f \in H^2 \text{ and } \|K_a\|^2 = K_a(a) = \left(1 - |a|^2\right)^{-1}.$$

These kernels K_a are often better adapted than the monomials z^n for the study of H^2 and its operators, but present the inconvenience of not being orthogonal. This inconvenience vanishes in certain cases, as shown by the following result.

12.5.2 Theorem *Let $(z_n)_{n \geqslant 1}$ be a sequence of distinct points of D.*

(1) *If $(z_n)_{n \geqslant 1}$ is an interpolating sequence of H^∞, with interpolation constant C, the system of vectors $(K_{z_n})_{n \geqslant 1}$ forms a Riesz system with $c_1 = C^{-1}$ and $c_2 = C$.*

(2) *Conversely, if the K_{z_n} form a Riesz system, $(z_n)_{n \geqslant 1}$ is an interpolating sequence of H^∞, with constant $C \leqslant \dfrac{c_2}{c_1}$.*

Proof If (1) is true, let $(\lambda_n)_{1 \leqslant n \leqslant N}$ be a finite sequence of scalars, and $S = \sum_{n=1}^{N} \lambda_n K_{z_n}$. Also, let $\gamma = (\gamma_n)_{1 \leqslant n \leqslant N}$ be a finite sequence of complex numbers of modulus 1 (complex signs), and $S_\gamma = \sum_{n=1}^{N} \gamma_n \lambda_n K_{z_n}$. Finally, let $g \in H^\infty$ be such that $g(z_n) = \overline{\gamma_n}$ for $n \geqslant 1$ and $\|g\|_\infty \leqslant C$. For $f \in H^2$ such that $\|f\| \leqslant 1$, we have

$$\langle S_\gamma, f \rangle = \sum_{n=1}^{N} \gamma_n \lambda_n \overline{f(z_n)} = \sum_{n=1}^{N} \lambda_n \overline{(fg)(z_n)} = \sum_{n=1}^{N} \lambda_n \langle K_{z_n}, fg \rangle = \langle S, fg \rangle.$$

It thus follows that

$$|\langle S_\gamma, f \rangle| \leqslant \|S\| \|fg\| \leqslant \|S\| \|g\|_\infty \|f\| \leqslant C \|S\|.$$

---

[19] Indeed, if $f(z) = \sum_{n=0}^{\infty} a_n z^n$, as $K_a(z) = \sum_{n=0}^{\infty} \overline{a}^n z^n$, we have $\langle f, K_a \rangle = \sum_{n=0}^{\infty} a_n a^n = f(a)$.

By passing to the upper bound on f, we obtain $\|S_\gamma\| \leqslant C\|S\|$ and hence, without effort, $\|S\| \leqslant C\|S_\gamma\|$ because of the arbitrary nature of the coefficients. Hence, finally,

$$C^{-1}\|S\| \leqslant \|S_\gamma\| \leqslant C\|S\|. \tag{12.27}$$

This unconditionality of the K_{z_n} is equivalent to the desired result. Indeed, if we square the inequality (12.27), replace $\gamma_1, \ldots, \gamma_N$ by $X_1(t), \ldots, X_N(t)$ where $X_j(t) = e^{2i\pi jt}$ (the functions X_j being orthonormal with respect to the Lebesgue measure on $[0, 1]$), and integrate over $[0, 1]$, we obtain the double inequality

$$C^{-2}\left\|\sum_{n=1}^{N} \lambda_n K_{z_n}\right\|^2 \leqslant \sum_{n=1}^{N} |\lambda_n|^2 \|K_{z_n}\|^2 \leqslant C^2 \left\|\sum_{n=1}^{N} \lambda_n K_{z_n}\right\|^2. \tag{12.28}$$

Conversely, if (2) holds, let $w = (w_k)_{k \geqslant 1} \in \ell^\infty$ be such that $\|w\|_\infty \leqslant 1$, and $\alpha = \frac{c_1}{c_2}$. Then, for all finite sequences $\lambda_1, \ldots, \lambda_N$ of scalars, we have

$$\sum_{1 \leqslant i, j \leqslant N} \frac{1 - \alpha^2 w_i \overline{w_j}}{1 - z_i \overline{z_j}} \lambda_i \overline{\lambda_j} = \sum_{1 \leqslant i, j \leqslant N} \left(1 - \alpha^2 w_i \overline{w_j}\right) \langle K_{z_j}, K_{z_i} \rangle \lambda_i \overline{\lambda_j}$$

$$= \left\|\sum_{j=1}^{N} \overline{\lambda_j} K_{z_j}\right\|^2 - \alpha^2 \left\|\sum_{j=1}^{N} \overline{w_j} \lambda_j K_{z_j}\right\|^2$$

$$\geqslant \sum_{j=1}^{N} \left(c_1^2 - \alpha^2 c_2^2 |w_j|^2\right) |\lambda_j|^2 \|K_{z_j}\|^2$$

$$\geqslant 0,$$

since

$$c_1^2 - \alpha^2 c_2^2 |w_j|^2 \geqslant c_1^2 - \alpha^2 c_2^2 = 0.$$

By the Schwarz–Pick Theorem 12.4.2, there exists an $f_N \in H^\infty$ such that

$$f_N(z_j) = \alpha w_j \text{ for } 1 \leqslant j \leqslant N \text{ and } \|f_N\|_\infty \leqslant 1.$$

By applying Montel's theorem for normal families to the sequence $\left(\alpha^{-1} f_N\right)_{N \geqslant 1}$, we obtain a function $f \in H^\infty$ such that

$$f(z_j) = w_j \text{ for } j \geqslant 1 \text{ and } \|f\|_\infty \leqslant \alpha^{-1} = \frac{c_2}{c_1}.$$

This proof is based on the Schwarz–Pick theorem. Here is another proof that is more self-contained and more in line with the spirit of this chapter. Suppose that $(K_{z_n})_{n \geqslant 1}$ is a Riesz system. In particular, for $n \geqslant 1$ fixed, we have

$$\left\| K_{z_n} - \sum_{j \neq n} \lambda_j K_{z_j} \right\|^2 \geq c_1^2 \left\| K_{z_n} \right\|^2$$

for all finite sums $\sum_{j \neq n} \lambda_j K_{z_j}$, which means[20] that the distance from K_{z_n} to the space generated by the other K_{z_j} is at least equal to $c_1 \left\| K_{z_n} \right\|$. But by the duality formula (12.8), applied to $X = X^* = H^2$, $E = \left(\text{vect} \left(K_{z_j} \right)_{j \neq n} \right)^\perp$, $E^\perp$ being the closed subspace generated by the K_{z_j} ($j \neq n$), we also have

$$\sup_{\substack{\|\varphi\| \leq 1 \\ j \neq n \Rightarrow \langle \varphi, K_{z_j} \rangle = 0}} \left| \langle \varphi, K_{z_n} \rangle \right| = \left\| K_{z_n} \right\|_{X^*/E^\perp} := \inf_{\psi \in E^\perp} \left\| K_{z_n} + \psi \right\| \geq c_1 \left\| K_{z_n} \right\|.$$

Decreasing c_1 if necessary, we can thus find $\varphi \in H^2$ orthogonal to the K_{z_j} of index $\neq n$, with norm ≤ 1, such that $|\varphi(z_n)| \geq c_1 \left\| K_{z_n} \right\|$. In particular, $(z_j)_{j \geq 1}$ is a Blaschke sequence.[21] But then (Theorem 12.2.10), we can write $\varphi = B_n g_n$ with $g_n \in H^2$ and $\|g_n\| = \|\varphi\| \leq 1$, so that

$$c_1 \left\| K_{z_n} \right\| \leq |\varphi(z_n)| = |B_n(z_n)| |g_n(z_n)| = |B_n(z_n)| \left| \langle g_n, K_{z_n} \rangle \right|$$
$$\leq |B_n(z_n)| \left\| K_{z_n} \right\| \|g_n\| \leq |B_n(z_n)| \left\| K_{z_n} \right\|.$$

Then, by simplification, $|B_n(z_n)| \geq c_1$, which is none other than the Carleson condition. $\qquad \square$

12.5.3 Hankel operators

12.5.3.1 Non-commutative approximation
According to the Weierstrass theorem, a function $f \in L^\infty([0, 1])$ is continuous on $[0, 1]$ if and only if it is a uniform limit on $[0, 1]$ of a sequence of polynomials, that is, if and only if

$$E_n(f) = \inf_{\deg P \leq n} \| f - P \|_\infty \searrow 0 \text{ as } n \nearrow +\infty.$$

According to some famous theorems of Bernstein ([46], pp. 332–339), the sequence $(E_n(f))_{n \geq 0}$ can decrease arbitrarily slowly to 0. The more quickly it tends to zero, the more regular is f, and vice versa. Similarly, if $T : H \to H'$ is a continuous operator between separable Hilbert spaces, T is compact if and only if the quantity

$$a_n(T) = \inf_{\text{rank}(R) < n} \| T - R \|,$$

---

[20] We say that the sequence $(K_{z_n})_{n \geq 1}$ is uniformly minimal.
[21] See the lines following (12.13).

called the *nth approximation number of* T, decreases to 0 as $n \to \infty$. The approximation numbers ([34], p. 41) have the ideal property, expressed by the inequality

$$a_n(ATB) \leqslant \|A\| \, a_n(T) \, \|B\|. \tag{12.29}$$

The speed at which $a_n(T)$ decreases to zero is an indicator of the degree of compactness of T. As in Bernstein's theorems, this sequence can decrease arbitrarily slowly to 0: given a sequence $(\varepsilon_n)_{n \geqslant 0}$ decreasing to 0, it suffices to consider a diagonal operator T such that $T(e_n) = \varepsilon_n e'_n$ for $n \geqslant 0$, where (e_n) and (e'_n) are orthonormal bases of H and H', respectively. But as soon as we restrict ourselves to operators T of a particular type, often associated with a function known as the *symbol* of the operator, the situation is much more complicated (study of the so-called composition operators is in progress). We examine here the case of operators known as *Hankel operators*.

12.5.3.2 Hankel operators

We are working on the unit circle $\mathbb{T}$. Set $e_n(z) = z^n$ for $n \in \mathbb{Z}$, and denote by P_+ the orthogonal projection from L^2 onto H^2. Then, $I - P_+$ is the orthogonal projection onto H^2_-, the orthogonal complement of H^2 in L^2. Orthonormal bases of H^2 and H^2_- are, respectively, $(e_n)_{n \geqslant 0}$ and $(f_n)_{n \geqslant 1}$, where $f_n = e_{-n}$. If $\varphi \in L^\infty$, the Hankel operator H_φ with symbol φ is defined by

$$H_\varphi : H^2 \to H^2_-, g \mapsto (I - P_+)(\varphi g).$$

We can immediately verify that the general term $a_{j,k}$ of the matrix of H_φ over the bases $(e_n)_{n \geqslant 0}$ and $(f_n)_{n \geqslant 1}$ *depends only on the sum* $j + k$, since

$$\langle H_\varphi(e_j), f_k \rangle = \widehat{\varphi}(-j - k) \text{ for } j \geqslant 0 \text{ and } k \geqslant 1.$$

A 1957 theorem of Nehari ([143], pp. 3–5) states that every operator whose matrix has the preceding property is of type H_φ for a certain function $\varphi \in L^\infty$, up to a unitary equivalence. Another theorem of Megretskii *et al.* ([143], pp. 520–522) states that the approximation numbers of a Hankel operator form an arbitrary non-increasing sequence. *The proof is extremely difficult.* But a weakened form of this result is accessible at the level of this book, with the Carleson interpolating sequences at its core. It is this form [142] that we will prove, after a few preliminaries on model spaces.

12.5.3.3 Model spaces

Let B be an infinite Blaschke product[22] with distinct zeros $(z_n)_{n \geqslant 1}$ forming an *interpolating sequence* with constant C. Then the space BH^2 is closed in

---

[22] In particular, B is an inner function.

H^2, and its orthogonal complement $K_B = (BH^2)^{\perp} = H^2 \ominus BH^2$ is called a *model space*. The importance of these spaces, odd at first sight, follows from the facts below (among others).[23]

(1) According to Beurling's theorem as proved above, the closed subspaces invariant under the unilateral shift S on H^2, defined by $Sf(z) = zf(z)$, are exactly the spaces uH^2 where u is an inner function. Their orthogonal complements $K_u = (uH^2)^{\perp} = H^2 \ominus uH^2$ are hence *exactly* the subspaces invariant under S^*, the "backward shift".

(2) One can show that any contraction of a certain type on a Hilbert space H^2 is unitarily equivalent to the restriction of S^* to a certain model space K_u. The understanding of these restrictions thus implies the understanding of (almost) all contractions.

Let us return to the case of our Blaschke product. The orthogonal projection on K_B is denoted by P_B and, with c_n a suitable constant, we set

$$e_n = \frac{B_n K_{z_n}}{B_n(z_n) \|K_{z_n}\|^2} = c_n \frac{B}{z_n - z}.$$

We can find our bearings in this model space thanks to the e_n, as shown by the following theorem.

12.5.4 Theorem (*1*) *The orthogonal projection* $P_B : H^2 \to K_B$ *is given by*

$$P_B(g) = g - BP_+ (\overline{B}g) \text{ for } g \in H^2.$$

(2) *The reproducing kernels* K_{z_n}, $n \geqslant 1$, *form a* Riesz basis *of* K_B, *of constant at most equal to* C.

(3) *We have* $\langle e_j, e_k \rangle = c_j \overline{c_k} \langle K_{z_k}, K_{z_j} \rangle$ *for* $j, k \geqslant 1$.

(4) *The functions* e_n, $n \geqslant 1$, *also form a* Riesz basis *of* K_B, *with constant at most equal to* C, *and bi-orthogonal to the* K_{z_n}, *in the sense that*

$$\langle e_j, K_{z_k} \rangle = \delta_{j,k} \text{ for } j, k \geqslant 1.$$

(5) *If* $f \in H^{\infty}$, *then* $P_B(fe_j) = f(z_j)e_j$ *for* $j \geqslant 1$.

Proof We will often use equation (12.9), due to the inner character of the Blaschke products B (or B_n):

$$\langle Bu, Bv \rangle = \langle u, v \rangle \text{ for } u, v \in L^2.$$

---

[23] The first fact was already pointed out in Section 12.3.

(1) Let $g \in H^2$. Note that $g - BP_+ (\overline{B}g) \in (BH^2)^\perp$ since, for $h \in H^2$,

$$\langle g - BP_+ (\overline{B}g), Bh \rangle = \langle \overline{B}g - P_+ (\overline{B}g), h \rangle = 0.$$

Moreover, as $BP_+ (\overline{B}g) \in BH^2$, we finally obtain $P_B(g) = g - BP_+ (\overline{B}g)$.

(2) Let V be the closed subspace generated by the K_{z_n} $(n \geqslant 1)$. By Theorem 12.2.10,

$$BH^2 = \{g \in H^2 / g(z_n) = \langle g, K_{z_n} \rangle = 0 \text{ for } n \geqslant 1\} = V^\perp.$$

With a passage to the orthogonal complements, we obtain $K_B = V$. Moreover, if C is the interpolation constant of $(z_n)_{n \geqslant 1}$, Theorem 12.5.2 tells us that $(K_{z_n})_{n \geqslant 1}$ is a Riesz system with constant at most equal to C, hence a Riesz basis of K_B.

(3) Via the unimodularity of B and of $\overline{z}$ on $\mathbb{T}$,

$$
\begin{aligned}
\langle e_j, e_k \rangle &= c_j \overline{c_k} \langle \frac{B}{z_j - z}, \frac{B}{z_k - z} \rangle \\
&= c_j \overline{c_k} \langle \frac{1}{z_j - z}, \frac{1}{z_k - z} \rangle \\
&= c_j \overline{c_k} \langle \frac{1}{1 - z_j \overline{z}}, \frac{1}{1 - z_k \overline{z}} \rangle \\
&= c_j \overline{c_k} \int_{\mathbb{T}} \frac{1}{1 - z_j e^{-i\theta}} \cdot \frac{1}{1 - \overline{z_k} e^{i\theta}} \, dm(\theta) \\
&= c_j \overline{c_k} \langle K_{z_k}, K_{z_j} \rangle.
\end{aligned}
$$

(4) First of all, $e_n \in K_B$. Indeed, let $h \in H^2$. We can write $Bh = B_n h_n$ with $h_n(z_n) = 0$, so that

$$\langle B_n K_{z_n}, Bh \rangle = \langle B_n K_{z_n}, B_n h_n \rangle = \langle K_{z_n}, h_n \rangle = \overline{h_n(z_n)} = 0,$$

hence $\langle e_n, Bh \rangle = 0$ by definition of e_n. Point (3) (preservation of inner products up to a factor $c_j \overline{c_k}$) shows that $(e_n)_{n \geqslant 1}$ is a Riesz system of K_B, with Riesz constant at most equal to C. Moreover, we have, by definition of the e_j: $\langle K_{z_k}, e_j \rangle = e_j(z_k) = \delta_{j,k}$. It remains to show that the e_n generate K_B, that is, a function $g \in K_B$ orthogonal to the e_n is identically zero. This is a (not so evident) consequence of the fact that the sequence $(K_{z_n})_{n \geqslant 1}$ is also a Schauder basis of K_B (see [120], p. 45). A function $g \in K_B$ can thus be written in the form of a convergent series $g = \sum_{n=1}^{\infty} \lambda_n K_{z_n}$. If moreover g is orthogonal to the e_n, the preceding bi-orthogonality gives

$$\lambda_n = \langle g, e_n \rangle = 0 \text{ for } n \geqslant 1,$$

hence $g = 0$.

(5) For $j \geqslant 1$, we clearly have $g := \left(f - f(z_j) \right) e_j \in BH^2$ since $g(z_k) = 0$ for $k \geqslant 1$. From this,

$$0 = P_B(g) = P_B(f e_j) - f(z_j) P_B(e_j) = P_B(f e_j) - f(z_j) e_j.$$

This completes the proof. $\square$

12.5.4.1 Approximation numbers of Hankel operators

We fix a non-increasing sequence $(\varepsilon_n)_{n \geqslant 1}$ of positive real numbers, with limit zero. Let $f \in H^\infty$, $(z_n)_{n \geqslant 1}$ an interpolating sequence to be adjusted later, and B the associated Blaschke product. We consider the operators $M_B : L^2 \to L^2$ of multiplication by B and $T_f : H^2 \to H^2$ of multiplication by f. We will see that the approximation numbers of the Hankel operator H_φ associated with the function $\varphi = f\overline{B} \in L^\infty$ tend to 0 at infinity like the ε_n, for a proper choice of f and B. Two more technical lemmas are necessary.

12.5.5 Lemma *Matching the arrows in the diagram below:*

$$H^2 \xrightarrow{H_{f\overline{B}}} H_-^2 \xrightarrow{M_B} L^2 \text{ and } H^2 \xrightarrow{P_B} K_B \xrightarrow{T_f} H^2 \xrightarrow{P_B} L^2,$$

we obtain the operator equality

$$M_B H_{f\overline{B}} = P_B T_f P_B.$$

Proof Let $g \in H^2$. Write $g = P_B(g) + v$, with $v \in BH^2$, and set $T = P_B T_f P_B$. By point (1) of Theorem 12.5.4,

$$\begin{aligned}
M_B H_{f\overline{B}}(g) &= B \left(f\overline{B}g - P_+ \left(f\overline{B}g \right) \right) \\
&= fg - B P_+ (\overline{B} fg) \\
&= P_B(fg) \\
&= P_B(f P_B(g))
\end{aligned}$$

since $fv \in BH^2$, and hence $P_B(fv) = 0$. $\square$

12.5.6 Lemma *Suppose that $f(z_n) = \varepsilon_n$ for $n \geqslant 1$, and set $T = P_B T_f P_B$. We then have the bounds*

$$C^{-2} \varepsilon_n \leqslant a_n(T) \leqslant C^2 \varepsilon_n \text{ for } n \geqslant 1,$$

where C is the interpolation constant of $(z_n)_{n \geqslant 1}$.

Proof (sound but pedestrian). We retain the notations of the preceding question, and in addition define $P_B(g) = \sum_{n=1}^{\infty} \tilde{g}(n)e_n$. Point (5) of Theorem 12.5.4 gives us

$$Tg = P_B\left(\sum_{j=1}^{\infty} \tilde{g}(j)fe_j\right) = \sum_{j=1}^{\infty} \tilde{g}(j)P_B(fe_j) = \sum_{j=1}^{\infty} \tilde{g}(j)f(z_j)e_j$$

$$= \sum_{j=1}^{\infty} \tilde{g}(j)\varepsilon_j\, e_j.$$

Hence, if R is the operator of rank $\leqslant n-1$ defined by $Rg = \sum_{j=1}^{n-1} \tilde{g}(j)\varepsilon_j\, e_j$, as the sequence $(\varepsilon_n)_{n \geqslant 1}$ is non-increasing, we have

$$\|(T-R)g\|^2 = \left\|\sum_{j=n}^{\infty} \tilde{g}(j)\varepsilon_j\, e_j\right\|^2$$

$$\leqslant C^2 \sum_{j=n}^{\infty} |\tilde{g}(j)|^2\, \varepsilon_j^2 \|e_j\|^2$$

$$\leqslant C^2\varepsilon_n^2 \sum_{j=n}^{\infty} |\tilde{g}(j)|^2\, \|e_j\|^2$$

$$\leqslant C^4\varepsilon_n^2\, \|P_B(g)\|^2$$

$$\leqslant C^4\varepsilon_n^2\|g\|^2.$$

This gives the right-hand inequality of Lemma 12.5.6.

On the contrary, we denote by E_n the subspace generated by $e_1, \ldots, e_n$. If $g = \sum_{j=1}^{n} \tilde{g}(j)e_j \in E_n$, we have $T(g) = \sum_{j=1}^{n} \tilde{g}(j)\,\varepsilon_j\, e_j$, so that

$$\|T(g)\|^2 \geqslant C^{-2} \sum_{j=1}^{n} |\tilde{g}(j)|^2\, \varepsilon_j^2 \|e_j\|^2 \geqslant C^{-2}\varepsilon_n^2 \sum_{j=1}^{n} |\tilde{g}(j)|^2\, \|e_j\|^2 \geqslant C^{-4}\varepsilon_n^2\|g\|^2.$$

The definition of the $a_n(T)$ as Bernstein numbers (see Exercise 12.15) thus gives the left-hand inequality of Lemma 12.5.6. $\qquad\square$

We can now state the principal result of this section.

12.5.7 Theorem [Hruscev–Peller] *Let $(\varepsilon_n)_{n \geqslant 1}$ be a non-increasing sequence of strictly positive real numbers, and $\rho > 0$. There exists a Hankel operator H_φ such that*

$$(1+\rho)^{-1}\varepsilon_n \leqslant a_n(H_\varphi) \leqslant (1+\rho)\varepsilon_n \text{ for } n \geqslant 1.$$

Proof Let $(z_n)_{n \geqslant 1}$ be an *interpolating sequence* with constant $C \leqslant \sqrt{1 + \rho}$. Such sequences exist, even if a clear reference on this point is not easy to find.[24] Let B be the Blaschke product associated with $(z_n)_{n \geqslant 1}$. We can thus find a function $f \in H^\infty$ such that $f(z_n) = \varepsilon_n$ for $n \geqslant 1$. By Lemma 12.5.6 we have

$$(1 + \rho)^{-1} \varepsilon_n \leqslant a_n(T) \leqslant (1 + \rho) \varepsilon_n,$$

where $T = P_B T_f P_B$. Now let $\varphi = f \overline{B} \in L^\infty$. By Lemma 12.5.5 and as $M_B : H^2_- \to L^2$ is an isometry, we also have (see Exercise 12.15)

$$a_n(H_\varphi) = a_n(M_B H_\varphi) = a_n(T),$$

which completes the proof. $\qquad\qquad\qquad\qquad\qquad\qquad\qquad\qquad\qquad$ □

12.6 Solution of the corona problem

Here is the context: as we know, $H^\infty = H^\infty(D)$ denotes the set of bounded holomorphic functions in the open unit disk D of $\mathbb{C}$. Equipped with the natural norm

$$\|f\|_\infty = \sup_{z \in D} |f(z)|,$$

it is a superb Banach algebra, commutative and unitary. It is even a uniform algebra, in the sense that $\|f^2\|_\infty = \|f\|_\infty^2$ if $f \in H^\infty$. It can thus be isometrically identified with a subalgebra of $C(K)$, where K is its spectrum. The algebra H^∞ is an involutive algebra for the involution $\tilde{f}(z) = \overline{f(\overline{z})}$, but not a C^*-algebra: indeed, $\|\tilde{f}\|_\infty = \|f\|_\infty$, but if, for example $f(z) = 1 + iz$, then

$$\tilde{f}(z) = 1 - iz \text{ and } (\tilde{f}f)(z) = \tilde{f}(z)f(z) = 1 + z^2, \text{ hence } \|\tilde{f}f\|_\infty = 2,$$

while $\|f\|_\infty^2 = 4$, so that

$$\|\tilde{f}f\|_\infty \neq \|f\|_\infty^2.$$

It is not separable (Exercise 12.1), hence its spectrum is not metrizable.[25] This spectrum contains a subset D' homeomorphic to D (that we identify with D), namely the set of evaluations $\delta_a : f \mapsto f(a)$ at the different points of D, $a \mapsto \delta_a$ being a homeomorphism of D to D' (see Proposition 11.3.3 in Chapter 11).

---

[24] See, for example, [136], p. 170 or the "thin" Blaschke products of [66], p. 430 or again Exercise 12.13, which completely details an example.

[25] See [152], p. 77.

We note that D' is exactly the set of $\varphi \in K$ such that $|\varphi(e)| < 1$, where e denotes the identity function

$$e : D \to \mathbb{C}, \; z \mapsto z.$$

Indeed, if $\varphi(e) = a \in D$, using the fact that all $f \in H^\infty$ can be written

$$f = f(a) + (e - a)g, \text{ where } g \in H^\infty,$$

we see that

$$\varphi(f) = f(a) + \big(\varphi(e) - a\big)\varphi(g) = f(a),$$

hence $\varphi = \delta_a$. The elements φ of $K \backslash D'$ thus satisfy $\varphi(e) = u \in \mathbb{T}$ (unit circle of $\mathbb{C}$), and we can write [66]

$$K = D' \cup \left(\bigcup_{u \in \mathbb{T}} K_u \right),$$

where

$$K_u = \{\varphi \in K / \varphi(e) = u\}$$

is the *fibre of K above u*.

This being said, the Banach algebra (and Banach space) H^∞ remains somewhat mysterious and monstrous. For example:

- we do not know if H^∞ has the approximation property[26] [66];
- H^∞ is not complemented in L^∞ (Exercise 12.2);
- H^∞ is not separable (Exercise 12.1), nor is L^∞ / H^∞ (Exercise 12.10);
- $H^\infty + \overline{H_0^\infty}$, where $H_0^\infty = \{f \in H^\infty / f(0) = 0\}$ and where the bar means conjugation, is not dense in L^∞ [90] – however, Bourgain [29] showed that if $h \in L^\infty$ and $\varepsilon > 0$, there exist $f \in H^\infty$ and $g \in H_0^\infty$ such that

$$\|h - f - \overline{g}\|_2 \leqslant \varepsilon \|h\|_\infty, \text{ with } \max(\|f\|_\infty, \|g\|_\infty) \leqslant C \ln \frac{1}{\varepsilon} \|h\|_\infty.$$

Our problem here is the evaluation of the place occupied by D (i.e., D') within K: Carleson's theorem (the corona theorem) says precisely that D is dense in K, that is, the corona $K \backslash \overline{D}$ is empty (where $\overline{D}$ denotes the closure of D in K). In other words, the corona theorem proves that there is **no** corona! But we will conform to the current terminology.

Here is a concrete reformulation of the theorem, *which we will apply later to* $X = D$ *and* $A = H^\infty$.

---

[26] See Chapter 13 for the definition.

12.6.1 Proposition *Let A be a uniform algebra, isometrically contained in the space $B(X)$ of bounded functions on a set X, $n \geqslant 1$ a fixed integer, K the spectrum of A, and δ a positive constant. The following assertions are equivalent.*

(i) If $f_1, \ldots, f_n \in A$ and

$$\sum_{j=1}^{n} |f_j(x)| \geqslant \delta > 0 \text{ for all } x \in X,$$

there exist $g_1, \ldots, g_n \in A$ such that we have a length-n Bézout identity

$$\sum_{j=1}^{n} f_j g_j = 1.$$

(ii) For any $\varphi \in K$ and any n-neighbourhood V of φ in K,

$$V = \{\psi \in K / |\varphi(f_j) - \psi(f_j)| < \varepsilon \text{ for } 1 \leqslant j \leqslant n\},$$

V meets X: there exists an $x \in X$ such that

$$|\varphi(f_j) - f_j(x)| < \varepsilon \text{ for } 1 \leqslant j \leqslant n.$$

Proof (ii) $\Rightarrow$ (i): Let $\varphi \in K$. By approaching φ with points of X as in the hypothesis, we see that $\sum_{j=1}^{n} |\varphi(f_j)| \geqslant \delta$. In particular, none of the $\varphi \in K$ is zero for all of the f_j. According to Gelfand theory, the f_j are not in a proper ideal of A, and the ideal that they generate is A itself; this translates exactly into a length-n Bézout identity $\sum_{j=1}^{n} f_j g_j = 1$.

(i) $\Rightarrow$ (ii): Suppose that there exist $\varphi \in K$ and an n-neighbourhood V of φ,

$$V = \{\psi \in K / |\psi(f_j) - \varphi(f_j)| < \varepsilon \text{ for } 1 \leqslant j \leqslant n\},$$

that does not intersect X. We set $\varphi(f_j) = c_j$. For all $x \in X$, there exists at least one $j_x \in [\![1, n]\!]$ such that $|\varphi(f_{j_x}) - f_{j_x}(x))| \geqslant \varepsilon$, and *a fortiori*

$$\sum_{j=1}^{n} |f_j(x) - c_j| \geqslant \varepsilon.$$

By the hypothesis, we have a Bézout identity $\sum_{j=1}^{n} (f_j - c_j) g_j = 1$, with $g_j \in A$. By taking the Gelfand transform of the two sides at φ, we obtain

$$1 = \sum_{j=1}^{n} (\varphi(f_j) - c_j)\varphi(g_j) = \sum_{j=1}^{n} 0 = 0 !.$$

This contradiction completes the proof. $\qquad\square$

12.6.2 Remark The condition $\sum_{j=1}^{n} |f_j(x)| \geqslant \delta$ is of course *necessary* to have a Bézout identity: indeed if $\sum_{j=1}^{n} f_j g_j = 1$, by setting $M = \sup_{1 \leqslant j \leqslant n} \|g_j\|_\infty$, we see that

$$1 \leqslant \sum_{j=1}^{n} |f_j(x)||g_j(x)| \leqslant M \sum_{j=1}^{n} |f_j(x)|,$$

hence

$$\sum_{j=1}^{n} |f_j(x)| \geqslant \delta := M^{-1}.$$

The last proposition (applied to $X = D$, $A = H^\infty$, and an arbitrary integer $n \geqslant 1$) means this: if we already know that D is dense in K (the spectrum of H^∞), by Gelfand theory, under the hypothesis contained in (i), we are assured of the existence of a Bézout identity. This is in fact the case for the separable subalgebra $A(D)$ of H^∞ made up of the functions holomorphic in D and continuous on Δ, the closure of D in the usual sense. We have seen in Chapter 11 that the spectrum of $A(D)$ is Δ, and this time the density of D in the spectrum of $A(D)$ is not a property, it is a definition!

The proof of (ii) $\Rightarrow$ (i) immediately implies the following fact: if $f_1, \ldots, f_n \in A(D)$ satisfy

$$\sum_{j=1}^{n} |f_j(z)| \geqslant \delta > 0 \text{ for all } z \in D,$$

then there exist $g_1, \ldots, g_n \in A(D)$ such that

$$\sum_{j=1}^{n} f_j g_j = 1. \tag{12.30}$$

Thus, a method of passing from $A(D)$ to H^∞ is obvious even to the least experienced analyst: let $f_1, \ldots, f_n \in H^\infty$ with $\sum_{j=1}^{n} |f_j(z)| \geqslant \delta$ for all $z \in D$. Set

$$f_j^{(r)}(z) = f_j(rz) \text{ for } z \in \overline{D} \text{ and } 0 < r < 1.$$

This time, $f_j^{(r)} \in A(D)$,

$$\sum_{j=1}^{n} |f_j^{(r)}(z)| = \sum_{j=1}^{n} |f_j(rz)| \geqslant \delta,$$

and (12.30) assures the existence of $g_1^{(r)}, \ldots, g_n^{(r)} \in A(D)$ such that

$$\sum_{j=1}^{n} f_j^{(r)}(z) g_j^{(r)}(z) = \sum_{j=1}^{n} f_j(rz) g_j^{(r)}(z) = 1 \text{ for all } z \in D. \qquad (12.31)$$

Provided that the $g_j^{(r)}$ are uniformly bounded as r varies, the theorem of normal families [160] guarantees that, modulo extraction,

$$g_j^{(r)} \to g_j \in H^\infty \text{ as } r \nearrow 1, \text{ uniformly on every compact subset of } D,$$

and the passage to the limit in (12.31), with z fixed, will give the coveted Bézout identity: $\sum_{j=1}^{n} f_j(z) g_j(z) = 1$. However, *everything lies in the "Provided that..."*. Gelfand theory gives the relatively spectacular result whereby an element x of a unitary Banach algebra B is invertible if and only if its Gelfand transform $\hat{x}$ is never zero. However, for example, the following plausible statement (quantitative version): "if $x \in B$ satisfies $\|x\| \leqslant 1$ and $|\hat{x}| \geqslant \delta > 0$, then $\|x^{-1}\| \leqslant C_\delta$, where C_δ depends only on δ (and on B)" is false.[27] Gelfand theory is here too soft and too general to give *quantitative* information on the $g_j^{(r)}$. We will sweat our way to such quantitative information in Section 12.6.

12.6.3 Remarks (1) Things are easier when we do not impose growth conditions: if $f_1, \ldots, f_n$ are holomorphic functions in D without common zeros, then there exist holomorphic $g_1, \ldots, g_n$ such that $\sum_{j=1}^{n} f_j g_j = 1$ (see Exercise 12.9).

(2) One difficulty of the problem is the multitude of choices for $g_1, \ldots, g_n$. On this subject, an instructive proof of the corona theorem can be found in [19], even if, in our opinion, it is not the most elementary.

The proof of the corona theorem that we will now present in detail is not the initial solution of Carleson (to which we will return in Section 12.7), but a solution no doubt more accessible for a non-specialist, due to the combined efforts of two mathematicians: L. Hörmander [92] and T. Wolff [189]. We take functions $f_1, \ldots, f_n$, holomorphic in a neighbourhood of $\overline{D}$ and additionally satisfying – without loss of generality – $\|f_j\|_\infty \leqslant 1$ for $1 \leqslant j \leqslant n$. Moreover, we suppose that the f_j satisfy a condition of the type $\sum_{j=1}^{n} |f_j| \geqslant \delta > 0$. Because of the equivalence of norms on $\mathbb{C}^n$, we can replace this condition by a condition of the type

$$\sum_{j=1}^{n} |f_j|^2 \geqslant \delta^2.$$

---

[27] Some interesting recent complements on this quantitative aspect and on the "invisible spectrum" can be found in an article of Nikolskii [137].

Under these conditions, we are certain of the existence of $g_1, \ldots, g_n \in H^\infty$ (and even in $A(D)$) such that $\sum_{j=1}^n f_j g_j = 1$. But in addition (*the whole difficulty lies here*) we aim for an *a priori* control of the type

$$\|g_j\|_\infty \leqslant C(n, \delta), \tag{12.32}$$

where the constant $C(n, \delta)$ may depend on n and δ, but is not allowed to depend on the f_j. The determination of suitable functions g_j entails two aspects: an algebraic aspect (Hörmander [92]) and an analytic aspect (Wolff [189]).

12.6.4 Algebraic aspect

Let us start with the case $n = 2$ (the case $n = 1$ is trivial, with $g_1 = f_1^{-1}$ and $\|g_1\|_\infty \leqslant \delta^{-1}$). The case $n \geqslant 3$ cannot in principle be deduced directly, but can be derived using the *ideas* put forward in this special case.

Following Hörmander, we set

$$h_1 = \frac{\overline{f_1}}{|f_1|^2 + |f_2|^2} \quad \text{and} \quad h_2 = \frac{\overline{f_2}}{|f_1|^2 + |f_2|^2}.$$

Thus, we have our boundedness $\left(\|h_j\|_\infty \leqslant \dfrac{1}{\delta^2}\right)$ and our Bézout identity $h_1 f_1 + h_2 f_2 = 1$. But of course, because of the symbols $\overline{}$ and $|\cdot|^2$, totally proscribed if we wish to remain within the holomorphic functions, the functions h_j thus formed are not at all holomorphic, and a conscientious student would never risk proposing h_1 and h_2 in this context, even if h_1 and h_2 are C^∞ with respect to $x = \operatorname{Re} z$ and $y = \operatorname{Im} z$. "No big deal!" basically replies Hörmander: we are going to adjust h_1 and h_2 to obtain a $\overline{\partial}$ null. *Here triumphs the point of view of partial differential equations*: as we have already said, a holomorphic function f is none other than a function satisfying the homogeneous partial differential equation $\overline{\partial} f = 0$. Thus, let g_1 and g_2 be "corrections" of h_1 and h_2. The algebraic aspect can then be separated into two sub-aspects.

(1) We want to preserve the Bézout identity $g_1 f_1 + g_2 f_2 = 1$. Then

$$f_1(g_1 - h_1) + f_2(g_2 - h_2) = 0,$$

in other words, if we ignore the problem of zeros,

$$\frac{g_1 - h_1}{f_2} = -\frac{g_2 - h_2}{f_1} =: v,$$

which gives us

$$\begin{cases} g_1 = h_1 + v f_2, \\ g_2 = h_2 - v f_1, \end{cases}$$

or, by introducing the functional column vectors $f = \begin{pmatrix} f_1 \\ f_2 \end{pmatrix}$, $g = \begin{pmatrix} g_1 \\ g_2 \end{pmatrix}$ and $h = \begin{pmatrix} h_1 \\ h_2 \end{pmatrix}$:

$$g = h + Af,$$

where A is the skew-symmetric functional matrix $\begin{bmatrix} 0 & v \\ -v & 0 \end{bmatrix}$. At this stage, we can observe that a skew-symmetric matrix is well adapted to the preservation of the Bézout identity. Indeed, by denoting $\langle \cdot, \cdot \rangle$ the symmetric bilinear functional defined on $\mathbb{C}^n$ by $\langle x, y \rangle = \sum_{j=1}^{n} x_j y_j$ and by the skew-symmetric nature of $A = [a_{jk}]_{1 \leqslant j, k \leqslant 2}$, we have

$$\langle Af, f \rangle = \sum_{j,k} a_{jk} f_j f_k = 0,$$

so finally

$$\langle g, f \rangle = \langle h, f \rangle + \langle Af, f \rangle = \langle h, f \rangle = 1.$$

(2) We want to establish that $\bar{\partial}$ is zero: $\bar{\partial} g_1 = \bar{\partial} g_2 = 0$. However, taking into account the fact that the f_j are holomorphic, and supposing the problem resolved, we have

$$\begin{cases} \bar{\partial} g_1 = \bar{\partial} h_1 + f_2 \bar{\partial} v = 0, \\ \bar{\partial} g_2 = \bar{\partial} h_2 - f_1 \bar{\partial} v = 0. \end{cases}$$

Multiplying the first equation by $\overline{f_2}$, the second by $-\overline{f_1}$ and adding, we obtain

$$\psi \, \bar{\partial} v = \overline{f_1} \, \bar{\partial} h_2 - \overline{f_2} \, \bar{\partial} h_1,$$

where $\psi = |f_1|^2 + |f_2|^2$, hence

$$\bar{\partial} v = h_1 \bar{\partial} h_2 - h_2 \bar{\partial} h_1.$$

Finally, by setting

$$h_{12} = h_1 \bar{\partial} h_2 - h_2 \bar{\partial} h_1,$$

the correction leads us to seek g_1 and g_2 of the form

$$\begin{cases} g_1 = h_1 + w_{12} f_2, \\ g_2 = h_2 - w_{12} f_1, \\ \bar{\partial} w_{12} = h_{12}. \end{cases}$$

This can also be written

$$
\begin{cases}
g_1 = h_1 + w_{11} f_1 + w_{12} f_2, \\
g_2 = h_2 + w_{21} f_1 + w_{22} f_2, \\
\overline{\partial} w_{jk} = h_{jk} := h_j \overline{\partial} h_k - h_k \overline{\partial} h_j, \\
\quad \text{the matrix } [w_{jk}]_{1 \leqslant j,k \leqslant 2} \text{ is skew-symmetric.}
\end{cases}
$$

For $n = 2$, we thus have no choice in the algebraic reduction. The immediate and naive generalisation (without a Koszul complex as in [66]) for the case of n functions would be to set

$$
\psi = \sum_{j=1}^{n} |f_j|^2, \quad h_j = \psi^{-1} \overline{f_j} \text{ and } h_{jk} = h_j \overline{\partial} h_k - h_k \overline{\partial} h_j,
$$

and then

$$
\begin{cases}
g_j = h_j + \sum_{k=1}^{n} w_{jk} f_k, \text{ with} \\
\overline{\partial} w_{jk} = h_{jk} \text{ and } [w_{jk}]_{1 \leqslant j,k \leqslant n} \text{ is skew-symmetric.}
\end{cases}
\tag{12.33}
$$

It is easy to acquire the skew-symmetry of the matrix $[w_{jk}]_{1 \leqslant j,k \leqslant n}$, since $h_{jk} = -h_{kj}$. For this, it suffices to solve the equation $\overline{\partial} w_{jk} = h_{jk}$ for $j < k$, and then to take $w_{jj} = 0$ and $w_{jk} = -w_{kj}$ if $j > k$. We will thus have, for such a pair (j, k):

$$
\overline{\partial} w_{jk} = -\overline{\partial} w_{kj} = -h_{kj} = h_{jk},
$$

and of course $\overline{\partial} w_{jj} = 0 = h_{jj}$. Moreover, if we control the norm (here the L^{∞}-norm) of the w_{jk} for $j < k$, with this construction we will control it for all pairs (j, k).

Finally, if the matrix $[w_{jk}]_{1 \leqslant j,k \leqslant n}$, solution of (12.33), is skew-symmetric, the Bézout identity is preserved, as we have already observed. Moreover, $\sum_{k=1}^{n} h_k f_k = 1$, hence $\sum_{k=1}^{n} f_k \overline{\partial} h_k = 0$ since the f_k are holomorphic, and

$$
\overline{\partial} g_j = \overline{\partial} h_j + \sum_{k=1}^{n} \left(h_j \overline{\partial} h_k - h_k \overline{\partial} h_j \right) f_k
$$

$$
= \overline{\partial} h_j + h_j \underbrace{\sum_{k=1}^{n} f_k \overline{\partial} h_k}_{=0} - \overline{\partial} h_j \underbrace{\sum_{k=1}^{n} h_k f_k}_{=1}
$$

$$
= \overline{\partial} h_j - \overline{\partial} h_j = 0.
$$

The algebraic aspect of the problem is thus resolved with (12.33).

12.6.5 Analytic aspect

The system (12.33) leaves a great deal of latitude in the choice of the w_{jk}; it remains to make a choice, if possible, providing a good control of the $\|w_{jk}\|_\infty$. In his article [92], Hörmander simply observes that this control is already present in Carleson's work, and he does not have an improvement to suggest on this point. Such an improvement in the proof was obtained by Wolff [189], and will now be presented.

Let us fix j and k in $[\![1, n]\!]$, and set

$$\varphi = h_{jk} = h_j \overline{\partial} h_k - h_k \overline{\partial} h_j.$$

The function φ is C^∞ in a neighbourhood of $\overline{D}$, hence (after multiplication, if necessary, by a C^∞ function with compact support that equals 1 in a neighbourhood of $\overline{D}$) we can replace it by a C^∞ function from $\mathbb{C}$ to $\mathbb{C}$, with compact support. The prerequisite of Section 12.2.6 allows us to claim that the equation $\overline{\partial} w = \varphi$ always has a solution w_0 that is C^∞ in a neighbourhood of $\overline{D}$.

Here, a clarification is required: in this section, we are going to sometimes work on the boundary of D, and in consequence we set

$$\|w\|_\infty = \sup_{|z|=1} |w(z)|,$$

concurrent with the notation for H^∞:

$$\|f\|_\infty = \sup_{z \in D} |f(z)| \text{ if } f \in H^\infty.$$

If we can have control of $\|w_{jk}\|_\infty$, with $\overline{\partial} w_{jk} = h_{jk}$, carrying this into (12.33) will give control of $\sup_{|z|=1} |g_j(z)|$. But the function g_j is holomorphic and bounded, and hence satisfies the maximum principle! We will thus also know how to control $\sup_{z \in D} |g_j(z)| = \|g_j\|_\infty$.

Let us return to $\overline{\partial} w = \varphi$. We have a solution w_0 given by the prerequisites, and we seek a solution w with norm $\|\cdot\|_\infty$ as small as possible, it being understood that another solution $w = w_0 + h$ will satisfy $\overline{\partial} w = \varphi$, so that $\overline{\partial} h = 0$ and hence $h \in H^\infty$. We thus see, where H^∞ appears here as $H^\infty(\mathbb{T})$, that

$$\inf_{\overline{\partial} w = \varphi} \|w\|_\infty = \inf_{\overline{\partial} h = 0} \|w_0 + h\|_\infty = \inf_{h \in H^\infty(\mathbb{T})} \|w_0 + h\|_\infty = \|w_0\|_{L^\infty / H^\infty}.$$

To estimate this quotient norm, we use the prerequisite (12.8), with

$$X = L^1 \text{ and } E = H_0^1 = \{f \in H^1 / \widehat{f}(0) = 0\},$$

the dual space of L^1 being L^∞ with the duality[28]

$$\langle f, g \rangle = \int_{\mathbb{T}} fg \, dm.$$

For this duality, $E^\perp$ is thus none other than H^∞. Indeed, thanks to the prerequisite of Section 12.2.8 (Theorem 12.2.9), we have

$$E^\perp = \left\{ f \in L^\infty / \int_{\mathbb{T}} f(e^{it})e^{int} \, dm = 0 \text{ for all } n \geqslant 1 \right\}$$
$$= \{ f \in L^\infty / \widehat{f}(n) = 0 \text{ for all } n \leqslant -1 \}$$
$$= H^\infty.$$

This prerequisite thus gives us

$$\overline{\partial} w_0 = \varphi \text{ and } \inf_{\overline{\partial} w = \varphi} \|w\|_\infty = \sup_{\substack{F \in H_0^1 \\ \|F\|_1 \leqslant 1}} \left| \int_{\mathbb{T}} w_0 F \, dm \right|. \qquad (12.34)$$

To estimate the right-hand side of (12.34), Wolff moves *to the interior of D* using formula (12.4) of the prerequisites (the Riesz formula):

$$\int_{\mathbb{T}} w_0 F \, dm = w_0(0)F(0) + \frac{1}{4} \int_D \Delta(w_0 F) d\lambda_1 = \frac{1}{4} \int_D \Delta(w_0 F) d\lambda_1,$$

since $F(0) = 0$. Let us calculate $\Delta(w_0 F)$. First of all, F being analytic,

$$\overline{\partial}(w_0 F) = F\overline{\partial} w_0 = \varphi F.$$

Next,

$$\partial\overline{\partial}(w_0 F) = \partial(\varphi F) = \varphi \, \partial F + F \partial \varphi$$

and finally, since $\partial F = F'$:

$$\boxed{\inf_{\overline{\partial} w = \varphi} \|w\|_\infty = \sup_{\substack{F \in H_0^1 \\ \|F\|_1 \leqslant 1}} \left(\left| \int_D F' \varphi \, d\lambda_1 + \int_D F \partial \varphi \, d\lambda_1 \right| \right).} \qquad (12.35)$$

We must now estimate the right-hand side of (12.35). Fortunately, in the calculation of $\varphi = h_j \overline{\partial} h_k - h_k \overline{\partial} h_j$, simplifications will help us in this task. Indeed, $h_j = \overline{f_j} \psi^{-1}$, hence

$$\overline{\partial} h_j = \overline{f_j'} \psi^{-1} - \overline{f_j} \psi^{-2} \overline{\partial} \psi,$$

so that

$$\varphi = \overline{f_j} \, \overline{f_k'} \psi^{-2} - \overline{f_j f_k} \psi^{-3} \overline{\partial} \psi - \left(\overline{f_k} \, \overline{f_j'} \psi^{-2} - \overline{f_k f_j} \psi^{-3} \overline{\partial} \psi \right),$$

---

[28] Different from that of Section 12.4.

thus finally

$$\varphi = \overline{f_j f_k' - f_k f_j'}\, \psi^{-2}. \tag{12.36}$$

We will now be able to complete the proof of the corona theorem, via two elegant lemmas.

12.6.5.1 Estimation of $\int_D F'\varphi\, d\lambda_1$

If we use brute force to bound $\left|\int_D F'\varphi d\lambda_1\right|$ by $\int_D |F'\varphi| d\lambda_1$, ψ^{-1} by δ^{-2} and the f_j by 1 (given the preceding expression for φ), we see that $\left|\int_D F'\varphi d\lambda_1\right|$ is bounded by two integrals of the form $\int_D |F'f'| d\lambda_1$, with $f \in H^\infty$ and $\|f\|_\infty \leqslant 1$. This corresponds to a certain duality (*the duality H^1-BMOA*, which we will discuss later). Next, we would like to bound these integrals, using only arguments from the prerequisites; Gamelin has done precisely that [65] in the following lemma (or rather in its proof).

12.6.6 Lemma *If $F \in H^1$ and $f \in H^\infty$, then*

$$\int_D |F'f'| d\lambda_1 \leqslant 4\|F\|_1 \|f\|_\infty. \tag{12.37}$$

Proof We can suppose that $\|f\|_\infty = 1$. Thanks to the factorisation Theorem 12.2.10, we can write

$$F = g_1 g_2, \text{ with } \|g_1\|_2^2 = \|g_2\|_2^2 = \|F\|_1.$$

Thus $F'f' = g_1'g_2 f' + g_1 g_2' f'$ and by Cauchy–Schwarz:

$$\int_D |g_1'g_2 f'| d\lambda_1 \leqslant \left(\int_D |g_1'|^2 d\lambda_1\right)^{1/2} \left(\int_D |g_2 f'|^2 d\lambda_1\right)^{1/2}.$$

According to the Littlewood–Paley identity (12.7), the first integral is bounded above by $\|g_1\|_2 = \|F\|_1^{1/2}$. For the second, we perform an integration by parts to focus the differentiation on g_2, that is, we write $g_2 f' = (g_2 f)' - g_2' f$. Thus, using the triangle inequality in $L^2(\lambda_1)$:

$$\left(\int_D |g_2 f'|^2 d\lambda_1\right)^{1/2} \leqslant \left(\int_D |(g_2 f)'|^2 d\lambda_1\right)^{1/2} + \left(\int_D |g_2' f|^2 d\lambda_1\right)^{1/2}$$

$$\leqslant \|fg_2\|_2 + \left(\int_D |g_2'|^2 d\lambda_1\right)^{1/2}$$

$$\leqslant \|g_2\|_2 + \|g_2\|_2 = 2\|F\|_1^{1/2}.$$

We can similarly bound the term corresponding to $g_1 g_2' f'$, and we finally obtain (12.37). $\qquad\square$

12.6.6.1 Estimation of $\int_D F \, \partial\varphi \, d\lambda_1$

Remember that

$$\varphi = \overline{(f_j f_k' - f_k f_j')} \, \psi^{-2}, \quad \text{where } \psi = \sum_{p=1}^n f_p \overline{f_p}.$$

Because $f_j f_k' - f_k f_j'$ is holomorphic, the calculation of $\partial\varphi$ is easy:

$$\partial\varphi = -2\overline{(f_j f_k' - f_k f_j')}\psi^{-3}\partial\psi = -2\overline{(f_j f_k' - f_k f_j')}\psi^{-3}\sum_{p=1}^n f_p' \overline{f_p}.$$

Here again, using brute force as in Section 12.6.5.1, we see that (up to an evident change of notation) $\left|\int_D F \, \partial\varphi \, d\lambda_1\right|$ is bounded by a sum of $2n$ integrals of the form $\int_D |Ff_1'f_2'|d\lambda_1$, with $f_1, f_2 \in H^\infty$ such that $\|f_j\|_\infty \leqslant 1$. Hence we must estimate these integrals; the differentiation has disappeared from the term in F but the price to pay is to have a product of two derivatives of functions in H^∞, with norms bounded by 1. This estimation also requires a lemma (whose elementary proof is again due to Gamelin [65]), but in fact, we have already proved this lemma while showing Lemma 12.6.6!

12.6.7 Lemma   *If $F \in H^1$ and $f_1, f_2 \in H^\infty$, we have*

$$\int_D |Ff_1'f_2'|d\lambda_1 \leqslant 4\|F\|_1\|f_1\|_\infty\|f_2\|_\infty.$$

Proof   We can suppose that $\|f_1\|_\infty = \|f_2\|_\infty = 1$. We factorise F as in Lemma 12.6.6 and again apply Cauchy–Schwarz. This gives

$$\int_D |Ff_1'f_2'|d\lambda_1 \leqslant \left(\int_D |g_1 f_1'|^2 d\lambda_1\right)^{1/2}\left(\int_D |g_2 f_2'|^2 d\lambda_1\right)^{1/2}.$$

However, the integrals on the right-hand side were already estimated in the proof of Lemma 12.6.6, each being bounded above by $2\|F\|_1^{\frac{1}{2}}$, hence the result. $\qquad\square$

Now, if we recapitulate, the proof of the corona theorem is complete.

12.6.8 Towards a version without dimension of the corona theorem

The key in Lemmas 12.6.6 and 12.6.7 is an inequality[29] of the form

$$\int_D |uf'|^2 d\lambda_1 \leqslant K^2\|f\|_\infty^2 \int_{\mathbb{T}} |u|^2 \, dm \quad \text{for } u \in H^2 \text{ and } f \in H^\infty. \quad (12.38)$$

---

[29] Itself deduced from the Littlewood–Paley identity and the fact that H^∞ is the set of multipliers of H^2.

This inequality means that if $\mu = |f'|^2 d\lambda_1$, the identity

$$H^2 \to L^2(\mu), u \mapsto u$$

is continuous with a norm bounded by $K \| f \|_\infty$, K being a numerical constant. That is (as we will see in Section 12.7) μ is a Carleson measure, with Carleson norm bounded by $K_1 \| f \|_\infty$, K_1 being a numerical constant. Moreover, if $a = |f|^2$, then $\Delta a = 4|f'|^2$, thus the inequality (12.38) can also be written

$$\int_D |u|^2 \Delta a \, d\lambda_1 \leqslant 4K^2 \|a\|_\infty \int_{\mathbb{T}} |u|^2 \, dm \text{ for all } u \in H^2,$$

and this new form can be generalised as follows [5, p. 136].

12.6.9 Lemma *Let a be a real subharmonic ($\Delta a \geqslant 0$) C^2 function in a neighbourhood of $\overline{D}$. Then we have the* a priori *inequality*

$$\int_D |u|^2 \Delta a \, d\lambda_1 \leqslant 8e \|a\|_\infty \int_{\mathbb{T}} |u|^2 \, dm \text{ for all } u \in H^2. \tag{12.39}$$

Proof We first suppose that $0 \leqslant a \leqslant 1$ on $\overline{D}$. Then, for all $u \in H^2$,

$$\Delta(|u|^2 e^a) \geqslant |u|^2 \Delta a. \tag{12.40}$$

Indeed, set $f = |u|^2$ and $g = e^a$. It is easy to see that

$$\Delta(fg) = f \Delta g + g \Delta f + 8 \operatorname{Re}(\partial f \, \overline{\partial g})$$

and that (chain rule)

$$\Delta f = 4|u'|^2, \ \Delta g = e^a \Delta a + 4|\partial a|^2 e^a, \ \partial f = u' \overline{u} \text{ and } \partial g = e^a \partial a,$$

hence

$$\begin{aligned}
\Delta(fg) &= |u|^2 e^a (\Delta a + 4|\partial a|^2) + 4|u'|^2 e^a + 8e^a \operatorname{Re}(u' \overline{u} \, \overline{\partial a}) \\
&= |u|^2 e^a \Delta a + 4e^a \big(|u|^2 |\partial a|^2 + |u'|^2 + 2 \operatorname{Re}(u' \overline{u} \, \overline{\partial a}) \big) \\
&= |u|^2 e^a \Delta a + 4e^a |u' + u \, \partial a|^2 \geqslant |u|^2 e^a \Delta a \geqslant |u|^2 \Delta a.
\end{aligned}$$

The positivity of Δa intervenes only in the last inequality written above. Next, (12.40) and the Riesz formula (12.4) show that

$$\begin{aligned}
\int_D |u|^2 \Delta a \, d\lambda_1 &\leqslant \int_D \Delta(|u|^2 e^a) d\lambda_1 = 4 \int_{\mathbb{T}} |u|^2 e^a \, dm - 4|u(0)|^2 e^{a(0)} \\
&\leqslant 4 \int_{\mathbb{T}} |u|^2 e^a \, dm \leqslant 4e \int_{\mathbb{T}} |u|^2 \, dm.
\end{aligned}$$

In the general case, if $C = \|a\|_\infty$ and $\alpha = \dfrac{a+C}{2C}$, we have $0 \leqslant \alpha \leqslant 1$ and $\Delta\alpha = \dfrac{\Delta a}{2C}$. Hence, by the preceding special case:

$$\int_D |u|^2 \frac{\Delta a}{2C} d\lambda_1 \leqslant 4e \int_{\mathbb{T}} |u|^2 \, dm,$$

which proves (12.39). □

The interest of Lemma 12.6.9 is considerable; in (12.32) it will lead to an estimation of the form

$$\sum_{j=1}^n |g_j(z)|^2 \leqslant C(\delta) \text{ for all } z \in D,$$

where the constant $C(\delta)$ is dimension-free, that is it does not depend on the number n of functions $f_1, \ldots, f_n$ considered; it is a first step towards vectorial versions of the corona theorem, which we will now examine.

But first of all, let us make an assessment of what we have just proved in this Section 12.6.

12.6.10 Theorem [corona theorem] *Let $\delta \in]0, 1[$, $n \geqslant 1$ and $f_1, \ldots, f_n \in H^\infty$ satisfying*

$$\delta^2 \leqslant \sum_{j=1}^n |f_j(z)|^2 \leqslant 1 \text{ for all } z \in D.$$

Then there exist $g_1, \ldots, g_n \in H^\infty$ such that

(i) $\sum_{j=1}^n f_j g_j = 1$,

(ii) $\sum_{j=1}^n |g_j(z)|^2 \leqslant C(n, \delta)$ *for all $z \in D$, where $C(n, \delta)$ is a positive constant depending only on n and δ.*

Here now is the improvement obtained by Tolokonnikov [136], which uses Lemma 12.6.9 instead of Lemmas 12.6.6 and 12.6.7, and a treatment *less brutal, and more vectorial* of the terms composing $F'\varphi$ and $F \partial\varphi$.

12.6.11 Theorem [dimension-free corona theorem] *In the preceding statement, one can replace conclusion (ii) by*

(ii)' $\left(\sum_{j=1}^n |g_j(z)|^2 \right)^{1/2} \leqslant C(\delta)$ *for all $z \in D$,*

where $C(\delta)$ depends only on δ. More precisely, we can take

$$C(\delta) = C\delta^{-4} \sqrt{\ln \frac{1}{\delta}},$$

where C is a numerical constant.

For the proof, we are going to work in the following vectorial context: the space B is a complex Hilbert space of finite dimension N, and $(u_1, \ldots, u_N)$ is a fixed orthonormal basis of B. If $1 \leqslant p \leqslant \infty$, we define $H^p(D, B)$ (abbreviated $H^p(B)$ when there is no risk of confusion) as the set of holomorphic maps[30] $f : D \to B$ such that

$$\|f\|_{H^p(B)} := \sup_{0 \leqslant r < 1} \left(\int_{\mathbb{T}} \|f(re^{it})\|^p \, dm(t) \right)^{1/p} < +\infty$$

(with the evident adaptation if $p = \infty$), where $\|\cdot\|$ denotes the norm associated with the Hilbertian inner product $(\cdot \mid \cdot)$ on B, that is

$$(x|y) = \sum_{j=1}^{N} x_j \overline{y_j} \text{ if } x = \sum_{j=1}^{N} x_j u_j \text{ and } y = \sum_{j=1}^{N} y_j u_j.$$

It will also be convenient to introduce the symmetric bilinear functional defined on B by the formula

$$\langle x, y \rangle = \sum_{j=1}^{N} x_j y_j.$$

This form depends of course on the choice of the basis $(u_1, \ldots, u_N)$, but this basis is destined to stay fixed. If $f \in H^p(D, B)$, each component f_j of f belongs to $H^p(D)$, and hence has m-almost everywhere a radial limit

$$f_j^*(e^{it}) = \lim_{r \nearrow 1} f_j(re^{it}),$$

which, for f, gives the radial limit $f^* = \sum_{j=1}^{N} f_j^* u_j$, and we have

$$\|f\|_{H^p(B)} = \|f^*\|_{L^p(\mathbb{T})}.$$

The prerequisite theorems are thus transformed as follows.

12.6.12 Theorem [factorisation]   *Let $f \in H^1(D, B)$. Then there exist $g_1 \in H^2(D)$ and $g_2 \in H^2(D, B)$ such that*

$$f = g_1 g_2 \text{ and } \|g_1\|_{H^2(D)} = \|g_2\|_{H^2(D,B)} = \|f\|_{H^1(D,B)}^{1/2},$$

in the sense that $f(z) = g_1(z)g_2(z)$ for $z \in D$.

Proof (shortened).   We can suppose $f \neq 0$; thus, for example, $f_1 \neq 0$, hence $\int_{\mathbb{T}} \ln |f_1^*| \, dm > -\infty$ according to a classical result [90], and *a fortiori* $\int_{\mathbb{T}} \ln \|f^*\|^{1/2} \, dm > -\infty$. Then, by a theorem of Szegö [90], there exists

---

[30] That is to say $f(z) = \sum_{j=1}^{N} f_j(z)u_j$, where each f_j is complex-valued and holomorphic in D.

an outer function $F \in H^2(D)$ such that $|F^*(e^{it})| = \|f^*(e^{it})\|^{1/2}$ almost everywhere on $\mathbb{T}$. We set $g_1 = F$ and $g_2 = \dfrac{f}{F}$. Thus

$$\|g_1\|^2_{H^2(B)} = \int_{\mathbb{T}} |F^*(e^{it})|^2 \, dm = \int_{\mathbb{T}} \|f^*(e^{it})\| \, dm = \|f\|_{H^1(B)},$$

and, F being an outer function, $\|f(z)\| \leqslant |F(z)|^2$ if $z \in D$, so that $\|g_2(z)\| \leqslant |F(z)|$ and $g_2 \in H^2(B)$, with

$$\begin{aligned} \|g_2\|^2_{H^2(B)} &= \int_{\mathbb{T}} \|g_2^*(e^{it})\|^2 \, dm \\ &= \int_{\mathbb{T}} \frac{\|f^*(e^{it})\|^2}{|F^*(e^{it})|^2} \, dm \\ &= \int_{\mathbb{T}} \|f^*(e^{it})\| \, dm = \|f\|_{H^1(B)}. \end{aligned}$$

Moreover, $f = g_1 g_2$ by construction. $\square$

12.6.13 Theorem [vectorial Littlewood–Paley identity]   *Let $f \in H^2(B)$. Then we have*

$$\|f\|^2_{H^2(B)} = \|f(0)\|^2 + \int_D \|f'(z)\|^2 d\lambda_1(z).$$

Proof   (easy). It suffices to write the scalar Littlewood–Paley identity for each component f_j of f, and sum up the results. $\square$

The resolution of $\bar\partial$ can also be done coordinate by coordinate. As for duality in Banach spaces, it suffices to observe the following facts.

Fact 1. If $X = L^1(B)$, $X^* = L^\infty(B)$ with the duality

$$\varphi(f, g) = \int_{\mathbb{T}} \langle f, g \rangle \, dm \text{ for } (f, g) \in L^1(B) \times L^\infty(B).$$

Fact 2. If $E = H_0^1(B) = \{f \in H^1(B)/f(0) = 0\}$, we have $E^\perp = H^\infty(B)$. Indeed,

$$E^\perp = \{g \in L^\infty(B)/\varphi(e^{int}x, g) = 0 \text{ for } x \in B \text{ and } n \geqslant 1\}.$$

But

$$\varphi(e^{int}x, g) = \int_{\mathbb{T}} \langle e^{int}x, g \rangle \, dm = \left\langle x, \int_{\mathbb{T}} e^{int} g(t) \, dm(t) \right\rangle = \langle x, \widehat{g}(-n) \rangle.$$

Hence

$$g \in E^\perp \Leftrightarrow \widehat{g}(-n) = 0 \text{ if } n \geqslant 1 \Leftrightarrow \widehat{g}(p) = 0 \text{ if } p \leqslant -1 \Leftrightarrow g \in H^\infty(B).$$

Let us revisit Lemmas 12.6.6, 12.6.7 and 12.6.9. First of all, we have the following result.

12.6.14 Lemma *Let a be a real subharmonic ($\Delta a \geqslant 0$) C^2 function in a neighbourhood of $\overline{D}$ and let $u \in H^2(B)$. Then we have the* a priori *inequality*

$$\int_D \|u\|^2 \Delta a \, d\lambda_1 \leqslant 8e\|a\|_\infty \int_{\mathbb{T}} \|u\|^2 \, dm.$$

Proof (easy). We apply (12.39) to each coordinate of u, and sum. $\qquad \square$

We proceed with the proof of the dimension-free corona theorem. We will work with *two* Hilbert spaces B_0 and B_1: the space B_0 will be the usual Hermitian space $\mathbb{C}^n$ with its canonical basis, and B_1 the space $\mathcal{M}_n(\mathbb{C})$ of matrices $n \times n$ with complex coefficients, with its canonical basis and associated Hermitian product[31]

$$(M|N) = \sum_{1 \leqslant j,k \leqslant n} m_{jk}\overline{n_{jk}}, \quad \text{if } M = [m_{jk}]_{1\leqslant j,k\leqslant n} \text{ and } N = [n_{jk}]_{1\leqslant j,k\leqslant n}.$$

If $f_1, \ldots, f_n$ are as in the statement of the theorem, we set $f = (f_1, \ldots, f_n) \in H^\infty(B_0)$ with

$$\delta^2 \leqslant \|f(z)\|_{B_0}^2 \leqslant 1 \quad \text{for all } z \in D.$$

With the same notations as in the scalar case, we set

$$\psi = \|f\|_{B_0}^2, h_j = \overline{f_j}\psi^{-1}, h_{jk} = h_j\overline{\partial}h_k - h_k\overline{\partial}h_j,$$

and

$$h = [h_{jk}]_{1\leqslant j,k\leqslant n} : D \to B_1.$$

We also consider a particular solution w_0 of the PDE

$$\overline{\partial}w = h, \text{ in the sense where } \overline{\partial}w_{jk} = h_{jk} \text{ for } 1 \leqslant j, k \leqslant n.$$

This time, as the orthogonal of $H_0^1(B_1)$ is $H^\infty(B_1)$, the duality in Banach spaces gives

$$\inf_{\overline{\partial}w=h} \|w\|_{L^\infty(B_1)} = \sup_{\substack{F \in H_0^1(B_1) \\ \|F\|_1 \leqslant 1}} \left| \int_{\mathbb{T}} \langle w_0, F \rangle \, dm \right|. \tag{12.41}$$

To estimate this lower bound, first we use the Riesz formula to move into D:

$$\int_{\mathbb{T}} \langle w_0, F \rangle \, dm = \frac{1}{4} \int_D \Delta(\langle w_0, F \rangle)d\lambda_1.$$

---

[31] In a sophisticated language, B_1 is the Schatten class $S_2(B_0)$ associated with B_0, or again the space of operators on B_0 with the Hilbert–Schmidt norm.

We calculate this Laplacian by writing $w_0 = [w_{jk}]$ and $F = [F_{jk}]$. We thus have

$$\langle w_0, F \rangle = \sum w_{jk} F_{jk} =: \chi,$$

so that

$$\bar{\partial} \chi = \sum F_{jk} \bar{\partial} w_{jk} = \sum F_{jk} h_{jk},$$

and then

$$\partial \bar{\partial} \chi = \sum F'_{jk} h_{jk} + \sum F_{jk} \partial h_{jk} = \langle F', h \rangle + \langle F, \partial h \rangle.$$

As $\partial \bar{\partial} = \dfrac{1}{4} \Delta$, relation (12.41) becomes

$$\inf_{\bar{\partial} w = h} \| w \|_{L^{\infty}(B_1)} = \sup_{\substack{F \in H_0^1(B_1) \\ \|F\|_1 \leqslant 1}} \left(\left| \int_D \langle F', h \rangle d\lambda_1 + \int_D \langle F, \partial h \rangle d\lambda_1 \right| \right). \quad (12.42)$$

It remains to estimate the right-hand side of (12.42). For this, the following identities will be useful.

12.6.15 Lemma *We have the two identities*

(1) $\| h \|_{B_1}^2 = \dfrac{1}{2} \psi^{-2} \Delta (\ln \psi)$,

(2) $\partial h = -2\psi^{-1}(f'|f) h$.

Proof For (1), according to a calculation already done in the scalar case,

$$h_{jk} = \overline{(f_j f'_k - f_k f'_j)} \psi^{-2},$$

hence

$$\| h \|_{B_1}^2 = \psi^{-4} \sum_{1 \leqslant j, k \leqslant n} \left(|f_j|^2 |f'_k|^2 + |f_k|^2 |f'_j|^2 - f_j f'_k \overline{f_k} \, \overline{f'_j} - f_k f'_j \overline{f_j} \, \overline{f'_k} \right)$$

$$= 2\psi^{-4} \left(\| f \|_{B_0}^2 \| f' \|_{B_0}^2 - \Big(\sum_{j=1}^n f_j \overline{f'_j} \Big) \Big(\sum_{k=1}^n f'_k \overline{f_k} \Big) \right)$$

$$= 2\psi^{-4} \left(\| f \|_{B_0}^2 \| f' \|_{B_0}^2 - |(f|f')|^2 \right).$$

Moreover, $\bar{\partial} \ln \psi = \dfrac{\bar{\partial} \psi}{\psi}$, hence $\partial \bar{\partial} \ln \psi = \dfrac{\partial \bar{\partial} \psi}{\psi} - \psi^{-2} \partial \psi \bar{\partial} \psi$, and, as $\psi = \sum f_j \overline{f_j} = \| f \|_{B_0}^2$, the f_j being holomorphic, we have

$$\bar{\partial} \psi = \overline{\partial \psi} = \sum f_j \overline{\partial f_j} = (f|f'),$$

thus

$$\partial\bar{\partial}\psi = \sum \partial f_j \overline{\partial f_j} = \sum |\partial f_j|^2 = \sum |f_j'|^2 = \|f'\|_{B_0}^2.$$

From all of this, we deduce that

$$\Delta(\ln\psi) = \frac{\Delta\psi}{\psi} - 4\psi^{-2}|\partial\psi|^2 = 4(\psi^{-1}\|f'\|_{B_0}^2 - \psi^{-2}|\partial\psi|^2)$$
$$= 4\psi^{-2}(\|f\|_{B_0}^2 \|f'\|_{B_0}^2 - |(f|f')|^2),$$

which gives (1).

For (2), we have already seen[32] that

$$\partial h_{jk} = -2\overline{(f_j f_k' - f_k f_j')}\psi^{-3}\partial\psi,$$

therefore

$$\partial h_{jk} = -2\psi^{-1}\overline{(f_j f_k' - f_k f_j')}\psi^{-2}\partial\psi = -2\psi^{-1}h_{jk}\partial\psi = -2\psi^{-1}h_{jk}(f'|f),$$

according to the calculations of point (1). This completes the proof of the lemma. □

We are almost ready to conclude, with the following variants of Lemmas 12.6.6 and 12.6.7.

12.6.16 Lemma *If $F \in H^1(B_1)$, we have*

$$I_1 := \int_D |\langle F', h\rangle|d\lambda_1 \leqslant C_\delta \|F\|_1, \text{ with } C_\delta = C\delta^{-2}\sqrt{\ln\frac{1}{\delta}},$$

C being a numerical constant.

Proof We factorise F as in Theorem 12.6.12: $F = g_1 g_2$, hence $|\langle F', h\rangle| \leqslant \|g_1' g_2\| \cdot \|h\| + \|g_1 g_2'\| \cdot \|h\|$, the norm being that of B_1. We then treat separately the two terms of the upper bound, using Cauchy–Schwarz:

$$\left(\int_D |g_1'| \cdot \|g_2\| \cdot \|h\| d\lambda_1\right)^2 \leqslant \int_D |g_1'|^2 d\lambda_1 \int_D \|g_2\|^2 \|h\|^2 d\lambda_1.$$

The first integral on the right-hand side is bounded above by $\|g_1\|_2^2 = \|F\|_1$ thanks to the scalar Littlewood–Paley identity. For the second, we use Lemmas 12.6.15 and 12.6.14:

---

[32] This results from (12.36) and the fact that the $f_j f_k' - f_k f_j'$ are holomorphic.

$$\int_D \|g_2\|^2 \|h\|^2 d\lambda_1 = \frac{1}{2} \int_D \psi^{-2} \|g_2\|^2 \Delta(\ln\psi) d\lambda_1$$

$$\leqslant \frac{1}{2} \delta^{-4} \int_D \|g_2\|^2 \Delta(\ln\psi) d\lambda_1$$

$$\leqslant 4e\delta^{-4} \|\ln\psi\|_\infty \int_{\mathbb{T}} \|g_2\|^2 dm \leqslant 8e\delta^{-4} \ln\frac{1}{\delta} \|F\|_1,$$

since $\delta^2 \leqslant \psi \leqslant 1$, hence $|\ln\psi| \leqslant 2\ln\frac{1}{\delta}$. Similarly,

$$\left(\int_D |g_1| \cdot \|g_2'\| \cdot \|h\| d\lambda_1\right)^2 \leqslant \int_D \|g_2'\|^2 d\lambda_1 \int_D |g_1|^2 \|h\|^2 d\lambda_1$$

$$\leqslant \|g_2\|_2^2 \int_D |g_1|^2 \frac{1}{2} \psi^{-2} \Delta(\ln\psi) d\lambda_1$$

$$\leqslant \|F\|_1 \frac{1}{2} \delta^{-4} \int_D |g_1|^2 \Delta(\ln\psi) d\lambda_1$$

$$\leqslant 8e\delta^{-4} \ln\frac{1}{\delta} \|F\|_1^2,$$

by the vectorial Littlewood–Paley identity and Lemmas 12.6.15 and 12.6.9. By summing the two bounds, we indeed obtain Lemma 12.6.16. □

12.6.17 Lemma *If $F \in H^1(B_1)$, we have*

$$I_2 := \int_D |\langle F, \partial h \rangle| d\lambda_1 \leqslant C_\delta' \|F\|_1, \text{ with } C_\delta' = C\delta^{-4}\sqrt{\ln\frac{1}{\delta}},$$

C being a numerical constant.

Proof We factorise F as in Lemma 12.6.16 and use point (2) of Lemma 12.6.15 to obtain, via Cauchy–Schwarz:[33]

$$I_2 \leqslant \int_D \|F\|_{B_1} \|\partial h\|_{B_1} d\lambda_1 \leqslant \int_D |g_1| \cdot \|g_2\|_{B_1} 2\psi^{-1} \|f'\|_{B_0} \|f\|_{B_0} \|h\|_{B_1} d\lambda_1$$

$$\leqslant 2\delta^{-2} \left(\int_D |g_1|^2 \|h\|_{B_1}^2 d\lambda_1\right)^{1/2} \left(\int_D \|g_2\|_{B_1}^2 \|f'\|_{B_0}^2 d\lambda_1\right)^{1/2} \text{ because } \|f\|_{B_0} \leqslant 1$$

$$\leqslant 2\delta^{-2} \left(\int_D |g_1|^2 \frac{1}{2} \delta^{-4} \Delta(\ln\psi) d\lambda_1\right)^{1/2} \left(\int_D \|g_2\|_{B_1}^2 \frac{1}{4} \Delta\psi \, d\lambda_1\right)^{1/2}$$

$$\text{because } \Delta\psi = 4\|f'\|_{B_0}^2$$

$$\leqslant \delta^{-4} \left(\int_D |g_1|^2 \Delta(\ln\psi) d\lambda_1\right)^{1/2} \left(\int_D \|g_2\|_{B_1}^2 \Delta\psi \, d\lambda_1\right)^{1/2}$$

$$\leqslant \delta^{-4} \sqrt{8e} \sqrt{2\ln\frac{1}{\delta}} \left(\int_{\mathbb{T}} |g_1|^2 dm\right)^{1/2} \sqrt{8e} \left(\int_{\mathbb{T}} \|g_2\|_{B_1}^2 dm\right)^{1/2}.$$

[33] Under the form $|(f'|f)| \leqslant \|f'\|_{B_0} \|f\|_{B_0}$.

We have used Lemmas 12.6.9 and 12.6.14, and again the fact that

$$\| \ln \psi \|_\infty \leqslant 2 \ln \frac{1}{\delta} \quad \text{and} \quad \|\psi\|_\infty \leqslant 1.$$

Finally,

$$I_2 \leqslant 8\sqrt{2}e\delta^{-4} \sqrt{\ln \frac{1}{\delta}} \|F\|_1,$$

which completes the proof. $\qquad\qquad\square$

Back to (12.42), we see that we can solve the equation

$$\overline{\partial} w = h$$

with the control

$$\|w\|_{L^\infty(B_1)} \leqslant \sup_{\substack{F \in H_0^1 \\ \|F\|_1 \leqslant 1}} (I_1 + I_2) \leqslant C_\delta + C'_\delta \leqslant C\delta^{-4} \sqrt{\ln \frac{1}{\delta}},$$

where C is a numerical constant. But this global resolution of $\overline{\partial} w = h$ makes us forget the skew-symmetry of $w = [w_{jk}]$, if indeed we had it for w_0. We re-establish this skew-symmety as in the scalar case, while keeping control of the norm: indeed, set $\chi_{jk} = w_{jk}$ if $j < k$, $\chi_{jj} = 0$ and $\chi_{jk} = -w_{kj}$ if $j > k$. Then, $\chi = [\chi_{jk}]$ is skew-symmetric by construction, and we have $\overline{\partial}\chi_{jk} = h_{jk}$ for all j and k. Finally, for $z \in D$,

$$\|\chi(z)\|_{B_1}^2 = 2 \sum_{j<k} |w_{jk}(z)|^2 \leqslant 2\|w(z)\|_{B_1}^2,$$

hence

$$\|\chi\|_{L^\infty(B_1)} \leqslant \sqrt{2}\|w\|_{L^\infty(B_1)}.$$

We can thus suppose that w is skew-symmetric in what follows.

If then

$$g = (g_1, \ldots, g_n) = h + Af,$$

with $A = [w_{jk}]$ skew-symmetric, we preserve our Bézout identity

$$\langle g, f \rangle = \sum_{j=1}^n g_j f_j = 1,$$

and moreover, for $z \in D$,

$$\|g(z)\|_{B_0} \leqslant \|h(z)\|_{B_0} + \|A(z)\|_{B_1}\|f(z)\|_{B_0},$$

because the norm of $A(z)$, seen as an operator on B_0, is bounded above by its Hilbert–Schmidt norm, if $A(z)$ is now seen as an element of the space B_1. However,

$$\|h(z)\|_{B_0} = \psi^{-1}(z)\|f(z)\|_{B_0} = \|f(z)\|_{B_0}^{-1} \leqslant \delta^{-1},$$

and

$$\|A(z)\|_{B_1} \leqslant C\delta^{-4}\sqrt{\ln\frac{1}{\delta}}$$

as we have just seen, whereas $\|f(z)\|_{B_0} \leqslant 1$ by hypothesis. Finally,

$$\|g(z)\|_{B_0} \leqslant C\delta^{-4}\sqrt{\ln\frac{1}{\delta}},$$

where C is an absolute constant (which changed over the last few lines!). This completes the proof of the dimension-free corona theorem: the constant $C(\delta)$ could be improved [136], nonetheless it deserves credit for not depending on the dimension n.

12.7 Carleson's initial proof and Carleson measures

Carleson obtained many superb results, among them the corona theorem (of course) but also, somewhat earlier, the characterisation, seen in Section 12.4, of interpolating sequences of H^∞. Remember that these are sequences $(z_n)_{n\geqslant 1}$ of points of D with the following property: for any bounded sequence $(w_n)_{n\geqslant 1}$ of complex numbers, there exists a *bounded* holomorphic function f in D realising the infinite Lagrange interpolation $f(z_n) = w_n$ for all $n \geqslant 1$.

Carleson's Theorem 12.4.4 characterises these sequences: let d be the pseudo-hyperbolic distance on D, defined by

$$d(a,b) = \left|\frac{a-b}{1-\bar{a}b}\right|.$$

Next, set

$$\delta_j = \prod_{k\neq j} d(z_k, z_j).$$

Then

$(z_n)_{n\geqslant 1}$ is an interpolating sequence if and only if $\inf_{j\geqslant 1} \delta_j = \delta > 0.$ (12.43)

12.7.1 Remark The interpolating sequences are closely related to the corona problem in the following manner: if $f_1, \ldots, f_n \in H^\infty$ satisfy $\sum_{j=1}^n |f_j(z)| \geqslant \delta$ and if one of these functions, for example f_n, is a Blaschke product whose zeros $(z_p)_{p\geqslant 1}$ satisfy (12.43), it is easy to construct a Bézout identity. Indeed, the hypothesis allows us to partition $\mathbb{N}^*$ as $N_1, \ldots, N_{n-1}$, N_j being a set of integers satisfying

$$|f_j(z_p)| \geqslant \frac{\delta}{n-1} \text{ for } p \in N_j.$$

By the equivalence (12.43), we can find $g_j \in H^\infty$ ($1 \leqslant j \leqslant n-1$) such that

$$g_j(z_p) = \begin{cases} \dfrac{1}{f_j(z_p)} & \text{if } p \in N_j, \\ 0 & \text{otherwise.} \end{cases}$$

But then, the function $1 - \sum_{j=1}^{n-1} f_j g_j$ is zero at each z_p, hence (see Theorem 12.2.10 or Exercise 12.5) it is of the form $g_n f_n$ with $g_n \in H^\infty$, so that $\sum_{j=1}^n f_j g_j = 1$.

As for the general case of the corona theorem, its proof, soon afterwards obtained by Carleson [36], relies on a very hard lemma concerning the Blaschke products (see [53], p. 203, lemma 1), whose proof can be considered as a (super-)elaboration of the preceding special case. While doing research on the interpolating sequences of H^∞, Carleson suppressed a parasitic additional condition[34] [90], and drew the following notion, which requires two preliminary definitions.

We denote by $\mathcal{M}^+$ the set of positive finite Borel measures on D. If $a \in \mathbb{T}$ and $h \in \ ]0, 1[$, the *Carleson window* $W(a, h)$ is the set of $z \in D$ such that $|z| \geqslant 1 - h$ and $|\arg(\overline{a}z)| \leqslant h$. See Figure 12.1.

We say that $\mu \in \mathcal{M}^+$ is a *geometric Carleson measure* if there exists a constant λ such that $\mu(W(a, h)) \leqslant \lambda h$ for all windows $W(a, h)$; the best constant is known as the Carleson norm of μ, and is denoted by $\|\mu\|_C$. We then have the following result.

12.7.2 Theorem [Carleson embedding theorem [66, 132]] *For $\mu \in \mathcal{M}^+$, the following properties are equivalent:*

(i) *μ is a geometric Carleson measure;*
(ii) *μ is a Carleson measure, that is, there exists some $p > 0$ such that the canonical inclusion $j : H^p \to L^p(\mu)$ is continuous,*

$$\int_D |f(z)|^p d\mu(z) \leqslant C_p^p \int_{\mathbb{T}} |f|^p \, dm \text{ for all } f \in H^p,$$

where C_p is a numerical constant;
(iii) *for all $p > 0$, the canonical inclusion $j : H^p \to L^p(\mu)$ is continuous.*

---

[34] Which happens to be an immediate consequence of (12.43).

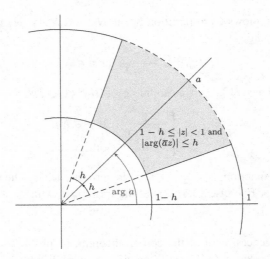

Figure 12.1

Moreover, in this case, for p fixed, C_p^p is equivalent to $\|\mu\|_C$, that is, their quotient is bounded above and below by positive numerical constants.

Another characterisation of interpolating sequences in terms of Carleson measures [66] is Theorem 12.4.13:

$(z_n)_{n \geqslant 1}$ *is an interpolating sequence if and only if it is separated and*

$$\mu = \sum_{n=1}^{\infty} (1 - |z_n|)\delta_{z_n}$$

is a Carleson measure

or again:

μ *is a Carleson measure if and only if* $(z_n)_{n \geqslant 1}$ *is a finite union of interpolating sequences*

(see Exercise 12.10 for an application of this result).

In short, as soon as the characterisation (12.43) is known, the Carleson measures are in the air. They notably reappeared in the 1970s, when Fefferman and Stein [61] introduced their famous bounded mean oscillation (BMO) space and identified it as the dual space of H^1. To remain in the context of this chapter, we will limit ourselves to the BMOA space of *holomorphic* functions of bounded mean oscillation. It can be described as the set of functions $f \in H^2$ whose values f^* at the boundary of D are in BMO, that is

$$\sup_I \frac{1}{m(I)} \int_I |f^* - f_I| \, dm =: [f] < +\infty,$$

where I runs over the set of non-trivial arcs of $\mathbb{T}$, and

$$f_I = \frac{1}{m(I)} \int_I f^* \, dm$$

is the mean of f^* over I. The BMOA space is a non-separable Banach space for the norm

$$\|f\|_* = [f] + \|f\|_2.$$

This definition makes obvious the continuous inclusion of H^∞ in BMOA (strict inclusion, as $\ln(1 - z) \in$ BMOA). One of the interests of this space lies in the following theorem [5].

12.7.3 Theorem [Fefferman–Stein duality theorem] *The dual space of H^1 is isomorphically identified with BMOA.*

Here is the meaning of this statement (see Exercise 12.11): if $g \in$ BMOA, the linear functional L_g, defined on H^2 by

$$L_g(f) = \int_\mathbb{T} f g \, dm,$$

can be extended to a continuous linear functional (still denoted by L_g) on H^1, in a necessarily unique manner since H^2 is dense in H^1. Moreover, we have

$$C_1 \|g\|_* \leqslant \|L_g\| \leqslant C_2 \|g\|_*,$$

C_1 and C_2 being numerical constants. Finally, $g \mapsto L_g$ is a surjection from BMOA onto the dual space of H^1. In fact, the identity (12.8) shows that the dual space of H^1 is isometric to L^∞/H^∞. We have here a more concrete description of this dual, avoiding the passage to the quotient, which again shows the non-separability of this quotient (see Exercise 12.10) as soon as we establish the (easy) fact that BMOA is non-separable.

The link between the BMOA space and the Carleson measures is contained in the following result.[35]

12.7.4 Theorem [equivalence theorem [5, 66]] *Let $f \in H^2$. The following two statements are equivalent:*

---

[35] Recall that $d\lambda_1$ is the probability $2 \ln \frac{1}{|z|} dA(z)$ on D.

(i) $f \in BMOA$,

(ii) the measure $|f'(z)|^2 d\lambda_1(z)$ is a Carleson measure on D.

Wolff knew all of this when he established (see [66], p. 312) his theorem on the solution of d-bar with L^∞ estimates; this theorem is not an immediate application of the Fefferman–Stein duality theorem, but uses the ideas involved with this duality, notably Green's formula!

Wolff's theorem can be stated as follows.

12.7.5 Theorem [Wolff [66, 189]]   *Let $G \in C^1(D)$ be such that the measures $d\mu_1 = |G|^2 d\lambda_1$ and $d\mu_2 = |\partial G| d\lambda_1$ are Carleson measures. Then, the PDE $\overline{\partial} w = G$ has a solution w such that $\|w\|_\infty \leqslant C_1 \|\mu_1\|_C + C_2 \|\mu_2\|_C$, C_1 and C_2 being numerical constants.*

To successfully apply this theorem to the corona problem, as we have already seen, Wolff had to solve, with an L^∞ estimate, the PDE

$$\overline{\partial} w = h_j \overline{\partial} h_k - h_k \overline{\partial} h_j =: h_{jk}.$$

If we were under the conditions of application of the preceding theorem with $G = h_{jk}$, we would have won. However, calculations close to those made in Section 12.6 show that

$$|h_{jk}|^2 \leqslant C \sum_{p=1}^{n} |f'_p|^2 \text{ and } |\partial h_{jk}| \leqslant C \sum_{p=1}^{n} |f'_p|^2,$$

where $C = C(n, \delta)$. But $f_p \in H^\infty$, hence $f_p \in BMOA$, and the equivalence theorem thus tells us that $|f'_p|^2 d\lambda_1$ is a Carleson measure, and hence so are $|h_{jk}|^2 d\lambda_1$ and $|\partial h_{jk}| d\lambda_1$. Wolff's method was thus crowned with success, and earned its author the prestigious Salem prize in 1985. The presentation "à la Gamelin" of Section 12.6 enables a reader unfamiliar with the notions of Section 12.7 to follow the entire proof of the corona theorem, while highlighting the crucial importance of Green's formula and its consequences (Riesz representation formula, Littlewood–Paley identity, etc.).

To conclude this section, let us point out that Carleson measures are still of considerable importance today, notably in the study of composition operators $C_\varphi : H^2 \to H^2$ defined by $C_\varphi(f) = f \circ \varphi$, the symbol φ being a holomorphic map from D to D. This symbol is associated with $\mu_\varphi = \varphi^*(m)$, the pullback measure of the Haar measure m on $\mathbb{T}$ by the boundary values φ^*; μ_φ is always a Carleson measure (corresponding to the bounded nature of C_φ), and the operator C_φ is compact if and only if μ_φ is a *vanishing* Carleson measure. This means that

$$\rho(h) = \sup_{a \in \mathbb{T}} \mu_\varphi(W(a, h)) = o(h) \text{ as } h \searrow 0.$$

For more details, refer to [43].

The notion of Carleson measures has been extended to spaces other than the Hardy H^p spaces, for example the Bergman B^p spaces; for the latter, the *geometric* condition $\mu(W(a, h)) \leqslant Ch^2$ is equivalent to the *analytic* condition of continuous inclusion of B^p in $L^p(\mu)$. For more details, refer to [54].

12.8 Extensions of the corona theorem

Many laymen honestly wonder what problems in mathematics are still to be resolved. In our opinion, the corona problem provides a fine example. This problem was brilliantly resolved by Carleson [36] in the early 1960s. Afterwards the question seemed closed, with nothing more to say. However, a parallel with the famous Hilbert's Nullstellensatz appears naturally. Let us recall its statement: if I is an ideal of the ring $A = \mathbb{C}[X_1, \ldots, X_d]$ of complex polynomials in d variables, and if the generators $f_1, \ldots, f_n$ of I do not have a common zero, we have a Bézout identity $\sum_{j=1}^n f_j g_j = 1$, in other words $I = A$. More generally, if $f \in A$ is dominated by I in the sense that its zeros contain those of I, then f belongs to the radical of I, that is a power of f belongs to I [8].

What happens in H^∞? Let $I = I(f_1, \ldots, f_n)$ be an ideal of this ring, of finite type and generated by $f_1, \ldots, f_n$. Here, the condition "do not have a common zero" is replaced by a condition of the type "cannot be simultaneously very small", that is by a condition $\sum_{j=1}^n |f_j(z)| \geqslant \delta$, and the Carleson theorem then tells us that $I = H^\infty$. By analogy with the Nullstellensatz, the following question arises naturally: this time $f \in H^\infty$ is said to be dominated by I (abbreviated $f \ll I$) if "f is small where I is small", that is if there exists a constant $C > 0$ such that

$$|f(z)| \leqslant C \sum_{j=1}^n |f_j(z)| \text{ for } z \in D. \tag{12.44}$$

Question 1. The corona theorem states that $1 \ll I \Rightarrow 1 \in I$. More generally, does $f \ll I$ imply $f \in I$, or at least $f^p \in I$ for a certain integer p?

Here are two elements for an answer.

Element 1 (Wolff [190]). If $f \ll I$, then $f^3 \in I$.

The proof, for which we refer to [66], is not much more difficult than that of the corona theorem, and is a bit of a fallout of the methods developed by Wolff himself to prove this theorem.

Element 2 (Rao [156]). $f \ll I$ does not always imply $f \in I$.

The example is simple and pleasing: let B_1 and B_2 be two Blaschke products such that[36]

$$|B_1(z)| + |B_2(z)| > 0 \text{ for } z \in D \text{ and } \inf_{z \in D}(|B_1(z)| + |B_2(z)|) = 0. \quad (12.45)$$

One way to prove the existence of B_1 and B_2 is the following: let $(z_n)_{n \geqslant 1}$ be a sequence of distinct points of D such that $\sum_{n=1}^{\infty}(1 - |z_n|) < \infty$, and let B_1 be the Blaschke product whose simple zeros are the z_n. By induction, we can find a sequence $(z'_n)_{n \geqslant 1}$ of distinct points of D satisfying the following conditions:

$$\begin{cases} |z_n - z'_n| & \leqslant 2^{-n}, \\ z'_n \neq z_m & \text{for } m, n \geqslant 1, \\ |B_1(z'_n)| & \leqslant 2^{-n}. \end{cases}$$

Then, $\sum_{n=1}^{\infty}(1 - |z'_n|) < \infty$. Let B_2 be the Blaschke product whose simple zeros are the z'_n. By construction, B_1 and B_2 have disjoint zeros, so that the first condition of (12.45) holds. Moreover,

$$\inf_{z \in D}(|B_1(z)| + |B_2(z)|) \leqslant \inf_{n \geqslant 1}(|B_1(z'_n)| + |B_2(z'_n)|)$$

$$= \inf_{n \geqslant 1} |B_1(z'_n)|$$

$$\leqslant \inf_{n \geqslant 1} 2^{-n} = 0,$$

and hence (12.45) holds. We then set

$$f_1 = B_1^2, \ f_2 = B_2^2, \ f = B_1 B_2 \text{ and } I = I(f_1, f_2).$$

It is clear that $|f| \leqslant |B_1|^2 + |B_2|^2 = |f_1| + |f_2|$, so that $f \ll I$. Let us suppose that f is an element of I, so it can be written

$$f = g_1 B_1^2 + g_2 B_2^2, \text{ with } g_1, g_2 \in H^\infty,$$

and fix a zero a of B_1. We have $0 = f(a) = g_2(a) B_2^2(a)$, hence $g_2(a) = 0$ since B_1 and B_2 do not have any common zeros. Therefore (Theorem 12.2.10 or Exercise 12.5), g_2 is divisible by B_1 in H^∞: $g_2 = B_1 h_2$ with $h_2 \in H^\infty$; similarly $g_1 = B_2 h_1$. Thus,

$$B_1 B_2 = f = B_2 h_1 B_1^2 + B_1 h_2 B_2^2 = B_1 B_2 (h_1 B_1 + h_2 B_2).$$

---

[36] For an explicit example, see Exercise 12.4.

In the integral domain H^∞, we can simplify by $B_1 B_2$, so that

$$h_1 B_1 + h_2 B_2 = 1.$$

But the second condition of (12.45) makes this Bézout identity impossible![37]
We have thus shown that $f \notin I$.

Again, this is not the last word, and two new questions arise. We replace the condition of domination by a stronger condition. Given $\alpha > 0$, we say that $f \in H^\infty$ is α-dominated by I (abbreviated $f \ll_\alpha I$) if there exists a constant $C > 0$ such that

$$|f(z)| \leqslant C \Big(\sum_{j=1}^n |f_j(z)| \Big)^\alpha \text{ for } z \in D.$$

Question 2. Does $f \ll_\alpha I$ imply $f \in I$?

Answer 2 (Bourgain, 1985 [28]). It is false for all $\alpha < 2$ (see Exercise 12.6). On the contrary, as soon as $\alpha > 1$, we have $f \in \overline{I}$, the closure of I in H^∞. In particular, $I = I(B_1, B_2)$ can be non-closed (see Exercise 12.7).

Answer 2' (Cegrell, 1990 [37]). It is true for all $\alpha > 2$. And this contains, with $\alpha = 3$, the result of Wolff, since $f \ll I \Rightarrow f^3 \ll_3 I$.

Of course, the following question remains.

Question 3. Does $f \ll I$ imply $f^2 \in I$?

Answer 3 (Treil, 2002 [179]). The answer is negative, even if I has only two generators.

For his counter-example, Treil also replies to another question.

Question 4. An improved proof [136] of the dimension-free corona theorem shows that if

$$\delta \leqslant \Big(\sum_{j=1}^n |f_j(z)|^2 \Big)^{1/2},$$

there exist $g_1, \ldots, g_n \in H^\infty$ with

$$\sum_{j=1}^n g_j f_j = 1 \text{ and } \Big(\sum_{j=1}^n |g_j(z)|^2 \Big)^{1/2} \leqslant C(\delta),$$

where

$$C(\delta) = C \delta^{-2} \ln \frac{1}{\delta},$$

and C is an absolute constant. Can we improve this dependence on δ?

---

[37] See Remark 12.6.2.

Answer 4 (Tolokonnikov [66]). In general, we cannot do better than

$$C(\delta) = C\delta^{-2}.$$

To obtain Answer 3, Treil showed the following.

Answer 4′ (Treil [179]). In general, we cannot do better than

$$C(\delta) = C\delta^{-2} \ln \ln \frac{1}{\delta}.$$

This suggested to Treil that an "operator theory" proof of the corona theorem is highly unlikely.

In short, one can easily raise multiple questions. What happens with other domains of the plane, for example an annulus?[38] What happens when we allow functions with values in a non-Hilbertian vector space?[39] What happens for functions of several complex variables?[40] In any case, each question, once answered, leads to at least two or three others. Mathematics is like the ocean: inexhaustible.

Exercises

12.1. For $\alpha \geqslant 0$, set

$$f_\alpha(z) = \exp\left(-\alpha \frac{1+z}{1-z}\right).$$

Show that $f_\alpha \in H^\infty$, and that if $\alpha \neq \beta$,

$$\|f_\alpha - f_\beta\|_\infty = \sup_{\mathrm{Re}\, w = 0} |e^{-\alpha w} - e^{-\beta w}| = 2.$$

Then deduce that H^∞ is non-separable. For a solution using Blaschke products, see the exercises of Chapter 9 of [120].

12.2. The functions considered in this exercise are defined on the circle $\mathbb{T}$. Let $P : L^\infty \to H^\infty$ be a continuous linear projection (if such exists), C the subspace of L^∞ of continuous functions, e_j the exponential $x \mapsto e^{ijx}$ ($j \in \mathbb{Z}$) and $T_a : L^\infty \to L^\infty$ the translation operator defined by

$$T_a f(t) = f(t+a), \quad \text{where } a \in \mathbb{T}.$$

---

[38] The answer is affirmative, cf. [5].

[39] For this, we refer to the book [136] of N. Nikolskii.

[40] The answer is negative in general [167] and the question remains open for the polydisk or the unit ball of $\mathbb{C}^d$ (see nonetheless [4]).

For $u \in L^\infty$ and $v \in L^1$, set

$$\langle u, v \rangle = \int_{\mathbb{T}} u(t) v(t) \, dm(t).$$

(a) Show that there exists a continuous linear map $Q : C \to L^\infty$ such that

$$\langle Qf, g \rangle = \int_{\mathbb{T}} \langle P T_a(f), T_a g \rangle \, dm(a) \text{ for } f \in C \text{ and } g \in L^1,$$

and that $\|Q\| \leqslant \|P\|$.

(b) Show that $Q(e_j) = e_j$ if $j \geqslant 0$ and $Q(e_j) = 0$ if $j < 0$.

(c) By testing on $f(t) = \sum_{n=1}^{N} \frac{\sin(nt)}{n}$, for which we know that $\|f\|_\infty \leqslant 2$, reach a contradiction. Thus, H^∞ is not complemented in L^∞.

12.3. Wild behaviour of H^∞ functions at the boundary. For $z \in D$, set

$$S(z) = \exp\left(-\frac{1+z}{1-z}\right).$$

The function S is an element of H^∞.

(a) Show that if $w \in \overline{D}$, there exists a sequence $(z_n)_{n \geqslant 1}$ of points of D such that $z_n \to 1$ and $S(z_n) \to w$.

(b) Let K_1 be the fibre of the spectrum of H^∞ above 1. Show that if $w \in \overline{D}$, there exists a $\varphi \in K_1$ $\left(\text{i.e., } \varphi(e) = 1\right)$ such that $\varphi(S) = w$.

12.4. A Blaschke product that tends to 0 on a radius. Let $\alpha > 1$ be a fixed real number. For $k \geqslant 1$, set $\varepsilon_k = k^{-\alpha}$ and consider the Blaschke product with zeros $z_k = 1 - \varepsilon_k$:

$$B_1(z) = \prod_{k=1}^{\infty} \frac{z_k - z}{1 - z_k z}.$$

(a) For $r = 1 - \varepsilon \in \,]0, 1[$, show that

$$|B_1(r)|^2 = \prod_{k=1}^{\infty} \left(1 - \frac{(1 - z_k)^2 (1 - r^2)}{(1 - z_k r)^2}\right) \leqslant e^{-g(\varepsilon)},$$

where

$$g(\varepsilon) = \sum_{k=1}^{\infty} \frac{\varepsilon_k \varepsilon}{(\varepsilon_k + \varepsilon)^2}.$$

(b) Show that $B_1(r) \to 0$ as $r \nearrow 1$.

(c) Let B_2 be the Blaschke product with zeros $z_k' = 1 - \left(k + \frac{1}{2}\right)^{-\alpha}$. Show that B_1 and B_2 do not have a common zero, but that

$$\inf_{z \in D} (|B_1(z)| + |B_2(z)|) = 0.$$

12.5. In this exercise, we study to what extent the usual results about factorisation of holomorphic functions hold for H^∞.

(a) Let B be a Blaschke product and let $f \in H^\infty$ be zero at the zeros of B, counted with multiplicity. Show that there exists an $h \in H^\infty$ such that $f = Bh$, and that $\|h\|_\infty = \|f\|_\infty$.

(b) Let $g \in H^\infty$ be such that its zeros are simple, and let $f \in H^\infty$ be such that any zero of g is a zero of f. Does there exist an $h \in H^\infty$ such that $f = gh$? Is this true if $|g(e^{it})| = 1$ almost everywhere?

12.6. Let r be an integer $\geqslant 2$, and let B_1 and B_2 be Blaschke products without common zeros such that

$$\inf_{z \in D} (|B_1(z)| + |B_2(z)|) = 0.$$

Set $f_1 = B_1^r$, $f_2 = B_2^r$, $f = (B_1 B_2)^{r-1}$ and $\alpha = 2(1 - r^{-1}) < 2$. Show that f is α-dominated by $I = I(f_1, f_2)$, but that $f \notin I$.

12.7. Let $f \in H^\infty$, $f \neq 0$. Show that the principal ideal generated by f is closed in H^∞ if and only if there exists a constant $c > 0$ such that $|f(e^{it})| \geqslant c$, m-almost everywhere. For the "if" part, apply Banach's isomorphism theorem to the operator of multiplication by f from H^∞ to fH^∞; next, test the inequality $\|fg\|_\infty \geqslant c\|g\|_\infty$ on $g \in H^\infty$ whose values at the boundary satisfy

$$|g| = 1 \text{ a.e. on } E, \text{ and } |g| = \varepsilon > 0 \text{ a.e. on } {}^c E,$$

where E is a Borel subset of the circle with positive measure and ε a well-chosen positive real number.

12.8. Let $g \in H(D)$. Suppose that $fg \in H^2$ for all $f \in H^2$.

(a) Show, by using the closed graph theorem, that there exists a constant $C > 0$ such that $\|fg\|_2 \leqslant C\|f\|_2$ for all $f \in H^2$.

(b) Show that for all integers $n \geqslant 0$,

$$\int_{\mathbb{T}} |g|^{2n} \, dm \leqslant C^{2n}.$$

(c) Finally, show that $g \in H^\infty$ and that $\|g\|_\infty \leqslant C$. Thus, the algebra of multipliers of H^2 can be identified isometrically with H^∞:

$$\|g\|_\infty = \sup_{\|f\|_2 = 1} \|fg\|_2.$$

12.9. Let $f_1, \ldots, f_n \in H(D)$ without common zeros. Show that there exist $g_1, \ldots, g_n$ in $H(D)$ such that $\sum_{j=1}^n f_j g_j = 1$. *Hint [5]*: prove that if $\varphi \in C^1(D)$, there exists a $w \in C^1(D)$ such that $\overline{\partial} w = \varphi$; and then apply Hörmander's algebraic method. *An alternative approach* could be the following:

first show that if $(z_n)_{n \geqslant 1}$ is a distinct sequence of points of D such that $|z_n| \to 1$, there exists a function $f \in H(D)$ whose (simple) zeros are exactly the z_n (see [160]); then, given a sequence $(w_n)_{n \geqslant 1}$ of complex numbers, solve in $H(D)$ the Lagrange interpolation problem $g(z_n) = w_n$ for all $n \geqslant 1$, in the form

$$g(z) = \sum_{n=1}^{\infty} w_n \frac{f(z)}{(z - z_n) f'(z_n)} e^{c_n(z-z_n)},$$

where the c_n are chosen with very large moduli and arguments opposite to those of the z_n, in order to force normal convergence of the series on every compact subset of D; next, proceed as in the remark of Section 12.7, by making the necessary adjustments in the case where some z_n are repeated.

12.10. Let $(z_n)_{n \geqslant 1}$ be a sequence of *uniformly separated* points of D (for example $z_n = 1 - 2^{-n}$). Fix $n \geqslant 1$ and set

$$\varphi_j(z) = \frac{z - z_j}{1 - \overline{z_j} z} \quad \text{and} \quad B_n = \varphi_1 \times \cdots \times \varphi_n.$$

(a) Show that there exists a $\delta > 0$, independent of n, such that

$$(1 - |z_j|) |B'_n(z_j)| \geqslant \delta \text{ for } 1 \leqslant j \leqslant n.$$

(b) Let $w_1, \ldots, w_n \in \mathbb{C}$, E_n the set of $f \in H^2$ such that $f(z_j) = w_j$ for $1 \leqslant j \leqslant n$ and K the unit ball of H^2. Show that E_n contains at least one element f_0, and that we have

$$\begin{aligned}
\inf_{f \in E_n} \|f\|_2 &= \inf_{g \in H^2} \left\| \frac{f_0}{B_n} - g \right\|_2 \\
&= \sup_{F \in K} \left| \frac{1}{2i\pi} \int_{\mathbb{T}} \frac{f_0(z)}{B_n(z)} F(z) \, dz \right| \\
&= \sup_{F \in K} \left| \sum_{j=1}^{n} \frac{w_j F(z_j)}{B'_n(z_j)} \right|.
\end{aligned}$$

Hint: as the orthogonal of H^2 in L^2 is $H_0^2 = z H^2$, for the duality

$$\langle f, g \rangle = \int_{\mathbb{T}} f g \, dm$$

use Remark 12.2.13, as well as the residue theorem.

(c) Show that there exists an $f \in E_n$ such that

$$\|f\|_2 \leqslant C \left(\sum_{j=1}^{n} (1 - |z_j|) |w_j|^2 \right)^{1/2},$$

where C is independent of n.

(d) Show that the map

$$H^2 \to \ell^2, \quad f \mapsto \left((1 - |z_j|)^{1/2} f(z_j)\right)_{j \geqslant 1}$$

is continuous and surjective.

(e) Show that the map

$$T : H^1 \to \ell^1, \quad f \mapsto \left((1 - |z_j|) f(z_j)\right)_{j \geqslant 1}$$

is continuous and surjective.

(f) Show that L^∞/H^∞ contains a subspace isomorphic to ℓ^∞, and in particular that L^∞/H^∞ is not separable.

12.11. Let $g : \mathbb{T} \to \mathbb{C}$ be a measurable function. Suppose that

$$\int_{\mathbb{T}} |fg| \, dm < +\infty \text{ for all } f \in H^1.$$

(a) By using the closed graph theorem, show that there exists a constant $C > 0$ such that $\|fg\|_1 \leqslant C\|f\|_1$ for all $f \in H^1$.

(b) Show that $\|hg\|_1 \leqslant C\|h\|_1$ for all $h \in L^1$, and then that $g \in L^\infty$. Why is this result not in contradiction to the fact that BMOA is the dual space of H^1?

12.12. Example of a separated Blaschke sequence that is not uniformly separated. In this exercise, d denotes the pseudo-hyperbolic distance on D.

(a) Let $\delta \in \,]0, 1[$ and $a, b \in D$ such that

$$|a - b| \geqslant \delta \max(1 - |a|, 1 - |b|).$$

By proving and using the identity

$$\frac{1}{d(a,b)^2} - 1 = \frac{(1 - |a|^2)(1 - |b|^2)}{|a - b|^2},$$

show that

$$d(a, b) \geqslant \delta(4 + \delta^2)^{-1/2}.$$

Conversely, show that

$$d(a, b) \geqslant \delta \Rightarrow |a - b| \geqslant \delta \max(1 - |a|, 1 - |b|).$$

(b) For $j \geqslant 1$, denote by Γ_j the following set, formed of 2^j points equidistributed on the circle $|z| = 1 - 2^{-j}$:

$$\Gamma_j = \left\{ z_{k,j} := (1 - 2^{-j}) \exp \frac{2ik\pi}{2^j}, 1 \leqslant k \leqslant 2^j \right\}.$$

Set $\Gamma = \bigcup_{j \geq 1} \Gamma_j$, and enumerate the elements of Γ in a sequence $(z_k)_{k \geq 1}$. Show that $(z_k)_{k \geq 1}$ is separated, but that

$$\sum_{k=1}^{\infty} (1 - |z_k|) \left(\ln \frac{1}{1 - |z_k|} \right)^{-1} = +\infty.$$

A fortiori, $\sum_{k=1}^{\infty} (1 - |z_k|) = +\infty$.

(c) In each Γ_j, we only keep the $z_{k,j}$ for which $k \leq 2^j j^{-2}$ to obtain a set Γ'_j, and then a new set $\Gamma' = \bigcup_{j \geq 1} \Gamma'_j$ and a new sequence $(z'_k)_{k \geq 1}$. Show that $(z'_k)_{k \geq 1}$ is separated *and* a Blaschke sequence.

Set

$$\mu = \sum_{k=1}^{\infty} (1 - |z'_k|) \delta_{z'_k}.$$

(d) For N integer ≥ 1, denote by W_N the Carleson window defined by

$$|z| \geq 1 - 2^{-N} 2\pi \quad \text{and} \quad |\arg z| \leq 2^{-N} 2\pi.$$

Show that W_N contains $\bigcup_{j \geq 2^{N/2}} \Gamma'_j$.

(e) Show that $\mu(W_N) \geq C 2^{-N/2}$, where $C > 0$ is a constant. Hence, μ is not a Carleson measure, and $(z'_k)_{k \geq 1}$ is not uniformly separated (it is not even a finite union of such sequences).

12.13. Quasi-isometric interpolation. Let $P = \{z \in \mathbb{C} / \operatorname{Im} z > 0\}$ be the upper half-plane, and let A be a positive constant. Set $z_n = i + A 2^n \in P$ for $n \geq 1$. Denote by $H^{\infty}(P)$ the space of bounded analytic functions on P, and $T : H^{\infty}(P) \to \ell^{\infty}$ the operator defined by $T(f) = (f(z_n))_{n \geq 1}$. Let $w = (w_n)_{n \geq 1} \in \ell^{\infty}$ such that $\|w\|_{\infty} \leq 1$, $\delta > 0$, and $f : P \to \mathbb{C}$ defined by

$$f(z) = (2i)^{\delta} \sum_{n=1}^{\infty} \frac{w_n}{(z - \overline{z_n})^{\delta}},$$

where we take the principal branch of the logarithm of $z - \overline{z_n} \in P$.

(a) Show that $f \in H^{\infty}(P)$, and more precisely

$$\|f\|_{\infty} \leq 2^{\delta} + A^{-\delta} \sum_{n=1}^{\infty} \frac{1}{2^{(n-3)\delta}} =: 2^{\delta} + A^{-\delta} C_{\delta}.$$

Hint: to bound $|f(Ax + iy)|$, first study the case $x \leq 1$, and then discuss according to whether x belongs to the left half or the right half of an interval of type $I_k = [2^{k-1}, 2^k]$, $k \geq 1$.

(b) Show that the function f of part (a) also satisfies

$$\|T(f) - w\|_\infty \leqslant A^{-\delta} \sum_{n=1}^{\infty} \frac{1}{2^{(n-2)\delta}} =: A^{-\delta} C'_\delta.$$

(c) Let $\varepsilon > 0$. By adjusting first δ and then A above, show that one can have

$$\|f\|_\infty \leqslant 1 + \varepsilon \text{ and } \|T(f) - w\|_\infty \leqslant \varepsilon.$$

(d) Show that P *contains interpolating sequences with constant arbitrarily close to* 1, and that it is the same for D. Recall that an almost surjective operator between Banach spaces is surjective (see [153], p. 201).

(e) Show nonetheless that D does not contain any interpolating sequence with constant exactly 1.

12.14. Naftalevitch's theorem. Let $(z_j)_{j \geqslant 1}$ be a Blaschke sequence of D. Set

$$\rho_j = |z_j|, \varepsilon_j = 1 - \rho_j, \theta_j = \sum_{n \geqslant j} \varepsilon_n \text{ and } z'_j = \rho_j e^{i\theta_j}.$$

(a) Let $j < k$ be two integers $\geqslant 1$. Show that

$$|z'_j - z'_k| \gg |\varepsilon_j - \varepsilon_k| + |\theta_j - \theta_k| \geqslant |\varepsilon_j - \varepsilon_k| + \varepsilon_j \geqslant \max(\varepsilon_j, \varepsilon_k)$$

(where $\gg$ means $>$ up to a positive constant). By using Exercise 12.12, show that the sequence $(z'_j)_{j \geqslant 1}$ is separated.

(b) Let μ be the measure $\mu = \sum_{j=1}^{\infty}(1 - |z'_j|)\delta_{z'_j}$ and $W(\xi, h)$ a Carleson window. Set $E = \{j \geqslant 1; z'_j \in W(\xi, h)\}$, where $\xi = e^{i\delta}, 0 \leqslant \delta < 2\pi$, and $j_0 = \min E$.

  (i) Show that if $\delta \leqslant 2h$, then

$$\mu(W(\xi, h)) = \sum_{j \in E} \varepsilon_j \leqslant \sum_{j \geqslant j_0} \varepsilon_j = \theta_{j_0} \leqslant 3h.$$

  (ii) Show that if $\delta > 2h$, then $j_1 = \max E < \infty$ and

$$\mu(W(\xi, h)) \leqslant \sum_{j_0 \leqslant j \leqslant j_1} \varepsilon_j \leqslant 2h + h = 3h.$$

(c) By using Theorem 12.4.13, show that $(z'_j)_{j \geqslant 1}$ is an interpolating sequence such that $|z'_j| = |z_j|$ for all $j \geqslant 1$. This proof of Naftalevitch's theorem is due to Vasyunin (see [136], p. 174).

What do you think of the sequence $z_j = \dfrac{j}{j+i}$?

12.15. Bernstein numbers and isometries. Let $T : X \to Y$ be an operator between Banach spaces, and $n \geqslant 1$ a fixed integer.

(a) Define the nth Bernstein number T by the formula

$$b_n(T) = \sup_{\dim E = n} \inf_{x \in S_E} \|Tx\|,$$

where S_E denotes the unit sphere of E. Show that $a_n(T) \geqslant b_n(T)$.

(b) Suppose X, Y, Z are Hilbertian, and that $U : Y \to Z$ is a linear isometry. Show that $a_n(UT) = a_n(T)$ for $n \geqslant 1$. *Hint:* use the fact that $U^*U = I_Y$ and the ideal property.

12.16. Malmqvist–Walsh basis. Let $(z_j)_{j \geqslant 1}$ be a Blaschke sequence and $B = \prod_{j=1}^{\infty} b_j$ the associated Blaschke product. Set $L_j = \dfrac{K_{z_j}}{\|K_{z_j}\|}$ for $j \geqslant 1$. Show that the sequence $(u_j)_{j \geqslant 1}$ defined by

$$u_1 = L_1 \text{ and } u_n = b_1 \cdots b_{n-1} L_n \text{ for } n \geqslant 2$$

is an *orthonormal basis* of the model space K_B associated with B.

12.17. Cyclic shift vectors. Consider the unilateral shift S on H^2.

(a) Let p be a fixed non-negative integer, and $f(z) = (1 - z)^p$. Let $g(z) = \sum_{k=0}^{\infty} a_k z^k \in H^2$ be such that $\langle g, z^n f \rangle = 0$ for $n \geqslant 0$.

  (i) Show that

$$\sum_{j=0}^{p} (-1)^j \binom{p}{j} a_{n+j} = 0 \text{ for } n \geqslant 0.$$

  (ii) Show that $g = 0$ and that f is cyclic for S.

(b) More generally, let f be a polynomial without zeros in D, or even a function analytic in a neighbourhood of $\overline{D}$ and without zeros in $\mathbb{D}$. Show that f is cyclic for S.

12.18. Interpolation sequences of the polydisk

(a) Show that the Cartesian product of an interpolating sequence A of the space $H^{\infty}(D^p)$ and an interpolating sequence B of the space $H^{\infty}(D^q)$ is an interpolating sequence of the space $H^{\infty}(D^{p+q})$. *Hint:* use Drury's method and normal families to show the existence of the P. Beurling functions associated with A and B, respectively.

(b) Consider the sequence $(p_{j,k})_{j,k \geqslant 0}$ of points of D^2 defined by

$$p_{j,k} = \left(\frac{j}{j+i}, \frac{k}{k+i} \right).$$

Show that it is an interpolating sequence for $H^\infty(D^2)$, yet does not satisfy the (sufficient) Carleson condition, since

$$\prod_{(j,k)\neq(0,0)} d(p_{j,k}, p_{0,0}) = 0,$$

where $d(a, b) := \max\left(\left|\dfrac{a_1 - b_1}{1 - \overline{a_1}b_1}\right|, \left|\dfrac{a_2 - b_2}{1 - \overline{a_2}b_2}\right|\right)$ is the distance between $a = (a_1, a_2)$ and $b = (b_1, b_2)$. This example is due to Berndtsson et al. [18], as is the sufficiency of the Carleson condition.

13

The problem of complementation
in Banach spaces

13.1 Introduction

This chapter is influenced by the study of an article by Lindenstrauss [121] to which we will often refer. Our principal aim is to study the phenomenon of *complementation* in Banach spaces within the general theory of these spaces, and within the problems of this theory, resolved or not.

We recall that a *Banach space* is a set X that satisfies the following three axioms:

(1) X is a linear space (over the field of real or complex numbers);
(2) X is equipped with a norm $\| \cdot \|$ verifying
   - $\|x + y\| \leqslant \|x\| + \|y\|$,
   - $\|\lambda x\| = |\lambda| \, \|x\|$,
   - $\|x\| \geqslant 0$, and $\|x\| = 0 \Leftrightarrow x = 0$;
(3) X is a *complete* metric space with respect to the distance associated with this norm.

From now on, B_X denotes the closed unit ball of X.

Recall that these spaces were first introduced to solve problems of hard analysis: convergence of Fourier series, integral equations, interpolation, existence of everywhere non-Hölderian continuous functions, etc.

Led by S. Banach, the Polish mathematicians who studied these spaces in the 1930s noted that their *completeness* automatically entailed a number of agreeable properties (thanks to Baire's theorem and to absolutely convergent series): the celebrated closed graph (or open mapping) theorem, the Banach–Steinhaus principle of uniform boundedness, etc.

Two other types of nice property also emerged: *convexity* (with the theorems of Hahn–Banach) and *compactness* for weakened topologies (with the theorems of Banach–Alaoglu, Eberlein–Smulyan, Krein–Milman, etc.). These properties were successfully reinvested in analysis: Tietze's theorem, the non-surjectivity of the Fourier transform from $L^1(\mathbb{T})$ to c_0, or from $L^1(\mathbb{R}^n)$

to $C_0(\mathbb{R}^n)$, the characterisation of $L^p([0, 2\pi])$ functions $(1 < p < \infty)$ by the boundedness in L^p of the sequence of partial sums of their Fourier series, etc. Hence, we start with analysis and return to analysis.

However, as observed by Lindenstrauss, the structure of Banach spaces is sufficiently interesting and mysterious to deserve study for its own sake, much as we study the theory of finite groups. Over the last 50 to 60 years, such studies produced results which were almost all *negative* properties (Enflo, Read, Gowers, Maurey, etc.), with the notable exception of the theorems of Dvoretzky (1961) and Rosenthal (1974). These results showed that, in a general Banach space, just about anything may occur, and finding a general structure common to all these spaces, other than their definition, is hopeless (see [120]).

These properties were discovered after the article of Lindenstrauss, which had an indisputable prospective value; a few of the questions posed in this article are cited below.

(1) Does every separable[1] Banach space have a Schauder basis?[2]
(2) Does every Banach space contain an unconditional basic sequence?
(3) Is every compact operator[3] between two Banach spaces always the limit in norm of operators with finite rank?
(4) Is every Banach space decomposable?
(5) Does every Banach space isomorphically contain either c_0, or an ℓ_p space?
(6) Is a separable Banach space which is isomorphic to its infinite-dimensional subspaces isomorphic to a Hilbert space?
(7) Is a Banach space always isomorphic to its hyperplanes?
(8) Does there exist a Banach space X with very few operators, such that every operator on X can be written $T = \lambda I_X + K$, where λ is a scalar, I_X the identity of X and K a compact operator of X?
(9) If every closed subspace of X is complemented, is X isomorphic to a Hilbert space?

Given the title of this chapter, our focus will especially be on question (9). Even if we have not yet specified all the definitions (this will be done by the end of the chapter), let us already indicate that:

- the answer is **no** to questions (1)–(5) and (7);
- the answer is **yes** to questions (6) and (9), question (9) having been solved by Lindenstrauss himself, in collaboration with Tzafriri;

---

[1] A normed space is said to be separable if it contains a countable dense subset.
[2] See the definition on p. 449.
[3] A linear operator $T : X \to Y$ between two normed spaces is said to be compact if $\overline{T(B_X)}$ is compact.

- problem (8), despite significant progress by Gowers and Maurey, remained open until very recently – it was answered in the affirmative by Argyros and Haydon [6].

In what follows, when we speak of a "subspace", we will always mean a "linear subspace"; the symbol $P : X \to E$ will always mean that P is a linear projection from X onto[4] E.

13.2 The problem of complementation

Let X be a Banach space and E a *closed* subspace of X. By the incomplete basis theorem, there exists a subspace F such that $X = E \oplus F$. The question that will interest us can be expressed in a number of ways.

(i) Can we always construct such an F that is also *closed*?
(ii) Does there always exist a *continuous* projection $P : X \to E$?
(iii) Does the identity $I_E : E \to E$ admit a continuous linear extension $P : X \to E$?

The equivalence of (i) and (ii) results from the closed graph theorem. As for statement (iii), it gives an idea of the difficulty of the problem.

If E satisfies one of the preceding equivalent properties, we say that E is *complemented* in X, and we write $E \subset\subset X$.

This is always the case if E is either finite-dimensional, or has finite codimension. The case where dim $E < \infty$ will be obtained as a consequence of the theorem of Kadeč–Snobar of Section 13.4, even if there exist much simpler qualitative proofs, for example an application of the Hahn–Banach theorem [120]. The case where codim $E < \infty$ is trivial, because a finite-dimensional subspace of X is always closed in X.

The formulation (iii) indicates that complementation is a *problem of operator theory*, and is likely to be difficult, since the (almost) only three general theorems known about the extension of a continuous linear operator f of a subspace X_0 of a Banach space X to a normed space Y are the following:

- such an extension exists, with preservation of norm, if the domain X is a Hilbert space;
- such an extension exists, with preservation of norm, if the range Y is the field of scalars (Hahn–Banach theorem);[5]

---

[4] In the sense of a surjection.
[5] Because of its generality and its optimality, this theorem is qualified by Zippin [194] as a *perfect* theorem.

- such an extension exists, with preservation of norm, if we accept enlarging the range, and we do not need to enlarge the range more than the domain, that is we can extend $f : X_0 \to Y$ to $g : X \to Z$, where Y is a subspace of Z such that Z/Y is isometric to X/X_0 (Kisliakov's theorem; see [107] or [48], p. 316, under "Kisliakov's lemma").

Note that in (iii), we are not allowed to enlarge the range, and thus must renounce using Kisliakov's theorem.

Another concern arises: the *utility* of the notion of complementation. We will examine this with three examples.

13.2.1 Example [the problem of linear lifting]   Let $T : X \to Y$ be a continuous linear surjection. Given $y \in Y$, the equation $T(x) = y$ always has a solution, but the larger the kernel of T, the more perplexing the selection of x (there are too many choices!). A natural question is whether we can smoothly navigate in this multitude of solutions, that is, does there exist a *continuous linear choice* $x = R(y)$ such that $T(x) = y$? Or, stated more abruptly, does T possess a right inverse? The answer is given in terms of complementation, as the following proposition shows.

13.2.2 Proposition   *Let X and Y be two Banach spaces.*

(1) *Let $T : X \to Y$ be a continuous linear surjection. The operator T possesses a right inverse R ($TR = I_Y$) if and only if $\ker T$ is complemented in X.*

(2) *Let $T : X \to Y$ be a continuous linear injection. The operator T possesses a left inverse L ($LT = I_X$) if and only if $\mathrm{Im}\, T$ is closed and complemented in Y.*

Proof   We restrict ourselves to point (1). First of all, if $TR = I_Y$, we set $Q = RT$; Q is a projection, since

$$Q^2 = R(TR)T = RT = Q.$$

Of course, $\ker T \subset \ker Q$. On the contrary, if $Qx = 0$, then $Tx \in \ker R = \{0\}$. Finally, $\ker T = \ker Q$, so that $\mathrm{Im}\, Q = \ker(I_X - Q)$ provides a closed complement of $\ker T$ in X.

Conversely, if $\ker T \oplus F = X$ with F a closed subspace, the equation $T(x) = y$ possesses a unique solution $R(y) \in F$, and the graph of R is closed. Indeed, if $y_n \to y$ and $R(y_n) \to x$, a passage to the limit in the equality $TR(y_n) = y_n$ gives $T(x) = y$, with $x \in F$ as F is closed, hence $x = R(y)$, and thus R is continuous. $\square$

A good reference for this result is the book of Meise and Vogt [128]. An interesting variant is Atkinson's theorem [9]: $T : X \to X$ is a Fredholm operator (i.e., ker T is finite-dimensional and Im T has finite codimension) if and only if T is invertible modulo the compact operators, meaning that there exists an $R \in \mathcal{L}(X)$ such that

$$T R = I_X + K_1 \text{ and } RT = I_X + K_2,$$

where $K_1, K_2 \in \mathcal{L}(X)$ are compact operators.

13.2.3 Example [the Cantor–Bernstein theorem for Banach spaces]   Let us introduce two useful notations. If X and Y are two Banach spaces:

- the notation $X \sim Y$ means that there exists a continuous linear isomorphism[6] from X onto Y;
- $X \oplus Y$ denotes the product space $X \times Y$ equipped with the norm

$$\|(x, y)\| = \max(\|x\|, \|y\|),$$

which defines the product topology on $X \times Y$ and turns $X \oplus Y$ into a Banach space.

We then have the following result.

13.2.4 Proposition [Pelczynski]   *Let X, Y be two Banach spaces. We suppose that:*

(i) X is isomorphic to a complemented *subspace of Y;*
(ii) Y is isomorphic to a complemented *subspace of X;*
(iii) X and Y are isomorphic to their squares

$$X \sim X \oplus X \text{ and } Y \sim Y \oplus Y.$$

Then, X is isomorphic to Y : $X \sim Y$.

Proof   The symbols guide us: there exists a Banach space R such that $Y \sim X \oplus R$, therefore

$$X \oplus Y \sim X \oplus (X \oplus R) \sim (X \oplus X) \oplus R \sim X \oplus R \sim Y.$$

As X and Y play symmetric roles, we also have $Y \oplus X \sim X$. But it is clear that $(x, y) \mapsto (y, x)$ is an isomorphism from $X \oplus Y$ onto $Y \oplus X$. Finally,

$$X \sim Y \oplus X \sim X \oplus Y \sim Y. \qquad \square$$

---

[6] Necessarily bi-continuous, according to the open mapping theorem.

13.2.5 Remark (1) The hypothesis (iii) is very often satisfied: if you are hunting for a Banach space non-isomorphic to its square, it's best to get up early (see the James space, for example)! That said, it has long been a question whether this hypothesis was superfluous. Fairly recently (1996), Gowers [70] showed that it was more or less necessary, by constructing a Banach space X not isomorphic to its square, but isomorphic to its cube. If we set $Y = X \oplus X$, the hypotheses (i) and (ii) are verified, but not the hypothesis (iii); and, in fact, X and Y are not isomorphic.

(2) The hypotheses (i) and (ii) (of complementation) are also more or less necessary: if $X = C([0, 1])$ and $Y = A(D)$ is the disk algebra, that is the algebra of functions analytic on the unit disk D and continuous on its closure, it is easy to verify that X is isomorphic to its square. Y is also isomorphic to its square (see [188], p. 190), and each of these spaces is isomorphic to a subspace of the other, as X and Y are isometrically universal for the class of separable Banach spaces, meaning that any separable Banach space is isometric to a subspace of X and to a subspace of Y. But X and Y are not isomorphic ([188], p. 191). More precisely, Y is not isomorphic to any complemented subspace of a space $C(K)$, where K is a compact topological space.

(3) In the hypothesis (iii), we can eliminate the condition $Y \sim Y \oplus Y$, but there is a price to pay: we must strengthen the hypothesis on X and suppose that either $X \sim \ell_p(X)$ for a $p \in [1, \infty[$ or else $X \sim c_0(X)$ (see [120], p. 62 or [188], p. 45). Here, $\ell_p(X)$ denotes the Banach space of sequences $x = (x_n)_{n \geqslant 1}$ of elements of X such that

$$\|x\|_{\ell_p(X)} := \Big(\sum_{n=1}^{\infty} \|x_n\|^p \Big)^{1/p} < \infty,$$

and $c_0(X)$ the Banach space of sequences of elements of X that converge to 0, with norm $\| \cdot \|_\infty$.

Here is the proof for the first case: as in the proof of Proposition 13.2.4, we have $X \oplus Y \sim Y$. Moreover, if $X \sim Y \oplus S$, we have $\ell_p(X) \sim \ell_p(Y) \oplus \ell_p(S)$. Hence, successively:

$$Y \oplus X \sim Y \oplus \ell_p(X) \sim Y \oplus \big(\ell_p(Y) \oplus \ell_p(S)\big) \sim \big(Y \oplus \ell_p(Y)\big) \oplus \ell_p(S)$$
$$\sim \ell_p(Y) \oplus \ell_p(S) \sim \ell_p(X) \sim X,$$

which completes the proof.[7]

(4) Some classic applications of proposition 13.2.4 are:

---

[7] We note the appearance of the idea of the Hilbert hotel: for any Y, we have $Y \oplus \ell_p(Y) \sim \ell_p(Y)$.

- the isomorphism between the space ℓ_∞ of bounded sequences and the space $L^\infty([0, 1])$ of bounded measurable functions on $[0, 1]$ (we will see this in the exercises);
- the "primary" nature of the Banach spaces $X = \ell_p$, $1 \leqslant p < \infty$. The closed infinite-dimensional subspaces of X, that are complemented in X, are isomorphic to X [2, 120]. The result is also true [123] for $X = \ell_\infty$.

13.2.6 Example [ideals of operators]   We fix a real number p such that $1 \leqslant p < \infty$; ℓ_p denotes the space of p-summable sequences

$$\ell_p = \left\{ f : \mathbb{N} \to \mathbb{C} : \|f\|_p = \left(\sum_{n=0}^{\infty} |f(n)|^p \right)^{1/p} < \infty \right\}.$$

$\mathcal{L}(\ell_p)$ denotes the algebra of continuous linear operators from ℓ_p to itself, and $K(\ell_p)$ the closed two-sided ideal of $\mathcal{L}(\ell_p)$ made up of the compact operators. We are going to prove the following result, due to Calkin in the Hilbertian case $p = 2$.

13.2.7 Proposition [the generalised Calkin theorem]   *Let I be a two-sided closed ideal of $\mathcal{L}(\ell_p)$, distinct from $\{0\}$ and $\mathcal{L}(\ell_p)$. Then I is equal to $K(\ell_p)$, the ideal of compact operators.*

Proof   Let us denote by X the Banach space ℓ_p. Since the ideal I is $\neq \{0\}$, it contains a non-zero operator T. Then, by the Hahn–Banach theorem, there exist an $a \in X$ and $b^* \in X^*$, the dual space of X, such that $b^*(Ta) = 1$. In general, for $y \in X$ and $x^* \in X^*$, we denote by $x^* \otimes y \in \mathcal{L}(X)$ the operator of rank $\leqslant 1$ defined by

$$x^* \otimes y(z) = x^*(z)y \ \text{ for } z \in X.$$

An immediate calculation shows that[8]

$$x^* \otimes y = (b^* \otimes y)T(x^* \otimes a),$$

so that $x^* \otimes y \in I$. The ideal I thus contains the operators of rank 1, hence the operators of finite rank, which are sums of operators of rank 1. As I is closed, it also contains the limits (in operator norm) of operators of finite rank, that is, the compact operators.[9] Hence, the ideal I contains $K(\ell_p)$. It remains to show that, if the inclusion was strict, we would have $I = \mathcal{L}(\ell_p)$, contrary to

---

[8]   The right-hand side of the equality is to be understood as the composition of three operators.
[9]   As the space ℓ_p has an evident Schauder basis, every compact operator from ℓ_p to ℓ_p is the limit (in operator norm) of a sequence of operators of finite rank (see Exercise 13.4).

the hypothesis. Thus, let $T \in I$ be non-compact. The theory of bases[10] shows that T is *not strictly singular* (see Exercise 13.3). That is, there exists a closed infinite-dimensional subspace $M \subset \ell_p$, such that the restriction of T to M is an isomorphism from M onto $T(M)$. To continue, we admit the following result (clear in the Hilbertian case $p = 2$), for which we refer to [2], p. 34 or to [120], pp. 61–62.

13.2.8 Lemma *If M_0 is a closed infinite-dimensional subspace of ℓ_p, then M_0 contains a closed subspace M_1 isomorphic to ℓ_p and complemented in ℓ_p.*

We apply this to $M_0 = T(M)$, observing that M_0 is a complete subset of ℓ_p, hence closed because of the completeness of M and the existence of a constant $\alpha > 0$ such that

$$\|T(x)\| \geqslant \alpha \|x\| \text{ for } x \in M.$$

Therefore, M_0 contains a closed subspace M_1 isomorphic to ℓ_p and complemented in ℓ_p. If necessary replacing M by $T^{-1}(M_1) \cap M$, we may suppose $T(M)$ isomorphic to ℓ_p *and* complemented in ℓ_p. We thus have at hand the following diagram:

$$\ell_p \xrightarrow{u} M \xrightarrow{T} T(M) \xrightarrow{v} \ell_p,$$

where u and v are isomorphisms. The composite $vTu : \ell_p \to \ell_p$ is thus an isomorphism, hence an invertible element of $\mathcal{L}(\ell_p)$, and if $vTu \in I$, we could conclude that $I = \mathcal{L}(\ell_p)$, since I contains an invertible element. However, at this stage, we cannot conclude this way, as v is not an element of $\mathcal{L}(\ell_p)$, but only a continuous linear map from $T(M)$ to ℓ_p. We need to be able to extend v continuously to ℓ_p, which can be done precisely because $T(M)$ is *complemented* in ℓ_p ($T(M) \subset\subset l_p$). Indeed, let $P : \ell_p \to T(M)$ be a continuous linear projection, then $w = vP$ defines a continuous linear extension of v, and we have $vTu = wTu$, with $wTu \in I$, since $u, w \in \mathcal{L}(\ell_p)$ and since I is a two-sided ideal. This time $vTu \in I$ (*a priori* not so evident), and, as we have already explained, the proof of Proposition 13.2.7 is now complete.   □

13.3 Solution of problem (9)

A year after the publication of Lindenstrauss' prospective article, major progress was made by Lindenstrauss and Tzafriri, with the following solution of problem (9) for Banach spaces all of whose subspaces are complemented.

---

[10] For the details, see for example [120], notably the exercises of Chapter 1.

13.3.1 Theorem [Lindenstrauss–Tzafriri] *Let X be a Banach space. The following statements are equivalent:*

(*i*) *X is isomorphic to a Hilbert space;*
(*ii*) *all closed subspaces of X are complemented in X.*

Proof Here we simply indicate the main ideas of the proof. For a detailed proof, we refer to Chapter 8 of [120].

(i) $\Rightarrow$ (ii) is clear.

For (ii) $\Rightarrow$ (i), we proceed in three steps.

Step 1. We show that there exists a constant C_1 such that every finite-dimensional subspace E of X is C_1-complemented, that is, there exists a continuous projection $P : X \to E$ such that $\|P\| \leqslant C_1$.

Step 2. Using step 1 and Dvoretzky's theorem [120], we show that there exists a constant C_2 such that, for every finite-dimensional subspace E of X, we have $d_E \leqslant C_2$, where d_E denotes the Banach–Mazur distance [178] between E and the Hilbert space with the same dimension as E:

$$d_E = d(E, \ell_2^{\dim E}).$$

Here, ℓ_2^n denotes the space $\mathbb{R}^n$ equipped with its canonical inner product. This is the crucial step that links the notions of Banach–Mazur distance and projection via Dvoretzky's theorem.

Step 3. Using step 2 and an argument of compactness, we construct an inner product on X such that the associated Hilbertian norm is equivalent to the initial norm on X. $\qquad\square$

This fundamental theorem of structure suggests that finding a non-complemented subspace E in X is going to be a *difficult and poorly rewarded* task, just like the task of deciding whether $x \in \mathbb{R}$ is transcendental. Recall that a complex number x is said to be transcendental if it resists all polynomials:

$$f \in \mathbb{Z}[X] \text{ and } f \neq 0 \Longrightarrow f(x) \neq 0.$$

In the same manner, $E \subset X$ is non-complemented if it resists all projections:

$$P : X \to E \text{ is a surjective linear projection} \Longrightarrow P \text{ is not continuous.}$$

As the set of algebraic numbers is at most countable, we surprise no one if we painfully show that x is transcendental ($x = e$, π, etc.). Of course x is transcendental, they (almost) all are. If it weren't, *that* would be a surprise... Similarly, Theorem 13.3.1 says that, if X is not Hilbertian, X

contains some non-complemented subspace E. The general rule is thus non-complementation. That said, one can give *explicit* examples of such pairs (E, X), comparable either to the Liouville numbers or to the numbers e, π, etc. in the theory of transcendental numbers. However, we will begin with a positive result, comparable to the fact that for every irrational x, one can find a second-order approximation (at least) by rationals, that is, there exists a sequence $\left(\dfrac{p_n}{q_n}\right)_{n \geqslant 0}$ of rationals converging to x such that

$$\left| x - \frac{p_n}{q_n} \right| \leqslant \frac{C}{q_n^2}.$$

13.4 The Kadeč–Snobar theorem

We are going to prove the following theorem, which is close to optimal.

13.4.1 Theorem [Kadeč–Snobar] *Let X be a Banach space, and E a subspace of X of finite dimension n. Then there exists a projection $P : X \to E$ such that $\|P\| \leqslant \sqrt{n}$.*

Kadeč–Snobar [99] initially gave a geometric proof, using ellipsoids "à la F. John"; *it will be presented here as a guided exercise*. The proof that follows is analytic, based upon the theory of 2-summing operators and the Pietsch factorisation theorem. Note first that one can suppose, without loss of generality, that $X = C(K)$, where K is a compact topological space. Indeed, if $K = B_X^*$ is the closed unit ball of the dual space X^* of X, equipped with the weak-* topology, then, by the Banach–Alaoglu theorem [120], K is compact, and by the Hahn–Banach theorem, the map $x \mapsto \widehat{x}$, where $\widehat{x}(\varphi) = \varphi(x)$ for $\varphi \in K$, is an isometry from X to a subspace of $C(K)$.

If we can find a projection $Q : C(K) \to E$, with norm $\leqslant \sqrt{n}$, then its restriction to X will provide a projection $P : X \to E$, with norm $\leqslant \sqrt{n}$. The fact that we take $X = C(K)$ will allow us to enhance the structure with the introduction of the Borel sets of K and of measure theory. As a matter of fact, this theory is already present in the initial proof, through considerations of the volume of convex bodies.

All will depend on the following crucial lemma, that will bring us to a Hilbertian situation. To simplify, we work in the field of *real numbers*.

13.4.2 Lemma *Let V be an n-dimensional subspace of a space $H = L^2(\Omega, \mathcal{A}, \mu)$, T a non-empty set, and $(F_t)_{t \in T}$ a bounded family of elements of V, indexed by T, such that the associated maximal function M, defined by*

$$M(x) = \sup_{t \in T} |F_t(x)| \text{ for } x \in \Omega,$$

is measurable. Then, we have

$$\|M\|_2 \leqslant \sqrt{n} \sup_{t \in T} \|F_t\|_2, \tag{13.1}$$

where $\| \cdot \|_2$ *denotes the norm of the Hilbert space H.*

Proof We set $C = \sup_{t \in T} \|F_t\|_2$, and fix an orthonormal basis $(\varphi_1, \ldots, \varphi_n)$ of V. We consider the function Δ defined by

$$\Delta = \left(\sum_{j=1}^{n} \varphi_j^2 \right)^{1/2}.$$

For $t \in T$, we can write

$$F_t = \sum_{j=1}^{n} c_j(t)\varphi_j, \quad \text{with} \quad \sum_{j=1}^{n} c_j(t)^2 = \|F_t\|_2^2 \leqslant C^2.$$

By the Cauchy–Schwarz inequality, for each $x \in \Omega$ we thus have

$$|F_t(x)|^2 \leqslant \left(\sum_{j=1}^{n} c_j(t)^2 \right) \left(\sum_{j=1}^{n} \varphi_j(x)^2 \right) \leqslant C^2 \sum_{j=1}^{n} \varphi_j(x)^2 = C^2 \Delta(x)^2,$$

so that

$$M(x)^2 \leqslant C^2 \Delta(x)^2.$$

By integrating with respect to μ, we obtain

$$\|M\|_2^2 \leqslant C^2 \|\Delta\|_2^2 = nC^2,$$

which proves (13.1). $\qquad\qquad\qquad\qquad\qquad\qquad\qquad\qquad\qquad\qquad\square$

13.4.3 Corollary *Let K be a compact topological space, with E an n-dimensional subspace of $C(K)$, $f_1, \ldots, f_N \in E$, and*

$$S = \left(\sum_{i=1}^{N} f_i^2 \right)^{1/2}$$

the associated square function. Then we have

$$\sup_{K} \left(nS^2 - \sum_{i=1}^{N} \|f_i\|_\infty^2 \right) \geqslant 0. \tag{13.2}$$

Proof We apply the lemma with $\Omega = \{1, \ldots, N\}$, $\mathcal{A} = \mathcal{P}(\Omega)$, $\mu = \delta_1 + \cdots + \delta_N$ the counting measure (where δ_j is the Dirac measure at the point j), and V the space generated by the F_t, where

$$F_t(i) = f_i(t) \text{ for } 1 \leqslant i \leqslant N \text{ and } t \in K.$$

Then, $\dim V \leqslant n$. Indeed, if $\gamma_1, \ldots, \gamma_n$ is a basis of E and if $f_i = \sum_{j=1}^n c_{ij}\gamma_j$, we have

$$F_t = \sum_{j=1}^n \gamma_j(t)\varphi_j, \text{ with } \varphi_j(i) = c_{ij} \text{ for } 1 \leqslant i \leqslant N,$$

so then

$$\text{Vect}(F_t)_{t \in K} \subset \text{Vect}(\varphi_1, \ldots, \varphi_n).$$

Moreover, for $t \in K$, we have

$$\|F_t\|_{L^2(\Omega)}^2 = \sum_{i=1}^N F_t(i)^2 = \sum_{i=1}^N f_i(t)^2 \leqslant \sum_{i=1}^N \|f_i\|_\infty^2,$$

therefore the family $(F_t)_{t \in K}$ is bounded in $L^2(\Omega)$. Given the choice of $\mathcal{A}$, the function M is automatically measurable, and this allows us to apply Lemma 13.4.2. Then, since

$$M(i) = \sup_{t \in K} |f_i(t)| = \|f_i\|_\infty \text{ and } \|F_t\|_{L^2(\Omega)} = S(t),$$

the inequality (13.1) can be read as

$$\left(\sum_{i=1}^N \|f_i\|_\infty^2\right)^{1/2} \leqslant \sqrt{n} \sup_{t \in K} S(t),$$

which proves (13.2).   □

Using a geometric version of the Hahn–Banach theorem, we formulate Lemma 13.4.2 and its corollary in a more manageable form, recognisable to an experienced reader as the Pietsch factorisation theorem.

13.4.4 Lemma [Pietsch] *Let K be a compact topological space and E an n-dimensional subspace of $C(K)$. Then there exists a Borel probability measure μ on K such that*

$$\int_K f^2 \, d\mu \leqslant \|f\|_\infty^2 \leqslant n \int_K f^2 \, d\mu, \quad \text{for } f \in E. \tag{13.3}$$

First let us show how this lemma enables us to complete the proof of the Kadeč–Snobar theorem. Let j denote the canonical inclusion $j : C(K) \to L^2(\mu)$ and Q the orthogonal projection from $L^2(\mu)$ onto $j(E)$, which is a closed subspace of $L^2(\mu)$ because of finite dimension. We can thus draw the following diagram:

$$C(K) \xrightarrow{\ j\ } L^2(\mu) \xrightarrow{\ Q\ } j(E) \xrightarrow{\ j^{-1}\ } E.$$

Relation (13.3), which can be written as

$$\|f\|_{L^2(\mu)} \leqslant \|f\|_{C(K)} \leqslant \sqrt{n}\, \|f\|_{L^2(\mu)},$$

shows that $\|j\| \leqslant 1$ $\big($as the left inequality of (13.3) holds for all $f \in C(K)\big)$ and that $\|j^{-1}\| \leqslant \sqrt{n}$. If we set $P = j^{-1}Qj$, P is a projection from $C(K)$ onto E, and we have

$$\|P\| \leqslant \|j^{-1}\|\, \|Q\|\, \|j\| \leqslant \sqrt{n} \times 1 \times 1 = \sqrt{n}.$$

To finish, we present the proof of Lemma 13.4.4.

Proof First recall a geometric version of the Hahn–Banach theorem [31]: let A and B be two disjoint convex subsets of a real normed space X, at least one of them being open. Then there exist a non-zero continuous linear functional L on X and a real constant α such that

$$\sup_{g \in B} L(g) \leqslant \alpha \leqslant \inf_{f \in A} L(f). \tag{13.4}$$

We suppose moreover that A and B are *cones*, that is, are invariant under multiplication by the positive real numbers. We thus have, for $g \in B$: $L(tg) = tL(g) \leqslant \alpha$ for all $t > 0$, which means that $L(g) \leqslant 0$. Similarly, $L(f) \geqslant 0$ for all $f \in A$. In short, the fact that A and B are cones *enables us to take $\alpha = 0$ in (13.4)*.

We apply this to the two convex cones of $C(K)$ defined by

$$A = \left\{ n \sum_{i=1}^{N} f_i^2 - \sum_{i=1}^{N} \|f_i\|_\infty^2,\ f_1, \dots, f_N \in E,\ N \in \mathbb{N}^* \right\},$$

$$B = \left\{ g \in C(K)/ \sup_K g < 0 \right\}.$$

The set A is definitely a convex cone, as it is invariant under multiplication by positive real numbers and under addition. We justify this last point: if

$$f = n \sum_{i=1}^{M} f_i^2 - \sum_{i=1}^{M} \|f_i\|_\infty^2 \text{ and } g = n \sum_{i=1}^{N} g_i^2 - \sum_{i=1}^{N} \|g_i\|_\infty^2$$

are two elements of A, with $f_1, \ldots, f_M, g_1, \ldots, g_N \in E$, then

$$f + g = n \sum_{i=1}^{M+N} h_i^2 - \sum_{i=1}^{M+N} \|h_i\|_\infty^2$$

by using the $(M + N)$-tuple $(h_1, \ldots, h_{M+N}) = (f_1, \ldots, f_M, g_1, \ldots, g_N)$.

According to the inequality of Corollary 13.4.3, if

$$f = n \sum_{i=1}^{N} f_i^2 - \sum_{i=1}^{N} \|f_i\|_\infty^2 \in A,$$

we have $\sup_K f \geqslant 0$, hence A and B are disjoint. Moreover, as B is open, the previously mentioned Hahn–Banach theorem for cones guarantees the existence of a continuous linear functional $L \neq 0$ on $C(K)$, given – by the Riesz representation theorem – by a Borel measure μ on K:

$$L(\varphi) = \int_K \varphi \, d\mu \text{ for } \varphi \in C(K),$$

such that (13.4) can be written as

$$\int_K g \, d\mu \leqslant 0 \leqslant \int_K f \, d\mu \text{ for } f \in A, g \in B. \tag{13.5}$$

The inequality on the left in (13.5) shows that μ is a positive measure.[11] If necessary replacing μ by $\dfrac{\mu}{\|\mu\|}$, we can suppose that μ is a probability measure. Now, if $f \in E$, the function $nf^2 - \|f\|_\infty^2$ is an element of A, and the inequality on the right in (13.5) gives

$$0 \leqslant n \int_K f^2 \, d\mu - \|f\|_\infty^2,$$

or

$$\|f\|_\infty^2 \leqslant n \int_K f^2 \, d\mu,$$

which completes the proof.   □

13.4.5 Remark   The estimation in $\sqrt{n}$ of the Kadec–Snobar theorem is almost optimal, as we will see in the following section and in the exercises.

[11]   Since, if $g \in C(K)$ is non-negative and $\varepsilon > 0$ then $-g - \varepsilon \in B$, thus

$$0 \geqslant \int_K (-g - \varepsilon) d\mu = -\int_K g \, d\mu - \varepsilon \mu(K)$$

and finally $\int_K g \, d\mu \geqslant 0$ by letting $\varepsilon \to 0$.

13.5 An example "à la Liouville"

We are going to give an ad hoc example of a pair (E, X) with X a Banach space and E a closed subspace of X non-complemented in X. In a way, this can be compared with the first examples of transcendental numbers constructed by Liouville. We proceed as follows: $I = \{1, 2, 4, \ldots, 2^k, \ldots\}$ denotes the set of powers of 2. Recall the following fact.

For $n \in I$, there exists an order-n square matrix $W_n = [w_{ij}]_{1 \leqslant i, j \leqslant n}$ (called a *Hadamard matrix*), with real coefficients, such that

$$\begin{cases} |w_{ij}| = n^{-1/2} \text{ for } 1 \leqslant i, j \leqslant n, \\ W_n = {}^t W_n = W_n^{-1}, \\ \operatorname{tr} W_n = 0 \text{ for } n \geqslant 2. \end{cases} \tag{13.6}$$

Thus, W_n is at the same time orthogonal, symmetric, with trace zero, and with coefficients of modulus $n^{-1/2}$. We can obtain W_n by $W_1 = [1]$ and the following recursive relation:

$$W_{2^{k+1}} = \frac{1}{\sqrt{2}} \begin{bmatrix} W_{2^k} & W_{2^k} \\ W_{2^k} & -W_{2^k} \end{bmatrix}.$$

Henceforth, we fix $n \in I$ such that $n \geqslant 2$. Since $W_n^2 = I$, $P_n := \dfrac{I + W_n}{2}$ is a projection of $\mathbb{R}^n$ (as usual, we identify the matrix W_n and the operator it represents in the canonical basis of $\mathbb{R}^n$). Let us consider the space ℓ_1^n, that is, by definition $\mathbb{R}^n$ equipped with the norm

$$\|(x_1, \ldots, x_n)\|_1 = \sum_{j=1}^n |x_j|,$$

denote by $(e_1, \ldots, e_n)$ its canonical basis, and set $E_n = \operatorname{Im} P_n$. We will consider E_n as a subspace of ℓ_1^n. Note right away that

$$\dim E_n = \operatorname{tr} P_n = \frac{1}{2}(\operatorname{tr} I + \operatorname{tr} W_n) = \frac{1}{2}\operatorname{tr} I = \frac{n}{2}.$$

We will see that E_n, whose dimension is half that of ℓ_1^n, is poorly complemented in ℓ_1^n, in the sense that

for every projection $P : \ell_1^n \to E_n$, we have $\|P\| \geqslant \dfrac{1}{2}\sqrt{n}$. $\tag{13.7}$

Moreover, this will show the optimality of $\sqrt{n}$, up to the factor 2, in the result of Kadeč–Snobar.

Indeed, let P be such a projection. Since E_n is none other than the set of fixed points of W_n, we have $P = W_n P$. Thus, by taking the traces and denoting by $\langle \cdot , \cdot \rangle$ the canonical inner product[12] of $\mathbb{R}^n$, we obtain

$$\frac{n}{2} = \dim E_n = \operatorname{tr} P = \operatorname{tr}(W_n P)$$

$$= \sum_{j=1}^{n} \langle W_n P e_j, e_j \rangle = \sum_{j=1}^{n} \langle P e_j, W_n e_j \rangle \text{ as } W_n \text{ is symmetric}$$

$$\leqslant \sum_{j=1}^{n} \| P e_j \|_1 \| W_n e_j \|_\infty \leqslant \| P \| \sum_{j=1}^{n} \| W_n e_j \|_\infty,$$

where

$$\| (x_1, \ldots, x_n) \|_\infty = \max_{1 \leqslant j \leqslant n} |x_j|.$$

However, $W_n e_j$ is none other than the jth column vector of W_n, so that $\| W_n e_j \|_\infty = n^{-1/2}$ according to (13.6). The preceding inequality can thus be read as

$$\frac{n}{2} \leqslant \| P \| \sum_{j=1}^{n} n^{-1/2} = \| P \| \, n^{1/2},$$

which proves (13.7).

Next, we "stack up" the E_n to obtain a non-complemented infinite-dimensional subspace, in the same way that we form a series to obtain a Liouville number. Let us now introduce a useful notation: let $p \geqslant 1$ be a fixed real number and let $(E_n)_{n \geqslant 1}$ be a sequence of Banach spaces. We denote by

$$\left(\bigoplus_{n=1}^{\infty} E_n \right)_p$$

the set of sequences $x = (x_n)_{n \geqslant 1}$ such that $x_n \in E_n$ for $n \geqslant 1$ and

$$\| x \| := \left(\sum_{n=1}^{\infty} \| x_n \|^p \right)^{1/p} < \infty.$$

One can verify that this defines a Banach space, known as the ℓ_p-*direct sum of the* E_n. Moreover, for a fixed $k \geqslant 1$, if we identify E_k with the set of elements $x = (x_n)_{n \geqslant 1}$ of $\left(\bigoplus_{n=1}^{\infty} E_n \right)_p$ such that $x_n = 0$ if $n \neq k$, E_k is a closed subspace of $\left(\bigoplus_{n=1}^{\infty} E_n \right)_p$.

---

[12] For which we have the following fundamental duality inequality

$$|\langle x, y \rangle| \leqslant \| x \|_1 \| y \|_\infty.$$

In our case, we consider the Banach space

$$X = \Big(\bigoplus_{n \in I} \ell_1^n \Big)_1,$$

the ℓ_1-direct sum of the spaces ℓ_1^n, $n \in I$. Within X, we consider

$$E = \Big(\bigoplus_{n \in I} E_n \Big)_1.$$

Easily, E is a closed subspace of X and, for any projection $P : X \to E$, we have

$$\|P\| \geqslant \|P_{|\ell_1^n}\|_{\ell_1^n} \geqslant \frac{1}{2} \sqrt{n} \text{ for all } n \geqslant 1,$$

so that $\|P\| = \infty$ and P is discontinuous.

For some variants of this construction, due to Sobczyk, see Chapter 8 of [178].

13.6 An example "à la Hermite"

In this section, we are going to consider imposed pairs (E, X) of Banach spaces with $E \subset X$ and show that E is not complemented in X, somewhat as Hermite or Lindemann showed that e or π were transcendental. Below are a few famous pairs (E, X).

- (c_0, ℓ_∞): Sobczyk, 1941 [123].
- $\big(K(H), \mathcal{L}(H) \big)$: Kalton [102].
- (H^1, L^1): Newman, 1961 [188]. Here,

$$L^1 = L^1(\mathbb{T}) \text{ and } H^1 = \Big\{ f \in L^1 / \widehat{f}(n) = 0 \text{ for } n < 0 \Big\}.$$

- (H^∞, L^∞): Rudin, 1961; also Curtis [188].

On the contrary, note that if $1 < p < \infty$, $H^p \subset\subset L^p$ according to a classic theorem of Riesz [188]. We also have $H^p \sim L^p$ according to a theorem of Boas (1955).

In what follows, the second example is detailed.

13.6.1 Theorem *Let H be a separable infinite-dimensional Hilbert space, $\mathcal{L}(H)$ the Banach space of all bounded operators[13] on H, and $K(H)$ the closed subspace consisting of all compact operators. Then, $K(H)$ is not complemented in $\mathcal{L}(H)$.*

---

[13] That is, the continuous linear maps from H to H.

Proof   We mimic Veech's proof of the non-complementation of c_0 in ℓ_∞: we need to eliminate all the continuous projections from $\mathcal{L}(H)$ on $K(H)$. We circumvent the difficulty by proving a stronger result, that will imply Theorem 13.6.1.

13.6.2 Theorem   *Let H be as in Theorem 13.6.1, and let $\Phi : \mathcal{L}(H) \to \mathcal{L}(H)$ be a continuous linear operator whose kernel contains the compact operators. Then there exists a projection P of infinite rank such that $\Phi(P) = 0$. In particular, we always have* $\ker \Phi \neq K(H)$.

Theorem 13.6.2 implies Theorem 13.6.1, because if Π was a continuous projection from $\mathcal{L}(H)$ on $K(H)$, the complementary projection $\Phi = I_{\mathcal{L}(H)} - \Pi$ would be an operator on $\mathcal{L}(H)$ with kernel $K(H)$.

Note now that a continuous projection $P : H \to F$ is compact if and only if $\dim F < \infty$. Indeed, if P is compact, $F = \ker(P - I)$ is finite-dimensional according to the Riesz theory of compact operators. To prove Theorem 13.6.2, it is necessary to "dig out" the projection P, which will require a few lemmas.

13.6.3 Lemma   *There exists an uncountable family $(A_i)_{i \in I}$ of infinite subsets of $\mathbb{N}^*$ such that*

$$|A_i \cap A_{i'}| < \infty \text{ if } i \neq i',$$

where $|\cdot|$ denotes cardinality.

Proof   Let $(p_k)_{k \geqslant 1}$ denote the increasing sequence of prime numbers, and I the uncountable set of infinite subsets of $\mathbb{N}^*$. For

$$i = \{j_1 < j_2 < \cdots < j_r < \ldots\} \in I,$$

we define

$$A_i = \{p_{j_1}, p_{j_1} p_{j_2}, p_{j_1} p_{j_2} p_{j_3}, \ldots\}.$$

Then, let $i' = \{j_1' < j_2' < \ldots\} \in I$ be distinct from i; denote by r the smallest index such that $j_r \neq j_r'$. We thus have, for example, $j_r < j_r'$, so that $j_r \notin i'$, and $A_i \cap A_{i'}$ is reduced to $\{p_{j_1}, p_{j_1} p_{j_2}, \ldots, p_{j_1} p_{j_2} \times \cdots \times p_{j_{r-1}}\}$. $\square$

In what follows, I will be as in the preceding lemma, and I_f will denote the set of *finite* subsets of I. We will say that a family $(T_i)_{i \in I}$ of $\mathcal{L}(H)$ is *bounded* if there exists a constant C such that

$$\left\| \sum_{i \in J} T_i \right\| \leqslant C \text{ for all } J \in I_f.$$

13.6.4 Lemma   *Let $(T_i)_{i \in I}$ be a bounded family of $\mathcal{L}(H)$. Then, the set of $i \in I$ such that $T_i \neq 0$ is at most countable.*

Proof Let $(e_n)_{n \geqslant 1}$ be a Hilbertian basis of H, and let $\langle \cdot , \cdot \rangle$ denote the inner product on H. For $u, v \in \mathbb{N}^*$ and $J \in I_f$, we have

$$\left| \sum_{i \in J} \langle e_v, T_i(e_u) \rangle \right| = \left| \left\langle e_v, \left(\sum_{i \in J} T_i \right)(e_u) \right\rangle \right|$$
$$\leqslant \|e_v\| \left\| \left(\sum_{i \in J} T_i \right)(e_u) \right\|$$
$$\leqslant \|e_v\| \|e_u\| \left\| \sum_{i \in J} T_i \right\|$$
$$\leqslant C \|e_v\| \|e_u\| = C.$$

As $J \in I_f$ is arbitrary, we easily deduce[14] that

$$\sum_{i \in J} |\langle e_v, T_i(e_u) \rangle| \leqslant 4C.$$

Then, if we set, for $u, v, n \geqslant 1$,

$$I_{u,v,n} = \left\{ i \in I \; / \; |\langle e_v, T_i(e_u) \rangle| \geqslant \frac{1}{n} \right\}$$

and

$$I_0 = \bigcup_{u,v,n \in \mathbb{N}^*} I_{u,v,n},$$

we see that the cardinality of $I_{u,v,n}$ is at most $4nC$, and hence I_0 is at most countable. Moreover, for $i \notin I_0$, we have $\langle e_v, T_i(e_u) \rangle = 0$ for $u, v \geqslant 1$, which implies $T_i(e_u) = 0$ for all $u \geqslant 1$, hence $T_i = 0$. $\qquad \square$

Now, we fix a Hilbertian basis $(e_n)_{n \geqslant 1}$ of H and denote by

- H_A, for $A \subset \mathbb{N}^*$, the closed subspace of H generated by the e_n, $n \in A$.
- P_A, the orthogonal projection operator from H on H_A.
- T_i, the operator $T_i = \Phi(P_{A_i})$, where the A_i are as in Lemma 13.6.3 and Φ as in Theorem 13.6.2.

Note that if A and A' are two *disjoint* subsets of $\mathbb{N}^*$, then

$$P_{A \cup A'} = P_A + P_{A'}.$$

13.6.5 Lemma *The family $(T_i)_{i \in I}$ is bounded in the sense given in Lemma 13.6.4.*

Proof The idea of the proof is the following: the A_i are almost disjoint by Lemma 13.6.3 so, if $i \neq i'$, $P_{A_i \cap A_{i'}}$ is a compact projection, because of finite rank. However, Φ does not see these projections as it is zero on the compact operators. From the point of view of Φ, everything happens as if the A_i were pairwise disjoint.

---

[14] See Exercise 13.12.

To be more precise, we fix $J \in I_f$. For $i \in J$, we define

$$
\begin{cases}
A'_i &= A_i \setminus \left(\bigcup_{j \in J, j \neq i} (A_i \cap A_j) \right), \\
A''_i &= A_i \setminus A'_i = \bigcup_{j \in J, j \neq i} (A_i \cap A_j), \\
B &= \bigcup_{i \in J} A'_i.
\end{cases}
$$

By construction, the A'_i, $i \in J$, are pairwise disjoint. By the properties of the A_i, the A''_i, $i \in J$, are finite sets, and we know that

$$
\sum_{i \in J} T_i = \sum_{i \in J} \Phi(P_{A'_i}) + \sum_{i \in J} \Phi(P_{A''_i}) = \Phi(P_B) + \sum_{i \in J} \Phi(P_{A''_i}) = \Phi(P_B).
$$

Indeed, the projections $P_{A''_i}$ are of finite rank, *a fortiori* compact, and by hypothesis Φ is zero on the compact operators. In other words, we have

$$
\sum_{i \in J} T_i = \Phi(P_B),
$$

so that

$$
\left\| \sum_{i \in J} T_i \right\| \leq \|\Phi\| \|P_B\| \leq \|\Phi\| =: C. \qquad \square
$$

The conclusion now ensues easily: by the preceding lemmas, the set of $i \in I$ such that $T_i \neq 0$ is at most countable, and as I is uncountable, there exists a $j \in I$ such that $T_j = \Phi(P_{A_j}) = 0$. However, P_{A_j} is a projection of infinite rank (as A_j is infinite), hence not compact, from the remark above. The kernel of Φ thus contains a non-compact operator: *quod erat demonstrandum*. $\qquad \square$

13.6.6 Remark (1) Let X be an infinite-dimensional Banach space. Is it always true that $K(X)$ is non-complemented in $\mathcal{L}(X)$? We can extend the preceding result to many spaces other than Hilbert spaces, as we will see in Exercise 13.6. But the general case remained an open problem up until the article of Argyros and Haydon [6]. They constructed a separable Banach space X having very few continuous operators, all of the form $\lambda I + K$, where K is compact: $K(X)$ is then trivially complemented in $\mathcal{L}(X)$ because it is of codimension one. The reader will find interesting complements in the excellent article [102] (see also [71]).

(2) An interesting positive result is Veech's theorem [123]: the space c_0 is *separably injective*, that is, complemented in every *separable* Banach space that contains it, even if it is not complemented in the non-separable space ℓ_∞. Another interesting positive result is Paley's theorem, to be seen in Exercise 13.8.

(3) Let λ_n denote the optimal constant such that, for every pair (E, X) of Banach spaces with $E \subset X$ and $\dim E = n$, there exists a projection $P : X \to E$ with norm $\leq \lambda_n$. We know that $\lambda_n \leq \sqrt{n}$ according to the

Kadeč–Snobar theorem. König and Tomczak-Jaegermann [178] refined this estimation by establishing the following bounds:

$$\sqrt{n} - \frac{\delta_1}{\sqrt{n}} \leqslant \lambda_n \leqslant \sqrt{n} - \frac{\delta_2}{\sqrt{n}},$$

where δ_1 and δ_2 are two positive constants. A less precise lower bound was seen with the construction of Sobczyk. A lower bound of the same type, using the Rudin–Shapiro polynomials, can be found in the exercises of Chapter 8 of [120].

An interesting result is due to Szarek [171]. One can construct (using high-level linear algebra) a Banach space X of dimension $2n$ in which *all the subspaces of dimension n* are as poorly complemented as possible: their complementation constant in X exceeds $c\sqrt{n}$, where c is a positive constant.

13.7 More recent developments

We enter here a rather specialised domain, which the reader can skip, but which helps situate the problem of complementation within the general theory of Banach spaces. First, let us give a few definitions to help understand the questions.

- $(e_n)_{n \geqslant 1}$ is said to be a *Schauder basis* of X if every $x \in X$ can be *uniquely* written as a convergent series

$$x = \sum_{n=1}^{\infty} x_n e_n,$$

where the x_n are scalars. The basis is said to be *unconditional* if in addition the preceding series is unconditionally convergent[15] for every $x \in X$.
- $(e_n)_{n \geqslant 1}$ is said to be an *unconditional basic sequence* if it is an unconditional basis of the closed subspace that it generates.
- The space X is said to have the *approximation property* (abbreviated as X has AP) if, for every compact subset $K \subset X$ and any $\varepsilon > 0$, there exists an operator $T \in \mathcal{L}(X)$ of finite rank such that

$$\|Tx - x\| \leqslant \varepsilon \text{ for every } x \in K.$$

---

[15] A series $\sum_{n \geqslant 1} u_n$ of vectors of a Banach space X is said to be unconditionally convergent if the family $(u_n)_{n \geqslant 1}$ is summable or, in other words, if the series $\sum u_{\sigma(n)}$ converges for every bijection $\sigma : \mathbb{N}^* \to \mathbb{N}^*$. In this case, we have automatically $\sum_{n=1}^{\infty} u_{\sigma(n)} = \sum_{n=1}^{\infty} u_n$. If X is *finite-dimensional*, unconditional convergence is equivalent to absolute convergence (see Chapter 2 of [120]).

- The space X is said to be *decomposable* if it can be written as the direct sum of E and F, where E and F are two closed infinite-dimensional subspaces of X. If not, X is said to be *indecomposable*.
- The space X is said to be *hereditarily indecomposable* if every closed infinite-dimensional subspace of X is indecomposable.
- The space X is said to be *homogeneous* if it is isomorphic to all its infinite-dimensional closed subspaces.

We note two implications between these different properties:

$$\text{if } X \text{ has a Schauder basis, then } X \text{ has AP.} \tag{13.8}$$

Indeed, if we set $P_n(x) = \sum_{j=1}^{n} x_j e_j$, the sequence $(P_n - I_X)_{n \geqslant 1}$ converges pointwise to 0 on X. However, it can be shown [120] that the P_n are continuous, and more precisely that

$$\lambda := \sup_{n \geqslant 1} \| P_n \| < \infty.$$

Thus, the $P_n - I_X$ are equi-$(1 + \lambda)$-Lipschitz, which is sufficient to pass from simple convergence to uniform convergence on all compact sets, thanks to the pre-compactness of the compact subsets of X.

On the contrary,

$$\text{if } X \text{ has an unconditional basis } (e_n)_{n \geqslant 1}, X \text{ is decomposable.} \tag{13.9}$$

Indeed, if E and F are the closed subspaces generated by the e_n with respectively odd and even indexes, the formulas

$$P(x) = \sum_{j=1}^{\infty} x_{2j-1} e_{2j-1} \quad \text{and} \quad Q(x) = \sum_{j=1}^{\infty} x_{2j} e_{2j}$$

define linear continuous (and complementary) projections from X onto E and F, respectively.

Now let us repeat the list of questions (1) to (9) from Section 13.1. Our spaces will all be infinite-dimensional.

(1) Does every separable Banach space have a Schauder basis?
(2) Does every Banach space contain an unconditional basic sequence?
(3) Is every compact operator between two Banach spaces always the limit in norm of operators with finite rank?
(4) Is every Banach space decomposable?
(5) Does every Banach space isomorphically contain either c_0, or some ℓ_p?
(6) Is a separable Banach space which is isomorphic to its infinite-dimensional subspaces isomorphic to a Hilbert space?
(7) Is a Banach space always isomorphic to its hyperplanes?

(8) Does there exist a Banach space X with very few operators, such that every operator on X can be written $T = \lambda I_X + K$, where λ is a scalar, I_X the identity of X and K a compact operator of X?

(9) If every closed subspace of X is complemented, is X isomorphic to a Hilbert space?

It is fascinating to see how the questions of Lindenstrauss have all been resolved after his 1970 article [121].

Question (1) was answered in the negative by Enflo in 1973: there exists a separable and reflexive[16] Banach space X that does not have AP [56] (see also [45]). Moreover, by (13.8), the space X does not have a Schauder basis either, which also answers question (3) in the negative, according to a theorem of Grothendieck detailed in [109], pp. 364–365 and 370.

Question (5) was answered in the negative by Tsirelson [180]. Today, his result is formulated as follows: there exists a reflexive space Banach T, possessing an unconditional basis, but containing neither c_0 nor any ℓ_p $(1 \leqslant p \leqslant \infty)$.

Questions (2), (4) and (7) were resolved all together by Gowers and Maurey in the 1990s [71]. By "sophisticating a complication" of the Tsirelson space by Schlumprecht, they constructed a Banach space X that is at the same time reflexive, separable, and hereditarily indecomposable. According to (13.9), the space X does not have an *unconditional* basic sequence although every Banach space is notoriously known to possess basic sequences [123]. Gowers and Maurey also showed that X is not isomorphic to its closed hyperplanes.[17]

By the way, we note that $C([0, 1])$ has a Schauder basis, does not have an unconditional basis ([120]), but contains unconditional basic sequences. Indeed, it isometrically contains all the separable Banach spaces, in particular those having an unconditional basis such as c_0 or ℓ_p.

Question (6) was answered in the affirmative by Gowers [69], who notably used the works of Tomczak and Komorowski: if X is homogeneous, X is isomorphic to a separable infinite-dimensional Hilbert space. The converse is evidently true.

Question (9) was answered positively by Lindenstrauss and Tzafriri [122], as we have seen.

Question (8), about the existence of a separable Banach space X such that all operators $T \in \mathcal{L}(X)$ can be written $T = \lambda I_X + K$, where λ is a scalar and K a

---

[16] If X is a normed space, the map $X \to X^{**}$, $x \mapsto \delta_x$ (where δ_x is the evaluation at x) is always injective (and even isometric) according to the Hahn–Banach theorem. If this map is surjective, the space X is said to be reflexive.

[17] Contrary to the fact that, if X is an infinite set and $a \in X$, $X \setminus \{a\}$ is equipotent to X.

compact operator, was still open in 2007, notwithstanding the substantial progress of Gowers and Maurey. They had constructed a separable space X such that any operator T on this space can be written $T = \lambda I_X + S$, where λ is a scalar and S a strictly singular operator, that is, an operator which in many ways behaves like a compact operator (see Exercise 13.3, showing that "compact" implies strictly singular and more). As we have already said, the question was resolved by Argyros and Haydon [6], who constructed a hereditarily indecomposable pre-dual of ℓ_1 that answers this question affirmatively. For some subtle variations on the theme of Gowers–Maurey spaces, see [62, 63].

Recently, Koszmider [115] constructed an infinite compact set K such that the Banach space $X = C(K)$ is indecomposable. On the contrary, according to a result of Pelczynski about the "(V) property" [188], such a space always contains a subspace isomorphic to c_0. As we have seen, this latter space is decomposable since it possesses an unconditional basis (its canonical basis). Thus, the Koszmider space provides a fine example of an indecomposable Banach space that is not hereditarily indecomposable.

Another interesting example is due to Pisier [146]: there exists a separable Banach space X (of cotype 2, as well as its dual) that, *among other properties*, satisfies:

(i) X does not have AP;
(ii) if $E \subset X$ and $\dim E = n$, every projection $P : X \to E$ verifies $\|P\| \geqslant c\sqrt{n}$, where c is a constant > 0.

Exercises

13.1. Let X, Y be Banach spaces.
(a) Let $(X \oplus Y)_{c_0}$ denote the product space $X \times Y$ equipped with the norm

$$\|(x, y)\| := \max(\|x\|, \|y\|) \text{ for } x \in X, y \in Y.$$

Show that it is a Banach space.
(b) Let $c_0(X)$ denote the space of sequences $x = (x_n)_{n \geqslant 0}$ of elements of X with limit zero, equipped with the norm

$$\|x\| = \sup_{n \geqslant 0} \|x_n\|.$$

Show that it is a Banach space.

13.2. An example of Jean Saint Raymond. We intend to show that the potential non-complementation of E in X does not result from the geometry of these spaces, but from their relative position. We start with a pair (A, B) of Banach spaces with $A \subset B$ and A non-complemented in B. With the notations of Exercise 13.1, set

$$
\begin{cases}
E = \left(A \oplus c_0(A) \oplus c_0(B)\right)_{c_0}, \\
X = \left(B \oplus c_0(A) \oplus c_0(B)\right)_{c_0}, \\
Y = \left(c_0(A) \oplus c_0(B)\right)_{c_0}.
\end{cases}
$$

(a) Show that E is a closed subspace of X and that E and X are both isometric to Y.

(b) Let Z be a Banach space. Show that if $(A \oplus Z)_{c_0} \subset\subset (B \oplus Z)_{c_0}$, then $A \subset\subset B$.

(c) Conclude that E is not complemented in X, in spite of being isometric to it.

13.3. (Super-) strictly singular operators. Let X and Z be two Banach spaces with $\dim X = \infty$, and $T : X \to Z$ a compact operator. For a closed subspace Y of X, set

$$
\rho(Y) = \inf_{y \in Y \setminus \{0\}} \frac{\|T(y)\|}{\|y\|}.
$$

Let $\varepsilon \in \,]0, 1[$. A subset C of X or Z is said to be ε-separated if we have $\|u - v\| > \varepsilon$ for distinct $u, v \in C$.

(a) Let A be a finite subset of the unit ball of X. Set $B = T(A)$. Show ([153], Chapter 7) that there exists an integer $P(\varepsilon) > 0$ such that, if B is ε-separated and if $T_{|A}$ is injective, then $|A| \leqslant P(\varepsilon)$.

(b) Suppose that $\dim Y = \infty$. Show that we always have $\rho(Y) = 0$. We say that T is *strictly singular*.

(c) Suppose $\dim Y = n$, and recall ([153], Chapter 7) that the unit ball of Y contains a subset A, of cardinality $p \geqslant \varepsilon^{-n}$, and ε-separated. Show that, if $\rho(Y) \geqslant \varepsilon$, then

$$
\varepsilon^{-n} \leqslant P(\varepsilon^2).
$$

(d) Show that, for any $\varepsilon > 0$, there exists an integer $N(\varepsilon) > 0$ such that, if $\dim Y \geqslant N(\varepsilon)$, then $\rho(Y) \leqslant \varepsilon$. We say that T is *super-strictly singular*. Note that the converse is false: the canonical injection

$$
j : L^\infty([0, 1]) \to L^2([0, 1])
$$

is super-strictly singular and non-compact.

(e) Consider the two Banach spaces[18]

$$X = \left(\bigoplus_{n=1}^{\infty} \ell_1^n \right)_1 \text{ and } Z = \left(\bigoplus_{n=1}^{\infty} \ell_1^n \right)_2,$$

and denote by j the canonical injection from X into Z. We admit here ([2], p. 38 or [120], p. 60) that $\ell_1 = \ell_1(\mathbb{N}^*)$ does not contain any reflexive infinite-dimensional closed subspace. Show that X is isomorphic to ℓ_1 and that Z is reflexive. Deduce that j is strictly singular without being super-strictly singular.

13.4. Let X be a Banach space possessing a Schauder basis $(e_n)_{n \geqslant 1}$. For $x = \sum_{j=1}^{\infty} x_j e_j \in X$, set $P_n(x) = \sum_{j=1}^{n} x_j e_j$. Recall [120] that $\sup_{n \geqslant 1} \| P_n \| < \infty$.

(a) Show that $P_n \to I_X$ as $n \to +\infty$, uniformly on all compact subsets of X.

(b) Let $T : X \to X$ be a compact operator. By considering the operators $P_n T$, show that T is a limit (in operator norm) of operators of finite rank.

13.5. Show that the Banach spaces $Y = \ell_\infty(\mathbb{N}^*)$ and $Y = L^\infty([0, 1])$ have the following property: for every pair X_0, X of Banach spaces with $X_0 \subset X$ and every operator $T : X_0 \to Y$, there exists a continuous extension of T, $\widehat{T} : X \to Y$, with the same norm as T. *Hint*: mimic the proof of the Hahn–Banach theorem by showing that the two spaces Y above have the *2-ball property*, that is, any family of closed balls of Y intersecting two by two has a non-empty intersection, as is the case for the segments of $\mathbb{R}$. For $\ell_\infty(\mathbb{N}^*)$, proceed coordinate by coordinate; for $L^\infty([0, 1])$, use the essential upper bounds.

13.6. Let X be a Banach space satisfying the following property: X contains an unconditional basic sequence $(e_n)_{n \geqslant 1}$ generating a closed space X_0 that is complemented in X.

(a) Show that $C([0, 1])$, equipped with its natural norm, satisfies this property. *Hint*: use the fact that c_0 is separably injective.

(b) Show that $K(X)$ is not complemented in $\mathcal{L}(X)$. *Hint*: mimic the proof of Theorem 13.6.2. Keeping the same notations, consider $T_i = \Phi(P_{A_i} Q)$, where $Q : X \to X_0$ is a continuous projection.

---

[18] See Section 13.5 for these notations.

13.7. Let $X = \ell_\infty(\mathbb{N}^*)$ and $Y = L^\infty([0, 1])$.

(a) Show that X and Y are isomorphic to their squares.

(b) For $n \geqslant 1$, let us denote by I_n the interval $\left[\dfrac{1}{n+1}, \dfrac{1}{n}\right[$, and by f_n its indicator function. Show that the map $T : X \to Y$ defined by

$$T(x) = \sum_{n=1}^\infty x_n f_n$$

is an isometry from X to Y.

(c) Show that there exists a sequence $(g_n)_{n \geqslant 1}$ of the unit ball of $L^1([0, 1])$ such that

$$\|y\|_\infty = \sup_{n \geqslant 1} \left| \int_0^1 y(t) g_n(t)\, dt \right| \quad \text{for } y \in Y.$$

(d) Show that the map $S : Y \to X$ defined by

$$S(y) = x, \quad \text{where } x_n = \int_0^1 y(t) g_n(t)\, dt \ \text{ for } n \geqslant 1,$$

is an isometry from Y to X.

(e) Show that X and Y are isomorphic.

13.8. The Paley projection

(a) Let $a = (a_n)_{n \geqslant 0}$ and $b = (b_n)_{n \geqslant 0}$ be two square-summable sequences of scalars, and $(c_n)_{n \geqslant 0}$ their Cauchy product:

$$c_n = \sum_{k=0}^n a_k b_{n-k}.$$

Show that

$$\sum_{n=0}^\infty |c_{2^n}|^2 \leqslant C \|a\|_2^2 \|b\|_2^2,$$

where C is a numerical constant.

(b) If $\Lambda \subset \mathbb{N}$, define H_Λ^1 by

$$H_\Lambda^1 = \{ f \in L^1(\mathbb{T}) / \widehat{f}(n) = 0 \text{ for all } n \notin \Lambda \}.$$

In particular, set $H^1 = H_\mathbb{N}^1$. Show that if Λ is the set of powers of 2, then H_Λ^1 is a closed subspace of H^1 and the map $P : H^1 \to H_\Lambda^1$, formally defined by

$$Pf = \sum_{n=0}^\infty \widehat{f}(2^n) e_{2^n}, \quad \text{with } e_j(t) = e^{ijt} \text{ for } j \in \mathbb{N},$$

is a continuous projection from H^1 onto H_Λ^1. *Hint*: use the fact that every function of H^1 is the product of two functions of H^2.

13.9. The nuclear norm. Let E and F be two real Banach spaces with the same dimension n, $G = \mathcal{L}(E, F)$ the space of linear maps from E to F, and E^* the dual space of E. If $e^* \in E^*$ and $f \in F$, define $e^* \otimes f \in G$ by

$$e^* \otimes f(x) = e^*(x)f \quad \text{for all } x \in E.$$

This is an operator of rank $\leqslant 1$. Denote by D the set of $e^* \otimes f$, where $\|e^*\| = \|f\| = 1$. For $u \in G$, set

$$N(u) = \inf\left\{ \sum_{j=1}^{m} \|e_j^*\|\|f_j\|, u = \sum_{j=1}^{m} e_j^* \otimes f_j \right\},$$

where the lower bound is taken over all decompositions of u as a finite sum of operators of rank $\leqslant 1$, with $e_j^* \in E^*$, $f_j \in F$ and $m \in \mathbb{N}^*$.

(a) Show that N is a norm on G; it is called the *nuclear norm*.

(b) Show that D is a compact subset of G.

(c) Show the inequality

$$|\mathrm{tr}\,(vu)| \leqslant \|v\|N(u) \quad \text{for } u \in G, v \in \mathcal{L}(F, E) =: H.$$

(d) Show that

$$\sup_{d \in D} |\mathrm{tr}\,(vd)| = \|v\| \quad \text{for all } v \in H.$$

(e) Show that

$$\sup_{u \in G, N(u) \leqslant 1} |\mathrm{tr}\,(vu)| = \|v\| \quad \text{for all } v \in H.$$

In other words, the operator norm $\|\cdot\|$ is the dual norm of the nuclear norm.

(f) Show that

$$\sup_{v \in H, \|v\| \leqslant 1} |\mathrm{tr}\,(vu)| = N(u) \quad \text{for all } u \in G.$$

(g) Show that B_N, the unit ball of G equipped with the nuclear norm, is none other than the convex hull of D.

For the following exercise, we recall the *Lewis theorem* ([120], p. 336 or [147], p. 28): if ℓ_2^n is the Euclidean space $\mathbb{R}^n$ equipped with its canonical inner product $\langle \cdot, \cdot \rangle$, and if E is as in Exercise 13.9, there exists an isomorphism $u_0 : \ell_2^n \to E$ such that $\|u_0\| = 1$ and $N(u_0^{-1}) = n$.

13.10. F. John's theorem [99]. The space E is as in Exercise 13.9. Define an inner product $(\cdot, \cdot)$ on E by the formula

$$(x, y) = \langle u_0^{-1}(x), u_0^{-1}(y) \rangle,$$

and denote by $|\cdot|$ the Euclidean norm on E associated with this inner product. Also, define an associated dual norm on E by the formula

$$\|x\|_* = \sup\{|(x, y)|, \|y\| \leqslant 1\}.$$

(a) Show that $\|x\| \leqslant |x| \leqslant \|x\|_*$ for $x \in E$.

(b) By using Exercise 13.9 and the fact that $N\left(\dfrac{u_0^{-1}}{n}\right) = 1$, show that there exist an integer $M \leqslant n^2 + 1$, real numbers $c_j \geqslant 0$ with sum n and unit vectors $e_j^* \in E^*$ and $f_j \in \ell_2^n$ such that

$$u_0^{-1} = \sum_{j=1}^M c_j e_j^* \otimes f_j.$$

(c) Let $u_j = u_0(f_j)$ satisfying $|u_j| = 1$, and $v_j \in E$ satisfying $(x, v_j) = e_j^*(x)$ for all $x \in E$ and $\|v_j\|_* = 1$. Show that the identity on E can be decomposed as follows:

$$I_E = \sum_{j=1}^M c_j(\cdot, v_j) u_j.$$

(d) By calculating the trace of I_E, show that $v_j = u_j$ and that $\|u_j\| = 1$.

(e) Conclude that we have the decomposition of F. John:

$$I_E = \sum_{j=1}^M c_j(\cdot, u_j) u_j, \text{ with } \|u_j\| = |u_j| = \|u_j\|_* = 1, c_j \geqslant 0, \sum_{j=1}^M c_j = n.$$

Thus, the u_j are situated at the confluence of three unit spheres.

(f) Show that, for all $x \in E$:

$$|x|^2 = \sum_{j=1}^M c_j |(u_j, x)|^2 \text{ and } |x| \leqslant \sqrt{n}\|x\|.$$

13.11. The Kadeč–Snobar theorem. Let E and X be two real Banach spaces, with $E \subset X$ and $\dim E = n$. Equip E with the inner product of Exercise 13.10, and keep the notations of that exercise.

(a) Show that there exist linear forms L_j on X, of norm 1, such that

$$L_j(x) = (x, u_j) \text{ for } x \in E.$$

(b) On X, define a semi-inner product[19] that extends $(\cdot, \cdot)$ on E by the formula

$$(x, y) = \sum_{j=1}^{M} c_j L_j(x) L_j(y) \text{ for } x, y \in X.$$

Set $M = \bigcap_{j=1}^{M} \ker L_j$, and denote by H_0 the pre-Hilbertian space X/M, by H its completion, and by

$$\rho : X \to X/M, x \mapsto x + M$$

the canonical surjection, considered naturally as a map from X to H. Note that if $y \in \rho(E)$, then y can be uniquely written in the form $y = x + M$, $x \in E$; indeed, if $x, x' \in E$ satisfy $x - x' \in M$, then $|x - x'| = 0$ hence $x = x'$ since $|\cdot|$ is a norm on E. This allows us to consider the following diagram:

$$X \xrightarrow{\rho} H \xrightarrow{Q} \rho(E) \xrightarrow{\rho^{-1}} E,$$

and we set $P = \rho^{-1} Q \rho$, where Q is the orthogonal projection from H onto $\rho(E)$. Show that

$$P(x) = \sum_{j=1}^{M} c_j L_j(x) u_j \text{ for all } x \in X.$$

(c) Show that P defines a projection from X onto E, of norm $\leqslant \sqrt{n}$.

13.12. A variant of Exercise 1.1 from Chapter 1. Let I be a finite set, and $(z_i)_{i \in I}$ a family of complex numbers indexed by I. Suppose that the real constant C satisfies

$$\left| \sum_{i \in J} z_i \right| \leqslant C \text{ for every subset } J \subset I.$$

(a) First suppose that the z_i are real numbers. Show that

$$\sum_{i \in I} |z_i| \leqslant 2C.$$

(b) In the general case, show that

$$\sum_{i \in I} |z_i| \leqslant 4C.$$

(c) Show that the constant 4 in the preceding inequality can be replaced by the constant π, this being optimal [120, 160].

---

[19] That is, a positive symmetric bilinear form.

13.13. Consider the Banach space c of convergent sequences of complex numbers $x = (x_n)_{n \geq 1}$, equipped with the natural norm defined by $\|x\| = \sup_{n \geq 1} \|x_n\|$, and c_0 the closed hyperplane of c consisting of the sequences with limit zero. If $x \in c$, $\ell(x)$ denotes the limit of x_n when $n \to \infty$.

(a) Define $P : c \to c_0$ by the formula $P(x) = x - \ell(x)e$, where $e \in c$ is the sequence identically equal to 1. Show that P is a projection from c onto c_0, with norm 2. Was the existence of P predictable?

(b) Show that, for any continuous projection $Q : c \to c_0$, we have $\|Q\| \geq 2$. *Hint*: consider $\|Q(e - 2e_n)\|$, where e_n is the nth vector of the canonical basis of c_0.

13.14. Let $X = H^\infty$ be the space of bounded analytic functions in the open unit disk D, equipped with the norm $\|f\|_\infty = \sup_{z \in D} |f(z)|$. For $n \in \mathbb{N}^*$, set $z_n = 1 - 2^{-n}$. Admit [188] the existence of a sequence $(f_n)_{n \geq 1}$ of elements of X satisfying the following properties:

$$f_n(z_j) = \delta_{n,j} \text{ for } n, j \geq 1 \text{ and } \sum_{n=1}^\infty |f_n(z)| \leq C \text{ for } z \in D,$$

where δ denotes the Kronecker symbol and C a constant. Finally, set $Y = \ell_\infty(\mathbb{N}^*)$, the space that we encountered in Exercise 13.5.

(a) Show that the operator $T : X \to Y$ defined by $T(f) = (f(z_n))_{n \geq 1}$ is a continuous linear surjection that admits a right inverse.

(b) Show that the closed subspace of X consisting of the functions that are zero on the sequence $(z_n)_{n \geq 1}$ is complemented in X.

13.15. Remark. The existence of a sequence $(f_n)_{n \geq 1}$ as in the statement of Exercise 13.14 holds for any interpolation sequence $(z_n)_{n \geq 1}$, thanks to a theorem of P. Beurling (see [66], as well as Section 12.7 of Chapter 12).

13.16. Justify the existence of a Banach space X, separable and infinite-dimensional, such that the space $\mathcal{L}(X)$ of continuous linear operators on X, equipped with the operator norm, is itself separable.

14

Hints for solutions

Exercises for Chapter 1

1.1. (a) Use the sequence $(\varepsilon_1, \ldots, \varepsilon_n)$ defined by

$$\varepsilon_k = \begin{cases} 1 & \text{if } \operatorname{Re} z_k \geqslant 0 \\ -1 & \text{otherwise} \end{cases} \quad \text{for } 1 \leqslant k \leqslant n.$$

1.2. (b) Use the non-trivial example of Hardy and Littlewood from Section 1.3.1.

1.3. (b) Start by showing $a_n = O(n^{-1})$.

1.4. (c) Recall that if a Dirichlet series $\sum a_n n^{-s}$ converges at s_0, then it converges for all $s \in \mathbb{C}$ such that $\operatorname{Re} s > \operatorname{Re} s_0$.

   (d) (i) Recall [153] that if $f : [1, +\infty[\to \mathbb{C}$ is C^1 and verifies $f' \in L^1$ $([1, +\infty[)$, then

$$\sum_{n=1}^N f(n) - \int_1^N f(t)\,dt \to \ell \in \mathbb{C}.$$

1.5. (b) Show that if $\sum a_n z^n$ is a power series with radius of convergence 1, with $a_n \to 0$, if $u \in \mathbb{T}$ and if

$$\sum_{n=0}^\infty b_n z^n = (z - u) \sum_{n=0}^\infty a_n z^n,$$

we have $b_n \to 0$, and the set of convergence at the boundary of $\sum b_n z^n$ is $E \cup \{u\}$ where E is the set of convergence at the boundary of $\sum a_n z^n$.

1.6. (c) Apply Littlewood's theorem to the series $\sum n^{-\alpha} a_n$.

1.7. (b) Show that $|\sin(\pi u)| \geqslant 2\|u\|$, $\|u\|$ being the distance from u to the closest integer.

(c) For $u = e^{2i\pi\alpha}$, we can conclude if

$$\left\| \frac{k}{\pi} + \alpha \right\| \geqslant c|k|^{-N} \text{ for all } k \in \mathbb{Z}^*.$$

The other cases appear less straightforward (resonance phenomena between π and α).

1.8. (a) Use the simple fact [76] that a series $\sum a_n$ that is Cesàro-summable of a certain order has coefficients with growth at most polynomial:

$$|a_n| \leqslant Cn^d \text{ for } n \geqslant 1.$$

(b) (ii) We can set $x = e^{-t}$ and use the Cauchy inequalities to bound the derivatives of ϕ on the positive real half-axis.

(b) (iii) Write $b = k - \alpha$, where $k \in \mathbb{N}$ and $0 \leqslant \alpha \leqslant 1$, and use Exercise 1.6.

1.9. Consider first the case $k = 2$, carry out a double Abel transformation in order to expose σ_n, and then use the estimation

$$\sum_{n=1}^{\infty} n^3 x^{n^2} = O((1-x)^{-2}) \text{ when } x \nearrow 1.$$

1.10. Make appropriate use of Theorems 1.3.11 and 1.3.12 of Littlewood, with the sequence $\varphi_n := n \dfrac{\mu_n}{\lambda_n}$.

1.12. (a) Start by showing that

$$f(x) = (1-x) \sum_{n=1}^{\infty} x^n \left(\sum_{d|n} (\Lambda(d) - 1) \right).$$

(c) Use the Kronecker lemma by which, if $\sum \dfrac{u_n}{n}$ converges, then

$$\frac{u_1 + \cdots + u_n}{n} \to 0.$$

1.13. (a) Use the fact that the dual of c_0 is isometric to ℓ_1.

(b) Use the principle of uniform boundedness.

1.14. Use Theorem 1.3.11.

Exercises for Chapter 2

2.2. (b) Using part (a) and Plancherel's theorem, show that $(1 + \|x\|^2)^{k/2} \widehat{f}(x) \in L^2(\mathbb{R}^d)$, where $\| \cdot \|$ denotes the standard Euclidean norm. Using spherical coordinates, conclude by showing that $(1 + \|x\|^2)^{-k/2} \in L^2(\mathbb{R}^d)$.

2.3. (b) Use Weierstrass' theorem.

2.5. (a) Use Theorem 2.6.3 and the fact that if $\xi \notin E$, there exists a $u \in A(\mathbb{R})$ such that $u = 1$ on E and $u(\xi) = 0$.

2.6. (a) Use Plancherel's theorem.

(b) A subspace of the Hilbert space $L^2(\mathbb{R})$ is dense if and only if its orthogonal is reduced to zero.

2.7. (a) As H^2 is a Hilbert space, the family $(g_a)_{a \in E}$ is complete if and only if any element of H^2 orthogonal to all of the functions g_a is zero.

(b) Show that, if $f \in W^+$, there exist u and $v \in H^2$ that verify the following conditions:

- $\widehat{u}(n) \neq 0$ for $n \geqslant 0$,
- $|\widehat{u}(n)| = |\widehat{v}(n)| = |\widehat{f}(n)|^{1/2}$ if $\widehat{f}(n) \neq 0$,
- $\widehat{f}(n) = \widehat{u}(n)\widehat{v}(n)$ for all $n \geqslant 0$.

(c) Thus, there exist generator sets of measure zero, which is quite startling, as Carleson himself points out: we do not need all of the translates, but only a "small number", to generate H^2.

2.8. (b) Bound *à la Tauber* (with $t = N^{-1}$)

$$\left| \sum_{n=1}^{\infty} a_n f(nt) - \sum_{n=1}^{N} a_n \right|,$$

by observing that f is Lipschitz and that $f(v) \leqslant Cve^{-v}$ if $v \geqslant 1$, where C is a positive constant.

(c) The ζ function has a meromorphic extension to $\mathbb{C}$, with a unique simple pole at $z = 1$, and $\zeta(1 + ix) \neq 0$ for all $x \in \mathbb{R}^*$.

2.9. (a) Observe that the function $\sigma : n \mapsto \sum_{d|n} \lambda(d)$ is multiplicative: $\sigma(mn) = \sigma(m)\sigma(n)$ if m and n are relatively prime.

(c) Obtain the estimate $L(x) = O\left((1-x)^{-1/2}\right)$ as $x \nearrow 1$ by comparing the series to an integral.

(e) By using the sieve properties of the Möbius function:

$$\sum_{d|n} \mu(d) = \begin{cases} 0 & \text{if } n \geqslant 2, \\ 1 & \text{if } n = 1, \end{cases}$$

show that, for all $n \geqslant 1$,

$$\sum_{d|n} \mu(d) S_d = \sum_{(m,n)=1} a_m,$$

where (m, n) is the GCD of m and n, and $S_d = \sum_{k=1}^{\infty} a_{kd}$. Deduce that $a_1 = 0$.

2.10. **(b)** To derive the integral expression for p_n, remark that

$$\varphi(t)^n = \mathbf{E}(e^{itS_n}) = \sum_{k\in\mathbb{Z}} \mathbf{P}(S_n = k)e^{ikt}.$$

Then, show that if t is a *non-zero* real number such that $|\varphi(t)| = 1$, there exists a real number α such that X_1 has values in $\alpha + \dfrac{2\pi}{t}\mathbb{Z}$ almost surely, and next that if t' is a non-zero real number such that $|\varphi(t')| = 1$, then $\dfrac{t'}{t} \in \mathbb{Q}$. Conclude that $|\varphi(t)| < 1$ almost everywhere, and that $p_n \to 0$.

In the example given, we have

$$\varphi(t) = \frac{1 + \cos t}{2} \geqslant 0,$$

so that, by Beppo Levi's theorem,

$$\sum_{n=0}^{\infty} p_n = \frac{1}{\pi} \int_{-\pi}^{\pi} \frac{dt}{1 - \cos t} = +\infty.$$

Deduce that

$$\lim_{x\nearrow 1} Q(x) = \lim_{x\nearrow 1} \left(1 - \frac{1}{P(x)}\right) = 1$$

and conclude.

(c) Observe that, by the Fubini–Tonelli theorem,

$$\sum_{n=0}^{\infty} r_n = \sum_{k=1}^{\infty}\sum_{n=0}^{k-1} q_k = \sum_{k=1}^{\infty} kq_k < \infty.$$

(d) If d is the GCD of the integers $n \geqslant 1$ such that $q_n > 0$, prove by induction on $N \geqslant 1$ that d divides all integers $n \in [\![1, N]\!]$ such that $p_n > 0$.

(e) Note that if $z = \rho e^{i\theta}$ ($0 \leqslant \rho \leqslant 1$ and $\theta \in \mathbb{R}$) is a zero of R in $\overline{D}$, then

$$\sum_{n=1}^{\infty} q_n \rho^n \cos(n\theta) = 1 = \sum_{n=1}^{\infty} q_n.$$

Conclude using part (d).

(g) Use parts (b) and (f) to conclude that such a random walk does not exist.

2.11. **(b)** If $f \in \mathcal{P}_A$ satisfies $\|f\|_\infty \leqslant 1$, use $g \in \mathcal{P}_A$ defined by $g(t) = f(st)$.

(c) Use the concavity inequality $|\sin u| \geqslant \dfrac{2}{\pi}|u|$ for $|u| \leqslant \dfrac{\pi}{2}$.

2.12. **(b)** See [40], p. 299.

Exercises for Chapter 3

3.1. (b) Use the PNT and Hadamard's formula for the radius of convergence.

3.2. (a) Observe that $\lambda_1, \ldots, \lambda_n$ are distinct divisors of P_n.

(b) We find, for $\lambda < \dfrac{1}{\ln 2}$ and n sufficiently large:

$$P(n) \geqslant \exp(\lambda \ln n \ln \ln n).$$

3.3. (a) Set $\Delta = m_1$. By hypothesis, there exist *rationals* v_j such that $m_j = \Delta^{v_j}$, that is $v_j = \dfrac{u_j}{b}$ with $u_1, \ldots, u_r$ and b mutually relatively prime after simplification. Let $\prod_p p^{\alpha_p}$ be the decomposition in prime factors of Δ. Use Bézout's identity to show that b divides α_p for all p, and conclude that $m_j = d^{u_j}$, with this time d and u_j integers.

(b) We find $b_p = \sum_{j=1}^{q} r_j b_{p-u_j}$, linear recurrence with characteristic equation $1 = \sum_{j=1}^{q} r_j r^{-u_j}$. The largest root of this equation is a simple root (easy), and satisfies $r = d^a$, with a as in Theorem 3.4.1. We thus have $b_p \sim c d^{ap}$, where c is a positive constant.

(d) The cluster set A of $\left(\dfrac{a_n}{n^a}\right)_{n \geqslant 1}$ is the segment $\left[\dfrac{c}{d^a}, c\right]$. In the special case outlined, we have $a = 1$, $b_0 = a_1 = 5$, $b_1 = a_3 = 13$ and $b_p = 2b_{p-1} + 3b_{p-2}$ for $p \geqslant 2$, so that $b_p = \dfrac{1}{2}(-1)^p + \dfrac{9}{2}3^p$, and then $A = \left[\dfrac{3}{2}, \dfrac{9}{2}\right]$.

3.4. (a) Use the equality $\varphi(n) = n \prod_{p \mid n} \left(1 - \dfrac{1}{p}\right)$ (p is a prime number), the inequality $1 - x \geqslant e^{-2x}$ which holds if $0 \leqslant x \leqslant \dfrac{1}{2}$ and the fact that $\sum_{p \leqslant x} \dfrac{1}{p} \sim \ln \ln x$ as $x \to +\infty$. We obtain an estimate of the form

$$\varphi(n) \geqslant n(\ln n)^{-C},$$

where C is a positive constant.

(b) Immediate consequence of (a). More simply, we note that if $\varphi(m) = n$ and if p^r (p prime, $r \geqslant 1$) divides m, then $p^r - p^{r-1} = p^{r-1}(p-1)$ divides n, which gives a finite number of possibilities for p and for r.

(d) Use Lemma 3.4.4.

(e) Show that $\dfrac{f(s)}{\zeta(s)}$ is defined by an infinite product absolutely convergent for $\operatorname{Re} s > 0$.

(f) Apply Ikehara's theorem.

3.5. (c) Apply the same method as in Exercise 3.4, with this time $\dfrac{f(s)}{\zeta(s)^{2^q}}$. Conclude here by applying Delange's Tauberian theorem.

3.6. (c) Select $y = \dfrac{x}{\ln^d x}$, with $d > 1$.

3.7. (b) Same method as in Exercises 3.4 and 3.5, using Ikehara's theorem.

Exercises for Chapter 4

4.1. (a) Given a pair (p, q) of integers that satisfies (4.2), first select an integer $n' \geqslant n$ such that $2^{-n'} \leqslant \left| x - \dfrac{p}{q} \right|$, and then (p', q') $(q' \geqslant 2)$ such that

$$\left| x - \frac{p'}{q'} \right| < \frac{1}{q'^n}.$$

(b) Observe that if P is the minimal polynomial of x and if $\dfrac{p}{q}$ is a rational number, then $\left| P\left(\dfrac{p}{q}\right) \right| \geqslant \dfrac{1}{q^d}$, and evaluate $\left| P\left(\dfrac{p}{q}\right) \right| = \left| P\left(\dfrac{p}{q}\right) - P(x) \right|$ using the mean value theorem.

(c) Write

$$\mathcal{L} = (\mathbb{R} \setminus \mathbb{Q}) \cap \underbrace{\bigcap_{n \geqslant 1} \bigcup_{p \in \mathbb{Z}} \bigcup_{q \geqslant 2} \left] \frac{p}{q} - \frac{1}{q^n}, \frac{p}{q} + \frac{1}{q^n} \right[}_{:= O_n}$$

and show that O_n is a dense open subset of $\mathbb{R}$.

(d) Verify that if $N \geqslant 1$ and $n \geqslant 3$, we have

$$\mathcal{L} \cap [-N, N] \subset \bigcup_{q \geqslant 2} \bigcup_{-qN \leqslant p \leqslant qN} \left] \frac{p}{q} - \frac{1}{q^n}, \frac{p}{q} + \frac{1}{q^n} \right[$$

and deduce that

$$\lambda(\mathcal{L} \cap [-N, N]) \leqslant 2 \sum_{q=2}^{\infty} \frac{2qN + 1}{q^n} \to 0 \text{ as } n \to +\infty.$$

4.2. (b) If x is a fixed real number, consider $X = x - \mathcal{L}$ and $Y = \mathcal{L}$.

4.3. (a) Copy the construction of the Cantor middle third set: select a sequence of positive numbers such that $\sum_{n=1}^{\infty} a_n = 1 - \alpha$, begin by removing from the segment $K_0 = [0, 1]$ a segment of length a_1 centred at $\dfrac{1}{2}$, which gives a set K_1, union of two disjoint segments, each of length $\dfrac{1 - a_1}{2}$. Continue in the same manner.

(b) Let $(K_n)_{n \geqslant 1}$ be a sequence of closed subsets of $[0, 1]$ with empty interiors, with $\lambda(K_n) \to 1$. Then apply Theorem 4.3.2 to the union of the K_n.

4.4. Using Baire's theorem, show that $[0, 1] \cap \mathbb{Q}$ is not a G_δ-set of $[0, 1]$.

4.5. If $x \in X$, then the singleton $\{x\}$ is a closed subset of X with empty interior, which implies that any countable subset of X is of first category. Conclude by applying Baire's theorem.

4.6. Show that the function

$$F : [0, 1] \to \mathbb{R}, x \mapsto \int_0^x f(t)\, dt$$

satisfies $F' = f$. Then use the estimates

$$\int_X^{+\infty} \frac{\sin u}{u^2}\, du \underset{X \to +\infty}{=} O\left(\frac{1}{X^2}\right)$$

and

$$\int_X^{+\infty} \frac{\sin^2 u}{u^2}\, du \underset{X \to +\infty}{\sim} \frac{1}{2X}.$$

4.7. Let Ω be an open subset of $\mathbb{R}$. Define $f : \Omega \to \mathbb{R}$ as follows:

$$f(x) = \begin{cases} 1 & \text{if } x \in \Omega, \\ 0 & \text{if } x \in \partial\Omega, \\ \mathbf{1}_{c\mathbb{Q}}(x) & \text{otherwise}, \end{cases}$$

where $\partial\Omega$ is the boundary of Ω and $\mathbf{1}_{c\mathbb{Q}}$ is the indicator function of the irrational numbers.

What are the points of continuity of f?

Now, if $E = \bigcap_{n=1}^\infty \Omega_n$ is a G_δ-set of $\mathbb{R}$, to each Ω_n we associate a function f_n as above; and we set $f = \sum_{n=1}^\infty a^n f_n$, with $0 < a < \frac{1}{2}$. What are the points of continuity of f? (Proceed as in the proof of Theorem 4.3.2.)

4.8. (a) (ii) Use the fact that any closed subset of a metric space is a G_δ-set.

(a) (iii) Show that, if $(x_n)_{n \geq 0}$ is a sequence of points of X converging to $x \in Y$, the sequence $(f(x_n))_{n \geq 0}$ is convergent. Verify that its limit $f(x)$ is independent of the choice of $(x_n)_{n \geq 0}$ and that the function $f : Y \to F$ thus defined is continuous.

(b) Given $y \in Y$, use a sequence of points in X that converges to y.

(c) Let E and F be two normed spaces, homeomorphic via $f : E \to F$. Using part (a), show that E is a G_δ-set that is dense in its completion $\widetilde{E}$. Deduce that if $x \in \widetilde{E}$, then $E \cap (x + E) \neq \varnothing$, and conclude.

4.9. (a) First show that F is dense in E, remembering that any proper subspace of E has empty interior. Then show that F cannot be written as a countable union of closed subsets of F with empty interiors, and then that if $(F_n)_{n \geqslant 0}$ is a sequence of closed subsets of F with empty interiors, $\bigcup_{n \geqslant 0} F_n$ has an empty interior in F.

(b) First show that an infinite set I is a countable union of an increasing sequence of proper subsets, using for example a countable subset of I.

4.10. (a) (i) Remark that $b_k \geqslant 2$ for $k \geqslant 1$, so we have the bound

$$b_k^{n+1-k} \leqslant 2^{n+1-k} \text{ for } n \geqslant 0 \text{ and } k \geqslant n + 1.$$

(ii) Show that a lower bound for $|f'(x)|$ is

$$b_1 - \left(1 + b_3^{-1} + b_4^{-2} + b_5^{-3} + \dots \right) \geqslant b_1 - \sum_{k=0}^{\infty} 2^{-k} = c_1.$$

(iii) Write

$$f^{(n)}(x) = i^n b_n e^{ib_n x} + \sum_{k<n} i^n b_k^{n+1-k} e^{ib_k x} + \sum_{k>n} i^n b_k^{n+1-k} e^{ib_k x},$$

so that

$$|f^{(n)}(x)| \geqslant b_n - \sum_{k<n} b_k^{n+1-k} - \sum_{k>n} b_k^{n+1-k}$$

$$\geqslant c_n + 2 - \sum_{k>n} 2^{n+1-k} = c_n.$$

(b) (i) To show that it is closed, use a sequence. To show that $F(b, p)$ has an empty interior in E, select $f \in F(b, p)$ and a neighbourhood $V = V_{n,\varepsilon}$ of 0 in E. According to Weierstrass' theorem, there exists a polynomial P belonging to $f + V_{n,\varepsilon/2}$. Then fix $c \in \mathbb{N}^*$ to be adjusted, and define the function

$$u : \mathbb{R} \to \mathbb{C}, x \mapsto \frac{\varepsilon}{2} c^{-n} e^{icx}.$$

Verify that $u \in V_{n,\varepsilon/2}$, so that $P + u \in f + V$. By noting that there exists a constant $M_p > 0$ such that

$$|P^{(k)}(x)| \leqslant M_p \text{ for } k \geqslant 0 \text{ and } |x| \leqslant p,$$

show finally that we can choose c in such a way that

$$|(P + u)^{(n+1)}(x)| > b^{n+1}(n + 1)! \text{ for } |x| \leqslant p,$$

and conclude that $P + u \notin F(b, p)$.

(ii) Apply Baire's theorem, and then show that a function $f \in E$ belongs to none of the $F(b, p)$ if and only if $R(f, x) = 0$ for all $x \in \mathbb{R}$. To conclude, use the following fact – already used in Exercise 4.2

and in part (c) of Exercise 4.8 –: if A is a residual subset of a Banach or Fréchet space E, then $A + A = E$.

Exercises for Chapter 5

5.1. (c) Use phases $\varphi_1, \ldots, \varphi_N$ such that

$$\left\| e^{iNt} \sum_{j=1}^{N} \cos(jt + \varphi_j) \right\|_\infty \leqslant C\sqrt{N}(\ln N)^{1/2},$$

so that

$$\left\| e^{iNt} \sum_{j=1}^{N} \cos(jt + \varphi_j) \right\|_U \leqslant C\sqrt{N}(\ln N)^{3/2}$$

from the properties of the L^1-norm of the Dirichlet kernel. Test then the *a priori* inequality of part (b) on this example to obtain a contradiction when $N \to +\infty$.

5.2. (a) Use the cotype 2 of $L^1(\mathbb{T})$ (Theorem 5.2.8).

(b) If $B^+ = \{n \in [\![1, N]\!]/\alpha_n = 1\}$ and $B^- = \{n \in [\![1, N]\!]/\alpha_n = -1\}$, show that either $A = B^+$ or $A = B^-$ is suitable. The size of $|A|$ follows immediately thanks to Parseval.

5.3. (b) (ii) We may use the fact that if $g \in C$, then

$$\sup_{f \in C, \|f\|_\infty \leqslant 1} \left| \int_\mathbb{T} fg \, dm \right| = \|g\|_1.$$

(iii) Using again the fact that L^1 is of cotype 2, show that

$$\left(\sum_{|n| \leqslant N} |c_n|^2 \right)^{1/2} \leqslant \sqrt{2}\|T\|.$$

5.5. Take an n-hypercube $Q_n = A_1 \times \cdots \times A_d$, and selectors $(X_i)_{i \in Q_n}$ with expectation $n^{\alpha-d}$, then define a notion of d-*disjoint* sets to pass from finite to infinite.

5.6. (a) First show, using characteristic functions, that if $T \in O(n)$, then $T(G)$ and G have the same distribution. Deduce that there exists an $O(n)$-invariant probability σ on S such that

$$\int_S f \, d\sigma = \mathbf{E}\left(f\left(\frac{G}{\|G\|} \right) \right) \text{ for all } f \in C(S).$$

Conclude by uniqueness that $\sigma = \mu$.

(b) Admit (see [120]) that $E(\sup_{1\leqslant j\leqslant n}|g_j|)$ behaves as $\sqrt{\ln n}$ while, by the law of large numbers, $\sqrt{\sum_{j=1}^{n}g_j^2}\sim\sqrt{n}$ with high probability. More precisely, if we set

$$Y_n = \sup_{1\leqslant j\leqslant n}|g_j|,\ X_n = \frac{Y_n}{\|G\|},\ I_n = E(X_n),$$

$$E_n = \left\{\|G\| \geqslant \frac{\sqrt{n}}{2}\right\},\ \text{and}\ F_n = \left\{\|G\| \leqslant 2\sqrt{n}\right\},$$

by remarking that $0 \leqslant X_n \leqslant 1$, we have

$$E(X_n) = E(X_n \mathbf{1}_{E_n}) + E(X_n \mathbf{1}_{^cE_n})$$
$$\leqslant \frac{2}{\sqrt{n}}E(Y_n) + P(^cE_n)$$
$$\leqslant C\left(\frac{\ln n}{n}\right)^{1/2} + \frac{C}{n},$$

from Tchebycheff's inequality for $\|G\|^2$.

Proceed similarly for the lower bound of I_n, by now using F_n. To find an upper bound for $E(Y_n \mathbf{1}_{^cF_n})$, use Cauchy–Schwarz and the fact that $E(Y_n^2) \leqslant C\ln n$.

5.8. (a) If $(x_n)_{n\geqslant 1}$ is α-distant, the inequality gives

$$\alpha^2 N(N-1) \leqslant 2N^2.$$

Simplify and let $N \to +\infty$. For the lower bound, use an orthonormal sequence.

(b) Use the canonical basis of ℓ_1.

(c) If $(r_n)_{n\geqslant 1}$ is a Rademacher sequence, then, for $i \neq j$, the random variable $|r_i - r_j|$ takes on values 0 or 2 with probability $\frac{1}{2}$. Moreover, $\|r_n\|_p = 1$ for all n.

Exercises for Chapter 6

6.1. (a) Use the continuity of the map

$$\tau : \mathbb{R} \to L^1(\mathbb{R}^n),\ a \mapsto f(\cdot + a).$$

(b) To show that the function $\mathbf{1}_A * \mathbf{1}_B$ is not identically zero, use the Fourier transform. In the case where $\lambda(A) = +\infty$, use the subset $A \cap [-N, N]^n$ for $N \in \mathbb{N}$ large enough.

6.2. (a) Use the Steinhaus theorem (Exercise 6.1) and the subsets

$$A_j = \{x \in \mathbb{R}^n / f(x) \leqslant j \text{ and } f(-x) \leqslant j\}, \ j \in \mathbb{N}$$

to show that there exists a j such that $A_j - A_j = A_j + A_j$ contains a neighbourhood of 0.

(b) Let $a, b \in \mathbb{R}^n$, distinct, such that $f(a) = f(b) = 0$, and $x \in [a, b]$. Note that x can be written as the midpoint of two points of $[a, b]$, one of them equal to a or b, and then that there exists $x_1 \in [a, b]$ such that $f(x) \leqslant \dfrac{f(x_1)}{2}$. Construct a sequence $(x_n)_{n \geqslant 1}$ of points of $[a, b]$ such that

$$f(x) \leqslant \frac{f(x_n)}{2^n} \text{ for all } n \geqslant 1.$$

By extracting a subsequence if necessary, we can suppose in addition that $x_n \to \ell \in [a, b]$. Explain why f is bounded above on a neighbourhood of ℓ, and thus deduce that $f(x) \leqslant 0$. Conclude then about the convexity of f.

6.4. (a) Use the metric defined by

$$d(x, y) = \sum_{n=1}^{\infty} 2^{-n} \frac{|x_n - y_n|}{1 + |x_n - y_n|} \text{ for } x, y \in X_0.$$

(b) Let X be a Polish space. As X is separable, we can write $X = \bigcup_{n_1=1}^{\infty} A(n_1)$, where the $A(n_1)$ are closed subsets of X, of diameter $\leqslant 1$. Step by step, we can write

$$A(n_1) = \bigcup_{n_2=1}^{\infty} A(n_1, n_2),$$

$$\vdots$$

$$A(n_1, \dots, n_k) = \bigcup_{n_{k+1}=1}^{\infty} A(n_1, \dots, n_k, n_{k+1}),$$

$A(n_1, \dots, n_k)$ being a closed subset of X of diameter $\leqslant \dfrac{1}{k}$. Show that if $x = (x_k)_{k \geqslant 1} \in X_0$, then $\bigcap_{k=1}^{\infty} A(x_1, \dots, x_k)$ is a singleton $\{T(x)\}$ and that the map $T : X_0 \to X$ so defined is a continuous surjection.

(c) Show that the class $\mathcal{C}$ of subsets B of $\mathbb{R}$ such that $B \in \mathcal{A}$ and $^cB \in \mathcal{A}$ is a σ-algebra containing the closed subsets of $\mathbb{R}$, by using the fact that any open subset of $\mathbb{R}$ is a countable union of closed subsets.

(d) Note that $\mathcal{A}$ contains the closed subsets of $\mathbb{R}$ because of part (b), as any closed subset of a Polish space is a Polish space. Let $(A_n)_{n \geqslant 1}$ be a sequence of non-empty elements of $\mathcal{A}$. To show that $\bigcap_{n=1}^{\infty} A_n \in \mathcal{A}$: for each $n \geqslant 1$ define a continuous surjection $f_n : X_0 \to A_n$, then set $X_1 = X_0^{\mathbb{N}^*}$ and

$$E = \{x \in X_1 / f_1(x_1) = f_2(x_2) = \cdots = f_n(x_n) = \cdots\}.$$

Show that X_1 (equipped with the product topology) is a Polish space and that E is a closed subset of X_1; finish by composing a continuous surjection from X_0 on E with the map $\psi : E \to A, x \mapsto f_1(x_1)$. To prove that $\bigcup_{n=1}^{\infty} A_n \in \mathcal{A}$, write $X_0 = \bigsqcup_{k=1}^{\infty} X(k)$, where $X(k)$ is the set of $x \in X_0$ such that $x_1 = k$. Verify that $X(k)$ is an open subset of X_0 homeomorphic to X_0, which shows the existence of a continuous surjection $f_k : X(k) \to A_k$. Finish by "gluing" the f_k to form a map from X_0 to $\bigcup_{n=1}^{\infty} A_n$.

(f) Given a countable dense subset D of X_0, prove the existence of an injection from the set of Borel sets to the set of functions from D to $\mathbb{R}$. The latter has the same cardinality as $\mathbb{R}$. Then, use the Cantor–Bernstein theorem to conclude.

6.5. Use Exercise 6.4.

6.6. Use the subsets $A = \{P(u), P \in \mathbb{N}[X] \text{ such that } P(0) = 0\}$ and $B = \{P(u), P \in \mathbb{N}[X] \text{ such that } P(0) \geqslant 1\}$.

6.10. (a) Note that $p(x) \geqslant -p(-x)$ so that $q(x) \geqslant -p(-x)$ for all $x \in X$.

(b) Let $x, y \in X$, $z = x + y$ and $u, v \in C$ be such that $p(ux)$ approaches $q(x)$ and $p(vy)$ approaches $q(y)$. Consider $w = v \circ u = u \circ v \in C$ and $p(wz)$.

(c) Set $C_n = \frac{1}{n} \sum_{k=0}^{n-1} A^k$ for $n \geqslant 1$, and start from the inequality

$$q(x - Ax) \leqslant p\big(C_n(x - Ax)\big).$$

6.11. (a) Use the inner product defined by

$$\varphi(x, y) = \int_G \langle \pi(t)x, \pi(t)y \rangle d\mu(t) \text{ for } x, y \in H,$$

μ being a G-invariant mean on G. To show that φ is well-defined and that it is an inner product, use the inequalities

$$|\langle \pi(t)x, \pi(t)y \rangle| \leqslant C^2 \|x\| \|y\|$$

and

$$\|\pi(t)x\| \geqslant \frac{1}{\|\pi(t^{-1})\|} \|x\| \geqslant \frac{1}{C} \|x\|.$$

Deduce ([118], p. 61) that there exists an $A \in \mathcal{L}(H)$, invertible, such that $\varphi(x, y) = \langle Ax, Ay \rangle$ for $x, y \in H$. To conclude, remark that φ is invariant under the $\pi(t)$, $t \in G$.

(b) The group $\mathbb{Z}$ is amenable.

(c) By part (b), one can assume that T is unitary. Hence, the spectrum of T is contained in the unit disk. Then, use the Gelfand–Naimark Theorem 11.4.14.

6.12. (a) Adapt to semigroups the method of Exercise 6.11, by defining an inner product on H by

$$\varphi(x, y) = L(f_{x,y}) \text{ for } x, y \in H,$$

where $f_{x,y}(t) = \langle \pi(t)x, \pi(t)y \rangle$ and L is a positive G-invariant linear functional on $\ell^{\infty}(G)$ such that $L(1) = 1$ (see Remark 6.2.15).

(b) Use the additive semigroup $\mathbb{N}^2$.

Exercises for Chapter 7

7.1. (a) Split the sum $F(x+h) - F(x)$ into two parts, and bound $|e^{i\pi n^2 h} - 1|$ by $\pi n^2 |h|$ if $n \leqslant N$ and by 2 otherwise.

(b) Choose N approximately equal to $|h|^{-1/2}$.

7.2. (a) Calculate the derivatives of $\widehat{\psi}$.

(b) Write $f(t) = f(t_0) + (t - t_0)f'(t_0) + (t - t_0)\varepsilon(t)$, where $\varepsilon(t) \to 0$ as $t \to t_0$, and then express $b^j f * \psi_j(t_0)$ as an integral using the function ε. Conclude by using the dominated convergence theorem.

(c) Show that $W * \psi_j(t_0) = \dfrac{1}{2} a^j e^{ib^j t_0}$.

(d) Show that $\widetilde{W} * \psi_j(t_0) = \dfrac{1}{2i} a^j e^{ib^j t_0}$.

7.3. First show that

$$\left| \sin\left(2^n(x + h)\right) + \sin\left(2^n(x - h)\right) - 2\sin(2^n x) \right| \leqslant 4^n h^2$$

for $x, h \in \mathbb{R}$ and $n \in \mathbb{N}$. To conclude, split the sum

$$g(x + h) + g(x - h) - 2g(x)$$

into two parts, using an integer N such that 2^N is approximately equal to $|h|^{-1}$.

7.4. (a) Observe that if g has a local minimum at $x \in \mathbb{R}$, then for $h > 0$ small enough we have

$$0 \leqslant \frac{g(x + h) - g(x)}{h} \leqslant \frac{g(x + h) + g(x - h) - 2g(x)}{h}.$$

This proves that $g'_d(x) = 0$. Proceed similarly for $h < 0$.

(b) If $a < b$ are fixed real numbers, and ℓ the linear function interpolating g at a and b, show that $g - \ell \in \lambda$ has a local extremum at some point of $]a, b[$. To show that the set of points where g is differentiable is uncountable, use the fact that if g is not linear, the set of slopes of the chords of the graph of g is an interval with non-empty interior.

(c) First show that if $x, h \in \mathbb{R}$ and $N \in \mathbb{N}$, we have

$$|\varphi(x+h) + \varphi(x-h) - 2\varphi(x)| \leqslant 2^{N+1} h^2 \sup_{n \geqslant 0} |\varepsilon_n| + 2^{-N+2} \sup_{n > N} |\varepsilon_n|.$$

Next, with $\alpha > 0$ fixed, choose N_0 such that $\sup_{n>N} |\varepsilon_n| \leqslant \alpha^2$ for $N \geqslant N_0$, and then $N \geqslant N_0$ such that 2^N is approximately equal to $\dfrac{\alpha}{|h|}$.

7.5. (a) Proceed by induction on n.

(b) Use the mean value theorem to show that

$$|P_n(x+h) + P_n(x-h) - 2P_n(x)| \leqslant h^2 \|P_n''\|_\infty$$

and conclude thanks to Bernstein's inequality.

(c) Proceed as in Exercise 7.3.

(d) The necessary and sufficient condition is $E_n(f) = o(n^{-1})$.

7.6. (a) Show that the rational number $\dfrac{p}{q}$ is an element of $\Omega_G(0)$ by reasoning by induction on $|q|$: suppose that $|p| < |q|$ and then consider $\sigma\left(\dfrac{p}{q}\right)$.

7.7. (b) Reason by induction on the denominator, as in Exercise 7.6.

(c) If $\varepsilon(g)$ is as we have supposed, show that there exists a $\delta \in \Gamma_0$ such that $g(\infty) = \delta(\infty)$, then that there exists an $n \in \mathbb{Z}$ such that $g = \delta\tau^n$. By using the homomorphism ε, show that n is even, and then that $g \in \Gamma_0$.

7.8. (a) Start with the case $b = 0$, by using the functional equation of θ_0. For the general case, use the bounds

$$na \leqslant na + b \leqslant (n+1)a \text{ for } n \in \mathbb{Z},$$

and distinguish the cases $n \geqslant 0$ and $n \leqslant -1$.

(b) (ii) To show that g is of type 1 or 5, use part (c) of Exercise 7.7. Then, explain why g can be written $g(z) = \dfrac{vz + w}{qz - p}$ and conclude by letting $\varepsilon \to 0$ in the equality

$$\sqrt{\varepsilon}\,\theta_0(r + i\varepsilon) = \frac{u\sqrt{\varepsilon}}{\sqrt{q(r + i\varepsilon) - p}}\theta_0\big(g(r + i\varepsilon)\big).$$

(iii) By using the inequality

$$|e^{ib} - e^{ia}| \leqslant |b - a| e^{-\operatorname{Im} a} \text{ if } 0 < \operatorname{Im} a \leqslant \operatorname{Im} b,$$

and the estimate

$$\sum_{n=1}^{\infty} n^2 e^{-n^2 t} = O(t^{-3/2}) \text{ as } t \searrow 0,$$

deduce that

$$\theta_0\left(-\frac{1}{r+i\varepsilon}\right) - \theta_0\left(-\frac{1}{r} + i\frac{\varepsilon}{r^2}\right) \to 0 \text{ as } \varepsilon \searrow 0.$$

To conclude in the case where $r > 0$, write

$$\sqrt{\varepsilon}\,\theta_0(r + i\varepsilon) = \frac{e^{i\pi/4}}{\sqrt{r+i\varepsilon}} \sqrt{\varepsilon}\,\theta_0\left(-\frac{1}{r+i\varepsilon}\right)$$

$$\simeq \frac{e^{i\pi/4}}{\sqrt{r+i\varepsilon}} \sqrt{\varepsilon}\,\theta_0\left(-\frac{1}{r} + \frac{i\varepsilon}{r^2}\right)$$

$$\simeq \frac{e^{i\pi/4}}{\sqrt{r+i\varepsilon}} r\sqrt{\frac{\varepsilon}{r^2}}\,\theta_0\left(-\frac{1}{r} + \frac{i\varepsilon}{r^2}\right),$$

and let $\varepsilon \to 0$.

(c) For example, let us treat the case where $q = 4n$, $n \geqslant 1$. In that case,

$$G_q = \sum_{k=0}^{4n-1} e^{ik^2\pi \frac{1}{2n}} = 2\sum_{k=0}^{2n-1} e^{ik^2\pi \frac{1}{2n}} = 2\sqrt{2n}\, S_{1/(2n)}.$$

Conclude, using the previous question.

7.11. (b) For $h > 0$, write

$$f\big(g(x+h)\big) = f\big(g(x) + r(h)\big)$$

$$= f\big(g(x)\big) + \sqrt{r(h)}\,\delta_+\big(f, g(x)\big) + o\big(\sqrt{r(h)}\big),$$

with $r(h) \geqslant 0$ because g is non-decreasing.

(d) Use the functional equation of F and the rules of computation (a) and (b) on $\delta_\pm$.

(f) Use the "little" functional equation $F(-x) = \overline{F(x)}$.

(g) Use (d), (f), the 2-periodicity of F, the fact that σ and τ^2 generate Γ_0 and finally that we can initially choose $\arg \delta_+(F, 0) = \dfrac{3\pi}{4}$ and $\arg \delta_-(F, 0) = \dfrac{5\pi}{4}$, that is, arguments differing by $\dfrac{\pi}{2}$.

(h) Use (d), (e) and (f). The answer is

$$\delta_+\left(F, \frac{1}{2}\right) = -\frac{\pi}{\sqrt{2}} \text{ and } \delta_-\left(F, \frac{1}{2}\right) = -i\frac{\pi}{\sqrt{2}}.$$

Exercises for Chapter 8

8.1. (a) Use Remark (8.3.7).

(b) Use the sum of the series as an upper bound for one term, then optimise in t.

(c) Use the increasing nature of the sequence $(p(n))_{n \geqslant 0}$ to obtain

$$\sum_{k=0}^{n} p(k)e^{-kt} \leqslant (n+1)p(n).$$

From there,

$$p(n) \geqslant \frac{1}{n+1}\left(f(e^{-t}) - \sum_{k=n+1}^{\infty} p(k)e^{-kt}\right).$$

To conclude, use the constants B and C, and choose $t = \dfrac{2B}{\sqrt{n}}$.

8.2. (a) On each of the intervals $I_1 = \mathbb{R}_-^*$ and $I_2 = \mathbb{R}_+^*$, make the (invertible) change of variable $x = t - \dfrac{1}{t} =: \phi_i(t)$. Observe that

$$\phi_1^{-1}(x) + \phi_2^{-1}(x) = x.$$

8.4. (a) Use partial fraction decomposition and observe (using the arithmetic hypothesis on the a_k) that the generating function of the sequence $(p_S(n))_{n \geqslant 0}$ has a unique pole of maximal multiplicity.

8.5. (b) Using (8.35), show that

$$g(e^{-2\pi z}) \simeq \frac{1}{\sqrt{2}}e^{\frac{\pi}{24z} + \frac{\pi z}{12}} \quad \text{when } z \text{ is close to } 0.$$

Exercises for Chapter 9

9.2. (a) Use Parseval's identity and the fact that the $|a_n|$ are non-decreasing.

(b) Consider the case where $a_n = n^{\delta + 1/2}e^{i\pi n^2 \sqrt{2}}$ and use Theorem 9.2.9.

9.3. (a) Use the Cauchy–Schwarz inequality and Parseval's identity.

(b) Consider the case where $a_n = r^n e^{i\pi n^2 \sqrt{2}}$.

9.4. (b) Select P such that $\widehat{P}(k) = \overline{c_k}$ for $-n \leqslant k \leqslant n$, and use the bound

$$\left|\int_{\mathbb{T}} P(-t)d\mu(t)\right| \leqslant \|P\|_\infty \|\mu\|.$$

9.5. (a) Apply Hölder's inequality to the functions $|X|^{r(1-\theta)}$ and $|X|^{r\theta}$, and to the conjugate exponents $\dfrac{p}{r(1-\theta)}$ and $\dfrac{q}{r\theta}$.

9.6. (b) Calculate $\mu(T^{-1}([0, x[))$ for $x \in [0, 1[$.

   (c) Show that

$$\mu(a_n > M_n) = \mu(a_1 > M_n) \leqslant \mu\left(\frac{1}{x} > M_n\right) = \mu\left(\left[0, \frac{1}{M_n}\right[\right)$$

by using the fact that T preserves μ. To conclude, use the Borel–Cantelli lemma.

   (d) If λ denotes the Lebesgue measure on $[0, 1[$, there exist positive constants C_1 and C_2 such that

$$C_1\lambda \leqslant \mu \leqslant C_2\lambda$$

(the Lebesgue measure and the Gauss measure on $[0, 1[$ are absolutely continuous with respect to one another).

9.7. (a) Recall that $q_1 = a_1 \geqslant 1$, $q_2 = a_2q_1 + q_0 = a_2a_1 + 1 \geqslant 2$ and

$$q_{n+1} = a_{n+1}q_n + q_{n-1} \text{ for } n \geqslant 1.$$

   (b) Use Exercise 9.6.
   (c) To establish the inequality, note that if $s \geqslant 2$, we have

$$q_s = a_sq_{s-1} + q_{s-2} \text{ and } q_{s-2} \leqslant q_{s-1}.$$

Then, observe that

$$q_s \leqslant 2a_sn \leqslant Cns(\ln s)^{1+\varepsilon},$$

and that $s \leqslant C' \ln n$ by part (a).
   (d) Use an Abel transformation.

9.8. (c) Use the fact that

$$\cot x = \cot(\pi - x) \leqslant \frac{1}{x} \text{ if } 0 < x < \pi.$$

9.9. (a) Use the equality

$$h_{n+1} - h_n = \iint_{0 \leqslant x, y \leqslant 1} f''(n-1+x+y)\, dx\, dy, \text{ with } f''(x) = \frac{1}{x}.$$

   (b) Add the inequalities of part (a).
   (c) Show that

$$p + \frac{c - a - 1}{3a} \leqslant h_c < p + \theta,$$

hence $|I_1| = c - a - 1 \leqslant C\theta a$.
   (d) Use Exercise 9.8.
   (e) $\frac{1}{\theta}$ and θa have the same order of magnitude.

(f) Write $z = e^{2i\pi t}$ and use part (e), as well as a decomposition of the sum in dyadic blocks of the form $\sum_{2^k < n \leqslant 2^{k+1}}$, and then use the maximum principle. For the lower bound, use Parseval's identity.

Exercises for Chapter 10

10.1. (a) Apply Parseval's identity to the function $x \mapsto f(x + 2h) - f(x)$.

(b) Use Fatou's lemma.

(c) For the first inequality, apply the result of part (a) to $h = \dfrac{\pi}{4N}$, where $N \in \mathbb{N}^*$.

(d) Use the Cauchy–Schwarz inequality.

(e) If the answer is yes, then by the closed graph theorem we have the *a priori* inequality

$$\sum_{n \in \mathbb{Z}} n^{1/2} |\widehat{f}(n)| \leqslant C \|f\|_{\mathrm{Lip}_1},$$

where

$$\|f\|_{\mathrm{Lip}_1} = |f(0)| + \sup_{0 \leqslant x \neq y \leqslant 1} \left| \frac{f(y) - f(x)}{y - x} \right|.$$

If we apply this inequality to the primitive with value zero at 0,

$$F(t) = \sum_{n \neq 0} \frac{c_n}{in} (e^{int} - 1)$$

of a trigonometric polynomial without a constant term $f(t) = \sum_{n \neq 0} c_n e^{int}$, since

$$\|F\|_{\mathrm{Lip}_1} \leqslant \|F'\|_\infty = \|f\|_\infty$$

we obtain the following inequality:

$$\sum_{n \in \mathbb{Z}} \frac{|\widehat{f}(n)|}{\sqrt{|n| + 1}} \leqslant (C + 1) \|f\|_\infty,$$

valid for all trigonometric polynomials f, hence also for all continuous functions $f : \mathbb{T} \to \mathbb{C}$. In other words, the sequence with general term $1/\sqrt{|n| + 1}$ is a multiplier of C in ℓ_1 in the sense of Exercise 5.3 in Chapter 5. This same exercise allows us to give a negative answer to the question.

10.2. (a) Observe that, for $|z| \leqslant 1$, we have

$$\sqrt{1 - z} = \sum_{n=0}^{\infty} \alpha_n z^n \quad \text{with } \alpha_n = O(n^{-3/2}),$$

and write

$$|1 - ae^{it}| = (1 - ae^{it})^{1/2}(1 - \overline{a}e^{-it})^{1/2}.$$

    (c) Factorise $|f|$ in a finite product of $|e^{it} - a|$ and use the fact that W is an algebra.

10.4. (b) Observe that

$$\frac{n_{k+1}}{n_k} \leqslant \frac{p_{k+2}}{p_k},$$

      so that

$$\varlimsup_{k \to +\infty} \frac{n_{k+1}}{n_k} \leqslant 1.$$

10.5. Combine Hardy's inequality and the Davenport–Erdős–Levêque lemma (Exercise 10.4).

10.6. If K_N denotes the Fejér kernel, we have

$$\sum_{n=1}^{N} \frac{|\widehat{K_N}(n)|}{n} \geqslant -1 + \sum_{n=1}^{N} \frac{1}{n},$$

    whereas $\|K_N\|_1 = 1$.

10.7. (b) We denote by $\mathcal{P}$ the space of trigonometric polynomials, $(c_n)_{n \in \mathbb{Z}}$ the indicator function of S, and σ the formal trigonometric series $\sum c_n e^{int}$. If $f(t) = \sum_{n \in \mathbb{Z}} a_n e^{int}$ is an element of $\mathcal{P}$, we set

$$\sigma * f(t) = \sum_{n \in \mathbb{Z}} c_n a_n e^{int} = \sum_{n \in S} a_n e^{int}$$

which is again an element of $\mathcal{P}$. Note that $\sigma * f = P(f)$, hence we have an inequality of the form

$$\|\sigma * f\|_1 \leqslant M \|f\|_1 \text{ for any } f \in \mathcal{P},$$

where $M = \|P\|$. By changing f to $f * K_n$ (K_n being the Fejér kernel), show that

$$\|\sigma * K_n\|_1 \leqslant M \text{ for all } n \geqslant 1.$$

By an argument of weak-* compactness that can be found, for example, in [90], there exists a measure μ on $\mathbb{T}$ such that $\widehat{\mu}(n) = c_n$ for all $n \in \mathbb{Z}$.

    (c) If μ is a measure on $\mathbb{T}$ such that $\widehat{\mu} = \mathbf{1}_S$, consider the operator of convolution by μ.

    (d) Use Hardy's inequality.

10.8. (b) See Exercise 9.5.

(c) Deduce from parts (a) and (b) that $\|\phi\|_1 \geqslant \delta \|\phi\|_2$, where $\delta = 2^{-\frac{\theta}{4(1-\theta)}}$.

10.10. Apply the generalised Hilbert inequality to $x_n = c_n e^{2i\pi\lambda_n a}$ and then $x_n = c_n e^{2i\pi\lambda_n b}$.

10.11. (a) Consider the triangle function Δ defined by $\Delta(t) = (1 - |t|)^+$, which satisfies

$$\widehat{\Delta}(x) = \left(\frac{\sin(\pi x)}{\pi x}\right)^2,$$

and take, for $\alpha \in \mathbb{R} \setminus \mathbb{Z}$ fixed,

$$\varphi(x) = \widehat{\Delta}(x) + \frac{1}{2}\left(\widehat{\Delta}(x - \alpha) + \widehat{\Delta}(x + \alpha)\right),$$

for which we have

$$\widehat{\varphi}(t) = (1 + \cos(2\pi\alpha t))\Delta(t),$$

and $\varphi > 0$ on $\mathbb{R}$. Then change φ to $\frac{\varphi}{c}$, where $c = \min_A \varphi$.

Exercises for Chapter 11

11.1. Observe that $\|e\|^{-1}\|x\| \leqslant \|x\|' \leqslant \|x\|$.

11.2. Show that the ideal generated by $f_1, \ldots, f_n$ is equal to A.

11.3. (b) (i) If $n \in \mathbb{Z} \setminus A$, there exists an $f \in I$ such that $\widehat{f}(n) = 1$. Remark then that if $e_n(t) = e^{int}$, we have $e_n * f = e_n$.

(ii) Use Fejér's theorem (which states that the Fourier series of $f \in L^1(\mathbb{T})$ converges to f in the sense of Cesàro in $L^1(\mathbb{T})$) to show that $I_A \subset I$. The other inclusion is obvious.

11.4. (b) (i) Take inspiration from the proof of Theorem 11.3.2 and use the compactness of $K \setminus U$.

(ii) If V is an open set containing $K \setminus U$, $\varphi \in I$ a function that does not vanish at any point of V, and $f \in A$ identically zero on U, note that the function $g : K \to \mathbb{C}$ defined by

$$g(x) = \begin{cases} \dfrac{f(x)}{\varphi(x)} & \text{if } x \in V, \\ 0 & \text{if } x \in U \end{cases}$$

is continuous and satisfies $f = g\varphi$.

(iii) Consider the closed disjoint sets

$$F = \{x \in K / |f(x)| \leqslant \varepsilon\} \text{ and } G = \{x \in K / |f(x)| \geqslant 2\varepsilon\}.$$

By Urysohn's lemma (see [152]), there exists a continuous function $u : K \to [0, 1]$ equal to 0 on F and 1 on G. Then define $g = uf$.

11.5. (b) Show that the spectrum of A can be identified with $[0, 1]$, by imitating the proof of Proposition 11.3.2 and using the fact that if $f \in A$, then $|f|^2 = f \overline{f} \in A$. More generally, if B is a self-adjoint Banach algebra continuously included in $C([0, 1], \mathbb{C})$, then any proper ideal of B possesses at least one zero in $[0, 1]$.

(c) If ever I is an intersection of maximal ideals of A, there exists a non-empty subset X of $[0, 1]$ such that $I = \{f \in A / f$ is zero on $X\}$. To conclude, consider the functions $x \mapsto (x - a)^2$ and $x \mapsto x - a$.

11.6. (b) In $\mathbb{C}$, the complement of a disk is connected, and the connected components of an open set are open.

(c) Use the maximum principle and the fact that a uniform limit of holomorphic functions is holomorphic.

(d) Apply Runge's theorem to the point a and to the compact set $\widehat{K}$.

(e) Let $u \in P(K)$ be defined by $u(z) = z$, and φ a character of A. Show that $a := \varphi(u)$ belongs to the polynomially convex hull of K and deduce that $a \in \widehat{K}$.

11.7. (c) By part (b), the kernel of Γ contains e, hence all the polynomials, hence A. Now, if φ is a multiplicative linear functional on A, it can be extended to an element of the spectrum of B. But then, if $f \in A$, we have

$$|\varphi(f)| = |\varphi(f^n)|^{1/n} \leqslant \|f^n\|^{1/n} \to 0 \text{ as } n \to +\infty.$$

11.8. (a) Note that I is a strict ideal of H^∞, and that $e - u$ belongs to I.

11.9. To show (c)$\Rightarrow$(a), use the fact that X is homeomorphic to the spectrum of $C(X, \mathbb{C})$.

11.10. The two-sided ideals of $\mathcal{M}_n(\mathbb{C})$ are $\{0\}$ and $\mathcal{M}_n(\mathbb{C})$.

11.11. (a) Use Parseval's identity.

(b) Use the maximum principle and Fejér's theorem.

11.12. If $\|e^{ita}\| = 1$ for all $t \in \mathbb{R}$, and if $x \in H$ is fixed, then the function $f(t) = \|x\|^2 - \|e^{ita}x\|^2$ attains its minimum at 0. To conclude, note that $f'(0) = 0$, that is, $-2 \operatorname{Im}\langle a(x), x \rangle = 0$.

11.13. (a) The differential equation $X' = (A + H)X$ can also be written $X' - AX = B$, with $B = HX$.

(b) Answer:

$$d(\exp)(A)(H) = \int_0^1 \exp((1-t)A)H\exp(tA)dt.$$

(c) Observe that the operators L and R commute, and hence so do their exponentials. Then deduce from the preceding question that

$$d(\exp)(A) = e^L \int_0^1 e^{t(R-L)}dt. \qquad (\star)$$

Next, show that

$$V := \int_0^1 e^{tU}dt = \Phi(U) \text{ for } U \in \mathcal{L}(\mathcal{M}_n(\mathbb{C})),$$

by observing that $UV = e^U - I_n = U\Phi(U)$. Variant: use $(\star)$ and a power series expansion.

(d) Show – or admit – that the spectrum of $R - L$ is the set of all $\lambda - \mu$, with λ and μ taken over the spectrum of A.

11.14. (a) First, show that if $g(z) = f(z)e^{ibz}$, we have

$$|g(z)| \leqslant \max(a, M) \text{ for all } z \in S_1 \cup S_2.$$

Next, *reapply* Phragmén–Lindelöf to g and to the convex hull of $S_1 \cup S_2$, which is a sector of angle π, with $\beta = 0$. One can proceed similarly with $S_3 \cup S_4$ and $h(z) = f(z)e^{-ibz}$.

11.16. (a) Use the fact that 1 is an extremal point of the unit ball of $\mathbb{C}$.

Exercises for Chapter 12

12.1. The map $z \mapsto w = \dfrac{1+z}{1-z}$ is a conformal mapping from D onto the right half-plane $\{\operatorname{Re} w > 0\}$, with boundary $\{\operatorname{Re} w = 0\}$.

12.2. (a) If $f \in C$ and $g \in L^1$ are fixed, the function $a \mapsto \langle PT_a f, T_a g \rangle$ is continuous, since translation is continuous on L^1 and on C; in particular, this function is measurable and the integral, which we denote by $L_f(g)$, is well-defined. L_f is a continuous linear functional on L^1, with norm bounded above by $\|f\|_\infty \|P\|$. Then use the fact that the dual space of L^1 is L^∞.

(b) See Exercise 10.7 of Chapter 10.

(c) Note that

$$\|Qf\|_\infty = Qf(0) = \frac{1}{2}\sum_{n=1}^N \frac{1}{n}.$$

12.3. (a) It suffices to consider the case $0 < |w| < 1$. We have

$$|S(re^{i\theta})| = \exp\left(-\frac{1-r^2}{1-2r\cos\theta + r^2}\right) \text{ for } 0 \leqslant r < 1 \text{ and } \theta \in \mathbb{R}.$$

Write $|w| = e^{-\varepsilon}$, with $\varepsilon > 0$. By the mean value theorem, for $r < 1$ close enough to 1, we can find an element θ_r of $]0, \pi[$ such that

$$\cos\theta_r = \frac{1}{2r}\left(1 + r^2 - \frac{1-r^2}{\varepsilon}\right).$$

Then, we have

$$|S(re^{i\theta_r})| = e^{-\varepsilon} \text{ and } \varphi_r := -\arg S(re^{i\theta_r}) = \frac{2r\sin\theta_r}{1-2r\cos\theta_r + r^2}$$
$$= \frac{2r\varepsilon\sin\theta_r}{1-r^2}.$$

Show that

$$\theta_r \sim \sqrt{\frac{2(1-r)}{\varepsilon}} \text{ and } \varphi_r \to +\infty \text{ as } r \nearrow 1.$$

Next, adjust r to have *exactly* $S(re^{i\theta_r}) = w$ and $z_r = re^{i\theta_r}$ close to 1. A less laborious method consists of using the conformal mapping $T : z \mapsto \dfrac{1+z}{1-z}$. The image by T of the lens $\{|z| < 1\} \cap \{|z-1| < \delta\}$ is the complement in the right half-plane of the disk $\overline{D}\left(-1, \dfrac{2}{\delta}\right)$, that is, the right half-plane deprived of a large cap.

(b) For φ, take a weak-* cluster point of the sequence $(\delta_{z_n})_{n \geqslant 1}$ in the spectrum of H^∞.

12.4. (a) If d denotes the pseudo-hyperbolic distance defined on p. 412, use the identity

$$1 - d(a,b)^2 = \frac{(1-|a|^2)(1-|b|^2)}{|1-\bar{a}b|^2},$$

as well as the inequality $1 - x \leqslant e^{-x}$.

(b) Observe that

$$g(\varepsilon) \geqslant \frac{1}{4\varepsilon}\sum_{\varepsilon_k \leqslant \varepsilon} \varepsilon_k.$$

12.5. (a) Denote by B_n the nth partial product of B. Use the maximum principle to show that

$$|f(z)| \leqslant \|f\|_\infty |B_n(z)| \text{ for } z \in D \text{ and } n \geqslant 1.$$

Then let $n \to +\infty$.

(b) Consider the example $f(z) = 1$ and $g(z) = \exp\left(-\dfrac{1+z}{1-z}\right)$.

12.6. Use the inequality

$$x^\beta + y^\beta \leqslant (x+y)^\beta,$$

valid for $x, y \geqslant 0$ and $\beta \geqslant 1$. Then, if $f = g_1 B_1^r + g_2 B_2^r$, show by using Exercise 12.5 that there exist $h_1, h_2 \in H^\infty$ such that $g_1 = B_1^{r-1} h_1$ and $g_2 = B_1^{r-1} h_2$, hence $h_1 B_1 + h_2 B_2 = 1$, which is absurd.

12.7. Suppose that the ideal generated by f is closed in H^∞ and fix $\gamma \in$ $]0, c[$. If ever we have $|f| \leqslant \gamma$ on a Borel set E of $\mathbb{T}$ of positive measure, use a function[1] $g \in H^\infty$ such that $|g| = 1$ a.e. on E and $|g| = \varepsilon$ a.e. on ^{c}E, where $0 < \varepsilon \leqslant \dfrac{\gamma}{\|f\|_\infty}$, to reach a contradiction.

12.8. (b) Proceed by induction on n.

(c) Show that, if $\lambda > 0$ and $n \in \mathbb{N}$,

$$m(|g| > \lambda) \leqslant \left(\frac{C}{\lambda}\right)^{2n}.$$

12.9. Set

$$M_n = \sup_{|z| \leqslant 2|z_n| - 1} \left| \frac{w_n f(z)}{(z - z_n) f'(z_n)} \right|.$$

Adjust first the argument of c_n to have $c_n z_n = |c_n z_n|$, then its modulus to obtain

$$M_n e^{-|c_n|(1 - |z_n|)} \leqslant 2^{-n}.$$

Finally note that if $r < 1$, then we have $r \leqslant 2|z_n| - 1$ for n large. The function g realises the desired interpolation.

12.10. (b) The existence of f_0 is evident, and hence $E_n = f_0 + B_n H^2$. Use the fact that

$$\|f_0 - g B_n\|_2 = \left\| \frac{f_0}{B_n} - g \right\|_2 \text{ for all } g \in H^2$$

and the duality formula in Banach spaces. The integral is calculated by the residue theorem, and its value is

[1] For the existence of such a function, see [90], p. 139.

$$\sum_{j=1}^{n} \frac{f_0(z_j) F(z_j)}{B_n'(z_j)}.$$

(c) Use part (a), the Cauchy–Schwarz inequality for $\sqrt{1 - |z_j|}|w_j|$ and $\sqrt{1 - |z_j|}|F(z_j)|$ as well as the Carleson embedding theorem.

(d) Part (c) provides an interpolating function for the n first points, with control of the H^2-norm. Next use an argument of normal families.

(e) Use the easy part of the factorisation theorem.

(f) The adjoint operator of a continuous linear surjection is an injection with a closed range, so that the dual space of H^1, isometric to L^∞ / H^∞, contains a copy of ℓ_∞, and hence is not separable.

12.11. (b) Apply the result of (a) to $f(t) = e^{iNt} p(t)$, where p is a trigonometric polynomial and N a large positive integer, to obtain $\|pg\|_1 \leqslant C\|p\|_1$. Next, if $h \in L^1$, use a sequence of trigonometric polynomials converging to h almost everywhere and in L^1 to show that $\|hg\|_1 \leqslant C\|h\|_1$ for any $h \in L^1$. The conclusion then follows easily.

There are functions g in BMOA$\setminus L^\infty$, but these functions operate continuously on H^1 otherwise than through an absolutely convergent integral $\int_{\mathbb{T}} fg \, dm$. This is analogous to the difference between Fourier in L^1 and Fourier–Plancherel in L^2.

12.12. (a) For the converse, remark that $|a - b| \geqslant \delta(1 - |a||b|)$.

(b) If a and b are two distinct elements of Γ, distinguish the cases where a and b belong to the same Γ_j or to two different Γ_j, and show that their Euclidean distance is always bounded below (up to a constant) by the sum of their distances to the boundary; then use part (a).

12.13. (d) The theorem cited shows that P has an interpolation constant $\leqslant \dfrac{1 + \varepsilon}{1 - \varepsilon}$.

(e) Consider the data $w_1 = 1$ and $w_j = 0$ for $j \geqslant 2$, and use the maximum principle.

12.15. (a) An operator of rank $< n$ cannot be injective on E if dim $E = n$.

12.16. Use the preservation of the inner product when u is an inner function: $\langle uf, ug \rangle = \langle f, g \rangle$.

12.17. (a) The characteristic equation associated with the linear recurrence of the preceding question is $(r - 1)^p = 0$. Thus there exists a polynomial P of degree $\leqslant p - 1$ such that $a_n = P(n)$ for $n \geqslant 1$. But $a_n \to 0$ since $g \in H^2$, hence $P = 0$ and $a_n = 0$ for $n \geqslant 0$.

Exercises for Chapter 13

13.2. (a) Use the fact that $(A \oplus c_0(A))_{c_0}$ is isometric to $c_0(A)$ (by shift).

(b) If $f : (B \oplus Z)_{c_0} \to (A \oplus Z)_{c_0}$ is a continuous projection, show that the map $B \to A, x \mapsto g(x)$, $g(x)$ being the component of $f(x, 0)$ on A, is a continuous projection.

(c) Use part (b), with $Z = (c_0(A) \oplus c_0(B))_{c_0}$.

13.3. (a) Use the pre-compactness of $T(B_X)$ and the injectivity of $T_{|A}$.

(b) By using the non-pre-compactness of the closed unit ball B_Y of Y, show that if $\rho(Y) > 0$ then $T(B_Y)$ is not pre-compact.

(c) Show that $T_{|A}$ is injective and that $T(A)$ is ε^2-separated.

(d) By part (c), if $0 < \varepsilon < 1$, $\dim Y = n$ and $\rho(Y) > \varepsilon$, then

$$n \leqslant \frac{\ln P(\varepsilon^2)}{\ln \dfrac{1}{\varepsilon}}.$$

For the canonical injection from $L^\infty([0, 1])$ into $L^2([0, 1])$, use a theorem of Grothendieck ([161], pp. 114–115) by which, if Y is a closed subspace of $L^\infty([0, 1])$ such that

$$\|f\|_\infty \leqslant C\|f\|_2,$$

C being a positive constant, then $\dim Y \leqslant C^2$.

(e) To show that j is not super-strictly singular, consider the subspace E_n of X consisting of the x whose coordinates are zero except perhaps the nth; then $\dim E_n = n$ and $\rho(E_n) = 1$. Next, to show that j is strictly singular, use the fact that a closed subspace of a reflexive space is itself reflexive, and that reflexivity is preserved under isomorphism. Thus, establish that if Y is an infinite-dimensional closed subspace of X, j cannot induce an isomorphism from Y to $j(Y)$.

13.4. (a) The sequence $(P_n)_{n \geqslant 1}$ is equicontinuous.

(b) $(P_n)_{n \geqslant 1}$ converges uniformly to the identity on the compact set $\overline{T(B_X)}$.

13.5. The fact that the 2-ball property implies Hahn–Banach can be found in [152]. To see that $Y = L^\infty([0, 1])$ has the 2-ball property, consider a family $(B_i)_{i \in I}$ of closed balls of Y, intersecting two by two, with $B_i = \overline{B}(f_i, r_i)$. For $i, j \in I$, we thus have, in the sense of the lattice Y:

$$f_i - r_i \leqslant f_j + r_j.$$

Then let φ denote the essential upper bound of the $f_i - r_i$, i running over I. By the preceding double inequality, we have

$$f_j - r_j \leqslant \varphi \leqslant f_j + r_j \text{ for all } j,$$

so that

$$\varphi \in \bigcap_{i \in I} B_i.$$

13.6. (a) $C([0, 1])$ is separable, and contains an isometric copy of any separable Banach space, hence in particular a copy of c_0, automatically complemented.

(b) If A is a subset of $\mathbb{N}^*$, consider

$$Y_A = \overline{\text{Vect}(e_n, n \in A)} \text{ and } P_A : X_0 \to Y_A$$

the natural projection. If Q is a continuous projection from X onto X_0, show that the $T_i := \Phi(P_{A_i} Q)$ form a bounded family and that, given the separability of X, there exist a sequence $(x_n)_{n \geqslant 1}$ dense in X and a sequence $(x_n^*)_{n \geqslant 1}$ pre-weakly total in X^*, that is,

if $x_n^*(x) = 0$ for all $n \geqslant 1$, then $x = 0$.

Deduce that the set of $i \in I$ for which $T_i \neq 0$ is at most countable.

13.7. (a) We have

$$X \sim \ell_\infty(2\mathbb{N}^*) \oplus \ell_\infty(2\mathbb{N} + 1)$$
$$\text{and } Y \sim L^\infty\left(\left[0, \tfrac{1}{2}\right]\right) \oplus L^\infty\left(\left[\tfrac{1}{2}, 1\right]\right).$$

(c) This is a general property of a separable Banach space and of its dual space.

(e) Use the injectivity[2] of X and Y, and the Pelczynski isomorphism Theorem 13.2.4.

13.8. (a) Split the sum c_{2^n} into two sums U_n and V_n corresponding respectively to $0 \leqslant k \leqslant 2^{n-1}$ and $2^{n-1} + 1 \leqslant k \leqslant 2^n$. We then have

$$|c_{2^n}|^2 \leqslant 2(|U_n|^2 + |V_n|^2)$$
$$\leqslant 2\|a\|_2^2 \sum_{2^{n-1} \leqslant j \leqslant 2^n} |b_j|^2 + 2\|b\|_2^2 \sum_{2^{n-1} < k \leqslant 2^n} |a_k|^2.$$

---

[2] Meaning that these spaces are complemented in all Banach spaces containing them.

By summing, deduce that

$$\sum_{n=0}^{\infty} |c_{2^n}|^2 \leqslant 6\|a\|_2^2\|b\|_2^2.$$

(b) If $f \in H^1$, write $f = gh$ with $g, h \in H^2$ such that $\|g\|_2 = \|h\|_2 = \|f\|_1^{1/2}$ and apply part (a) to $a_n = \widehat{g}(n)$, $b_n = \widehat{h}(n)$ and $c_n = \widehat{f}(n)$. Then deduce that $Pf \in L^2$ and that

$$\|Pf\|_1 \leqslant \|Pf\|_2 \leqslant \sqrt{6}\|g\|_2\|h\|_2 = \sqrt{6}\|f\|_1.$$

13.9. (a) Show that $\|e^* \otimes f\| = \|e^*\| \|f\|$, and then that $N(u) \geqslant \|u\|$.

(c) Show that $\mathrm{tr}\,(e^* \otimes f) = e^*(f)$ and that $v(e^* \otimes f) = e^* \otimes v(f)$.

(f) By reflexivity, if the operator norm is the dual of the nuclear norm, then the latter is the dual of the former.

(g) For the inclusion $B_N \subset \mathrm{co}\,D$, observe that $\mathrm{co}\,D$ is compact by Carathéodory's theorem, and then show that, with $u \in B_N$, we have

$$|\mathrm{tr}\,(vu)| \leqslant \sup_{d \in D} |\mathrm{tr}(vd)|.$$

Then use the Hahn–Banach theorem.

13.10. (d) We have

$$\mathrm{tr}\,I_E = \sum_{j=1}^{M} c_j(u_j, v_j), \quad \sum_{j=1}^{M} c_j = n \text{ and } (u_j, v_j) \leqslant |u_j||v_j| \leqslant 1,$$

hence $(u_j, v_j) = 1$ for all j. From the case of equality in Cauchy–Schwarz, $u_j = v_j$ for all j. But then, the equality

$$1 = (u_j, v_j) \leqslant \|u_j\|\|v_j\|_* = \|u_j\|$$

shows that $\|u_j\| = 1$ for all j.

(e) Use the fact that $u_j = v_j$.

(f) Take the inner product with x in the equality

$$x = \sum_{j=1}^{M} c_j(x, u_j)u_j.$$

13.11. (a) Use the Hahn–Banach theorem and the fact that $\|u_j\|_* = 1$.

(b) Set

$$S(x) = \sum_{j=1}^{M} c_j L_j(x)u_j$$

and show that $(x - S(x), u_k) = 0$ for all k.

(c) Part (a) of Exercise 13.10 shows that $\|\rho^{-1}\| \leqslant 1$ and part (f) of the same exercise shows that $\|\rho\| \leqslant \sqrt{n}$.

13.12. (a) Use $I^+ = \{i \in I / z_i \geqslant 0\}$ and $I^- = \{i \in I / z_i < 0\}$.

13.13. (a) Use the sequence x defined by $x_1 = -1$ and $x_n = 1$ for $n \geqslant 2$ to show that $\|P\| = 2$.

(b) Observe that $\|e - 2e_n\| = 1$; then, if $f = Q(e)$, we have

$$|f_n - 2| \leqslant \|Q\| \text{ for all } n \geqslant 1.$$

13.14. (a) Use the Cauchy integral formula to show that the series of general term f_n converges normally on every compact subset of the unit disk; next show that with $w \in Y$ and $R(w) = \sum_{n=1}^{\infty} w_n f_n$, then $R(w) \in X$ and $T R(w) = w$.

(b) Use Proposition 13.2.2.

13.16. Use the Argyros–Haydon space and the following fact: if the dual space X^* of a Banach space X has the approximation property (AP), then X itself has the AP ([123], p. 34).

References

[1] N. H. Abel, Recherches sur la série $1 + \frac{m}{1}x + \frac{m(m-1)}{1\times2}x^2 + \frac{m(m-1)(m-2)}{1\times2\times3}x^3 + \ldots$, *J. für Math.* (1826) pp. 311–319.

[2] F. Albiac and N. Kalton, *Topics in Banach Space Theory*, Berlin, Springer, 2006.

[3] M. Alessandri, *Thèmes de Géométrie*, Paris, Dunod, 1999.

[4] E. Amar, On the corona problem, *J. Geometric Anal.* 1, 4 (1991) pp. 291–305.

[5] M. Andersson, *Topics in Complex Analysis*, Berlin, Springer, 1997.

[6] A. Argyros and R. Haydon, A hereditary indecomposable $\mathcal{L}_\infty$-space that solves the scalar-plus-compact problem, *Acta Math.* 206, 4 (2011) pp. 1–54.

[7] J. Arias, *Pointwise Convergence of Fourier Series*, Lecture Notes in Mathematics No. 1875, Berlin, Springer, 2002.

[8] M. Artin, *Algebra*, Englewood Cliffs, NJ, Prentice-Hall, 1991.

[9] W. Arveson, *A Short Course on Spectral Theory*, Berlin, Springer, 2002.

[10] *Autour du Centenaire Lebesgue*, Panoramas et Synthèses no. 18, Paris, Société Mathématique de France, 2004.

[11] J. Bak and D. J. Newman, *Complex Analysis*, Berlin, Springer, 1999.

[12] S. Banach, Sur le problème de la mesure, *Fund. Math.* 4 (1923) pp. 7–33.

[13] S. Banach, Un théorème sur les transformations biunivoques, *Fund. Math.* 6 (1924) pp. 236–239.

[14] S. Banach and A. Tarski, Sur la décomposition des ensembles de points en parties respectivement congruentes, *Fund. Math.* 6 (1924) pp. 244–277.

[15] N. Bary, *A Treatise on Trigonometric Series*, Vols. 1 & 2, Oxford, Pergamon Press, 1964.

[16] D. Bellay, La résolution de la conjecture de Littlewood, Mémoire de DEA (avec A. Bonami), Université d'Orléans, 2002.

[17] L. Bernal-Gonzàlez, Lineability of sets of nowhere analytic functions, *J. Math. Anal. Appl.* 340 (2008) pp. 1284–1295.

[18] B. Berndtsson, A. Chang and K. C. Lin, Interpolating sequences in the polydisk, *Trans. Amer. Math. Soc.* 302, 1 (1987) pp. 161–169.

[19] B. Berndtsson and T. Ransford, Analytic multifunctions, the $\bar{\partial}$-equation, and a proof of the corona theorem, *Pacific J. Math.* 124, 1 (1986) pp. 57–72.

[20] P. Billinglsey, *Ergodic Theory and Information*, New York, John Wiley & Sons, Inc., 1965.

[21] P. Billingsley, *Probability and Measure*, 3rd edn, New York, John Wiley & Sons, Inc., 1995.

[22] R. Blei, *Analysis in Integer and Fractional Dimensions*, Cambridge, Cambridge University Press, 2001.

[23] R. Blei and T. W. Kőrner, Combinatorial dimension and random sets, *Israel J. Math.* 47 (1984) pp. 65–74.

[24] R. P. Boas, *A Primer of Real Functions*, Carus Mathematical Monographs No. 13, Washington, D.C., Mathematical Association of America, 1996.

[25] S. V. Bočkarëv, Logarithmic growth of arithmetic means of Lebesgue functions of bounded orthonormal systems, *Dokl. Akad. Nauk SSSR* 223 (1975) pp. 16–19. English translation in *Soviet Math. Dokl.* 16 (1975).

[26] P. du Bois-Reymond, Versuch einer Classification der willkürlichen Functionen reeller Argumente nach ihren Änderungen in den kleinsten Intervallen, *J. für Math.* 79 (1875) p. 28.

[27] E. Bombieri and J. Bourgain, A remark on Bohr's inequality, *Int. Math. Res. Notes* (2004) pp. 4307–4330.

[28] J. Bourgain, On finitely generated closed ideals in $H^\infty(D)$, *Ann. Inst. Fourier* 35, 4 (1985) pp. 163–174.

[29] J. Bourgain, A problem of Rudin and Douglas on factorization, *Pacific J. Math.* 121, 1 (1986) pp. 47–50.

[30] J. Bourgain, On $\Lambda(p)$-subsets of squares, *Israel J. Math.* 67, 3 (1989) pp. 291–311.

[31] H. Brézis, *Analyse Fonctionnelle*, Paris, Dunod, 1994.

[32] A. M. Bruckner, J. B. Bruckner and B. S. Thomson, *Real Analysis*, Englewood Cliffs, NJ, Prentice-Hall, 1997.

[33] A. M. Bruckner and G. Petruska, Some typical results on bounded Baire one functions, *Acta Math. Hung.* 43, 3&4 (1984) pp. 325–333.

[34] B. Carl and I. Stephani, *Entropy, Compactness and the Approximation of Operators*, Cambridge, Cambridge University Press, 1990.

[35] L. Carleson, An interpolation problem for bounded analytic functions, *Amer. J. Math.* 80 (1958) pp. 921–930.

[36] L. Carleson, Interpolation by bounded analytic functions and the corona problem, *Ann. Math.* 76 (1962) pp. 547–559.

[37] U. Cegrell, A generalization of the corona theorem in the unit disk, *Math. Z.* 203 (1990) pp. 255–261.

[38] Ch. Cellérier, Note sur les principes fondamentaux de l'analyse, *Bull. Sci. Mathématiques* 14 (1890) pp. 143–160.

[39] K. Chandrasekharan, *Classical Fourier Transforms*, Berlin, Springer, 1989.

[40] Y. S. Chow and H. Teicher, *Probability Theory*, 3rd edn. Berlin, Springer, 2003.

[41] K. L. Chung, *A Course in Probability Theory*, New York, Academic Press, 2001.

[42] P. Cohen, On a conjecture of Littlewood and idempotent measures, *Amer. J. Math.* 82 (1960) pp. 191–212.

[43] C. Cowen and B. McCluer, *Composition Operators on Spaces of Analytic Functions*, Boca Raton, FL, CRC Press, 1995.

[44] H. Davenport, On a theorem of P. Cohen, *Mathematika* 7, 2 (1960) pp. 93–97.

[45] A. M. Davie, The approximation problem for Banach spaces, *Bull. London Math. Soc.* 5 (1973) pp. 261–266.

[46] Ph. J. Davis, *Interpolation and Approximation*, London, Dover Publications, 1975.

[47] H. Delange, Généralisation du théorème de Ikehara, *Ann. Sc. Éc. Norm. Sup.* 3, 71 (1954) pp. 213–242.

[48] J. Diestel, H. Jarchow and A. Tonge, *Absolutely Summing Operators*, Cambridge Studies in Advanced Mathematics No. 43, Cambridge, Cambridge University Press, 1995.

[49] J. Dixmier, Les moyennes invariantes dans les semi-groupes et leurs applications, *Acta Sci. Math. (Szeged)* 12 (1950) pp. 213–227.

[50] W. F. Donoghue, The lattice of invariant subspaces of a completely continuous quasinilpotent transformation, *Pacific J. Math.* 7 (1957), pp. 1031–1935.

[51] J. Dugundji, *Topology*, Boston, Allyn and Bacon, 1966.

[52] J. J. Duistermaat, Selfsimilarity of Riemann's nondifferentiable function, *Nieuw Archief voor Wiskunde* 9 (1991) pp. 303–337.

[53] P. Duren, *Theory of H^p Spaces*, 2nd edn, London, Dover Publications, 2000.

[54] P. Duren and A. Schuster, *Bergman Spaces*, Mathematical Surveys and Monographs No. 100, Providence, RI, AMS, 2004.

[55] A. Dvoretsky and P. Erdős, On power series diverging everywhere on the circle of convergence, *Michigan Math. J.* 3 (1955–1956) pp. 31–35.

[56] P. Enflo, A counterexample to the approximation property in Banach spaces, *Acta Math.* 130 (1973) pp. 309–317.

[57] P. Erdős, On a problem of Sidon in additive number theory, *Acta Sci. Math. (Szeged)* 15 (1953–1954) pp. 255–259.

[58] P. Erdős, A. Hildebrand, A. Odlyzko, P. Pudaite and B. Reznick, A very slowly convergent sequence, *Math. Mag.* 58 (1985) pp. 51–52.

[59] P. Erdős, A. Hildebrand, A. Odlyzko, P. Pudaite and B. Reznick, The asymptotic behaviour of a family of sequences, *Pacific J. Math.* 126, 2 (1985) pp. 227–241.

[60] P. Eymard and J. P. Lafon, *Autour du nombre π*, Paris, Hermann, 1999.

[61] C. Fefferman and E. Stein, H^p spaces of several variables, *Acta Math.* 129 (1972) pp. 137–193.

[62] V. Ferenczi, A uniformly convex hereditarily indecomposable space, *Israel J. Math.* 102 (1997) pp. 199–225.

[63] V. Ferenczi, Operators on subspaces of hereditarily indecomposable spaces, *Bull. London. Math. Soc.* 29 (1997) pp. 338–344.

[64] T. Gamelin, *Uniform Algebras*, Englewood Cliffs, NJ, Prentice-Hall, 1969.

[65] T. Gamelin, Wolff's proof of the corona theorem, *Israel J. Math.* 37 (1980) pp. 113–119.

[66] J. Garnett, *Bounded Analytic Functions*, revised first edition, Berlin, Springer, 2007.

[67] I. Gelfand, D. Raikov and G. Shilov, *Commutative Normed Rings*, Providence, RI, AMS Chelsea, 1964.

[68] J. Gerver, The differentiability of the Riemann function at certain rational multiples of π, *Amer. J. Math.* 92 (1970) pp. 33–55.

[69] T. Gowers, A new dichotomy for Banach spaces, *Geom. Funct. Anal.* 6 (1996) pp. 1083–1093.

[70] T. Gowers, A solution to the Schröder–Bernstein problem for Banach spaces, *Bull. London Math. Soc.* 28 (1996) pp. 297–304.

[71] T. Gowers and B. Maurey, The unconditional basic sequence problem, *J. Amer. Math. Soc.* 6 (1993) pp. 851–874.

[72] V. Gurariy and W. Lusky, *Geometry of Müntz Spaces and Related Questions*, Lecture Notes in Mathematics No. 1870, Berlin, Springer, 2005.

[73] G. H. Hardy, Theorems relating to the summability and convergence of slowly oscillating series, *Proc. London Math. Soc.* 2, 8 (1910) pp. 301–320.

[74] G. H. Hardy, Weierstrass's non-differentiable function, *Trans. Amer. Math. Soc.* 17 (1916) pp. 301–325.

[75] G. H. Hardy, Some famous problems of the theory of numbers, and in particular Waring's problem; an Inaugural Lecture delivered before the University of Oxford at the Clarendon Press, 1920.

[76] G. H. Hardy, *Divergent Series*, Oxford, Clarendon Press, 1949 (3rd edn, 1967).

[77] G. H. Hardy, *Collected Papers*, Oxford, Clarendon Press, 1974.

[78] G. H. Hardy, *L'Apologie d'un Mathématicien*, Paris, Belin, 1985.

[79] G. H. Hardy, *Ramanujan*, Providence, RI, AMS Chelsea, 1991.

[80] G. H. Hardy, *Divergent Series*, Providence, RI, AMS Chelsea, 1991.

[81] G. H. Hardy and J. E. Littlewood, Some problems of Diophantine approximation (II): The trigonometrical series associated with the elliptic θ-functions, *Acta Math.* 37 (1914) pp. 193–238.

[82] G. H. Hardy and J. E. Littlewood, A new solution of Waring's problem, *Quart. J. Math.*, 48 (1920) pp. 272–293.

[83] G. H. Hardy and J. E. Littlewood, A new proof of a theorem on rearrangements, *J. London Math. Soc.* 23 (1948) pp. 163–168.

[84] G. H. Hardy and S. Ramanujan, Asymptotic formulæ for the distribution of integers of various types, *Proc. London Math. Soc.* 2, 16 (1917) pp. 112–132.

[85] G. H. Hardy and S. Ramanujan, Asymptotic formulæ in combinatory analysis, *Proc. London Math. Soc.* 2, 17 (1918) pp. 75–115.

[86] G. H. Hardy and M. Riesz, *The General Theory of Dirichlet Series*, Cambridge, Cambridge Tracts in Mathematics and Mathematical Physics No. 18, 1915.

[87] G. H. Hardy and E. M. Wright, *An Introduction to the Theory of Numbers*, 5th edn, Oxford, Oxford University Press, 1979.

[88] F. Hausdorff, *Grundzüge der Mengenlehre*, Berlin, Veit, 1914.

[89] E. Hlawka, J. Schoissengeier and H. Taschner, *Geometric and Analytic Number Theory*, Berlin, Springer, 1991.

[90] K. Hoffman, *Banach Spaces of Analytic Functions*, Englewood Cliffs, NJ, Prentice-Hall, 1962.

[91] M. Holschneider and Ph. Tchamitchian, Pointwise analysis of Riemann's non differentiable function, *Invent. Math.* 105 (1991) pp. 157–175.

[92] L. Hörmander, Generators for some rings of analytic functions, *Bull. Amer. Math. Soc.* 73 (1967) pp. 943–949.

[93] V. Hugo, *Oeuvres Poétiques*, Vol. 2, Paris, Bibliothèque de la Pléiade, 1967.

[94] A. E. Ingham, On Wiener's method in Tauberian theorems, *Proc. London Math. Soc.* 2, 38 (1933–1934) pp. 458–480.

[95] S. Itatsu, Differentiability of Riemann's function, *Proc. Japan. Acad. Ser. A* 57 (1981) pp. 492–495.

[96] S. Jaffard, The spectrum of singularities of Riemann's function, *Revista Matematica Iberoamericana* 12, 2 (1996) pp. 441–460.

[97]   S. Jaffard, Y. Meyer and R. D. Ryan, *Wavelets, Tools for Science and Technology*, Philadelphia, Society for Industrial and Applied Mathematics, 2001.

[98]   F. John, Extremum problems with inequalities as subsidiary conditions, *Courant Anniversary Volume*, New York, Wiley-Interscience, 1948, pp. 187–204.

[99]   M. I. Kadeč and M. G. Snobar, Some functionals over a compact Minkowski space, *Math. Notes* 10 (1971) pp. 694–696.

[100] J. P. Kahane, Sur certaines classes de séries de Fourier absolument convergentes, *J. Math. Pures et Appliquées* 35 (1956) pp. 249–259.

[101] J. P. Kahane, *Séries de Fourier Absolument Convergentes*, Berlin, Springer, 1970.

[102] N. Kalton, Spaces of compact operators, *Math. Annal.* 208 (1974) pp. 267–278.

[103] Y. Katznelson, Sur les fonctions opérant sur l'algèbre des séries de Fourier absolument convergentes, *C.R.A.S.* 247 (1958) pp. 404–406.

[104] Y. Katznelson, *An Introduction to Harmonic Analysis*, 3rd edn, Cambridge, Cambridge University Press, 2004.

[105] Y. Katznelson and K. Stromberg, Everywhere differentiable, nowhere monotone, functions, *Am. Math. Monthly* 81 (1974) pp. 349–354.

[106] P. B. Kennedy and P. Szüsz, On a bounded increasing power series, *Proc. Amer. Math. Soc.* (1966) pp. 580–581.

[107] S. V. Kisliakov, On spaces with small annihilators, *Sem. Leningrad. Otdel. Mat. Inst. Steklov (LOMI)* 73 (1978) pp. 91–101.

[108] S. V. Kisliakov, Fourier coefficients of boundary values of functions analytic in the disc and in the bidisc, *Trudy Matem. Inst. im. V.A. Steklova* 155 (1981) pp. 77–94 (in Russian).

[109] I. Klemes, A note on Hardy's inequality, *Canad. Math. Bull.* 36 (1993) pp. 442–448.

[110] H. Koch, *Number Theory: Algebraic Numbers and Functions*, Graduate Studies in Mathematics No. 24, Providence, RI, American Mathematical Society, 2000.

[111] S. V. Konyagin, On the Littlewood problem, *Izv. A. N. SSSR, ser. mat.* 45, 2 (1981) pp. 243–265. English translation in *Math. USSR-Izv.* 18, 2 (1982) pp. 205–225.

[112] P. Koosis, *Lectures on H^p Spaces*, London, London Mathematical Society Lecture Notes Series 40, 1980.

[113] J. Korevaar, *Tauberian Theory, a Century of Developments*, Berlin, Springer, 2004.

[114] T. Kőrner, *The Behaviour of Power Series on their Circle of Convergence*, Lecture Notes in Mathematics No. 995, Berlin, Springer, 1983, pp. 56–94.

[115] P. Koszmider, Banach spaces of continuous functions with few operators, *Math. Annal.* 330 (2004) pp. 151–183.

[116] L. Kronecker, Quelques remarques sur la détermination des valeurs moyennes, *C.R.A.S.* 103 (1887) pp. 980–987.

[117] S. Lang, *Real and Functional Analysis*, 3rd edn, Berlin, Springer, 1993.

[118] P. D. Lax, *Functional Analysis*, Chichester, John Wiley & Sons, 2002.

[119] H. Lebesgue, Sur une généralisation de l'intégrale définie, *C.R.A.S.* 132 (1901) pp. 1025–1031.

[120] D. Li and H. Queffélec, *Introduction à l'Étude des Espaces de Banach, Analyse et Probabilités*, Cours spécialisé de la SMF No. 12, 2004.

[121] J. Lindenstrauss, Some aspects of the theory of Banach spaces, *Adv. Math.* 5 (1970) pp. 159–180.

[122] J. Lindenstrauss and L. Tzafriri, On the complemented subspaces problem, *Israel. J. Math.* 9 (1971) pp. 263–269.

[123] J. Lindenstrauss and L. Tzafriri, *Classical Banach Spaces I, Sequence Spaces*, Berlin, Springer, 1977.

[124] J. E. Littlewood, The converse of Abel's theorem on power series, *Proc. London Math. Soc.* 2, 9 (1911) pp. 434–448.

[125] J. E. Littlewood, *Littlewood's Miscellany*, Cambridge, Cambridge University Press, 1986.

[126] G. G. Lorentz, *Approximation of Functions*, Providence, RI, AMS Chelsea, 1986.

[127] O. C. McGehee, L. Pigno and B. Smith, Hardy's inequality and the L^1-norm of exponential sums, *Annals of Math.* 113 (1981) pp. 613–618.

[128] R. Meise and D. Vogt, *Introduction to Functional Analysis*, Oxford, Clarendon Press, 1997.

[129] H. L. Montgomery, *Ten Lectures on the Interface between Analytic Number Theory and Harmonic Analysis*, CBMS No. 84, Providence, RI, American Mathematical Society, 1994.

[130] L. Mordell, The approximate functional formula, *J. London Math. Soc.* 1 (1926) pp. 68–72.

[131] F. Nazarov, Some remarks on the Smith–Pigno–McGehee proof of the Littlewood's conjecture, *St. Petersburg Math. J.* 7, 2 (1996) pp. 265–275.

[132] C. W. Neville, A short proof of an inequality of Carleson, *Proc. Amer. Math. Soc.* 65 (1977) pp. 131–132.

[133] D. J. Newman, A simple proof of Wiener's $\frac{1}{f}$ theorem, *Proc. Amer. Math. Soc.* 48, 1 (1975) pp. 264–265.

[134] D. J. Newman, Simple analytic proof of the prime number theorem, *Amer. Math. Monthly* 87, 9 (1980) pp. 693–696.

[135] D. J. Newman, *Analytic Number Theory*, Berlin, Springer, 1998.

[136] N. Nikolskii, *Treatise on the Shift Operator*, Berlin, Springer, 1986.

[137] N. Nikolskii, In search of the invisible spectrum, *Ann. Inst. Fourier* 49, 6 (1999) pp. 1925–1998.

[138] I. Niven, *Irrational Numbers*, Carus Mathematical Monographs No. 11, Washington, D.C., Mathematical Association of America, 2006.

[139] A. M. Olevskii, Fourier series and Lebesgue functions, *Usp. Mat. Nauk* 22, 3, 135 (1967) pp. 237–239.

[140] B. Osofsky and S. Adams, Some rotations of $\mathbb{R}^3$, solution of problem 6102, *Amer. Math. Monthly* 85, 6 (1978) pp. 504–505.

[141] J. Partington, *Interpolation, Identification and Sampling*, Oxford, Oxford University Press, 1997.

[142] V. Peller and S. V. Khruscev, Hankel operators, best approximation and stationary Gaussian processes, *Russian Math. Surveys* 37 (1982), pp. 61–144.

[143] V. Peller, *Hankel Operators and their Applications*, Berlin, Springer, 2003.

[144] S. K. Pichorides, *Norms of Exponential Sums*, Publications Mathématiques d'Orsay No. 77-73, 1977.

[145] S. K. Pichorides, On the L^1-norm of exponential sums, *Ann. Inst. Fourier* 30, 2 (1980) pp. 79–89.

[146] G. Pisier, Counterexamples to a conjecture of Grothendieck, Séminaire de l'École Polytechnique, 1982, pp. 1–35.

[147] G. Pisier, *The Volume of Convex Bodies and Banach Space Geometry*, Cambridge, Cambridge University Press, 1989.

[148] D. Pompeiu, Sur les fonctions dérivées, *Math. Ann.* 63 (1907) pp. 326–332.

[149] K. Prachar, *Primzahlverteilung*, Berlin, Springer, 1957.

[150] H. Queffélec, Dérivabilité de certaines sommes de séries de Fourier lacunaires, *C.R.A.S.* 273 série A (1971) pp. 291–293.

[151] H. Queffélec, L'inégalité de Vinogradov et ses conséquences, Publications Mathématiques d'Orsay 81-08, exposé No. 4, pp. 1–15.

[152] H. Queffélec, *Topologie*, 4th edn, Paris, Dunod, 2012.

[153] H. Queffélec and C. Zuily, *Analyse pour l'Agrégation*, 4th edn, Paris, Dunod, 2013.

[154] J. Racine, *Oeuvres Complètes,* Vol. 1, Bibliothèque de la Pléiade, 1999.

[155] H. Rademacher, On the partition function $p(n)$, *Proc. London Math. Soc* 2, 43 (1937) pp. 241–254.

[156] K. Rao, On a generalized corona problem, *J. Analyse Math.* 18 (1967) pp. 277–278.

[157] R. Remmert, *Classical Topics in Complex Function Theory*, Berlin, Springer, 1998.

[158] L. Rodriguez-Piazza, On the mesh condition for Sidon sets, pre-print.

[159] W. Rudin, *Fourier Analysis on Groups*, New York, Wiley-Interscience, 1962.

[160] W. Rudin, *Real and Complex Analysis*, 3rd edn, New York, McGraw-Hill, 1987.

[161] W. Rudin, *Analyse Fonctionnelle*, Paris, Ediscience International, 1995.

[162] W. Rudin, A converse to the high indices theorem, *Proc. Amer. Math. Soc.* 17 (1966) pp. 434–435.

[163] V. Runde, *Lectures on Amenability*, Lecture Notes in Mathematics No. 1774, Berlin, Springer, 2002.

[164] R. Salem, On a problem of Littlewood, *Amer. J. Math.* 77 (1955) pp. 535–540.

[165] D. Sarason, A remark on the Volterra operator, *J. Math. Anal. Appl.* 12 (1965), pp. 244–246.

[166] A. Selberg, Reflections around the Ramanujan centenary, in *Collected Papers*, Vol. 1, Berlin, Springer, 1989, pp. 695–701.

[167] N. Sibony, Prolongement analytique des fonctions holomorphes bornées, *C.R.A.S.* 275 (1972) pp. 973–976.

[168] E. M. Stein and R. Shakarchi, *Complex Analysis*, Princeton, NJ, Princeton University Press, 2003.

[169] E. M. Stein and R. Shakarchi, *Real Analysis*, Princeton, NJ, Princeton University Press, 2005.

[170] K. Stromberg, The Banach–Tarski paradox, *Amer. Math. Monthly* 86 (1979) pp. 151–161.

[171] S. Szarek, The finite-dimensional basis problem with an appendix on nets of Grassmann manifolds, *Acta Math.* 151 (1983) pp. 153–179.

[172] J. Tannery and J. Molk, *Traité des Fonctions Elliptiques*, Paris, Gauthier-Villars, 1893.

[173] A. Tauber, Ein Satz aus der Theorie der unendlichen Reihen, *Monatsh. f. Mathematik u. Physik* 8 (1897) pp. 273–277.

[174] G. Tenenbaum, *Introduction à la théorie analytique et probabiliste des nombres*, Cours spécialisé de la SMF, 1995.

[175] G. Tenenbaum and M. Mendès-France, *Les nombres premiers*, PUF, Collection Que sais-je?, 1997.

[176] E. C. Titchmarsh, *The Theory of Functions*, Oxford, Oxford Science Publications, 1932.

[177] E. C. Titchmarsh, *The Theory of the Riemann Zeta Function*, revised by D. R. Heath-Brown, Oxford, Oxford Science Publications, 1986.

[178] N. Tomczak-Jaegermann, *Banach–Mazur Distances and Finite-Dimensional Operator Ideals*, Pitman Monographs and Surveys in Pure and Applied Mathematics, Oxford, Longman-Wiley, 1983.

[179] S. Treil, Estimates in the corona theorem and ideals of H^∞, a problem of T. Wolff, *J. Analyse Math.* 87 (2002) pp. 481–4950.

[180] B. S. Tsirelson, Not every Banach space contains ℓ_p or c_0, *Funct. Anal. Appl.* 8 (1974) pp. 139–141.

[181] S. Uchiyama, On the mean modulus of trigonometric polynomials whose coefficients have random sign, *Proc. Amer. Math. Soc.* 16 (1965) pp. 1185–1190.

[182] W. A. Veech, *A Second Course in Complex Analysis*, San Francisco, CA, Benjamin, 1967.

[183] S. Wagon, *The Banach–Tarski Paradox*, Cambridge, Cambridge University Press, 1985.

[184] C. E. Weil, On nowhere monotone functions, *Proc. Amer. Math. Soc.* 56 (1976) pp. 388–389.

[185] A. Wells and B. Williams, *Embeddings and Extensions in Analysis*, Berlin, Springer, 1970.

[186] N. Wiener, Tauberian theorems, *Annals of Math.* 33 (1932) pp. 1–100.

[187] N. Wiener, *The Fourier Integral and Certain of its Applications*, Cambridge, Cambridge University Press, 1933 (republished by Dover in 1958).

[188] P. Wojtaszczyk, *Banach Spaces for Analysts*, Cambridge Studies in Advanced Mathematics, Cambridge, Cambridge University Press, 1991.

[189] T. Wolff, Some theorems on vanishing mean oscillation thesis, University of California, Berkeley, CA 1979.

[190] T. Wolff, A refinement of the corona theorem, in *Linear and Complex Analysis Problem Book, 199 Research Problems*, Lecture Notes in Mathematics No. 1043, Berlin, Springer, 1984, pp. 399–400.

[191] K. Yosida, *Functional Analysis*, Berlin, Springer, 1980.

[192] D. Zagier, Newman's short proof of the prime number theorem, *Amer. Math. Monthly* 104, 8 (1997) pp. 705–708.

[193] Z. Zalcwasser, Sur les polynômes associés aux fonctions modulaires θ, *Stud. Math.* 7 (1937) pp. 16–35.

[194] M. Zippin, Extension of bounded linear operators, in *Handbook of the Geometry of Banach Spaces*, Vol. 2, Amsterdam, North-Holland, 2003, pp. 1703–1741.

[195] A. Zygmund, Smooth functions, *Duke Math. J.* 12 (1945), pp. 47–76.

[196] A. Zygmund, *Trigonometric Series*, 2nd edn, Cambridge, Cambridge University Press, 1959.

Notations

Index

Printed in the United States
by Baker & Taylor Publisher Services